NanoAgroceuticals & NanoPhytoChemicals

NanoAgroceuticals & NanoPhytoChemicals

Editor-in-Chief
Bhupinder Singh

Editors
Om Prakash Katare
Eliana B. Souto

CRC Press
Taylor & Francis Group
Boca Raton London New York

CRC Press is an imprint of the
Taylor & Francis Group, an **informa** business

CRC Press
Taylor & Francis Group
6000 Broken Sound Parkway NW, Suite 300
Boca Raton, FL 33487-2742

Printed on acid-free paper

International Standard Book Number-13: 978-0-8153-8977-4 (Hardback)

Library of Congress Cataloging-in-Publication Data

Names: Bhoop, Bhupinder Singh, editor.
Title: Nanoagroceuticals & nanophytochemicals / editor: Bhupinder Singh Bhoop.
Other titles: Nanoagroceuticals and nanophytochemicals
Description: Boca Raton, FL : CRC Press, Taylor & Francis Group, 2019. |
Includes bibliographical references and index.
Identifiers: LCCN 2018046889| ISBN 9780815389774 (hardback : alk. paper) |
ISBN 9781351139281 (ebook)
Subjects: LCSH: Nanobiotechnology. | Nanostructured materials. |
Pharmaceutical biotechnology. | Phytochemicals.
Classification: LCC TP248.25.N35 N264 2019 | DDC 620.1/15--dc23
LC record available at https://lccn.loc.gov/2018046889

Visit the Taylor & Francis Web site at
http://www.taylorandfrancis.com

and the CRC Press Web site at
http://www.crcpress.com

Contents

Foreword

Nanotechnology has been expanding its horizons to embrace all the vital sectors influencing human life, including agriculture and herbal technologies. Today, it is considered as a pivotal tool to meet various challenges posed by agriculture, herbiculture, and horticulture, including sustainability, plant protection, and early diagnosis of any bug(s) in the crops. Various surrogates to synthetic antifeedants and fertilizers have recently been investigated for effective pest control and enhanced crop yield in different agro-systems. The plant-derived products, also known as phytochemicals, in this context, have proved to be quite eco-friendly, residue-free, biodegradable, and cost-effective.

The major aims of various nanodelivery systems of phytochemicals in agriculture is to improve the storage shelf-life of crops, reduce the amount of sprayed chemicals through smart delivery of phytochemicals, obtain sustained delivery of pesticides and fertilizers, decrease the loss of nutrient(s) in fertilizers, and augment percentage yield of bioactives through optimized water and nutrient management. Also, efficacy of food processing has been known to be augmented through intelligent nanodelivery of phytochemicals, bioseparation of proteins, and rapid sampling of biological and chemical contaminants. Nanotechnology, overall, offers phenomenal applications for developing novel and innovative products out of agricultural products and herbal supplements, and their subsequent production, processing, storage, and packaging. Apart from this, nanotechnology-derived devices, such as DNA microarrays, microelectro-mechanical systems, and microfluidics, have exhibited tremendous potential to facilitate the insights of nanostructures in plant breeding and genetic transformation.

The escalating trend in scientific publications and patents in the wider domain offers a definitive testimony to the increasing potential of nanotechnology, especially for plant disease management and crop fortification, in spearheading the agrochemical, herbal, and food industries. Nevertheless, the domain is still in its infancy, and the information available needs consolidation, accordingly.

This book volume, *NanoAgroceuticals & NanoPhytoChemicals*, encompasses the recent advances made on nanoscale tools for delivery of diverse constituents of plants and agriculture in order to address the critical issues related to their safety, efficacy, and efficient and cost-efficient development and production. The editors have brought together opinion leaders in the domains of nanoagriculture and nanophytomedicine to give a 15-chapter overview, bringing forth tangible benefits to the farmers, agricultural scientists, food industry, and consumers, especially patients.

The editors and publishers need to be applauded for bringing forth a magnificent reference book comprising the latest advances in the field. I sincerely hope and wish that this volume will be one of the greatest treasures for researchers in times to come.

Mandip Sachdeva
Professor and Section Leader, Pharmaceutics
Editor in Chief, CRC Critical Reviews in Therapeutic Drug Carrier Systems
AAPS Fellow, 2007
Fulbright Fellow, 2014
College of Pharmacy and Pharmaceutical Sciences
Tallahassee, FL A&M University

About the Series

Implying miniaturization, the term "nano" has become a household term. Nanotechnology encompasses the study and application of functional nanomaterials in diverse fields even at their molecular levels. Virtually, nothing remains to be "nanonized" at the moment, whatever the domain may be. Traversing a voyage from the origin of this idea by Richard Feynman in 1959, nanotechnology has now entered into our day-to-day living, society, and culture, shaping our future in a categorical manner. On one hand, nanotechnology is being explored to develop newer nanostructures, and on the other hand, the horizons of their promising applications are being widened manifold. Not merely an "evolution," this has lately been turning out to be a "revolution" across different scientific disciplines and industrial sectors across the globe.

Notwithstanding the promising benefits of the cutting-edge technology, its applications in healthcare are considered as paramount. This perspective assumes immense significance, as the whole world is experiencing unprecedented biomedical challenges and changes, particularly in context with "3-**D**" succession of **D**iseases, **D**iagnosis, and **D**rugs. In line with the futuristic predictions made by Robert Freitas in 1999 in his book *Nanomedicine*, today this is ostensibly the fastest emerging offshoot of nanotechnology. Further integration of nanomedicine with the biological sciences, such as biotechnology, biochemistry, pharmaceutical sciences, biomedical engineering, bioimaging, nanomedical robotics, and biomedical technology, has spurred the genesis of a whole contemporary field of "nanobiomedicine." Employing various nanostructures and nanodevices, the science of nanobiomedicine primarily endeavors to improve upon the patient safety, efficacy, economy, and compliance of varied therapeutic, preventive, and diagnostic strategies. Not merely dealing with the nuances of interactions of nanostructures with biomolecular receptors, the modern discipline of nanobiomedicine also embraces the methods of design, engineering, and technological transfer of nanoformulations of drugs and biomacromolecules, as well as the fabrication of advanced biomedical appliances. Drawing apt inputs from almost all the frontiers of science and technology, nanobiomedicine today is heralding a new era in medical care with lots of hopes and promises for the suffering society. Nanobiomedicine, has been making rapid strides recently, eventually providing significant impetus toward scientific enthusiasm, research efforts, industrial anticipation, and regulatory attention.

Of late, several nanomedicinal products have been introduced into commercial circulation, with a few thousand more in the pipeline and yet many more in the early stages of developmental research. These have been documented to exhibit stellar benefits not only of enhanced surface-area-per-unit volume, but also of improved aqueous solubility, target specificity, controlled-release potential, biocompatibility, stealth characteristics, precise control of particle size, and ability for drug combination therapy. Such nanoscaled products principally encompass different types of nanoparticles, nanocrystals, nanoemulsions, nanoconjugates, liposomes, dendrimers, nanogels, nanocosmeceuticals, nanocapsules, nanophytomedicines, lipid nanospheres, and nanocomposites. A huge proportion of these innovative nanotech-enabled products have demonstrated enormous patient benefits and plausible solutions to the erstwhile insurmountable medical problems. Alarmingly high incidences of inadequate efficacy, safety, and patient compliance reported with the current crop of drugs, have been calling for developing such newer and therapeutically superior nanomedicinal formulations to address the present patient needs and market demands. Driven by the technological updates, increased gubernatorial support, recent notifications of newer regulatory guidance for drug products containing nanomaterials and enhanced preference for targeted therapeutics, the world has now been witnessing a rapid surge in such nanomedicines. Nanopharmaceuticals, therefore, are an outsized industry today, with its market crossing a few hundreds of billions of US dollars, over 500 companies and over 80 products worldwide. In the forthcoming decades too, these nanomedicines are anticipated to massively impact almost all the terrains of patient care, personalized medicine, clinical and veterinary practice, and above all, the pharmaceutical market. The ardent need in the current hour of shifting paradigms, accordingly, is to acquaint the professionals on advances in this vibrant science of nanobiomedicine, growing at a phenomenally striking pace. A recent literature survey of the leading journals revealed a phenomenally high rate of publications in this multi- and interdisciplinary domain of nanobiomedicine. For a realm advancing as fast as this one, it is quite challenging to obtain a complete overview of the exciting developments, as most updated information on such nano-bio interfaces lies scattered in diverse journals. Owing to nonavailability of the standard treatise(s) on the subject, therefore, we have undertaken a grand task of bringing about a book series covering a gamut of subject topics spanning nanobiomedicine. Accordingly, the current book series, "**Emerging Trends in NanoBioMedicine**," an anthology of four volumes, aspires to dexterously present the pertinent facts and figures on an array of novel subject topics. Each individual volume of the compendium is an assortment of choicest reviews and research articles, contributed by illustrious scientists and experts selectively drawn from across the world. The primary motivation behind this Herculean effort has been to bring forth high quality insight into latest research in the field as a multi-volume digest. It comprises four volumes:

Volume 1: **NanoBioMaterials**
Volume 2: **NanoBioEngineering**
Volume 3: **NanoNutraceuticals**
Volume 4: **NanoAgroceuticals & NanoPhytoChemicals**

NanoBioMaterials, the first volume of the series, maps out an exhaustive overview of the contemporary research accomplishments of variegated nanomaterials exhibiting the profound impact in improving treatment efficiency of various diseased states. This volume tacitly introduces and demonstrates the close association of biology with material science, with intent for effective and safe use of nanomaterials for different biological applications. Being a promising, though burgeoning, realm of science, the treatise accentuates diverse applications of nanomaterials in the fields of drug delivery, bioimaging, tissue engineering, and many more. It primarily focuses on innovations based on various nanobiomaterials, such as polymeric, metallic, lipidic, peptidal, carbon-based, and prodrugconjugate-based, leading finally to produce next-generation biomedicinal products.

Highlighting challenges for fundamental research and opportunities for ground-breaking advances in nano-enabled bioengineering, the second volume of this series successfully unfolds the topic of **NanoBioEngineering**. Being a rational fusion of biology, engineering and nanotechnology, it presents breakthrough techniques and tools amid the existing armamentarium to utilize novel phenomena like microfluidics and nanotooling and nanomaterials such as aptamers, nanocargos, nanofibers, nanocrystals, hyperbranched dendrimers, surface-engineered and superparamagnetic nanoparticles for meeting the desired healthcare needs. The integrated volume, herein, provides a lucid account on far-reaching topics from diverse disciplines of life sciences and technology, including physics, chemistry, biology, engineering, biomedical technology, and drug development. The enormous potential of nutraceuticals in fostering human healthcare, the burning issues confronting their poor bioavailability and usage of nanotechnology principles for resolving these issues, have been adroitly deliberated upon in the third volume of the book series, **NanoNutraceuticals**. Provision of the requisite knowledge on a gamut of modern hi-tech approaches to improve their activity, especially using nanotechnology through a single textual source enshrined, this book volume can be instrumental in ameliorating the subdued popularity of nutraceuticals and lack of serious consideration by health regulatory agencies. Thus, the contents of this book volume have the capability to metamorphose the biological performance of these functional foods through their skilful delivery to the specific target using apt nanocarriers for safer and more efficacious management of diseased states such as cancer and neurodegenerative diseases and for benefitting the human healthcare *per se*.

Volume four, **NanoAgroceuticals & NanoPhytoChemicals**, is a testament to the sophisticated and unconventional extrapolation of nanotech applications in the field of agroceuticals and phytochemicals. Albeit still in its infancy, the subject matter encompassed in the book volume holds immense potential to bring forth tangible benefits to the farmers, food industry, patients, and consumers. This book presents a cogent account on various nano-enabled delivery technologies of phytochemicals in agriculture to improve the storage shelf-life of crops, reduce the amount of sprayed chemicals through smart delivery of phytochemicals, sustain delivery of fertilizers, decrease the loss of nutrient(s) in fertilizers, and augment the percent harvesting yield through optimized water and nutrient management. The role of nanotechnology in the development of novel and innovative products out of agriculture products and herbal supplements, and their subsequent production, processing, storage, and packaging, has also been explicitly highlighted in this book volume.

All these volumes put together highlight some of the most remarkable advances that have taken place recently in the vast field of nanobiomedicine and reflect new challenges and issues that confront their further progress. The series seeks to offer the readers an apt tête-à-tête with fresh and original perspectives, epitomizing a vast multiplicity of experience, wisdom, and vision, which is quite implausible to be made available collectively anywhere else.

Verily, in this context, I feel quite satiated and fascinated that the long-felt void of a repository on the much-sought-after discipline of nanobiomedicine is being filled up by spearheading this current multi-volume book series. Although the field of nanobiomedicine has become too vast today, meticulous endeavor has been made to customize the contents of the book series and steer the novitiates as well as experts on the recent and upcoming trends. Not only does this treatise embrace holistic applications of a plethora of promising nanoconstructs, especially in the fields of medical therapeutics and diagnostics, biotechnology, pharmacogenomics, human gene biology, clinical pharmacology, and so on, but it also touches upon their regulatory, safety, and industrial scale-up issues.

At this juncture, I would like to take this opportunity to express my heart-felt gratitude to several people whose pooled efforts brought forth this dream of a mega-series into reality. The list includes all the authors, editors, the publishing house team, my research group, and research funding agencies, besides others.

First and foremost, I put across my heart-felt gratitude to the eminent and erudite authors for dexterously working on the chapters by bringing together the most comprehensive and updated information on wide-ranging subject(s). Not merely did these contributors meet with the stringent timelines, they also abided by a battery of editorial suggestions sportingly. They garnished and flavoured these chapters with apt perspectives, emanating from their own experiential wisdom. Earnest efforts have been made by them to encompass a diversity of graphic illustrations, presented immaculately as expository graphs, photographs, methodology flow charts, hierarchical tabulations, bar charts, and so on. Each chapter has been duly referenced to current literature, also citing vital texts for further reading.

As the designated editor-in-chief of the book series, I fall short of apposite words to express my deep appreciation to the distinguished editors of the four volumes for pouring in their invaluable and pertinent inputs from time to time. The sagacious, prompt, and critical feedback provided by them, as well as by the worthy reviewers of the individual chapters, all of whom spent their priceless time, helped a great deal in ameliorating the quality standards of the manuscripts.

I put into record the highly painstaking, persistent, and meticulous efforts of the management and staff members of the publisher, *viz.*, M/s CRC Press (Taylor & Francis Group), USA, in bringing about this mega-compilation. Without their immense support and direction, it could not have been possible to bring forth this colossal series in a timely manner.

Our well-knit team of exceptionally diligent, efficient, and perseverant researchers and post-docs deserve special mention, as they were verily the spine of this massive and momentous project. Naming a few, while leaving many others, may demean their stellar contribution, for which I would always remain beholden.

I also seize this opportunity in acknowledging the University Grants Commission (UGC), New Delhi, India, for awarding this prestigious National UGC Centre of Excellence in “Applications of Nanomaterials, Nanoparticles & Nanocomposites" with focus area as "Biomedical Sciences" to the Panjab University, Chandigarh, India. In fact, the drive to take up this marathon task for writing the explicit treatise was initiated after my assuming the charge of this Centre of Excellence, dedicated to explore and establish various nanobiomedicinal precepts. Thanks are also due for the tangible financial support provided by several other governmental agencies and pharma industrial houses to enable us to bring out tangible outcomes on diverse innovative nanostructured systems and in characterizing them. Several of such systematic works on nanopatterned structures, primarily employing Quality by Design (QbD)-based paradigms, have fetched exalted global recognitions by the American Association of Pharmaceutical Sciences (USA), Select Biosciences (UK), StateEase Inc. (USA), Evonik AG (Germany), Minitab Ltd. (UK), CPhI-Asia, EBA-Oxford (UK), APTI (India), and many more.

With this, I invite and incite you to start this amusing expedition to the salubrious and picturesque world of nanobiomedicine, with a missionary message, "Think Tiny and Act Big." I earnestly wish, hope, and pray that this four-volume book series on **Emerging Trends in NanoBioMedicine** will serve as a classic repertoire of essential knowledge and know-how in the domain to address the unmet needs of the scientists and practitioners from current and future generations.

Bhupinder Singh

Preface

The world of nanoscience has emerged to a very high level of advancements today, showing its omnipresence in the field of science and technology. Making materials in nanoforms has already brought a tangible difference into the life of human beings. The intrinsic benefits of the enlarged surface area associated with improved functionality and deviated behavior has found several applications in the domains of agriculture and herbal pharmaceuticals. The plants, seen as the greatest source of nutritional and therapeutic agents, when explored through the disruptive nanotechnological tools, have found newer applications with newer dimensions, hitherto unraveled. Any phyto-entity brought about out of the plant source, when framed in apposite nanostructures, has demonstrated exciting potential to make things better, simpler, safer, more effective and cost-effective.

This book, entitled *NanoAgroceuticals & NanoPhytoChemicals*, is a valuable compilation of the treatises written by the eminent scientists and experts in their respective realms of super-specialty. The success of the research work, so far, has been attained by taking different plant bioactive materials and generating nanostructures useful for a variety of applications using suitable methods to transform various agroceuticals and phytochemicals for agricultural and therapeutic outcomes, respectively. The work encompassed in this domain affirms the utility of nanomode in the metamorphosis of agroceuticals from hazardous to eco-friendly ones. Use of nanophytochemicals in therapeutics, which is one of the most prominent areas of interest for all stakeholders in health sciences, has also been aptly discussed in this book.

Briefly, Chapter 1 successfully elaborates the synergistic interplay of two assorted disciplines, *nanotechnology* and *agrochemicals,* to grossly benefit the growth of agriculture, particularly related to pesticides and fertilizers. This chapter includes a discussion of how the approach of nanoformulation in agrochemicals has let to phenomenal returns towards improving their aqueous solubility, controlling their release and toxicological profile, widening the ambit of their future applications in agriculture, and accomplishing disease-free crop production. Chapter 2 reports the production and characterization of nanonutrients and their applications to various agricultural crops. A definitive insight is provided on the biosynthesis of nanonutrients based on phytochemicals with documented agricultural applications, testifying substantial improvement in yield through biomobilization of nutrients as biofertilizers in nanostructured forms.

Chapter 3 describes the green synthesis of nanomaterials and their major role as agroceuticals in reducing the environmental impact of pesticides. This chapter also deals with the transition-metal nanoparticles, which can provide cheap and eco-friendly pesticides with significantly reduced side effects to the food chain. Chapter 4 addresses different types of nanonutrient fertilizers that can be used to improve plant nutrition by means of mineral biofortification and biotic or abiotic stress tolerance. An overview on nanofertilizers involving nanocarriers based on phytochemicals and their use has been furnished, emphasizing their eco-friendly characteristics.

Chapter 5 focuses on the structure and activity of a paramount class of pesticides, organophosphorus-based compounds, and their different detection methods employing nanomaterials. As such, organophosphorus pesticides are well-known for their potential hazards to human health and environmental pollution; their precise detection using nanotech-enabled biosensors has been the major highlight of the chapter. Chapter 6 provides updated cognizance into various nanofomulations including polymeric, metallic, and silica nanoparticles; self nanoemulsifying drg delivery systems (SNEDDs); and phytosomes and other vesicular systems for efficacious delivery of a diversity of phytopharmaceuticals. It also provides a holistic overview on the scientific and patent literature, along with the regulatory and clinical perspectives of such nanophytopharmaceuticals.

Chapter 7 discusses the current developments in nanocarrier-loaded plant bioactives and their potential uses, in general, and in chemotherapy, in particular. Such phytobioactives invariably encounter the physiochemical challenges during their gastrointestinal absorption and/or metabolism, thus calling for the need to develop their appropriate nanoconstructs. Chapter 8 pays special and focussed attention on the useful anticancer potential of neem phytoconstituents, especially on neem leaf glycoprotein (NLGP), with its proven immunomodulatory activities and potential immunity against tumor growth. Besides, this exclusive work dwells on the strategic nanocarriers to improve the stability, sustainability, and controlled release of the NLGP.

Chapter 9 discusses orally administered nanocarriers for the delivery of taxanes as anticancer therapy, hitherto administered via parenteral administration. Specific attention is imparted to undesirable side effects of the existing injectable formulations of taxanes, formulation strategies for developing technologies using novel modes of nanocouturing, and compilation of successful outcomes of oral taxanes. Chapter 10 reports plant-based bioactives (e.g., quercetin, rutin, resveratrol, curcumin, etc.) for management of various neurodegenerative diseases employing promising tools of nanotechnology. The chapter also unearths salient advantages of such nanoscale systems, majorly owing to their remarkable resemblance with the biomolecules.

Chapter 11 fundamentally describes the preparation, characterization, and application of natural and artificial celluloses, as well as their derivatives. It further touches upon the recent developments in nanostructured systems of cellulosic biomaterials, such as implants and antibacterial non-woven multilayered dressings. Chapter 12 reviews the mechanisms, risk factors, and study models related to aged macular degeneration (AMD), a multifactorial disease responsible for legal blindness among the elderly population, and potential nanotherapeutic strategies employed for the improved management of AMD. Chapter 13 discusses the use of chitosan-based nanocarriers for effective delivery of bioactives for different routes of adminstration—namely, colon, nasal, dental, and ocular. Developments in the areas of chitosan-based biomaterials and additive substances in biomedicine are elaborated in the form of nanoparticles, nanofibrils, and nanofilaments, made of nanostructured cellulose, chitin, chitosan, and so forth.

Chapter 14 addresses the use of lipid-based nanovesicular delivery systems for topical delivery of various natural bioactives. Through their uniqueness of interactions at the skin barrier interface, such vesicular systems have demonstrated enormous benefits in overcoming serious hiccups encountered by bioactives during permeation across skin barriers. Chapter 15 provides a terse account on the challenges and opportunities of using various nanostructured systems for biopharmaceutical applications and the subsequent clinical outcomes. The relevant real-time benefits of nanotechnology have also been lucidly discussed using case studies on vital plant-based biopharmaceuticals.

Conclusively, the facts and findings presented here from different perspectives not only successfully enlighten and inspire the readership, but also enkindle their intuitive power for developing novel nanotech strategies to alleviate their plausible untoward characteristics. Experimentally and experientially, an amazing power has been perceived when certain active biomolecules get into association with appropriate functional excipients at the nanoscale. According to the adage, "A needle cannot be picked by a crane," subtler means are invariably required to attain subtler objectives. Nevertheless, such means need to be further explored and accomplished in arriving at the meaningful and reproducible outcomes to promote the growth of nanoagroceuticals & nanophytochemicals.

Bhupinder Singh
Om Prakash Katare
Eliana B. Souto

Editors

Professor Bhupinder Singh Bhoop, MPharm, PhD, DSt, FPAS, is globally acclaimed for his extensive research work on diverse nanostructured drug-delivery systems, developed primarily using Quality-by-Design (QbD) paradigms. His work encompasses variegated nanoengineered systems, naive as well as functionalized, such as SLNs, NLCs, SNELs, MWCNTs, polymeric nanoparticles, lipid-polymer hybrid nanoparticles, mixed nanomicelles, nanocapsules, nanosponges, nanobilosomes, liposomes and other vesicular systems, for a gamut of disease states, particularly cancers and cardiovascular, immunodeficiency and neurodegenerative disorders.

A prolific writer, he has over 280 original journal publications, 16 books, one directory, 67 book chapters, five patents and three technology transfers to his credit. He is on editorial and review boards of scores of journals. A widely travelled scientist, he has delivered more than 280 invited talks across the globe and has duly trained tens of hundreds of scientists from the pharmaceutical industry on nanomedicines and QbD. He has completed 16 research projects and guided 31 PhD (previously and currently), 3 postdoctoral and 60 postgraduate scholars. Currently, he is serving as founder and coordinator of UGC Centre for Excellence in Nano Biomedical Applications, and coordinator, UGC Centre of Advanced Studies (Pharmaceutical Sciences), both at Panjab University, Chandigarh, India. He has been Chairman, University Institute of Pharmaceutical Sciences, and Dean, Faculty of Pharm Sciences, as well as a Fellow of Panjab University. His work has earned Professor Singh numerous awards and accolades at global and national levels.

Om Prakash Katare, MPharm, PhD, is a world-renowned scientist on nano-enabled, phospholipid-based topical drug delivery who has contributed quite significantly towards scientific research, particularly through industry-academia collaborations. More than 220 publications in peer-reviewed journals, 20 book chapters, over 75 plenary and invited talks, 13 patents filed (6 granted), 3 successful industrial tech-transfers of topical nanomedicines, supervision of 17 PhD and 46 MPharm candidates, and numerous national/international awards and recognitions (including the DBT Technology National Award by the president of India and the OPPI Award) are all testimony to his proficiency in technological research. Professor Katare has also handled key administrative posts, including the Founder Director of the Research Promotion Cell of Panjab University, India.

Eliana B. Souto, PharmD, MSc, PhD (pharmaceutical technology, biopharmaceutics and nanotechnology) is a Faculty of Pharmacy at University of Coimbra (UC, Portugal). Her current research topics include design, development, and characterization of new drug-delivery systems, based on innovative nanomaterials. Dr. Souto has published more than 150 papers in scientific peer-reviewed, international journals, edited 5 books, and contributed over 40 book chapters, 60 conference proceedings, and 120 abstracts in international conferences to date. She has guided 19 PhD students and is guiding 4 more candidates.

Contributors

Afrah Jalil Abd
Nanomedicine-Laboratory of Immunology and Molecular Biomedical Research NLIMBR)
School of Medicine (SoM)
Centre for Molecular and Medical Research (C-MMR)
Deakin University
Victoria, Australia

Anchal
University Institute of Pharmaceutical Sciences
UGC Centre of Advanced Studies
Panjab University
Chandigarh, India

Umesh Banakar
Banakar Consulting Services
Carmel, Indiana

Sarwar Beg
University Institute of Pharmaceutical Sciences
UGC Centre of Advanced Studies
Panjab University
Chandigarh, India

Suman Chaudhary
Nanomedicine-Laboratory of Immunology and Molecular Biomedical Research (NLIMBR)
School of Medicine (SoM)
Centre for Molecular and Medical Research (C-MMR)
Deakin University
Victoria, Australia

Ganesh Elango
Chemistry of Heterocycles and Natural Product Research Laboratory
Department of Chemistry, School of Advanced Sciences
Vellore Institute of Technology
Vellore, Tamilnadu

and

Department of Bio-Medical Engineering
Saveetha School of Engineering
Saveetha University
Thandalam, Tamilnadu, India

Rajashree Gude
Goa College of Pharmacy
Panaji, Goa, India

Michael Ioelovich
Designer Energy Limited
Rehovot, Israel

Ashay Jain
UGC-Centre of Excellence in Applications of Nanomaterials, Nanoparticles and Nanocomposites (Biomedical Sciences)
Panjab University
Chandigarh, India

Atul Jain
UGC-Centre of Excellence in Applications of Nanomaterials, Nanoparticles and Nanocomposites (Biomedical Sciences)
Panjab University
Chandigarh, India

Anu Kalia
Electron Microscopy and Nanoscience Laboratory
Department of Soil Science
College of Agriculture
Punjab Agricutural University
Ludhiana, Punjab, India

Jagat Rakesh Kanwar
Nanomedicine-Laboratory of Immunology and Molecular Biomedical Research (NLIMBR)
School of Medicine (SoM)
Centre for Molecular and Medical Research (C-MMR) Strategic Research Centre
Faculty of Health Deakin University
Victoria, Australia

Rupinder Kaur Kanwar
Nanomedicine-Laboratory of Immunology and Molecular Biomedical Research (NLIMBR)
School of Medicine (SoM)
Centre for Molecular and Medical Research (C-MMR) Strategic Research Centre
Faculty of Health Deakin University
Victoria, Australia

Om Prakash Katare
University Institute of Pharmaceutical Sciences
UGC Centre of Advanced Studies
Panjab University
Chandigarh, India

Harleen Kaur
Department of Microbiology
College of Basic Sciences and Humanities
Punjab Agricultural University
Ludhiana, Punjab, India

Harmanjot Kaur
University Institute of Pharmaceutical Sciences
UGC Centre of Advanced Studies
Panjab University
Chandigarh, India

Harpreet Kaur
GHG Khalsa College
Ludhiana, India

M. Kaur
Department of Chemistry
Punjab Agricultural University
Ludhiana, Punjab, India

Ripandeep Kaur
UGC-Centre of Excellence in Applications of Nanomaterials, Nanoparticles and Nanocomposites (Biomedical Sciences)
Panjab University
Chandigarh, India

Satvinder Kaur
GHG Khalsa of Pharmacy
Ludhiana, India

Shimayali Kaushal
National Agri Food Biotechnology Institute
Manauli
Punjab, India

Rajneet Kaur Khurana
University Institute of Pharmaceutical Sciences
UGC Centre of Advanced Studies
Panjab University
Chandigarh, India

Shikha Lohan
UGC-Centre of Excellence in Applications of Nanomaterials, Nanoparticles and Nanocomposites (Biomedical Sciences)
Panjab University
Chandigarh, India

Akanksha Mahajan
University Institute of Pharmaceutical Sciences
UGC Centre of Advanced Studies
Panjab University
Chandigarh, India

Selvaraj Mohana Roopan
Chemistry of Heterocycles and Natural Product Research Laboratory
Department of Chemistry, School of Advanced Sciences
Vellore Institute of Technology
Vellore, India

Misbah Momin
Bombay College of Pharmacy
Kalina, Santacruz (East)
Mumbai, India

Alka Mukne
Bombay College of Pharmacy
Kalina, Santacruz (East)
Mumbai, India

Swapna Nair
Bombay College of Pharmacy
Kalina, Santacruz (East)
Mumbai, India

Indira Rathore
Central Arid Zone Research Institute
Jodhpur, Rajasthan, India

Kaisar Raza
Department of Pharmacy
School of Chemical Sciences and Pharmacy
Central University of Rajasthan
Ajmer, India

Sumant Saini
University Institute of Pharmaceutical Sciences
UGC Centre of Advanced Studies
Panjab University
Chandigarh, India

Amita Sarwal
University Institute of Pharmaceutical Sciences
UGC Centre of Advanced Studies
Panjab University
Chandigarh, India

Shahdeep Kaur
UGC-Centre of Excellence in Applications of Nanomaterials, Nanoparticles and Nanocomposites (Biomedical Sciences)
Panjab University
Chandigarh, India

Gajanand Sharma
University Institute of Pharmaceutical Sciences
UGC Centre of Advanced Studies
Panjab University
Chandigarh, India

P. Sharma
Department of Chemistry
Punjab Agricultural University
Ludhiana, Punjab, India

Sumit Sharma
University Institute of Pharmaceutical Sciences
UGC Centre of Advanced Studies
Panjab University
Chandigarh, India

Rahul Shukla
University Institute of Pharmaceutical Sciences
UGC Centre of Advanced Studies
Panjab University
Chandigarh, India

Bhupinder Singh
University Institute of Pharmaceutical Sciences
and
UGC-Centre of Excellence in Applications of Nanomaterials, Nanoparticles and Nanocomposites (Biomedical Sciences)
Panjab University
Chandigarh, India

Charan Singh
University Institute of Pharmaceutical Sciences
UGC Centre of Advanced Studies
Panjab University
Chandigarh, India

Guneet Singh Randhawa
University Institute of Pharmaceutical Sciences
UGC Centre of Advanced Studies
Panjab University
Chandigarh, India

S. Singh
Department of Pharmacy
College of Pharmacy
Dongguk University-Seoul
Republic of Korea

Vishal Singh
National Agri Food Biotechnology Institute
Punjab, India

Nitin Kumar Singhal
National Agri Food Biotechnology Institute
Punjab, India

V. R. Sinha
University Institute of Pharmaceutical Sciences
UGC Centre of Advanced Studies
Panjab University
Chandigarh, India

Eliana B. Souto
Department of Pharmaceutical Technology
and
REQUIMTE/LAQV
Group of Pharmaceutical Technology
Faculty of Pharmacy
University of Coimbra (FFUC)
Coimbra, Portugal

Jagadish Tarafdar
Central Arid Zone Research Institute
Jodhpur, Rajasthan, India

Teenu Sharma
University Institute of Pharmaceutical Sciences
UGC Centre of Advanced Studies
Panjab University
Chandigarh, India

Kanika Thakur
University Institute of Pharmaceutical Sciences
UGC Centre of Advanced Studies
Panjab University
Chandigarh, India

Divya Utreja
Department of Chemistry
Punjab Agricultural University
Ludhiana, Punjab, India

1 Applications of Nanoagrochemicals
Precepts and Prospects

Nitin Kumar Singhal, Vishal Singh, and Shimayali Kaushal*

CONTENTS

1.1 INTRODUCTION

Ensuring food availability is one of the burning issues in the contemporary era. Conservative agricultural strategies, in this regard, must be improved using novel and emerging techniques, which are relatively fast and cost-effective. Nowadays, nanotechnology is one of the foremost strategies of research and development, which has constantly been helping mankind to flourish, even in unfavorable situations. Nanotechnology can play a vital role in the agricultural field too, and many research trials, accordingly, have been conducted in the last few years (Ditta, 2012). There should be expansion in the agricultural yield using diverse modern techniques to tackle global struggles such as a steeply growing human population, changing climatic conditions, and reducing approachability of vital plant macronutrients and other nutrients. A nanomaterial, as per the European Commission, is a natural, accompanying, or manufactured material, including particles, in a progressive state or as agglomerate, where more than half of the population of particles size lies in 1–100 nm (Rauscher et al., 2017). Due to the environmental issues, human well-being, and effectiveness, the number-size distribution threshold of 50% may be altered to be between 1% and 50% (Jampílek and KráL'Ová, 2015). Considering the aforementioned definition, graphene flakes, chemically modified graphenes, fullerenes, and single-walled carbon nanotubes (SWCNTs) should fall under the category of nanomaterials, having an external diameter of less than 100 nm (Potočnik, 2011).

A major difficulty, hitherto inadequately addressed, is the possible cytotoxicity of these nanoparticles (NPs). The cytotoxicity mainly comes from the shape and surface of NPs, as these parameters are mainly responsible for NP-cell interactions, as well as transfection to cells. Nanotubes are found to be the most toxic among the NPs, ostensibly because of their pinpoint edge (Nalwa,

* Corresponding author.

2014). By improvement of management and maintenance of efforts in agriculture, animal science, and the fish industry, nanotechnology can notably improve its existing techniques (Sekhon, 2014). Nanotechnology also has the admirable prospects to augment food production globally, improve the nutritional content of foods, and trim down the leftovers intended for "sustainable intensification" of agricultural production (Mousavi and Rezaei, 2011; Sonkaria et al., 2012; Pérez-de-Luque and Carmen Hermosín, 2013). The extensive exercise of pesticidal chemicals and fertilizers are the main cause for environment pollution, carnage of flora and fauna diversity in a particular habitat, and the manifestation of pests and pathogens. Overcoming the listed shortcomings, different nano-based formulations offer routes to deliver fertilizer in effective and target-specific manners, thus leading to improved yield and reduced damage. Principal thrust areas of exploration with nanotechnology in food segments are food safety and security. For instance, nanobiosensors of various types and smart, direct, and effective food storage systems increase the wealth gain on food items by incorporating bioactive food constituents via nanoencapsulation or nanodelivery mechanism(s) (Joseph and Morrison, 2006; Sekhon, 2014). Nanotechnology also provides solutions in the area of agriculture, and, more precisely, to optimize the conditions of supply of agricultural plants. Nanopreparations, which can provide increased crop harvesting capacity by 1.5–2 times (Baianu et al., 2004), will be significantly helpful for mankind, by and large. Nanoagrochemicals differ widely from conventional agrochemicals, as the former contain NPs. The number of research attempts in nano-based applications increased considerably recently, due to the discovery of their unusual physicochemical attributes, which are often totally different from those of the same material in the continuous phases or macroscopic dispersions (Figure 1.1). Formulations can be a combination of inorganic moieties, a

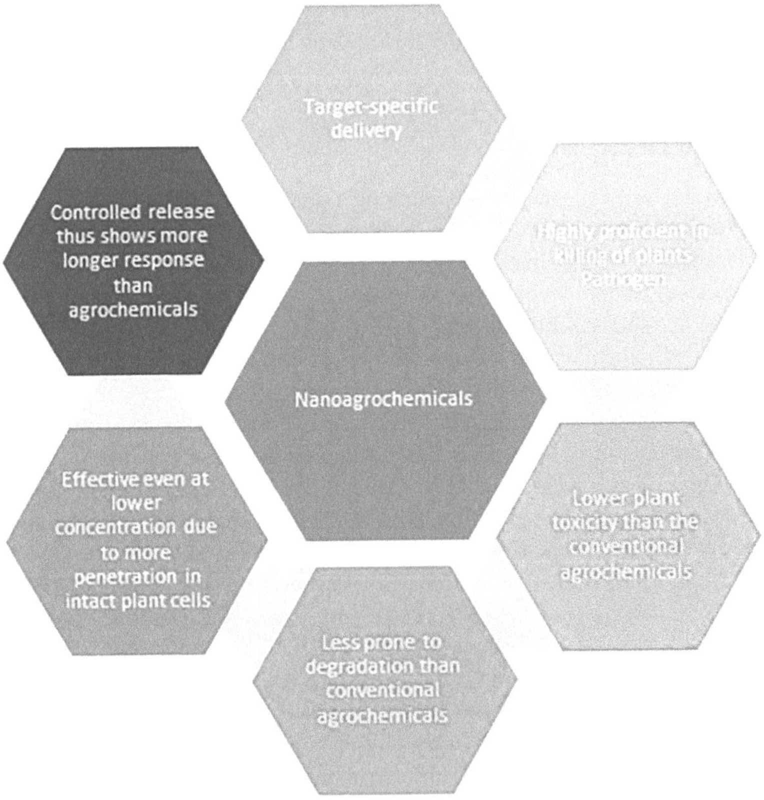

FIGURE 1.1 Nanoagrochemical classification on the basis of their functions and comparison with conventional methods of agrochemical usage.

number of different inorganic and organic surfactants, and an active constituent, all mixed to form a nanometer-sized NP (Perlatti et al., 2013).

Current NP systems can be arranged into three main classes:

i. Nonbiodegradable solid inorganic-based NPs
ii. Commonly biodegradable organic-based NPs
iii. Hybrid NPs

Nanoparticles derived from organic-based systems mostly exploit organic materials in agricultural fields and are usually composed of hydrophilic decomposable environment-responsive polymers, such as lignin, phospholipids, chitosan, lecithin, starch, cellulose derivatives, alginates, lactalbumin polylactides, poly(propylene glycol), polyacrylamide, and polysorbate. Encapsulation technology, generally grounded on biocompatible organic-based matrices, permits design of sustained-release nanocargos, wherein the shell surface can also be altered by various other site-specific targeting molecules (Kaushik et al., 2013). Using encapsulation, the nanoformulation approach not only tends to modify the aqueous solubility of the active ingredient (i.e., lipophilicity or hydrophilicity), but can also ensure its targeted biodistribution, thus reducing the dosage of agrochemicals, since nanoencapsulation offers safety of the agrochemicals from the surrounding environment and vice versa (Perlatti et al., 2013).

Bioavailability of all agricultural chemicals changes with their miscibility and absorptivity parameters. Traditionally, different methods are available to increase the solubility of some poorly soluble agrochemicals, but these methods do not work with certain structures with explicit properties in regard to their chemistry. Common alterations to enhance penetrability are grounded on a few vibrant thoughts (namely, decrease in the ionizability and upsurge in hydrophobicity) (Kerns and Di, 2013), but all have some restrictions. One of the approaches is targeted at refining solubility, permeability, and biological accessibility—for example, permeability accompaniments, surfactants, or complexing means can be coined as formulation and used through various means. Another possibility to enhance bioavailability of agrochemicals along with targeted delivery is to employ NPs. This targeted delivery of NPs allows many biological agrochemicals to be specifically deposited on the desired location and accordingly act as per requirement. There are myriad advantages of nanotechnology in agrochemicals: (i) augmented bioavailability (swift dissolution, better diffusion or infiltration through the membranes of the cell), (ii) reduced quantities, (iii) concentration-dependent toxicity, (iv) controlled release, (v) site-specific delivery, and (vi) reduced environmental influence on bioavailability irregularity (Rao et al., 2004; Bhushan, 2013; Nalwa, 2014) (Table 1.1).

1.1.1 Exploitation of Nano-Based Formulations for Plant Safety and Their Biological Activities

1.1.1.1 Nanofertilizers

The term *nanofertilizers* can be defined as a large collection of materials covering the nature and balanced growth, and even implication for agricultural exercises (Mattiello and Marchiol, 2017). A vital role is played by the fertilizers to increase the agricultural production up to 35%–40%. Mainly, two types of fertilizers—controlled-release fertilizers (CRFs) and slow-release fertilizers (SRFs)—have become highly widespread, because CRFs comprise plant nutrients which cannot be taken up by the plant readily; rather, their uptake is delayed after application. CRFs deliver nutrients to the plants for a longer duration of time as compared to the quicker release of fertilizers. The CRF term has become well-accepted, the main aspects considered being monitoring the rate of release, duration of release, and release time lag, all of which were quite conventional regarding CRF research (Trenkel, 2010). The rate, pattern, and duration of plant nutrient release of nutrients are controlled with encapsulation materials (Figure 1.2), by slow hydrolysis of polymeric materials. The release and requirement can be adjusted using appropriate coating materials for the encapsulation of the nanofertilizers. In nanofertilizers, important nutrient components are distributed to crops

TABLE 1.1
Holistic Advantages of Nanoagrochemicals over Conventional Agrochemicals

Types of Nanoagrochemicals	Subtypes	Target Plant Species	Comparison to Conventional Agrochemicals	References
Nanoherbicides	Poly(ε-caprolactone) containing herbicide atrazine	*Brassica* sp.	Lower toxicity to nontargeted plant species	Sousa et al. (2018), Pereira et al. (2014)
	Glyphosate isopropylamine (GIPA) nanoemulsion formulation	Goose grass (*Eleusine indica*)	Enhanced bioactivity and significantly lower spray deposition	Lim et al. (2013)
	Clodinafop-propyrgyl loaded carboxymethyl cellulose nanoparticles	Unwanted grasses in wheat crops	Improving yield of crop than other conventional chemicals	Davidson et al. (2013)
	Chitosan and sodium alginate polymer	Annual grasses and broad-leaved weeds	Reduced environmental impact and causes less chromosome damage in onion (*Allium cepa*)	Silva et al. (2011)
	Phenoxy-herbicides intercalated into the Zn-Al-layered double hydroxide inorganic interlayers	Broad-leaf weeds	Improved herbicide encapsulation and controlled release	Sarijo et al. (2010)
	Light-responsively controlled-release herbicide particle (LCHP)	Target multiple species	Facile approach to control release of pesticide and reduce pollution risk on environment	Chen et al. (2018)
Nanofungicides	*Zataria multiflora* essential oil encapsulated in chitosan NPs	*Botrytis cinerea*	Controlled and sustained release of essential oil and decreased disease severity	Mohammadi et al. (2015)
	Chitosan-lactide copolymer as hydrophobic carrier for pyraclostrobin	*Colletotrichum gossypii*	Better fungicidal activity and sustained release	Jampílek and KráL'Ová (2015)
	Hexaconazole NPs stabilized by PEG	*Rhizoctonia solani*	More potent and exhibited more stability than conventional formulation	Jampílek and KráL'Ová (2015)
	ZnO and TiO_2 NPs	*Macrophomina phaseolina*	More efficient prevention of fungal growth than conventional agrochemicals	Navale et al. (2015), Jamdagni et al. (2018)
	Cu-chitosan nanoparticles	*A. alternate*, *M. phaseolina*, *R. solani*, *Fusarium oxysporum*	More efficient suppression of phytopathogenic fungus	Saharan et al. (2015), Mohamed et al. (2018)

(*Continued*)

TABLE 1.1 (*Continued*)
Holistic Advantages of Nanoagrochemicals over Conventional Agrochemicals

Types of Nanoagrochemicals	Subtypes	Target Plant Species	Comparison to Conventional Agrochemicals	References
	Sulfur NPs	*Fusarium solani*, *Venturia inaequalis*, *Aspergillus niger*	Efficient prevention of fungal growth and significantly reduce phospholipid content	Choudhury et al. (2010)
	Silver NPs	*Alternaria citri*	Amount of inoculums in the field is reduced; eco-friendly method for management of fungal plant pathogens	Farooq et al. (2018)
	Nanoformulated heteroleptic metal complexes	*Fusarium oxysporum*	Better alternative due to lower toxicity and water dispersibility	Gumber et al. (2018)
Nanoinsecticides	Nanoencapsulated essential oil of *Carum copticum*	*Plutella xylostella* larvae	Higher toxicity than the oil alone	Jampílek and KrāL'Ová (2015)
	α-Pinene and linalool with SiO_2 NPs	*Spodoptera litura*, *Achaea janata*	Enhanced anti-feedant potential, longer shelf life of terpenes, and increased biological activity against *S. litura*	Jampílek and KrāL'Ová (2015)
	Zein nanoparticles loaded with the main constituents of the essential oil of citronella	*Tetranychus urticae*	Formulations are environmentally friendly and also contribute to the goal of sustainability	Oliveira et al. (2018)
	NPs loaded with *Azadirachta indica* (neem) extracts	*Plutella xylostella*	Showed 100% larval mortality	Forim et al. (2013)
	Fipronil nanoformulations	*Nilaparvata lugens*	Nanoformulations showed improved insecticidal activity against Brown plant hopper (BPH)	Kumar et al. (2018)
	Water dispersible amorphous permethrin NPs	*Culex quinquefasciatus*	Increased larvicidal activity relative to the conventional one and noncytotoxicity to nontarget organisms	Anjali et al. (2010)
	Lantana camara nano-extract	*Spodoptera litura*	More effective in controlling *S. litura*	Kasmara et al. (2018)

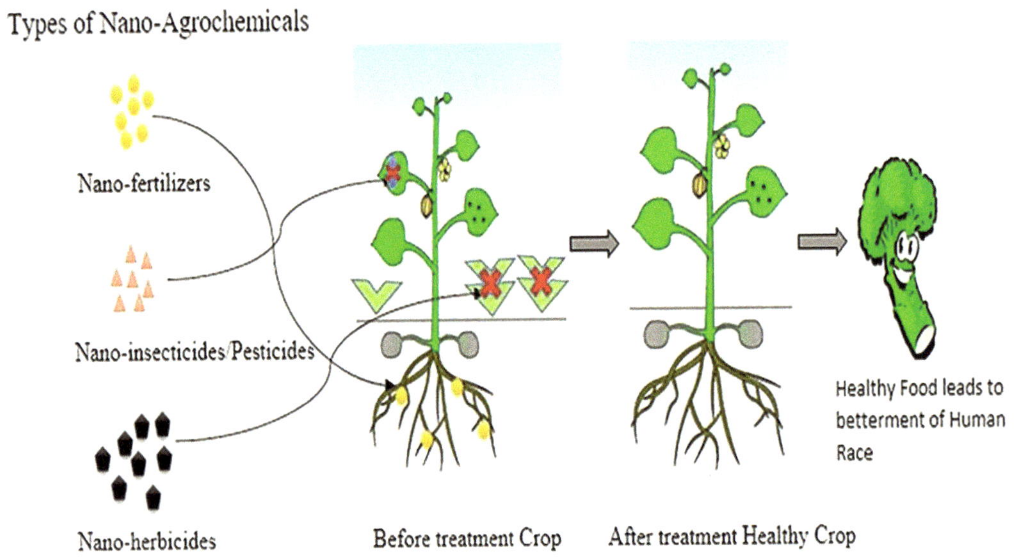

FIGURE 1.2 Types of nanoagrochemicals: (a) nanofertilizers, (b) nanoinsecticides, and (c) nanoherbicides, and their beneficial effects on crop development.

within coated nanoparticles, where inorganic and polymeric organic materials were used to entrap nutrients, which can be transported as particles or emulsions of sizes in the nanometric scale (Rai, 2012). Nanofertilizers have various benefits after application, such as lowering the nitrogen loss due to leaching, controlled release, and long-term integration of soil by microbes.

Novel types of NPs, which can infiltrate seeds or move in the root tissue, have a future to fabricate novel nutrient cargo systems, which will utilize the permeable surface of the nano-sized area of the plant and show continued release of nutrients on command (Zafar et al., 2016). Amid the number of different micronutrients, iron is the micronutrient element that plays a vital role in many oxidizing and reducing enzymes and has been found to be indispensable for chlorophyll synthesis. Fertilizers based on the nanosystems that bind with the iron characterize a rich and key source of bivalent iron for plants because of the higher stability and controlled release of iron in a wide pH range (3–11) and confirm a greater Fe^{2+} to Fe^{3+} ion ratio, which results in increasing synthesis of chlorophyll in the plants. In one study, after more than two months of treating the *Catharanthus roseus* plant with magnetic nanoparticles, it developed significant growth parameters of the leaf and the quantity of carbohydrate as paralleled to control plants, but the amounts of proline amino acid, the stomatal density of the leaf, and antioxidants did not show any major change (Jampílek and KráL'Ová, 2015).

1.1.1.2 Nanoherbicides

Sustainable agriculture requires the use of the least number of agricultural chemicals in order to defend the environment and diverse species. Encapsulating the herbicide in polymeric core shell nanoparticles results in harmless and appropriate control of herbicides that ensure environmental security (Jampílek and KráL'Ová, 2015). Parasitic weeds with nanoencapsulated herbicides were discovered to alleviated the phytotoxic effects on crops that was caused by the herbicides (Pérez-de-Luque and Rubiales, 2009). Encapsulated, targeted herbicide molecules in nanoparticles are intended for an exact receptor position in the roots of chosen weeds, thereafter translocating to the weed root system and finally to parts that delay the glycolysis of food assets in the root system, eventually making the specific weed plant starved for food and ablated (Jampílek and KráL'Ová, 2015).

Among the first studies evaluating the effectiveness of microencapsulated herbicides, Buhler et al. (1994) studied the loss of atrazina due to lixiviation and concluded that microencapsulation diminishes the contamination of underground waters in the context of low levels of herbicide lixiviation. Targeted formulations of poly(ε-caprolactone) in nanoform having entrapped atrazine have been discovered to be efficient for the management of the *Brassica* species and also displayed lesser harm to off-target species—for example, alga *Pseudokirchneriella subcapitata*—in contrast to the herbicide only (Pereira et al., 2014).

1.1.1.3 Nanofungicides

Fungal diseases cause a significant decrease in crop yield, which eventually has a noteworthy impact on a country's economics. Therefore, it is essential to carry out research on efficient fungicides that can check the fungal infections, particularly by preventing or assassinating the fungal source of infections. Though, it is noteworthy to mention here that fungicides cannot control all types of fungal diseases, such as the vascular wane fungal diseases, Fusarium wilt and Verticillium wilt. Recently, the chemicals against the fungus are commonly applied in farming to check the soil, seed, or airborne fungal pathogens (Correia et al., 2016), since they can direct an infection during its occurrence and growth, enhance yield of crops, lessen imperfection, and increase the shelf life and the quality of harvested plants. Nanofungicide usage needs awareness and proper knowledge to avoid likely interactions with off-target species by conserving soil fertility directly or indirectly (Jampílek and KráL'Ová, 2015). Agents with fungicidal characteristics comprise natural or artificial molecules, elements, inorganic-based compounds, and metal complexes.

Nanosized sulfur nanoparticles have been discovered to be very efficient in inhibiting the progression of fungal species such as *Fusarium solanii* and *Venturia inaequalis*, responsible for causing Fusarium wilt disease and apple scab disease, respectively. Sulfur NPs have fungicidal efficacy, mainly via their accumulation on the cell wall and its consequent destruction (Jampílek and KráL'Ová, 2015). Overall lipid amount of the treated fungal species, such as *Aspergillus niger*, by spherical and cylindrical sulfur NPs resulted in the marked lowering of the expression of vital enzymes such as desaturases (linoleoyl-CoA desaturase, stearoyl-CoA 9-desaturase, and phosphatidylcholine desaturase). Also, increase in buildup of saturated fatty acids and worn-out lipid layers can be considered to be major factors of sulfur NP-enabled mycostasis. Besides inhibitory effects of sulfur NPs on fungal development and sporulation of *A. niger* and *Fusarium oxysporum*, they can also considerably decrease the phospholipid content (Roy Choudhury et al., 2011).

Commercially available fungicidal preparations showed sustain release only for seven days, while controlled release of the systemic fungicide carbendazim, from nanoformulations of polyethylene glycol (PEG)-based functionalized amphiphilic, was around 10–35 days. Though, discharge of the greatest quantity of carbendazim from the fabrication was reliant on the PEG's molecular mass, as it was increasing with increases in molecular weight. Effectiveness of these nano-based formulations against the plant pathogenic fungi *R. solani* was expressed using an ED_{50} value ranging between 0.40 and 0.74 mg/dm^3 (Jampílek and KráL'Ová, 2015). The antimycotic property of silver is much more superior to that of other metals, since the Ag(I) ion results in deactivation of SH groups of the cell wall, leading to disintegration of trans-membrane energy metabolism and membrane electron transport chain (ETC), alterations in fungal nucleic acid, alienation of the enzymatic substrate complex (which is very important for the respiratory system), and reduced membrane penetrability, along with cellular destruction (Chan and Don, 2012; Durán and Marcato, 2013). Effectiveness of silver nanoparticles was reliant on particle size and shape and decreased with growing particle size. In addition, the shortened trilateral particle form presented a more "cidal" effect than spherical and rod shaped particles (Kvi and Vec, 2006). Antifungal properties of silver NPs have been exploited to a great extent against an array of human pathogens (Morones et al., 2005; Tian et al., 2007; Sarsar et al., 2015). At present, there is growing interest in utilizing the antifungal properties of silver NPs for plant disease management (Singh et al., 2014).

However, pre- and postharvest scaling of plant essential-oil application might be a choice to manipulate the chemically synthesized fungicides to dodge fruit postharvest worsening, as well as enhance the shelf life, while holding the overall superiority of a wide variety of the fresh produce. Encapsulation of the essential oil of *Zataria multiflora* in chitosan nanoparticles of 125–175 nm in size exhibited a continuous and systemic release of the essential oil for more than a month (Mohammadi et al., 2015). Further, the scientists observed that dose of the nanocapsules, at a concentration of 1500 ppm, considerably decreased the infection harshness and occurrence of *Botrytis*-injected strawberries for seven days at 4°C and two to three days more when kept at 20°C.

1.1.1.4 Nanoinsecticides

The term *insecticides* can be defined as molecules or agents of natural, biological, or chemical origin which assist in checking the insect growth and spread by killing or ruining its action, believed to be damaging. Insecticides are exceptionally dynamic materials that can menace the environmental integrity of marine and terrestrial communities. Prevalent use of insecticides has resulted in polluted croplands and surface water and leads to extreme danger to marine life, in addition to destruction of harvests and invasion of insect pests in stored food products (Neethirajan and Jayas, 2011). There is a need for an alternate efficient strategy, such as integrated pest utilization, for eradicating pests in agriculture without any harm to the environment due to the reason that many synthetic chemical insecticides were found to exert toxic and fatal effects on off-target organisms, leading to development of physiological resistance in the target and causing adverse environmental effects. Nano-based technologies have an alternate solution for pest utilization (Zhang et al., 2014). However, for revealing and physical obliteration of small-sized organisms and transporters of pathogens, such as malaria vectors, fleas, ants, and other arthropod pests, photoacoustic-photothermal nanotheranostic platforms based on lasers were used (Foster et al., 2014), with laser wavelengths that are not dangerous for humans, and also the delivery from nanoparticles can be controlled by using the photoacoustic effect. Significant potential has been exhibited to enhance crop production, not only by nano-based artificial insecticides, but also by nanotechnology-enabled botanical insecticides that are active mediators extracted from plants or are crucial oils isolated from specific plants (Singh, 2014; Oliveira et al., 2015). In general, nanoencapsulated inherent insecticides for crops show increased effectiveness due to their improved absorption into plants and less washing away. However, nanoencapsulation methods can be applied as developing tools to manage the insecticides liberation and its consequent accessibility (Scrinis and Lyons, 2007; Chhipa, 2017).

1.2 EMERGENCE OF NANOAGROCHEMICALS IN THE IMMEDIATE FUTURE

Most of the nanoagrochemicals mentioned in this chapter do not accommodate well within current market limits. Few of them showed little agronomic relevance, and some are connected with markedly undesirable dangers without strong compensation. Modified NPs, for instance, which have been well-recognized in other divisions, show very little potential for substantial agricultural applications (Ingle et al., 2017). Comparably, as per the expansions observed in other divisions of the food chain over the decade, curiosity has shifted from synthetic to organic-based nanomaterials (Amenta et al., 2015). Nanoagrochemicals, based on organic delivery systems, are popularly synthesized for food or therapeutic applications. However, these are quite costlier than other agrochemicals. A large gap in the research, which should be treaded on war footing, lies in measuring whether the products presented in the literature are competent or not to contest with the conventional materials in terms of cost and representation. Hence, it is easily visible that nanoagrochemicals will arise shortly as "nano" formulations, which mainly include constituents already being used or accepted worldwide.

1.3 FUTURE OF NANOAGROCHEMICALS

A possible strategy of nanoagrochemicals that might be applicable in the near future is represented in Figure 1.3. Guidelines are highly feasible in the near future for nanoagrochemicals, and other developments are being carried out, besides the current ongoing trials. Eventually, nanoagrochemicals are predictable to become the subsequent promising group of pollutants associated with agricultural applications. Also, nanotechnology has immense promise to turn out as a dormant basis of developing

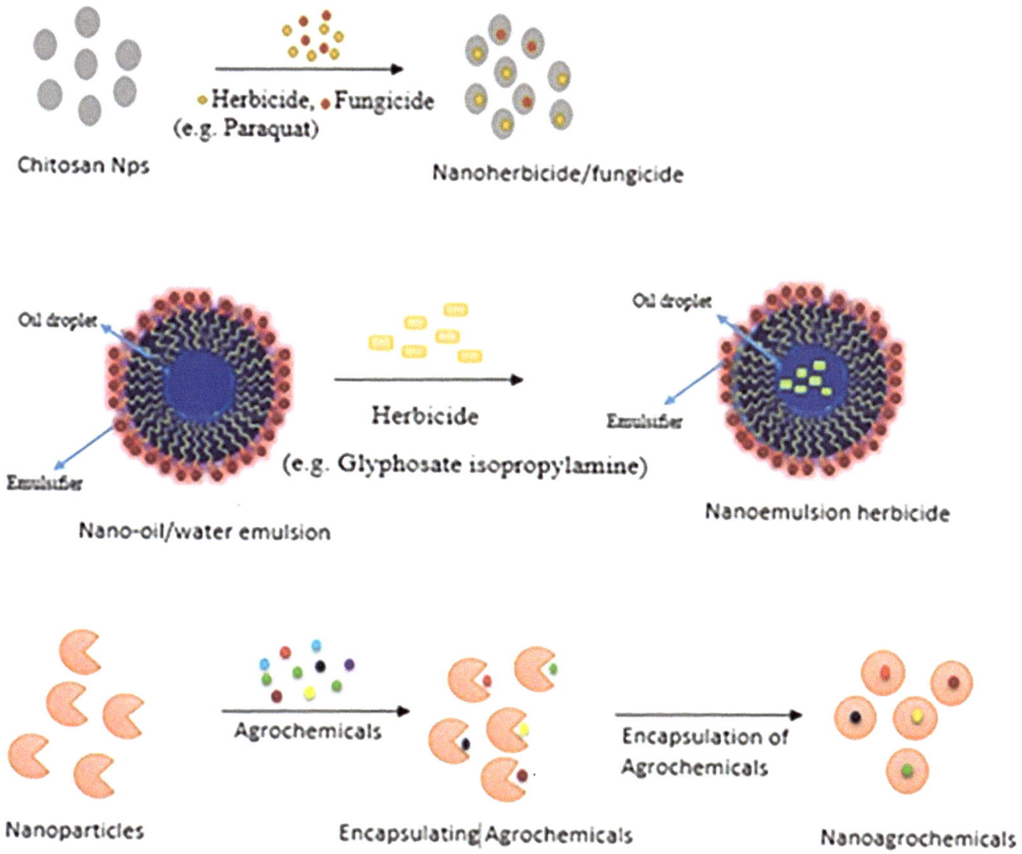

FIGURE 1.3 Classification of nanoagrochemicals based upon the type of nanocarriers being employed for the delivery of agrochemicals.

various solutions, which can alleviate the adulteration caused by pesticides and fertilizers. Sudden changes in production will be mandatory to achieve the second course for researchers and supervising organizations to propose guidelines such as

a. The EU authority suggested modern guidelines relating to a superior monitoring of preparations that are characteristically experiential as additional restraints are urgently required. Regular current experiments to chemical controlling assessment usually encompass applying increased security features to explain the growing level of ambiguity. Alternatively, a unique scientific-based device ought to be recognized to adjudge and take benefits out of formulation, to extend based on the danger factors, and to be reimbursed over the whole life cycle of the products.
b. Direct valuation of nanoagrochemicals characteristically considers that of all the products discussed in literature, many have significant toxicity and other shortcomings, relevant to agriculture, directly or indirectly. Augmentation among disciplines that is included at all phases of the growth and valuation of agrochemicals will hold up the development of fitting goods contained within several restraints of the sector of agrochemicals and are expected to bring an additional value comparative to the existing goods.
c. With rising regulatory demands and risks of reprehension, inducements are fascinating to endorse innovations that may lead to the advancement of the smarter elucidations for plant safety and nourishment. Endorsement of an alliance within different sectors and

incorporation of communal science and law will confirm public and consumer recognition and the development of apposite legal bases.

d. Developing a magnitude threshold to differentiate a "nano" product from "non-nano" product is immensely important for determining efficacy and cost-efficacy of agrochemicals. Touching a wider thought of nano-enabled knowledge and construction on the skill from other segments will be more precious to hold the expansion of more sustainable agrochemicals.

1.4 ASSESSMENT OF NANOAGROCHEMICAL RISKS AND UTILIZATION OF MANAGEMENT

Risk evaluation and organization in the agriculture sector is of huge significance, as it affects the biological organisms at the inhabitant, society, and bionetwork levels. Most significant aspects of nanotechnology should be improved quality and good dietary worth. Particles can exhibit noxious effects when they reach the nanosize (Borm et al., 2006). Even though nanotechnology presents enormous potential in the agri-food sector, appropriate knowledge about their safety is not aptly known yet. NP toxicity is characterized by the size, appearance, and perishability of the particles, which can be divided into four classes: (i) biodegradable (>100 nm), (ii) nonbiodegradable (>100 nm), (iii) biodegradable (<100 nm), and (iv) nonbiodegradable (<100 nm) (Nehoff et al., 2014; Ahamed et al., 2015). Accumulation of nonbiodegradable materials in the body can lead to stimulation of the immune system, which could lead to further greater risks of toxicity in the body (Keck and Müller, 2013). NPs can easily make their means in the human system through diverse routes such as the respiratory system, oral route, or the epidermis. Inhalation of these toxic NPs and their contact with the skin pose major risks of toxicity in the agriculture field, such s damaging membranes, producing reactive oxygen species (ROS), and causing genotoxicity (Hoet et al., 2004; Emanuel et al., 2015).

Safety factors have to be prioritized in order to ensure safer nano-based agrochemicals. The risk factors linked with their usage need total pruning by using different investigational procedures, analytical practices (Herges, 2009), and concepts that are dissimilar from the conventional theories that are typically applied to agrochemicals (Kah, 2015).

1.5 CONCLUSIONS

Agrochemicals with the involvement of nanotechnology have achieved enhanced effectiveness and reduced toxicity burden on the environment, as well as on the human race. Agrochemicals loading on a wide variety of nanoparticles have the potential to combat the key problems associated with the conventional agricultural chemicals, such as lower efficacy, higher toxicity, nonspecific delivery, and the requirement of higher dosage to achieve the desired results. Based on the afore-cited facts in this chapter, it can be clinched that nanoagrochemicals possess several advantages over the conventional agrochemicals, including target-specific delivery resulting in a lower dosage requirement, protection of the active ingredient from the physical and chemical factors, controlled release to have longer duration of response, enhanced bioavailability of the active ingredient resulting in reduced harmfulness for off-target organisms, and also decreased environmental burden. This chapter also accentuates different nanoagrochemicals like nanoherbicides, nanofungicides, and nanoinsecticides, and their synthesis using different nanoparticles and nanoemulsions. Table 1.1 vividly mentions the composition of different nanoagrochemicals, their advantages over the conventional agrochemicals, and their target plant, fungus, and insect species.

Despite the enormous advantages of using nanoagrochemicals over conventional agrochemicals, there is the serious issue of food safety and toxicity. The acute need of the hour, therefore, is the development of more eco- and human-friendly nanotechnology-based agrochemicals; safety measurements of nanoagrochemicals and various risk factors associated with their use and their impact on the environment, nontarget organisms, and the human race; and, above all, proper regulatory guidelines.

REFERENCES

Ahamed M, Akhtar MJ, Alhadlaq HA, Khan MA, Alrokayan SA. Comparative cytotoxic response of nickel ferrite nanoparticles in human liver HepG2 and breast MFC-7 cancer cells. *Chemosphere.* 2015; 135: 278–88.

Amenta V, Aschberger K, Arena M, Bouwmeester H, Botelho Moniz F, Brandhoff P et al. Regulatory aspects of nanotechnology in the agri/feed/food sector in EU and nonEU countries. *Regul. Toxicol. Pharmacol.* 2015; 73: 463–76.

Anjali CH, Sudheer Khan S, Margulis-Goshen K, Magdassi S, Mukherjee A, Chandrasekaran N. Formulation of water-dispersible nanopermethrin for larvicidal applications. *Ecotoxicol. Environ. Saf.* 2010; 73: 1932–6.

Baianu IC, Lozano PR, Prisecaru VI, Lin HC. *Applications of Novel Techniques to Health Foods, Medical and Agricultural Biotechnology.* Cornell Univ. Libr., 2004, p. 38.

Bhushan B. Handbook springer of nanotechnology. *J. Chem. Inf. Model.* 2013; 17–45.

Borm PJA, Robbins D, Haubold S, Kuhlbusch T, Fissan H, Donaldson K et al. The potential risks of nanomaterials: A review carried out for ECETOC. *Part. Fibre Toxicol.* 2006; 3: 11.

Buhler DD, Koskinen WC, Schreiber MM, Gan J. Dissipation of alachlor, metolachlor, and atrazine from starch-encapsulated formulations in a sandy loam soil. *Weed Sci.* 1994; 42: 411–7.

Chan YS, Don MM. Characterization of Ag nanoparticles produced by white-rot fungi and its *in vitro* antimicrobial activities. *Int. Arab J. Antimicrob. Agents.* 2012; 2: 1–8.

Chen C, Zhang G, Dai Z, Xiang Y, Liu B, Bian P. Fabrication of light-responsively controlled-release herbicide using a nanocomposite. *Chem. Eng. J.* 2018; 349: 101–10.

Chhipa H. Nanopesticide: Current status and future possibilities. *Agricultural Res. & Tech.* 2017; 5(1): 555651.

Choudhury SR, Nair KK, Kumar R, Gogoi R, Srivastava C, Gopal M et al. Nanosulfur: A potent fungicide against food pathogen, *Aspergillus niger. AIP Conference Proceedings.* 2010, pp. 154–7.

Correia M, Rodrigues M, Paíga P, Delerue-Matos C. Fungicides. In: Caballero B, Finglas PM, Toldrá F (eds.) *Encyclopedia of Food and Health.* Academic Press, Elsevier Inc., Waltham, USA, 2016, pp. 169–76.

Davidson DW, Verma MS, Gu FS. Controlled root targeted delivery of fertilizer using an ionically crosslinked carboxymethyl cellulose hydrogel matrix. *Springerplus.* 2013; 2: 318.

Ditta. How helpful is nanotechnology in agriculture? *Adv. Nat. Sci. Nanosci. Nanotechnol.* 2012; 3: 033002.

Durán N, Marcato P. Nanobiotechnology perspectives. Role of nanotechnology in the food industry: A review. *Int. J. Food Sci. Technol.* 2013; 48: 1127–34.

Emanuel H, Grundschober AF, Leuthold S, Meier PJ, St-pierre MV, Perfetto SP et al. Functional analysis of the extracellular cysteine residues in the human organic anion transporting polypeptide, OATP2B1. *Drug Metab. Dispos.* 2015; 806–17.

Farooq M, Ilyas N, Ilyas N, Khan I, Saboor A, Khan S et al. Antifungal activity of plant extracts and silver nano particles against citrus brown spot pathogen (*Alternaria citri*). *IJOEAR.* 2018; 4: 118–25.

Forim MR, Costa ES, Da Silva MFDGF, Fernandes JB, Mondego JM, Boiça J et al. Development of a new method to prepare nano-/microparticles loaded with extracts of *Azadirachta indica*, their characterization and use in controlling *Plutella xylostella. J. Agric. Food Chem.* 2013; 61: 9131–9.

Foster SR, Galanzha EI, Totten DC, Beneš H, Shmookler Reis RJ, Zharov VP. Photoacoustically-guided photothermal killing of mosquitoes targeted by nanoparticles. *J. Biophotonics.* 2014; 7: 465–73.

Gumber K, Sidhu A, Vk S. Evaluation of nano-formulated heteroleptic metal complexes as potential antifungals against *Fusarium oxysporum. Der Pharma Chemica.* 2018; 10: 144–7.

Herges R. Computational nanoscience. *Chemistry Central Journal.* 2009; 3.

Hoet PH, Brüske-Hohlfeld I, Salata OV. Nanoparticles—Known and unknown health risks. *J. Nanobiotechnology.* 2004; 2: 12.

Ingle PK, Deshmukh AG, Padole DA, Dudhare MS, Moharil MP, Khelurkar VC. Silver nanoparticles: A new Era of nanobiotechnology. *J. Pharm. Innov.* 2017; 6(1): 01–04.

Jamdagni, P, Rana JS, Khatri P, Nehra K. Comparative account of antifungal activity of green and chemically synthesized zinc oxide nanoparticles in combination with agricultural fungicides. *Int. J. Nano Dimens.* 2018; 9: 198–208.

Jampílek J, KráL'Ová K. Application of nanotechnology in agriculture and food industry, its prospects and risks. *Ecol. Chem. Eng. S.* 2015; 22: 321–61.

Joseph T, Morrison M. Nanotechnology in agriculture and food. *Communications.* 2006; 1–15.

Kah M. Nanopesticides and nanofertilizers: Emerging contaminants or opportunities for risk mitigation? *Front. Chem.* 2015; 3: 1–6.

Kasmara H, Melanie, Nurfajri DA, Hermawan W, Panatarani C. The toxicity evaluation of prepared *Lantana camaranano* extract against *Spodoptera litura* (Lepidoptera: Noctuidae). *AIP Conf. Proc.* 2018; 030046.

Kaushik P, Shakil NA, Kumar J, Singh MK, Singh MK, Yadav SK. Development of controlled release formulations of thiram employing amphiphilic polymers and their bioefficacy evaluation in seed quality enhancement studies. *J. Environ. Sci. Health. B.* 2013; 48: 677–85.

Keck CM, Müller RH. Nanotoxicological classification system (NCS)—A guide for the risk-benefit assessment of nanoparticulate drug delivery systems. *Eur. J. Pharm. Biopharm.* 2013; 84: 445–8.

Kerns EH, Di L. Drug-like properties: Concepts, structure design and methods from ADME to toxicity optimization, climate change 2013. *The Physical Science Basis.* 2013; 2: 1–580.

Kumar N, Kumar R, Shakil NA, Sarkar DJ. Evaluation of fipronil nanoformulations for effective management of brown plant hopper (*Nilaparvata lugens*) in rice. *Int. Jour. of Pest Management.* 2018; 874.

Kvi L, Vec R. Silver colloid nanoparticles: Synthesis, characterization, and their antibacterial activity. *J. Phys. Chem. B.* 2006; 110: 16248–53.

Lim CJ, Basri M, Omar D, Abdul Rahman MB, Salleh AB, Raja Abdul Rahman RN. Green nanoemulsion-laden glyphosate isopropylamine formulation in suppressing creeping foxglove (*A. gangetica*), slender button weed (*D. ocimifolia*) and buffalo grass (*P. conjugatum*). *Pest Manag. Sci.* 2013; 69: 104–11.

Mattiello A, Marchiol L. Application of nanotechnology in agriculture: Assessment of TiO2 nanoparticle effects on barley. *Intech.* 2017; 953–78.

Mohamed A, Gaber H, Elsharabasy F. Evaluating the *in vivo* efficacy of copper- chitosan nanocomposition for treating vascular wilt disease in date palm. *IJEAB.* 2018; 2: 447–54.

Mohammadi A, Hashemi M, Hosseini SM. Nanoencapsulation of *Zataria multiflora* essential oil preparation and characterization with enhanced antifungal activity for controlling *Botrytis cinerea*, the causal agent of gray mould disease. *Innov. Food Sci. Emerg. Technol.* 2015; 28: 73–80.

Morones JR, Elechiguerra JL, Camacho A, Holt K, Kouri JB, Ramírez JT et al. The bactericidal effect of silver nanoparticles. *Nanotechnology.* 2005; 16: 2346–53.

Mousavi Sr, Rezaei M. Nanotechnology in agriculture and food production. *J. Appl. Env. Biol. Sci.* 2011; 1: 414–19.

Nalwa HS. *Encyclopedia of Nanoscience and Nanotechnology*, vol. 3. Ceramic Nanopowders, USA, 2014. p. 200. ISBN: 1-58883-001-2

Navale GR, Thripuranthaka M, Late FJ. Antimicrobial activity of ZnO nanoparticles against pathogenic bacteria and fungi. *JSM Nanotechnol. Nanomedicine.* 2015; 3(1): 1033.

Neethirajan S, Jayas DS. Nanotechnology for the food and bioprocessing industries. *Food Bioprocess Technol.* 2011; 4: 39–47.

Nehoff H, Parayath NN, Domanovitch L, Taurin S, Greish K. Nanomedicine for drug targeting: Strategies beyond the enhanced permeability and retention effect. *Int. J. Nanomedicine.* 2014; 22(9): 2539–55.

Oliveira HC, Stolf-Moreira R, Martinez CBR, Grillo R, De Jesus MB, Fraceto LF. Nanoencapsulation enhances the post-emergence herbicidal activity of atrazine against mustard plants. *PLoS One.* 2015; 10: 7.

Oliveira JLD, Campos EVR, Pereira AES, Pasquoto T, Lima R, Grillo R et al. Zein nanoparticles as eco-friendly carrier systems for botanical repellents aiming sustainable agriculture. *J. Agric. Food Chem.* 2018; 66: 1330–40.

Pereira ES, Grillo R, Mello NFS, Rosa AH, Fraceto LF. Application of poly(epsilon-caprolactone) nanoparticles containing atrazine herbicide as an alternative technique to control weeds and reduce damage to the environment. *J. Hazard. Mater.* 2014; 268: 207–15.

Pérez-de-Luque A, Carmen Hermosín M. Nanotechnology and its use in agriculture. In: *Bio-Nanotechnology. A Revolution in Food, Biomedical and Health Sciences.* 2013; pp. 383–98.

Pérez-de-Luque A, Rubiales D. Nanotechnology for parasitic plant control. *Pest Manag. Sci.* 2009; 65: 540–45.

Perlatti P, Bergo PDS, Fernandes JB, Forim MR. Polymeric nanoparticle-based insecticides: A controlled release purpose for agrochemicals. *Insectic.-Dev. Safer More Eff. Technol.* 2013; 523–50.

Potočnik J. Commission recommendation of 18 October 2011 on the definition of nanomaterial (2011/696/EU). *Off J. Eur. Union L.* 2011; 275: 38–40.

Rai V. Implications of nanobiosensors in agriculture. *J. Biomater. Nanobiotechnol.* 2012; 03: 315–24.

Rao CNR, Müller A, Cheetham K. *The Chemistry of Nanomaterials Synthesis, Properties and Applications.* Willey Publications Ltd., 2004, pp. 1–761.

Rauscher H, Rasmussen K, Sokull-Klüttgen B. Regulatory aspects of nanomaterials in the EU. *Chem. Ing. Tech.* 2017; 89: 224–31.

Roy Choudhury S, Ghosh M, Mandal A, Chakravorty D, Pal M, Pradhan S et al. Surface-modified sulfur nanoparticles: An effective antifungal agent against *Aspergillus niger* and *Fusarium oxysporum. Appl. Microbiol. Biotechnol.* 2011; 90: 733–43.

Saharan V, Sharma G, Yadav M, Choudhary MK, Sharma SS, Pal A et al. Synthesis and *in vitro* antifungal efficacy of Cu-chitosan nanoparticles against pathogenic fungi of tomato. *Int. J. Biol. Macromol.* 2015; 75: 346–53.

Sarijo SH, Bin M, Zobir H, Yahaya AH, Zainal Z, Yarmo MA. Synthesis of phenoxyherbicides-intercalated layered double hydroxide nanohybrids and their controlled release property. *Curr. Nanosci.* 2010; 6: 199–205.

Sarsar V, Selwal MK, Selwal KK. Biofabrication, characterization and antibacterial efficacy of extracellular silver nanoparticles using novel fungal strain of Penicillium atramentosum KM. *J. Saudi Chem. Soc.* 2015; 19: 682–8.

Scrinis G, Lyons K. The emerging nano-corporate paradigm: Nanotechnology and the transformation of nature, food and agri-food systems. *Int. J. Sociol. Food Agric.* 2007; 15: 22–44.

Sekhon BS. Nanotechnology in agri-food production: An overview. *Nanotechnol. Sci. Appl.* 2014; 7: 31–53.

Silva M dos S, Cocenza DS, Grillo R, de Melo NFS, Tonello PS, de Oliveira LC et al. Paraquat-loaded alginate/chitosan nanoparticles: Preparation, characterization and soil sorption studies. *J. Hazard. Mater.* 2011; 190: 366–74.

Singh D. *Advances in Plant Biopesticides.* Springer Publications Ltd, 2014, pp. 279–93.

Singh K, Panghal M, Kadyan S, Chaudhary U, Yadav J. Green silver nanoparticles of *Phyllanthus amarus*: As an antibacterial agent against multi drug resistant clinical isolates of *Pseudomonas aeruginosa. J. Nanobiotech.* 2014; 12: 40.

Sonkaria S, Ahn SH, Khare V. Nanotechnology and its impact on food and nutrition: A review. *Recent Pat. Food. Nutr. Agric.* 2012; 4: 8–18.

Sousa GFM, Gomes DG, Campos EVR, Oliveira JL. Post-emergence herbicidal activity of nanoatrazine against susceptible weeds. *Front. Env. Sci.* 2018; 6: 1–6.

Tian J, Wong KKY, Ho CM, Lok CN, Yu WY, Che CM et al. Topical delivery of silver nanoparticles promotes wound healing. *Chem. Med. Chem.* 2007; 2: 129–36.

Trenkel ME. *Slow- and Controlled-Release and Stabilized Fertilizers: An Option for Enhancing Nutrient Use Efficiency in Agriculture.* 2nd edition, IFA, Paris, France, 2010, pp. 1–164.

Zafar H, Ali A, Ali JS, Haq IU, Zia M. Effect of ZnO nanoparticles on *Brassica nigra* seedlings and stem explants: Growth dynamics and antioxidative response. *Front. Plant Sci.* 2016; 7: 535.

Zhang Q, Wu C, Ren F, Zhang YL, Yan J, Warburton ML et al. Role of nanotechnology in agriculture with special reference to management of insect pests. *Genetics.* 2014; 3: 714–8.

2 Biosynthesis of Nanonutrients for Agricultural Applications

Jagadish Tarafdar, Indira Rathore, Ripandeep Kaur, and Ashay Jain*

CONTENTS

2.1 INTRODUCTION

Nanotechnology accomplishes its magnificent objectives by controlling the particle size of matter, by keeping at least one dimension between the size range of 1 and 100 nanometers (Jain et al., 2010). Nanotechnology has now started emerging out of the confines of laboratories and is conquering almost all the spheres of our lives with its novel applications (Kesharwani et al., 2016). A series of general methods for nanoparticle synthesis have been developed and are well-documented in the literature as well (Jain et al., 2016; Garg et al., 2017). The physical methods involve generation of nanoparticles in the solid or gas phase, whereas the chemical methods often generate nanoparticles in the solution phase (Iravani et al., 2014;Kumar et al., 2017). Currently, the novel techniques have been employing various apt biological sources, that is, the biological methods are being explored for the synthesis of nanoparticles (Tarafdar and Adhikari, 2015).

Living organisms have huge potential for the production of nanoparticles/nanodevices with desirable shapes and sizes for varied applications (Tarafdar et al., 2012b; Jeevanandam et al., 2016;

* Corresponding author.

Bindraban et al., 2018). Of late, the researchers have stated being inclined towards extensively exploring the biological systems for the production of nanofractions, it being a clean, reliable, and biocompatible process (Pantidos and Horsfall, 2014). Recently, microbial synthesis of nanoparticles has emerged as a promising field of research (Li et al., 2011; Saklani et al., 2013; Fariq et al., 2017). Microorganisms such as bacteria, fungi, actinobacteria, yeasts, and viruses are considered as potential biofactories for the production of metal nanoparticles, either intra-or extracellularly (Narayanan and Sakthivel, 2010; Tarafdar and Adhikari, 2015; Abdel-Raouf et al., 2017; Saravanan et al., 2018; Vishnukumar et al., 2018). Numerous efforts, therefore, have been undertaken for the biosynthesis of metallic nanoparticles employing bacteria (Klaus et al., 1999; Husseiny et al., 2007), fungi (Tarafdar et al., 2012a; Tarafdar and Raliya, 2013), actinobacteria (Ahmad et al., 2003; Sastry et al., 2003), yeast (Kowshik et al., 2003; Banerjee et al., 2018), and viruses (Lee et al., 2002; Merzlyak and Lee, 2006; Krajina et al., 2018). Further, another advantageous and profitable approach of biogenic synthesis is the under-exploitation of resources, which utilizes parts of the entire plant(s) as the biological factories to synthesize metallic nanoparticles (Akhtar et al., 2013).

2.1.1 Biosynthesis Process

The major reaction involved in biosynthesis of nanoparticles is oxidation/reduction. The microbial enzymes or the plant phytochemicals are usually responsible for reduction of the salts/metal compounds to their respective nanoparticles (Kaul et al., 2012). The biosynthesized nanoparticles are generally more stable due to natural encapsulation by the mother protein. Several organisms have been identified for production of different nanonutrients. Some well-known examples of bacteria synthesizing inorganic materials include megatotactic bacteria (synthesizing magnetic nanoparticles) and S-layer bacteria, which produce gypsum and calcium carbonate layers (Noda et al., 2018; Wacławek et al., 2018). Table 2.1 lists some microorganisms explored in our laboratories for the production of different nanoparticles from the respective soil solution, along with their average sizes and their NCBI (National Center for Biotechnology Information) GenBank Accession number.

Generally, the microorganisms are made to grow in the respective salt solutions for the preparation of nanoparticles. Selected organisms, which grow well, are kept for fungal ball preparation, allowing the release of enzymes for the breakdown of their salts into nanoform (Figure 2.1).

The factors responsible for biosynthesis of nanoparticles include salt concentration, pH, temperature, stirring speed, and protein-to-salt ratio. A salt concentration of 0.1 mM at a pH of

TABLE 2.1
A List of Organisms Producing 100% Particles in Nanoform from the Precursor Salts

Name of the Organism	NCBI Genebank Accession No.	Type of Nanoparticle	Average Size of Nanoparticle (nm)
Aspergillus terrus CZR1	JF681300	Ag	2.5
		Zn	6.8
		Mg	3.9
Aspergillus tubingenesis TFR5	JQ675292	P	28.2
		Ti	44.8
Rhizoctonia bataticola TFR6	JQ675307	Zn	18.5
		Au	6.2
Aspergillus fumigatus TFR8	JQ675291	ZnO	3.8
Aspergillus tubingensis TFR3	JN126255	ZnO	2.8
Aspergillus Oryzae TFR9	JQ675292	Fe	17.3
Aspergillus brasiliensis TFR23	JX999490	Mg	5.9
Aspergillus tubingenesis TFR29	KT213693	N	1.4
Aspergillus ochraceus TFR23	KC806053	K	2.2

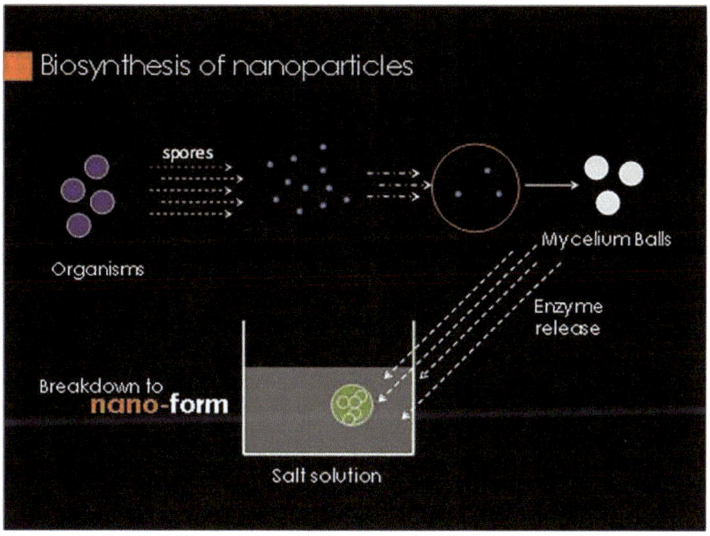

FIGURE 2.1 Biosynthesis of nanoparticle showing natural encapsulation by mother protein.

TABLE 2.2
Optimum Salt Concentration Required for Production of Nanonutrients

Metal	Concentration (mM)	Compound
Mg	0.1, 0.5	MgO, $MgNO_3$, $MgSO_4$
Zn	0.1	ZnO, $ZnCl_2$, $ZnSO_4$, $ZnNO_3$
Fe	0.1	FeO_3, $FeCl_3$, $K_3Fe(CN_6)_6$
P	0.1	$Ca_3(PO_4)_2$, Phytin
Ti	0.1	TiO_2
Ag	0.1	$AgNO_3$
Au	0.1	$HAuCl_3$
K	0.1	KNO_3
N	0.1	$NH4(NO_3)_2$
Pt	0.1	$Pt(OH)_2$
B	0.1–0.5	H_3BO_3
Mo	0.1–0.5	$(NH4)_2MoO_4$

5.5–7.0 and temperature of 28°C, with stirring at 60–80 rpm and a protein-to-salt ratio of 1:1 were found to be ideal for the production of nanoparticles, as mentioned in Table 2.2. The extracellular fungal protein was responsible for the synthesis of different monodispersed nanonutrients, such as N, P, K, Mg, S, B, Fe, Mo, and Zn. It has been found that the protein band of 32 kDa weight was mainly responsible for the breakdown of salt into nanoform.

2.1.2 Uptake of Nanoparticles in Plants

Nanoparticles on spraying get adsorbed on plant surfaces and are taken up through natural nanometer- or micrometer-scale plant openings. Several existing and proposed pathways for nanoparticle movement and uptake in plants are depicted in Figure 2.2. The rate of uptake depends on the size and surface properties of the nanoparticles; that is, nanoparticles with a very small size can penetrate through the cuticle, while the bigger ones move through cuticle-free areas, such as

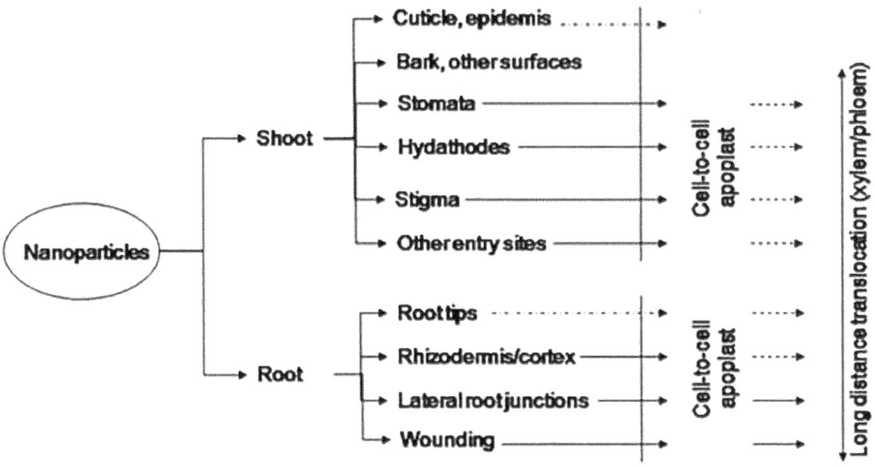

FIGURE 2.2 Pathways of nanoparticles uptake and translocation in plants. Broken lines indicate the assumption of very low rates of transport.

TABLE 2.3
Optimum Concentration Standardized to Different Nanonutrients as Foliar Application

	Optimum Concentration (ppm)	
Nanonutrient	**Cereals**	**Legumes**
N	80	80
P	40	40
K	40	20
Mg	20	20
Zn	10	10
Fe	30	30
B	4	10
Mo	6	2

hydathodes, the stigmata of flowers, and the stomata (Diet and Herth, 2011). Nanoparticles must traverse the cell wall before entering the intact cell protoplast of a plant. Results suggest that nanoparticles of less than 5 nm in size were able to traverse efficiently across the cell wall of the undamaged cell (Schreiber, 2011).

It was observed that cube-shaped nanoparticles of less than 20 nm in size, when supplied by foliar spray to plants of two weeks of age, showed maximum efficiency (Tarafdar et al., 2012b). The optimum concentration of some of the nanonutrients is summarized in Table 2.3.

2.1.3 Characterization of Biosynthesized Nanoparticles

Nanoparticles were characterized using particle-size analysis (PSA)/dynamic light scattering (DLS), transmission electron microscopy (TEM), scanning electron microscopy (SEM), atomic force microscopy (AFM), fourier transform infrared spectroscopy (FTIR), X-ray diffraction (XRD), ultraviolet-visible absorption spectroscopy, energy dispersive X-ray spectroscopy, lithography, inductively coupled plasma mass spectrometry (ICPMS), and inductively coupled optical emission spectrometry (ICPOES). The properties of nanoparticles depend upon a variety of parameters,

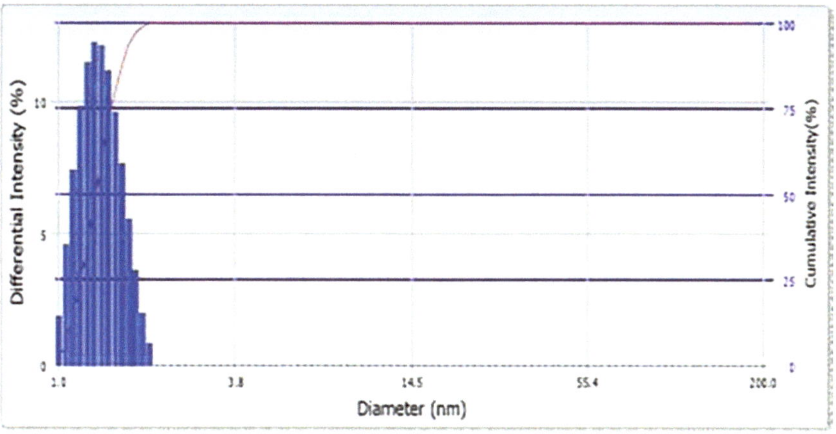

FIGURE 2.3 Particle size distribution of biosynthesized nitrogen nanoparticles.

such as particle size, dispersity, surface area, porosity, solubility, aggregation, and zeta potential. Different techniques, which have been employed to characterize these biosynthesized nanoparticles, are discussed below.

2.1.3.1 Particle-Size Analysis (PSA)

PSA measures the particle size distribution in terms of the polydispersity index (PDI) and zeta potential. The basic underlying principle of the instrument is that the light hitting the nanoparticles gets scattered in all directions. Larger particles exhibit more scattering of light than smaller particles owing to their larger area of contact. DLS measures the time-dependent fluctuations in the scattering of light intensity to determine the translational diffusion coefficient and, subsequently, the hydrodynamic diameter from the Stokes-Einstein equation. The concentration of the particles determines the probability of light scattering of the given solution. This technique also measures the PDI of the particle and zeta potential of the solution. If the PDI value is less than 1, then the particle is considered as monodispersed. The zeta potential of the solution determines the degree of aggregation of nanoparticles: the higher the zeta potential, the less the aggregation. It also indicate the degree of stability of the particle. Figure 2.3 depicts the particle-size distribution graph of nitrogen nanoparticles.

2.1.3.2 Transmission Electron Microscopy (TEM)

TEM is a microscopic technique, transmitting a beam of electrons through an ultrathin specimen, facilitating measurement of all three dimensions of the nanoparticles, along with their composition. The TEM images of the biosynthesized nanoparticles are shown in Figure 2.4.

2.1.3.3 Scanning Electron Microscopy (SEM)

SEM is a microscopic technique imaging the sample's surface by using a high energy beam of electrons in a raster scan pattern. This produces signals that reconstruct the sample using information provided (topography, composition, conductivity, etc.). The SEM furnishes information on surface structure of the nanoparticles, as shown in Figure 2.5.

2.1.3.4 Atomic Force Microscopy (AFM)

AFM is a very high-resolution type of scanning probe microscopy. AFM is considered as one of the most accurate tools for nanoparticle analysis with precise imaging, measurement, and manipulation of nanoparticles. The forces inside the nanoparticles—such as mechanical contact force; van der Waals force; and magnetic, electrostatic and chemical bondings—are measured by this technique.

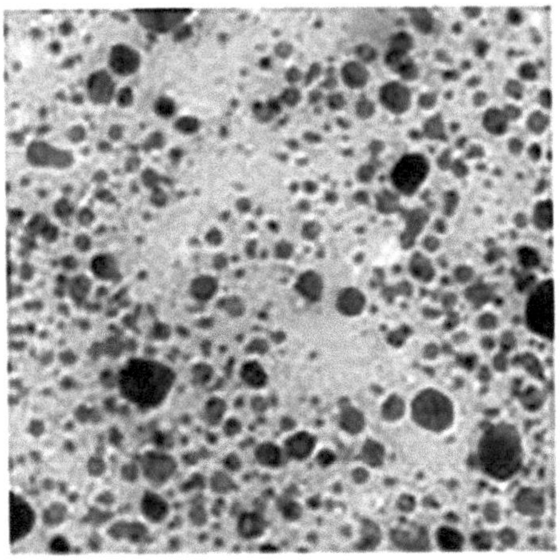

FIGURE 2.4 TEM image of biosynthesized nanoparticles.

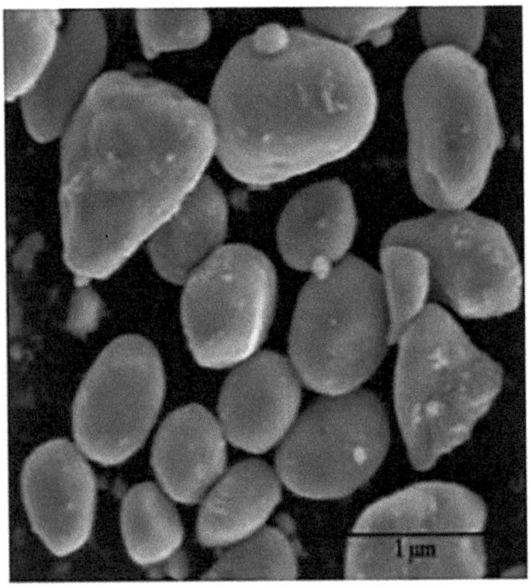

FIGURE 2.5 SEM image of biosynthesized nanoparticles.

Figure 2.6 depicts a high-resolution three-dimensional (3-D) AFM image of the nanoparticles with their detailed structural information.

2.1.3.5 Fourier Transform Infrared (FTIR) Spectroscopy

In FTIR, the sample is analyzed by placing it in front of an infrared (IR) beam. This technique determines the chemical functional groups present in the sample. Different functional groups absorb at different IR frequencies. The radiation is then mathematically converted to nanoparticle structure using the Fourier transform. This offers advantages over other methods, as it is a nondestructive and environmentally friendly technique, ensuring better speed and more reliable outputs. It also suggests as to which functional group is involved in breakdown or absorption of particles (Figure 2.7).

2.1.3.6 X-ray Diffraction (XRD)

XRD determines the arrangement of atoms within a nanoparticle. On striking the crystal, the X-ray beam gets diffracted into different directions. From the diffraction angles and intensities

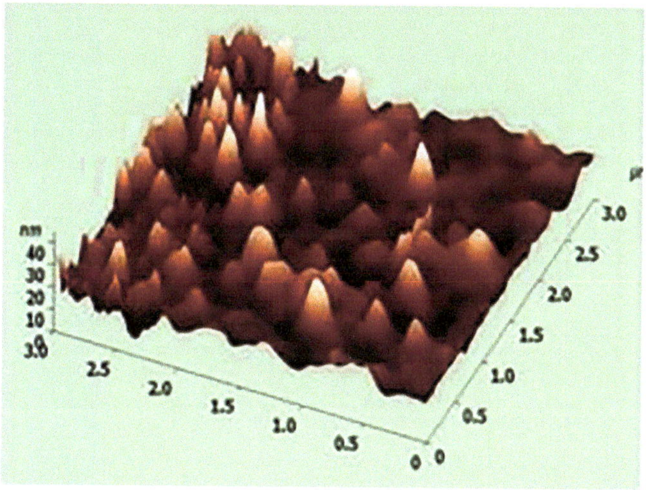

FIGURE 2.6 AFM image of biosynthesized nanoparticles.

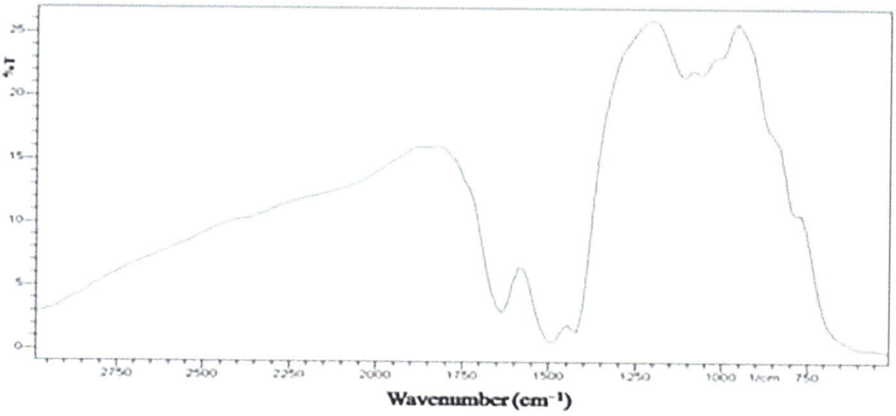

FIGURE 2.7 FTIR image of biosynthesized magnesium nanoparticles.

of these beams, we can get a 3-D picture of the atoms inside the nanoparticles. There are many methods through which crystals of nanoparticles can be created, including the hanging drop method, the sitting drop method, and microdialysis. When the X-ray pattern is recorded (Figure 2.8), the relative peaks are formed by the mathematical interpretation of the data, and the peak provides us 3-dimensional structure which further facilitates identification of the particles.

2.1.3.7 Lithography

Figure 2.9 portrays the lithography technique, which is meant for defining shapes to the surface of a semiconductor wafer. There are many lithographic techniques—such as, photolithography, electron beam lithography, X-ray lithography, and so forth—which employ different lens systems and for different durations of radiation exposure. This technique is used to transfer a previously determined pattern on a surface and is a crucial technique in the characterization of nanoparticles.

2.1.3.8 Ultrasonication

Sonication uses sound energy to agitate particles and break intermolecular interactions, if any. It is also used to initiate many chemical reactions. It can be used to remove dissolved gases from liquid.

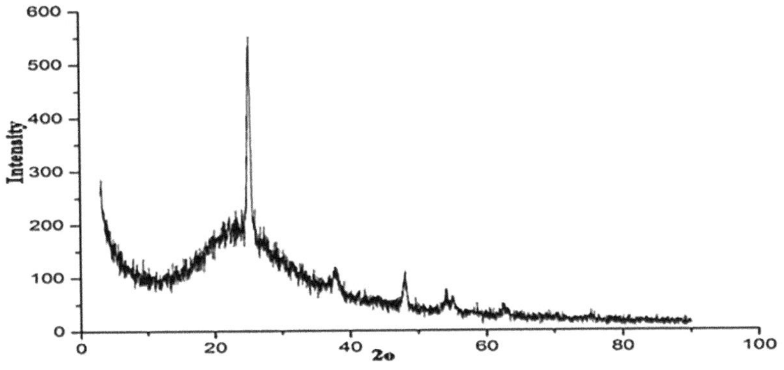

FIGURE 2.8 XRD spectra of biosynthesized titanium nanoparticles.

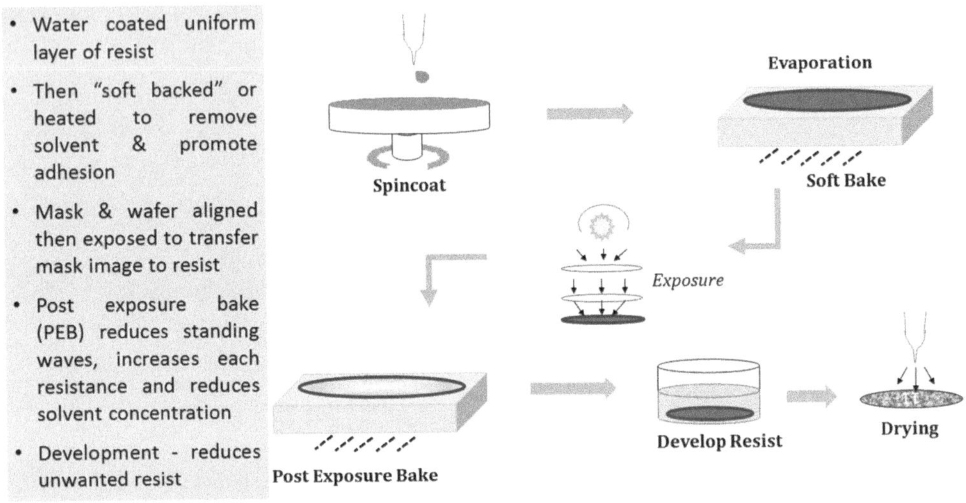

FIGURE 2.9 Conceptual optical lithography process.

It disperses the nanoparticles in liquid, which breaks the interparticulate agglomeration. However, the concentration of solution plays a crucial role during sonication.

2.1.3.9 Inductively Coupled Plasma Mass Spectroscopy (ICPMS)

ICPMS is a very sophisticated technique, which uses plasma to dissociate ions contained in a sample. These ions are then extracted from plasma and analyzed in a mass spectrometer. This technique is very sensitive and capable of detecting nanoparticles at extremely low concentrations, that is, on the order of 10^{-12} g (pg). Coupling of plasma with mass spectrometry provides an insight into nanoparticle structure, but at same time increases the complexity.

2.1.4 Application of Biosynthesized Nanoparticles

As mentioned earlier, the effectiveness of biosynthesized nanoparticles is maximum when sprayed in the recommended doses on two-week-old plants. Aerosol spray (with the help of a nebulizer) was found to be superior to traditional spray, as there was loss of only 15% nanoparticles *vis-à-vis* 33% loss by traditional spray (Tarafdar et al., 2012b). The nanoparticles of <20 nm showed best penetration capability through plants and microorganisms. Cubic geometry of the nanoparticles was

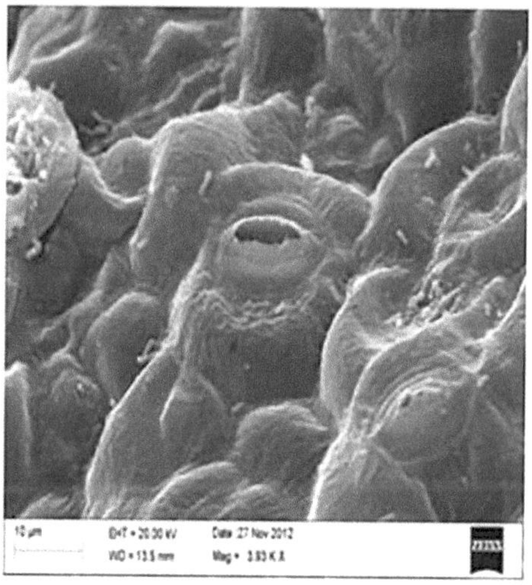

FIGURE 2.10 Penetration and movement of nanoparticle in mung bean under arid environment.

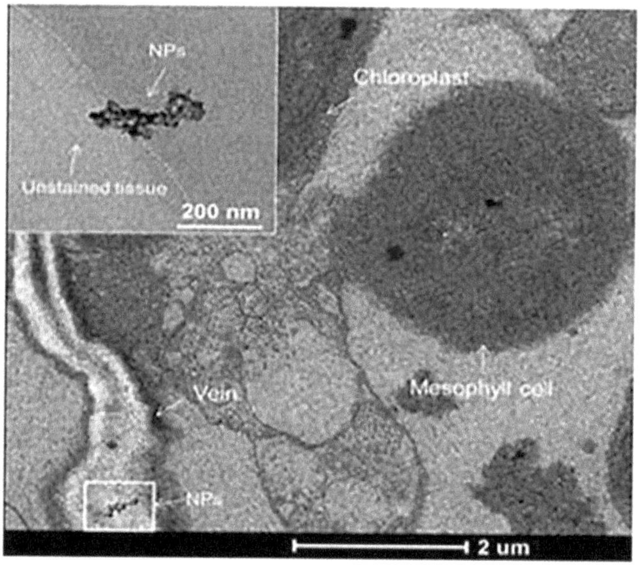

FIGURE 2.11 Exuded carbon in pearl millet and cluster bean.

found to be an apt shape for better penetration in both plants and microorganisms as compared to the other shapes (plate, wire, cage, etc.) of the particles.

Nanoparticles are adsorbed on the plant surface and taken up through natural nanometer- or micrometer-scale plant tissue openings (Figure 2.10). It was predicted that they may enter through the cuticle, stomata, hydathodes, stigma, and other parts of the shoot, as well as through the tips, rhizodermis/cortex, lateral root junctions, and wounding of roots (Tarafdar and Rathore, 2016). Uptake rates normally depend on the size of nanoparticles. A smaller size of nanoparticles can help penetratation through cuticle-free areas, such as hydathodes, the stigma of flowers, and stomata (Figure 2.11).

The nanoparticles, after entering, may travel through a sugar solution or cell sap, and become large particles with time due to agglomeration and start accumulating in the vacuole (Tarafdar and Rathore, 2016).

TABLE 2.4
Percentage Improvement in Organic Acid Concentration[a] in the Rhizosphere and Nanophosphorus Uptake by the Plants

Crops	Organic Acid Concentration	P-uptake
Clusterbean	23.2	27.2
Moth bean	19.5	23.5
Mung bean	20.7	22.7
Pearl millet	15.5	17.3

[a] Nano-P application 640 mg ha^{-1}.

2.1.5 Effect of Biosynthesized Nanoparticles on Enzyme Exudation, Nutrient Use Efficiency, and Plant Metabolism

The yield of beneficial enzymes from plants may vary from 18% and 283% due to nanoinduction. Activities of enzymes such as esterase, acid phosphatase, alkaline phosphatase, phytase, aryl-sulphatase, nitrate reductase, urease, cellulase, hemicellulose, and lignase in the biosynthesized nanoparticle-sprayed plants were observed to be better than control plants. Application of nanophosphorus also helps to improve the organic acid concentration in the rhizosphere and phosphorus uptake by the plants (Table 2.4).

Carbon leaching through roots may be reduced due to the application of nanoparticles, resulting in more biomass production of cluster bean and moth bean. Nanozinc (nano-Zn) and nanoiron (nano-Fe) application was found to be associated with high protein content and low superoxide dismutase (SOD) activity, further resulting in more stress tolerance by the plant. The relevant data are presented in Table 2.5.

It has also been found that application of nano-Zn and nano-Fe on plant leaves increases chlorophyll content further, resulting in better prevention of membrane against damage. Data are tabulated in Table 2.6.

A similar experiment with magnesium (Mg) nanoparticles showed higher light absorption and chlorophyll and leaf-protein content of cluster bean and wheat (Raliya et al., 2014; Rathore and Tarafdar, 2015). Response of nanoparticles on root growth and development was studied for several crops. The results ratified the efficiency of biosynthesized nanophosphorus (nano-P) with an average improvement of the root length by 32%, root area by 21%, root biomass by 10%, and root nodulation under legumes by 68%. Similar results were observed with the application of nano-Zn, nano-Fe, nano-Mg, and nano-N, where increase in root length varies between 2% and 12%, root area between 4% and 20%, dry biomass between 2% and 6%, and nodulation between 5% and 58%.

TABLE 2.5
Effect of Nanozinc and Nanoiron on Protein Content and Superoxide Dismutase (SOD) Activity in Chickpea (% Change over Control)

Treatments	Protein (mg/mL)	SOD Activity (unit/mg Protein)
10 ppm mega-Zn	+13.5	−28.7
10 ppm nano-Zn	+30.4	−38.7
1.5 ppm mega-Fe	+20.6	−33.2
1.5 ppm nano-Fe	+32.9	−39.7

TABLE 2.6
Effect of Nanozinc and Nanoiron Nanoparticles on Chlorophyll and Malondialdehyde Content (% Change over Control)

Treatments	Chlorophyll (mg/g FW)	Malondialdehyde (mM/g FW)
10 ppm mega-Zn	+31.6	−4.3
10 ppm nano-Zn	+68.1	−5.7
1.5 ppm mega-Fe	+20.7	−1.2
1.5 ppm nano-Fe	+41.5	−4.0

An improvement of 4–20 times in the nutrient use efficiency (NUE) was noticed with the application of nano-P, nano-Zn, and nano-Fe (Tarafdar et al., 2015) compared to the chemical fertilizer used. Nanosized polysaccharide powder resulted in 33%–82% more soil aggregation, 10%–14% more moisture retention, and 2%–5% more carbon buildup in the soil (Tarafdar et al., 2012c; Tarafdar and Rathore, 2016).

2.1.6 Assessment of Biosynthesized Nanonutrient Potential on Crop Yield

Multilocation field trials were conducted with different nanonutrients under a research field and a farmer's field. The results obtained for 11 different tested crops are summarized in Table 2.7. With the application of critical doses of nanonutrients, 12%–48% improvement of crop yield was observed under research field conditions, while in the farmer's field, it varied between 18% and 54%. Moreover, almost 10–21 days of advancement in crop maturity was noticed in plants supplemented with nanonutrients applied *vis-à-vis* the control plants.

2.1.7 Safety Assessment of Biosynthesized Nanonutrient-Derived Food

A number of experiments have been conducted to date to screen out any health hazards owing to the use of nanofoods. The results clearly established the safety of nanonutrients with hardly any adverse effects on percentage of seed germination, as well as on soluble seed protein content in vital arid crops of cluster bean, mung bean, moth bean, and pearl millet (Table 2.8).

TABLE 2.7
Yield Increase (%) due to Nanonutrient Application under Different Crops

Crops	Nanonutrient Applied	Research Field (%)	Farmer's Field (%)
Capsicum	P, Zn	19–24	–
Castor	P, Zn	–	24–37
Clusterbean	P, Zn, Fe, Mg	14–36	21–37
Cauliflower	P, Zn	–	47–54
Maize	P	23–32	–
Moth bean	P, Zn, Fe	12–31	18–27
Mung bean	P, Zn, Fe, Mg	17–42	23–38
Pearl millet	P, Zn, Fe, Mg	13–48	18–43
Rice	P	22–28	–
Tomato	P, Zn	18–23	25–29
Wheat	P	21–32	–

Doses applied: P-40 ppm; Fe-30 ppm; Mg-20 ppm; Zn-10 ppm.

TABLE 2.8
Effect on Seed Germination and Soluble Protein Content of Arid Crops

	Seed Germination (%)				Soluble Protein ($\mu g\ g^{-1}$ seed)			
Treatment	Cluster Bean	Moth Bean	Mung Bean	Pearl Millet	Cluster Bean	Moth Bean	Mung Bean	Pearl Millet
Control	90	82	89	80	105.6	98.5	78.5	70.1
Ordinary Zn	93	82	92	80	105.0	99.2	79.1	74.3
Nano-Zn	92	84	91	82	108.1	106.9	79.8	76.1
Ordinary Fe	89	83	90	81	106.3	108.5	78.6	75.0
Nano-Fe	97	84	93	84	115.4	105.3	78.9	75.4
Ordinary P	94	82	90	81	105.3	100.4	79.0	78.2
Nano-P	93	85	93	85	110.2	103.6	81.1	77.9
LSD ($p = .05$)	NS	NS	NS	NS	NS	NS	NS	NS

NS: Not Significant; LSD: Least Significant Difference.

Furthermore, application of a nano-Zn concentration up to 10 ppm increased the microbial population significantly, as shown in Table 2.9. Also, no adverse effect was noticed in body weight, grain consumption rate, and blood pH of mice feeding on nanoparticle-sprayed plant grains as compared to the control (Table 2.10).

A preclinical safety evaluation of pearl millet and mung bean grown with biosynthesized nanonutrients showed

i. Absence of any preterminal death in all groups of mice
ii. Lack of any abnormal clinical signs, behavioral activity, and so forth in animals receiving test nanomaterials
iii. Insignificant difference in feed intake and gain in body weight between the treated animals and the control group
iv. No change(s) in individual organ weights and gross necropsy
v. No adverse effects in rats, even after feeding pearl millet and mung bean grown with nanofertilizer more than two and one half times higher than the dose limit

The health effects of biosynthesized nanofoods were evaluated in female Wistar rats, employing repeated subacute toxicity studies. Histopathological analysis of liver, kidney, and spleen tissues

TABLE 2.9
Effect on Microbial Population with the Application of Nanozinc at 10 ppm Concentration

Treatment	Fungi (cfu $\times 10^4\ g^{-1}$)	Bacteria (cfu $\times 10^6\ g^{-1}$)	Actinobacteria (cfu $\times 10^6\ g^{-1}$)
Master Control[a]	9	18	6
Control[b]	14	24	8
Ordinary Zn	17	32	12
Nano-Zn	23	39	16
LSD ($p = .05$)	3.2	7.1	2.4

[a] Crop not grown.
[b] Crop grown without treatment.

TABLE 2.10
Weight, Grain Consumption Rate, and Blood pH of Mice at Different Time Intervals with the Feeding of Nanozinc (10 ppm) Treated Material

	Days Feeding											
	Body Weight (g)				Consumption Rate (g/100 g Body wt/day)				Blood pH			
Treatment(s)	7	14	21	28	7	14	21	28	7	14	21	28
Control	4.9	6.3	11.8	13.6	12.3	13.2	13.9	14.6	7.2	7.2	7.3	7.2
Ordinary Zn	5.2	6.8	12.4	14.0	12.4	13.8	14.2	14.8	7.2	7.3	7.2	7.2
Nano-Zn	5.8	7.1	12.9	14.4	13.6	14.3	14.8	15.2	7.2	7.2	7.3	7.4
LSD ($p = .05$)	0.3	0.4	6.5	0.7	0.6	0.7	0.7	NS	NS	NS	NS	NS

NS: Not Significant.

TABLE 2.11
Summary of Histopathological Analysis for Estimating Toxicological Effects of the Test Substance (–ve Indicates No Toxicity Observed)

Tissue/Groups	Group A	Group B	Group C	Group D	Group E
Liver	–ve	–ve	–ve	–ve	–ve
Kidney	–ve	–ve	–ve	–ve	–ve
Spleen	–ve	–ve	–ve	–ve	–ve

Group A: Normal diet.
Group B: Normal diet supplemented with control (unsprayed) wheat plant.
Group C: Normal diet supplemented with Fe nanoparticle (sprayed) wheat plants.
Group D: Normal diet supplemented with 100 mg per kg Fe nanoparticles.
Group E: Normal diet supplemented with ferrous sulphate (100 mg per kg) sprayed wheat plants.

TABLE 2.12
Gene Ontology Distribution for Unigenes in Pearl Millet (Six-Week-Old Plant)

Sample Name	Biological Processes	Molecular Functions	Cellular Components
Control	9222	12,044	6223
Nano-P	10,902	14,435	9415

revealed that biosynthesized nanonutrients produced no significant adverse effects, as evidenced by the normal tissue architecture observed in the exposed animals at the postinstillation time period of 90 days in comparison to the standard diet-exposed controls. In a nutshell, the gross architecture of tissues was intact with no noticeable necrosis or fibrosis (Table 2.11).

A complete bioinformatics study was conducted after the application of nanonutrients to six-week-old cereal (pearl millet) and legume (mung bean) plants. Gene ontology (GO) sequence distributions help in specifying all the annotated nodes comprising GO functional groups. Unigenes associated with similar functions are assigned to the same GO functional groups. The GO sequence distribution is shown in Table 2.12 for all the three GO domains—that is, biological processes, molecular functions, and cellular components.

TABLE 2.13
Gene Ontology Distribution for Unigenes of Mung Bean (Six-Week-Old Plant)

Sample Name	Biological Processes	Molecular Functions	Cellular Components
Control	3926	4660	2277
Nano-P	8395	9862	4481

The results showed 6542 unigenes, expressed differently in pearl millet due to nanophosphorus application, out of which 3918 were significantly downregulated and 2624 were significantly upregulated. An enzyme known as endo-β-mannase (EC 3.2.1.78) was encountered in a nanoparticle-treated pearl millet sample and not in respective control sample. This enzyme plays a key role in plant growth and development, including embryogenesis, seed germination, shoot growth, leaf formation, flower development, and fruiting. The GO distribution of mung bean (control and treated) plants (Table 2.13) showed predominantly more biological process, molecular function, and cellular component terms of the mung bean plant. Comparative accounting of the distribution of GO terms between mung bean (control) and mung bean (nanoparticles-treated plants) showed 2488 unigenes were exclusively present in nanoparticle-treated mung bean, helping in different metabolic activities like carbohydrate metabolism (605), lipid metabolism (237), mucleotide metabolism (323), amino acid metabolism (314), and biosynthesis of secondary metabolism (110). In general, 3792 unigenes expressed differently, out of which, 2751 were significantly downregulated and 1041 were significantly upregulated. The results showed that application of recommended doses of biosynthesized nanonutrients had a positive influence on different metabolic activities, leading eventually to sustainable development and higher yield.

2.2 CONCLUSIONS

An ecofriendly and low-cost protocol for biosynthesis and characterization of nanoparticles for plants has been employed and discussed. The results from application of recommended doses clearly demonstrated their positive effect on plant growth and development, including crop yield, nutrient mobilization, and use efficiency with no adverse effects. Thus, the biosynthesis technique exhibits impressive potential to minimize or omit the use of chemical fertilizers.

ACKNOWLEDGMENTS

The authors acknowledge the Indian Council of Agricultural research (ICAR) and the University Grant Commission (UGC) for financial assistance, under which the present investigations were carried out.

REFERENCES

Abdel-Raouf N, Al-Enazi N M, Ibraheem I B M. Green biosynthesis of gold nanoparticles using Galaxaura elongata and characterization of their antibacterial activity. *Arab J Chem*. 2017; 10: S3029–S3039.

Ahmad A, Senapati S, Khan M I, Kumar R, Sastry M. Extracellular biosynthesis of monodisperse gold nanoparticles by a novel extremophilic actinomycete, *Thermomonospora sp. Langmuir*. 2003; 19: 3550–3553.

Akhtar M S, Panwar J, Yun Y S. Biogenic synthesis of metallic nanoparticles by plant extracts. *ACS Sustain Chem Eng*. 2013; 1: 591–602.

Banerjee K, Rai V R. A review on mycosynthesis, mechanism, and characterization of silver and gold nanoparticles. *BioNanoScience*. 2018; 8: 17–31.

Bindraban P S, Dimkpa C O, Angle S, Rabbinge R. Unlocking the multiple public good services from balanced fertilizers. *Food Security*. 2018; 1: 1–3.

Diet K J, Herth S. Plants nanotoxicology. *Trends Plant Sci.* 2011; 16: 582–589.
Fariq A, Khan T, Yasmin A. Microbial synthesis of nanoparticles and their potential applications in biomedicine. *J Appl Biomed.* 2017; 15(4): 241–248.
Garg N K, Tyagi R K, Sharma G, Jain A, Singh B, Jain S, Katare O P. Functionalized lipid polymer hybrid nanoparticles mediated co-delivery of methotrexate & aceclofenac: A synergistic effect in breast cancer with improved pharmacokinetics attributes. *Mol Pharm.* 2017; 14: 1883–1897.
Husseiny M I, El-Azia M A, Badr Y, Mahmoud M A. Biosynthesis of gold nanoparticles using *Pseudomonas aeruginosa. Spectrochim Acta.* 2007; Part A 67: 1003–1006.
Iravani S, Korbekandi H, Mirmohammadi S V, Zolfaghari B. Synthesis of silver nanoparticles: Chemical, physical and biological methods. *Res Pharm Sci.* 2014; 9: 385–406.
Jain A, Agarwal A, Majumder S, Lariya N, Khaya A, Agrawal H, Majumdar S, Agrawal GP. Mannosylated solid lipid nanoparticles as vectors for site-specific delivery of an anti-cancer drug. *J Control Release.* 2010; 148: 359–367.
Jain A, Garg N K, Jain A, Kesharwani P, Jain A K, Nirbhavane P, Tyagi R K. A synergistic approach of adapalene-loaded nanostructured lipid carriers, and vitamin C co-administration for treating acne. *Drug Dev Ind Pharm.* 2016; 42: 897–905.
Jeevanandam J, Chan Y S, Danquah M K. Biosynthesis of metal oxide nanoparticles. *Chem Bio Eng Rev.* 2016; 3(2): 55–67.
Kaul R K, Kumar P, Burman U, Joshi P, Agrawal A, Raliya R, Tarafdar J C. Magnesium and iron nanoparticles production using microorganism and various salts. *Material Sci-Poland.* 2012; 30: 254–258.
Kesharwani P, Jain A, Jain A, Jain A K, Garg N K, Tekade R K, Raj Singh T R, Iyer A K. Cationic bovine serum albumin (CBA) conjugated poly lactic-co-glycolic acid (PLGA) nanoparticles for extended delivery of methotrexate into brain tumors. *RSC Adv.* 2016; 6: 89040–89050.
Klaus T, Joerger R, Olssen E, Granqvist C G. Silver based crystalline nanoparticles, microbically fabricated. *Proc Natl Acad Sci.* 1999; 96: 13611–13614.
Kowshik M, Arhtaputre S, Kharrazi S, Vogel W, Urban J, Kulkarni S K, Paknikar K M. Extracellular synthesis of silver nanoparticles by a silver- tolerent yeast strain MKY3. *Nanotech.* 2003; 14: 95–100.
Krajina B A, Proctor A C, Schoen A P, Spakowitz A J, Heilshorn S C. Biotemplated synthesis of inorganic materials: An emerging paradigm for nanomaterial synthesis inspired by nature. *Prog Mater Sci.* 2018; 91: 1–23.
Kumar R, Ghoshal G, Jain A, Goyal M. Rapid Green synthesis of silver nanoparticles (AgNPs) using (Prunus persica) plants extract: Exploring its antimicrobial and catalytic activities. *J Nanomed Nanotechnol.* 2017; 8: 2.
Lee S W, Mao C, Flynn C, Belcher A M. Ordering of quantum dots using genetically engineered viruses. *Sci.* 2002; 296: 892–895.
Li X, Xu H, Chen Z S, Chen G. Biosynthesis of nanoparticles by microorganisms and their applications. *J Nanomater.* 2011; 2011: 1–16.
Merzlyak A, Lee S W. Phage as template for hybrid materials and mediators for nanomaterials synthesis. *Curr Opin Chem Biol.* 2006; 10: 246–252.
Narayanan K B, Sakthivel N. Biological synthesis of metal nanoparticles by microbes. *Adv Colloid Interface Sci.* 2010; 156: 1–13.
Noda Y, Asaka T, Fudouzi H, Hayakawa T. Accumulation and interparticle connections of triangular Ag-coated Au nanoprisms by oil-coating method for surface-enhanced Raman scattering applications. *Appl Surf Sci.* 2018; 435: 687–698.
Pantidos N, Horsfall L E. Biological synthesis of metallic nanoparticles by bacteria, fungi and plants. *J Nanomed Nanotechnol.* 2014; 5.
Raliya R, Tarafdar J C, Singh S K, Gautam R, Choudhary K, Maurino V G, Saharan V. MgO nanoparticles biosynthesis and its effect on chlorophyll contents in the leaves of clusterbean (*Cyamopsis tetragonoloba L.*). *Adv Sci Engineer Med.* 2014; 6: 538–545.
Rathore I, Tarafdar J C. Prospective of biosynthesized magnesium nanoparticles in foliar application of wheat plant. *J Bionanosci.* 2015; 9: 209–214.
Saklani V, Majumder S, Jain V K. Microbial synthesis of silver nanoparticles: A review. *J Biotechnol Biomater.* 2013; S13: 007.
Saravanan M, Barik S K, MubarakAli D, Prakash P, Pugazhendhi A. Synthesis of silver nanoparticles from *Bacillus brevis* (NCIM 2533) and their antibacterial activity against pathogenic bacteria. *Microb Pathog.* 2018; 116: 221–226.
Sastry M, Ahmad A, Khan M I, Kumar R. Biosynthesis of metal nanoparticles using fungi and actinomycete. *Curr Sci.* 2003; 85: 162–170.

Schreiber L. Transport barriers made of catin, suberin and associated waxes. *Trends in Plant Sci.* 2011; 15: 546–553.

Tarafdar J C, Adhikari T. 2015. Nanotechnology in soil science: *Soil Science: An Introduction* (ISBN 81-90 3797-7-1). In: R K, Rattan, J C, Katyal, B S, Dwivedi, A K, Sarkar, T, Bhattacharyya, J C, Tarafdar, S S, Kukal, eds., *Indian Society of Soil Science*, New Delhi, pp. 775–807.

Tarafdar J C, Agrawal A, Raliya R, Kumar P, Burman U, Kaul RK. ZnO nanoparticles induced synthesis of polysaccharides and phosphatases by *Aspergillus* fungi. *Adv Sci Eng Med.* 2012c; 4: 1–5.

Tarafdar J C, Raliya R. Rapid low-cost and ecofriendly approach for iron nanoparticle synthesis using *Aspergillus oryzae* TFR9. *J Nanopart Res.* 2013; 2013: 1–4.

Tarafdar J C, Raliya R, Rathore I. Microbial synthesis of phosphorus nanoparticles from Tri-calcium phosphate using *Aspergillus tubingensis* TFR-5. *J Bionanosci.* 2012a; 6: 84–89.

Tarafdar J C, Rathore I. 2016. Microbial synthesis of nanoparticles for use in agriculture ecosystem: *Microbes for Plant Stress Management.* In: D J Bagyaraj, Jamaluddin, eds., *New India publishing Agency*, New Delhi, India, pp. 105–118.

Tarafdar J C, Rathore I, Thomas E. Enhancing nutrient use efficiency through nanotechnological interventions. *Ind J Fert.* 2015; 11: 46–51.

Tarafdar J C, Xiang Y, Wang WN, Dong Q, Biswas P. Standardization of size, shape and concentration of nanoparticle for plant application. *Appl Biol Res.* 2012b; 14: 138–144.

Vishnukumar P, Vivekanandhan S, Misra M, Mohanty A K. Recent advances and emerging opportunities in phytochemical synthesis of ZnO nanostructures. *Mater Sci Semicond Process.* 2018; 80: 143–161.

Wacławek S, Gončuková Z, Adach K, Fijałkowski M, Černík M. Green synthesis of gold nanoparticles using Artemisia dracunculus extract: Control of the shape and size by varying synthesis conditions. *Environ Sci Pollut Res.* 2018; 1–10.

3 Scientific Role of Transition-Metal Nanoparticles as Agrochemicals

Selvaraj Mohana Roopan and Ganesh Elango*

CONTENTS

* Corresponding author.

3.1 INTRODUCTION

Agriculture has been the backbone of the Indian economy since ancient times. Agriculture in India supports almost 17% of the world population, but with only 2.3% of the world's geographical area and 4.2% of the world's water. Indian agriculture is characterized by agro-ecological characteristics such as soil, rainfall, temperature, and cropping systems. For instance, Tamilnadu is the second largest state in India in economic activity, after Maharashtra. The state possesses various types of agro-climatic conditions, ranging from hot tropical to temperate climates; major soil types like red, black, alluvial, loamy, and sandy loam; and several crop varieties like cereals, pulses, oilseeds, fruits, vegetables, and other crops. The well-developed, agro-based cotton textiles and sugar industries are the added strengths for the rapid growth of the economy. But a weaker point is that it has an uneven rainfall pattern and demand for quality agricultural products due to urbanization and industrialization. Apart from these, crop productivity is facing a great threat in the form of pests and disease (Pandey, 2009; Sivagnanam, 2011).

Several pesticides and insecticides have been used in order to annul the harmful effects of these pests, such as chemical and organic fertilizers. But, these pesticides and insecticides have been found to cause harmful effects, not only to the pests, but also to the environment (Pereira et al., 2017, Madhumitha et al., 2016). The environmental impact caused by the chemical pesticides turned the view of agriculturalists and scientists towards natural and organic fertilizers. Natural and organic fertilizers are those that use plants or extracts from plants and microorganisms as pesticides, respectively. Biotechnology, in this context, has played a vital role in using microorganisms as a whole, or a part of a microorganism, as pesticides.

To date, usage of these organic and natural pesticides has not been able to satisfy the farmers in clearing out pests because of lack of awareness and high costs. At present, the scientists from the pertinent field of nanotechnology have undertaken efforts and extended the applications of pesticides to protect the crop growth in the form of nanopesticides, produced using silver, zinc, palladium, and others (Hemalatha and Madhumitha, 2015; Fowsiya et al., 2016; Aupama and Madhumitha, 2017). The current book chapter deals with notable nanopesticides, their agricultural applications, and the research endeavors underway to develop nanopesticides further. (Pandey, 2009; Quinones et al., 2017a,b).

3.2 AGRICULTURE IN INDIA

The agro-based industries of India play a major role in Indian economy. About 80% of the Indian population depends on agricultural sources. India has diversified ecological factors—such as soil, temperature, monsoon, and irrigation—and, thus, crops. But, the pressure on agriculture has been increasing since 1950, and the struggle to feed the rapidly building population is still increasing. Indian agriculture has been facing a formidable threat due to degradation of land and surface, loss of ground-water resources, insect pests, diseases, weeds, drought, salinity, heat, cold, and so forth. Moreover, there is a dire need for storage improvement, marketing, and transportation to attain security of livelihood (Pandey, 2009; Duhan et al., 2017). But agriculture has to face many obstacles—such as pollution, industrialization, urbanization, and also pests—to feed the population. Many of these vital factors have been discussed in brief (Pandey, 2009; Sivagnanam, 2011).

3.2.1 Industrialization and Urbanization

Urbanization and industrialization have created severe loss of fertile lands, since these are converted into nonagricultural lands for their use as industries and residential areas. Increased urbanization and industrialization during recent years has also created an increased market demand for quality agricultural products, such as milk, fruits, and also vegetables. This indicates an ardent need for increasing incentives in order to increase the yield per acre in field crops, vegetables, and fruits production, as well as milk and meat production (Pandey, 2009).

3.2.2 Pests

The outbreaks of pests and diseases pose tremendous threats to the crops, leading to complete loss to many farmers. There are several pests present around the globe, some important ones being the caterpillar, beetle, bollworm, and prodenia.

3.2.3 Irrigation

Uncertain monsoon and irrigation are among the major causes for this extreme reduction in the growth rates of agriculture. The irrigation process mostly consists of ground water extractions, which leads to lowering of the ground water level, and in some cases, to extinction of wells. In addition, the interstate disputes on major irrigation systems, such as Cauvery, Mullai Periyar, Palar, and others, also significantly affect the agricultural growth (Pandey, 2009).

3.3 PESTS AND PESTICIDES

3.3.1 Pests

Pests have been a problem to farmers during both the preharvesting and poststage of agriculture (Pandey, 2009). The micro- and macroorganisms (rodents, fungi, nematodes, viruses, etc.), commonly called pests, are illustrated in Figure 3.1. These pests damage the crops and reduce crop yield, not only causing economic loss to the farmers, but being hazardous to human health too (Paul and Das, 1994). Pests include several insects such as cow pea seed beetles, Colorado potato beetles, cabbage loopers, caterpillars, thrips, aphids, whiteflies, ants, termites, crane fly larvae, fleas, cutworms,

FIGURE 3.1 Pest that affect several crops.

FIGURE 3.2 Chemical organic pesticides.

grasshoppers, cabbage worms, diamond-back moths, armyworms, fruit flies, leaf rollers, mealy bugs, spider mites, and others. Several formulations have been produced in order to control these pests and to protect crop from damage (Rouhani et al., 2012; Murray et al., 2013).

3.3.2 Pesticides

Often, chemical mixtures utilized for reducing, killing, and repelling pests are known as pesticides (Figure 3.2), thereby enhancing the growth of crops (Rouhani et al., 2012). The pesticides are classified based on different criteria, listed in Table 3.1.

3.3.2.1 Organic Pesticides

Organic pesticides are the chemical substances that are used to attack and kill pests. The organic pesticides are classified into several types (acetamides, ureas, carbamates, organochlorines, thiocarbamates, organophosphates, etc.) (Rouhani et al., 2012). These include copper, copper sulfate, lime sulfur, sulfur, adjuvants, aluminum phosphate, barium carbonate, barium fluorosilicate, barium polysulfide, cadmium arsenate, calcium arsenate, and others. (Rouhani et al., 2012; Mader and Adamson, 2017). If used improperly, organic pesticides can harm people and the environment, as these mix with soil and water. Soil and water, when contaminated by these pests, affect plants, animals, and human beings too. This also affects the food chain and the food web of living organisms (Rouhani et al., 2012).

Federal Insecticide, Fungicide, and Rodenticide Act has delegated authority to the US Environment Protection Agency (EPA) to ensure that the usage of pesticides does not result in unreasonable adverse effects on humans and the environment. Also, the pesticide must be registered, furnishing

TABLE 3.1
Different Types of Pesticides are Classified Based on Various Criteria

S. No	Classification Criteria	Classification
1	Based on insects	Fungicides, algaecides, insecticides, miticides, etc.
2	Based on chemical nature	Organic and inorganic pesticides
3	Based on site of action	Stomach poison, contact poison, systemic poison, and fumigants
4	Based on formulation	Sprays, dry powder, gas/vapor, and other formulations

proper data for its acceptance to be used in the field. Based on the data, the pesticide is classified as highly toxic, toxic, moderately toxic, low toxic, and safe. It also is classified as a carcinogen or a noncarcinogen. With this classification, certain chemical pesticides are approved for limited usage and some are even banned. Though several acts and organizations have been created to regularly check the pesticides in action, the effects still could not be controlled, and alternatives for these chemical organic fertilizers is required.

3.3.2.2 Natural Pesticides

Natural pesticides, popularly known as chemicals, minerals, or biological forms, are produced by natural sources. In other words, these are also called biopesticides. Sometimes, these natural pesticides are also classified under organic-based pesticides on the basis of their specificity and toxicity. The natural pesticides include *Bacillus thuringiencis*, diatomaceous earth, neem, pepper, tobacco, castor bean, garlic, dahlia, mineral oil, papaya, onion, and others. These natural pesticides are considered safe and nontoxic since they are derived from natural sources. These pesticides have virulence similar to those of the organic pesticides and kill pests by attacking their physiological systems holistically. Still, the farmer and scientific communities feel the constant need of an improved alternative, which has now turned their interest towards nanotechnology and initiated the production of nanopesticides (Majumdar et al., 2013).

3.4 NANOTECHNOLOGY PERSPECTIVES

Nanoparticles are a unique subset of the broad field called nanotechnology. Nanoparticles are those unagglomerated particles, ranging in size from 1 to 100 nm. These nanoparticles are gaining importance in present-day technology due to their use in various fields such as medicine, drug delivery, information, energy, and the environment (Okuyama et al., 2004; Scalf and West, 2006). Nanoparticles can be mainly classified into two types: engineered nanoparticles and nonengineered nanoparticles. Engineered nanoparticles are those created or synthesized artificially (silver nanoparticles, gold nanoparticles, etc.) for their use in several techniques, whereas nonengineered nanoparticles are freely available in the environment (atmospheric nanoparticles that are produced during combustion, aerosols, etc.). Both engineered and nonengineered nanoparticles have demonstrated their uses in several industries. Several novel methods have lately been reported for nanoparticle synthesis, including physical, biological, and chemical techniques (Okuyama et al., 2004).

3.4.1 Two Approaches of Synthesis of Nanoparticles

Top-down and bottom-up are the two major approaches utilized for nanomaterial synthesis. The top-down approach follows decomposition of larger particles into smaller ones, whereas the bottom-up approach includes assembly of nanoparticles from molecules or atoms. The top-down method includes mechanical milling, ablation, and etching, and the bottom-up method includes electrochemical precipitation, vapor deposition, the sol–gel process, and green synthesis, which has gained increasing attention recently due to its relatively lower cost and less-toxic nature making it safer to use (Singh et al., 2015).

3.4.2 Green Synthesis

Biosynthesis of nanoparticles is one of the methods that can act as an alternative and eco-friendly approach for physical- and chemical-mediated synthesis, due to its cost-effectiveness. Further, chemical and physical synthesis of nanoparticles was found to involve hazardous materials (Basavegowda et al., 2013; Khan et al., 2018; Puji et al., 2018). Green synthesis using plant extracts is increasing in popularity as compared to synthesis mediated by microorganisms such as bacteria, algae and fungi, because of their pathogenicity and elaborative cell culture maintaining process (Scalf and West, 2006; Basavegowda et al., 2013). While comparing to microorganism-mediated synthesis of nanoparticles, the plant-mediated synthesis plays a major role due to its various advantages like extracellular synthesis and process ease. Production of nanoparticles is accomplished by mixing metal salt and various plant extracts to maintain the exact size, shape, and morphological natures by varying several conditions such as pH, temperature, concentration of metal salt, reaction time, reaction medium, and others. By maintaining these parameters, one can synthesize metal nanoparticles in a required morphological formation (Madhumitha et al., 2013). Currently, synthesis of nanoparticles extracellularly is carried out by utilizing most of the plant parts such as *Elaeagnus latifolia* (Phanjom et al., 2012), *Tinospora crispa* (Sarada et al., 2015), *Phyllanthus niruri* (Kathireswari et al., 2014), *Amaranthus spinosus* (Das et al., 2012), *Ananus comosus* (Basavegowda et al., 2013), *Cyphomandra betacea* (Elango and Roopan, 2016), *Persia Americana,* and also several green sources (Shah et al., 2018; Shao et al., 2018; Vedelago et al., 2018).

Apart from plants, microorganisms have also been used in the synthesis of nanoparticles. In this process, pigment or the whole cell extract is used as the source for synthesis of nanoparticles. After synthesis, the nanoparticles are characterized by several methods to find their appropriate shape, size, and chemical nature, for example, for establishing their use in a particular field or application.

3.4.3 Characterization of Nanoparticles

The synthesized nanoparticles are analyzed by various microscopical techniques such as transmission electron microscopy (TEM), scanning electron microscopy (SEM), dynamic light scattering (DLS), Brownian motion analysis, and analytical disc centrifugation. These methods are explained in brief.

3.4.3.1 Scanning Electron Microscopy (SEM)

In this method, the sample surface is scanned with a beam of electrons and the electrons that are back-scattered from the surface are measured, by which an image is obtained. The obtained image provides information on the size and shape of the synthesized nanoparticles.

3.4.3.2 Transmission Electron Microscopy (TEM)

Herein, imaging of the sample is accomplished by transmission and diffraction of an electron beam in high vacuum. It gives better resolution than SEM and also provides particle-size distribution with high accuracy (Wang et al., 2006).

3.4.3.3 Dynamic Light Scattering (DLS)

The sample is scattered by a laser light beam at its surface of dispersed nanoparticles, and the back-scattered light from the surface is measured. This method is used to find particle size, particle-size distribution, and also zeta potential (Wang et al., 2006).

3.4.3.4 Brownian Motion Analysis

The visual tracking of particle movement can be conducted using this method, which is examined by a laser light beam at the surface of the sample. This method is used to find the dispersed and agglomerated nanoparticles, zeta potential, and hydrodynamic diameter of the synthesized nanoparticles (Wang et al., 2006).

3.4.3.5 Analytical Disc Centrifugation

This involves sedimentation of the dispersed nanoparticles due to centrifugal force on a rotating disc in a gradient sugar solution. This method is used to find the polydispersed and agglomerated particles and their respective particle size. After characterization of these synthesized nanoparticles, they are used for many applications in various fields (Wang et al., 2006).

3.4.4 Applications of Nanoparticles

The synthesized nanoparticles have been used to study various applications, including antibacterial, antifungal, antihelminthic, antidiabetic, antialgal, insecticidal, pesticidal, and anticancer properties, as well as drug delivery in the field of biology.

3.4.4.1 Antimicrobial Activity

It is known that inorganic nanomaterials can be used as better bactericidal agents. These days, most of the nanomaterials can act as bactericidal agents, which play a major role in several industries, particularly in pharmaceutical industries. Mostly, silver (Ag) is the metal that plays a superior role as a bactericidal agent. Moreover, in several cases, most of the bacteria have developed their resistance against Ag metal. To overcome this, most metal nanoparticles have been prepared by much simpler and more eco-friendly methods, and using these metal nanoparticles with oligodynamic effect, most of the microbes are inactivated without causing any side effects (Khatami et al., 2018; Mohammed et al., 2018).

3.4.4.2 Medicine

The major responsibility of medical and biological research committees is to identify the solutions for designed and accidental problems, with the help of nanoparticles. The outcome of this has resulted in an amalgamation of nanotechnology with the fields of biotechnology, biomedical sciences, nanomedicine, drug design, and others. Most of the surgical tools for diagnostic, analytical, and therapeutic applications have been reported using nanotechnology.

3.4.4.3 Diagnostics

The magnetic nanoparticles, subjected towards suitable antibodies are used to label specific molecules. DNA embedded with gold (Au) nanoparticles can identify the samples for genetic sequence, where conversion of nucleotides to nucleic acids is performed by technology of nanopores, a useful nanoelectronic tool.

3.4.4.4 Drug Delivery

With the help of active agents placed in a specific region, one can decrease the side effects in case of prolonged intake or overconsumption of drugs. The nanoparticles, when attached with biological molecules, were found to reach the site accurately and to have a profound effect on the damaged or diseased area. The nanoparticles in drug delivery, especially in cancer treatment, could be targeted at the site of the tumor nearly perfectly. Also, nanoparticles have natural oxidizing activity, and along with the drug, these are found to be more effective than the conventional drug-delivery systems (Abhilash, 2010).

3.4.4.5 Stem Cell Therapy

Nanoparticles can be used as an effective tool in several fields, mainly in the field of stem cell diagnosis. A recent report from researchers confirms that the nanoparticles succeeded in enhancing the properties of stem cells with the help of mouse modelling. It was also found to stimulate blood vessel formation and renew tissue(s) effectively (Abhilash, 2010).

3.4.4.6 Other Uses

In addition, nanoparticles have been successfully used in the production of polymer composites, high-tech foams, gels, paints, and cosmetics (Abhilash, 2010).

3.5 NANOTECHNOLOGY IN AGRICULTURE

Recently, researchers have reported the use of nanoparticles in various fields of agriculture. Nanoparticles have been found to be used as growth promoters, nanopesticides, nanoherbicides, nanofertilizers, nanobiosensors, disease mitigators, and photosynthesis facilitators. The role of nanoparticles in each field has been explained in the following text:

3.5.1 Growth Promoters

In the field of agriculture, most researchers identified nanoparticles as growth promoters towards germination of seeds, growth of the plants, and so forth, and also they also observed various physicochemical alterations, which may be utilized for various purposes in in the agricultural field. Various interactions among greeneries and nanomaterials have resulted in enhancement of properties. Among the nanoparticles synthesized, carbon nanomaterials play a major role in the field of enhancing seed germination due to their improved chemical, mechanical, thermal, and electrical properties. Multiwalled carbon nanotubes (MWCNTs), when reacted with tomato seeds, lead to faster generation for biomass vegetation (Khodakovskaya et al., 2009).

3.5.2 Disease Suppression

Most of the reports state that the nanomaterials have the ability to eliminate the disease-causing pathogens, which act against growth of the plants. It has been reported that silver nanoparticles (Ag NPs) significantly reduced disease-causing pathogens in grass varieties (Jo et al., 2009) and those affecting the growth of various plant species (Lamsal et al., 2011). In another study, fluconazole tagged with Ag NPs resulted in notably superior activity against various bacterial and fungal pathogens (Gajbhiye et al., 2009). Several other reports also suggest that other nanoparticles can be effectively used as disease suppressors.

3.5.3 Nanopesticides

Of late, several drawbacks of using pesticides have been observed as usage of these toxic pesticides leads to adverse effects on plants as well as the environment, which results in elimination of nitrogen fixation, disturbing the bird habitat, farmer economy, decreased land remuneration, and so forth. Therefore, to overcome the aforementioned drawbacks, the nanoparticles play a major role without any toxic effects to the environment (eco-friendly synthesized metal nanoparticles) (Ghormade et al., 2011). Various studies on nanopesticides and their efficiency have been explained in detail in this chapter.

3.5.4 Nanofertilizers

Nanofertilizers are nanoparticle formulations that act similar to the conventional fertilizers, providing nutrients to plants. Exchange of old fertilizers with recently developed fertilizers embedded with nanoparticles due to the characteristic feature of the latter in releasing their nutrients and maintaining water flow in a controlled manner have been reported to result in notable gains in the agricultural fields (Moaveni and Kheiri, 2011; Naderi and Danesh-Shahraki, 2013). A study on cooper nanoparticles embedded with soil resulted in tremendous increase of seedling growth, from 40% to 95% (Shah and Belozerova, 2009). Similarly, a study on hydroxyapatite, $Ca_5 (PO_4)_3OH$, nanoparticles exhibited a significant fertilizing effect on soybeans (Liu and Lal, 2014). Thus, from the said findings, it was found that nanoparticles can play a prominent role due to their eco-friendly nature in reducing or eradicating the toxicity levels present in the soil (Singh et al., 2015).

3.5.5 Nanobiosensors

Nanoparticles have been found to be used as biosensors in detecting plant pathogens. In addition, they are also found to diagnose metabolites and chemical materials such as pesticides, herbicides, and so forth. Though this research is in the initial stage in the agricultural field, it can be used as an effective investigating tool to detect compounds that indicate disease. Studies on fluorescent nanoparticles have been designed in combination with antibodies and are used for diagnosis of disease in the plant *Xanthomonas axonopodis* (family Solanaceae) (Yao et al., 2009). Another study revealed the detection of metabolites such as glucose, urea, and others and suggested that nanobiosensors can be effectively used in detecting pesticides, insecticides, and the disease too (Rai et al., 2012).

3.5.6 Photosynthesis

Currently, most scientists focus towards increasing the effort towards augmenting photosynthesis with the help of nanotechnology. In this regard, nanotechnology has its distinct contribution in manipulating genes by increasing the efficiency of the synthesized materials. A recent report states that the single-walled carbon nanotubes (SWCNTs) significantly increased the activity of photosynthesis (Giraldo et al., 2014). Also, a study on gene-induced nano-anatase resulted in increased carboxylation, thus unravelling a promising aspect of nanoparticles in increasing the photosynthetic efficiency (Ma et al., 2008).

3.6 NANOPESTICIDES

Nanopesticides can be defined as any formulation that is produced in the range of the nanometer and to be used against insect pathogens in agriculture. Nanopesticides can be of different types, such as organic forms and inorganic forms, and also they exist in various formulations. The main function of the nanoformulation is to gain the tendency of ingredients that are unsoluble in nature and to maintain the exposure of active soluble ingredients to save the active ingredient from degradation (Ragaei and Sabry, 2014).

3.6.1 Nanoformulation

Various types of nanoformulations that are majorly utilized as nanopesticides (Abhilash, 2010; Ragaei and Sabry, 2014) are clearly listed in Table 3.2.

3.6.2 Works on Nanopesticides

Several studies have been performed for the use of nanoparticle as nanopesticides, of which few are discussed as follows. A study on CuO and ZnO nanoparticles has been found to be active against the Pythium isolates *Pythium ultimum* and *Pythium aphanidermatum*. They were found to affect the Fe-chelating ability of these two pests and were found to affect their growth (Zabrieski et al., 2015).

Another study on silver nanoparticles synthesized by an aqueous extract of *Euphorbia prostrate* was found to act against *Sitophius oryzae L* (Zahir et al., 2012). Eco-friendly, synthesized Ag NPs from various plant sources have bactericidal effects on *E. Coli*, *Staphylococcus*, *Streptococcus*, *Shigella*, *Salmonella*, *Vibrio*, and *Enterobacteria* and pesticidal activity on the pest *Sitophius oryzae* (Banerjee et al., 2014; Sankar and Abideen, 2015). Another study (Devi et al., 2014) on Ag NPs synthesized by *Euphoria hirta* was found to act against larvae and pupae of the cotton boll worm, *Helicoverpa armigera*. It affects the pest by exhibiting larval toxicity and affecting longevity of males and females and also fecundity of the pest.

A study on bactericidal activity on plant pathogens by pine cone extract showed that pine cone extract can effectively kill the plant pathogens *X. Oryzae*, *B. megaterium*, *B. glumae*, and others (Velmurugan et al., 2013). Ag NPs also act as pathogens for plant fungus (Marimuthu et al., 2011). The fruit peel extract of *Annona squamosa* was found to have adulticidal activity against several flies, mites, bugs, mosquito larvae, and others. Moreover, most of the scientists have proven that

TABLE 3.2
Some Types of Nanoformulations Used as Nanopesticides

S. No	Nanoformulation	Definition	Morphology
1	Nanospheres	Aggregate in which active compound is homogenously distributed into polymeric matrix	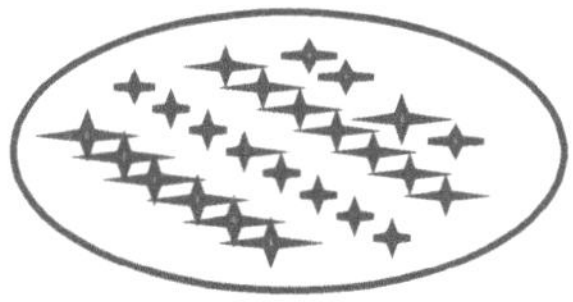
2	Nanocapsules	Aggregate in which active compound is concentrated near the centre core and lined by the matrix polymer	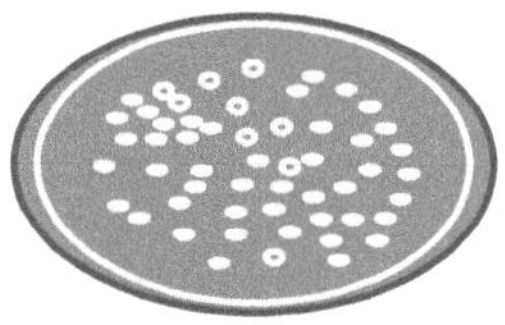
3	Nanogels	Hydrophilic polymers that absorb high volumes of water	
4	Micelles	Aggregate formed in aqueous solutions by molecules containing hydrophilic and hydrophobic moieties	

the synthesized metal nanoparticles and various fruit extracts also can be utilized as larvicidal and pesticidal agents, which are cheaper and nontoxic to nature (Madhumitha et al., 2012).

Another study on various pests such as *Rhipicephalus microplus*, capitis, cattle kick, mites, and others are studied against green synthesized ZnO nanoparticles and the results are significant. In addition, green-synthesized ZnO NPs were subjected to act as larvicidal agents against malarial and dengue vectors, which were stated to play a prominent role as larvicidal and pesticidal agents (Kirthi et al., 2011). The Ag NPs synthesized from *Hibiscus rosasinensis* tend to act has larvicidal agents and they were found to effective against the larvae of the *Aedes albopictus* mosquito reared in a rubber plantation (Sarah et al., 2012).

Similarly, a study on silica nanoparticles (SNPs) and Ag NPs found that they affect the larval stage and adults of *Callosobruchus maculatus* on cowpea seed. Also a study on aluminium and zinc particles was found to affect the larvae of cereal grains, for example, *Tribolium castaneum* (Deka and Sinha, 2015). In another case, the synthesized material from *Penicillium pinophilum* was found to effectively remove the pesticides from plant as well as the environment (Abd-El-Salam et al., 2015). From these studies, it is evident that nanoparticles can be effectively used as pesticides to nullify the action of pesticides on plants and also the environment.

3.7 CONCLUSIONS

Nanoparticles can also play a role as agrochemicals which will facilitate more researchers to focus their work progress towards nanoagrochemicals. Moreover, in this chapter, the authors discussed some basics of industrialization and urbanization of agricultural pests, which can play a role against various pesticides by leading those to mortality rate. These types of nanoparticles were majorly termed as nanoagrochemicals and also this chapter deals with nanoparticles synthesis and characterization techniques, which can be of great help to some basic researchers at the starting potential for synthesis of nanoagrochemicals.

ACKNOWLEDGMENTS

The authors are thankful to VIT management and DBT (No.BT/PR6891/GBT/27/491/2012) for providing financial support to carry out this research progress.

REFERENCES

Abd-El-Salam SA, Hamzah AM, El-Taweelah NM. Aluminum and zinc oxides nanoparticles as a new method in controlling the red flour beetle, *Tribolium castaneum* (herbest) compared to malathion insecticide. *Int J Scientific Res. Agri. Sci.* 2015; 2: 1–6.

Abhilash M. Potential applications of nanoparticles. *Inter. J. Pharma. Biosci.* 2010; 1(1): 1–12.

Aupama N, Madhumitha G. Green synthesis and catalytic application of silver nanoparticles using *Carissa carandas* fruits. *Inorg. Nano-Metal. Chem.* 2017; 47: 116–120.

Banerjee P, Satapathy M, Mukhopahayay A, Das P. Leaf extract mediated green synthesis of silver nanoparticles from widely available Indian plants: Synthesis, characterization, antimicrobial property and toxicity analysis. *Biores. Bioprocess.* 2014; 1(3): 1–10.

Basavegowda N, Sobczak-Kupiec A, Malina D, Yathirajan HS, Keerthi VR, Chandrashekar N et al. Plant mediated synthesis of gold nanoparticles using fruit extracts of *Ananas comosus* (L.) (pineapple) and evaluation of biological activities. *Adv. Mat. Lett.* 2013; 4(5): 332–337.

Das RK, Gogoi N, Babu PJ, Sharma P, Mahanta C, Bora U. The synthesis of gold nanoparticles using *Amaranthus spinosus* leaf extract and study of their optical properties. *Adv. Mater. Phy. Chem.* 2012; 2: 275–281.

Deka AC, Sinha SK. Mycogenic silver nanoparticle biosynthesis and its pesticide degradation potentials. *Int. J. Tech. Enhancement Emerg. Eng. Res.* 2015; 3(5): 108–113.

Devi D, Murugan K, Selvam CP. Green synthesis of silver nanoparticles using *Euphorbia hirta* (Euphorbiaceae) leaf extract against crop pest of cotton bollworm, *Helicoverpa armigera* (Lepidoptera: Noctuidae). *J. Biopest.* 2014; 7: 54–66.

Duhan JS, Kumar R, Kumar N, Kaur P, Nehra K, Duhan S. *Nanotechnology: The new perspective in precision agriculture.* Biotech. Report. 2017; pp. 11–23.

Elango G, Roopan SM. Efficacy of SnO_2 nanoparticles toward photocatalytic degradation of methylene blue dye. *J. Photochem. Photobiol. B.* 2016; 155: 34–38.

Fowsiya J, Madhumitha G, Al-Dhabi NA, Arasu MV. Photocatalytic degradation of Congo red using *Carissa edulis* extract capped zinc oxide nanoparticles. *J. Photochem. Photobio. B.* 2016; 162: 395–401.

Gajbhiye M, Kesharwani J, Ingle A, Gade A, Rai M. Fungus-mediated synthesis of silver nanoparticles and their activity against pathogenic fungi in combination with fluconazole. *Nanomed-Nanotechnol.* 2009; 5(4): 382–386.

Ghormade V, Deshpande MV, Paknikar KM. Perspectives for nanobiotechnology enabled protection and nutrition of plants. *Biotech. Adv.* 2011; 29: 792–803.

Giraldo JP, Landry MP, Faltermeier SM, McNicholas TP, Iverson NM, Boghossian AA et al. Plant nanobionics approach to augment photosynthesis and biochemical sensing. *Nat, Mater.* 2014; 13: 400–408.

Hemalatha K, Madhumitha G. Eco-friendly synthesis of palladium nanoparticles, environmental toxicity assessment and its catalytic application in Suzuki miyaura coupling. *Res. J. Pharm. Technol.* 2015; 8: 1691–1700.

Jo YK, Kim BH, Jung G. Antifungal activity of silver ions and nanoparticles on phytopathogenic fungi. *Plant Dis.* 2009; 93(10): 1037–1043.

Kathireswari P, Gomathi S, Saminathan K. Plant leaf mediated synthesis of silver nanoparticles using *Phyllanthus niruri* and its antimicrobial activity against multi drug resistant human pathogens. *Int. J. Curr. Microbiol. App. Sci.* 2014; 3(3): 960–968.

Khan MZH, Tareq FK, Hossen MA, Roki MNAM. Green synthesis and characterization of silver nanoparticles using *Coriandrum sativum* leaf extract. *J. Eng. Sci. Technol.* 2018; 13: 158–166.

Khatami M, Iraj S, Nobre MAL, Niloofar Z, Aflatoonian MR. Waste-grass-mediated green synthesis of silver nanoparticles and evaluation of their anticancer, antifungal and antibacterial activity. *Green. Chem. Lett. Rev.* 2018; 11: 125–134.

Khodakovskaya M, Dervishi E, Mahmood M, Xu Y, Li Z, Watanabe F et al. Carbon nanotubes are able to penetrate plant seed coat and dramatically affect seed germination and plant growth. *ACS Nano.* 2009; 3: 3221–3227.

Kirthi AV, Rahuman AA, Rajakumar G, Marimuthu S, Santhoshkumar T, Jayaseelan C et al. Acaricidal, pediculocidal and larvicidal activity of synthesized ZnO nanoparticles using wet chemical route against blood feeding parasites. *Parasitol. Res.* 2011; 109: 461–472.

Lamsal K, Kim SW, Jung JH, Kim YS, Kim KS, Lee YS. Application of silver nanoparticles for the control of *Colletotrichum* species *in vitro* and pepper anthracnose disease in field. *Mycobiology.* 2011; 39(3): 194–199.

Liu R, Lal R. Synthetic apatite nanoparticles as a phosphorus fertilizer for soybean (Glycine max). *Sci. Rep.* 2014; 4: 5686–5691.

Ma L, Liu C, Qu C, Yin S, Liu J, Gao F et al. Rubisco activase mRNA expression in spinach: Modulation by nanoanatase treatment. *Biol. Trace. Elem. Res.* 2008; 122(2): 168–178.

Mader E, Adamson NL. Organic approved pesticides, minimizing risks to bees. *The Xerces Soc. Invertebr. Conserv.* 2017; 10: 1–6.

Madhumitha G, Elango G, Roopan SM. Biotechnological aspects of ZnO nanoparticles: Overview on synthesis and its applications. *Appl. Microbiol. Biotechnol.* 2016; 100: 571–581.

Madhumitha G, Rajakumar G, Roopan SM, Rahuman AA, Priya KM, Saral AM et al. Acaricidal, insecticidal, and larvicidal efficacy of fruit peel aqueous extract of *Annona squamosa* and its compounds against blood-feeding parasites. *Parasitol. Res.* 2012; 111: 2189–2199.

Madhumitha G, Roopan SM. Devastated crops: Multifunctional efficacy for the production of nanoparticles. *J. Nano.* 2013; 2013: 1–12.

Majumdar A, Miles J, Reeves M, Gray G, Kelly N, East W, Chapman L, Rear BO. Insecticides for organic commercial & backyard vegetable production, Alabama cooperative extension system, ANR1428, 2013: pp. 1–8.

Marimuthu S, Rahuman AA, Rajakumar G, Santhoshkumar T, Kirthi AV, Jayaseelan C et al. Evaluation of green synthesized silver nanoparticles against parasites. *Parasitol Res.* 2011; 108: 1541–1549.

Moaveni P, Kheiri T. *2nd International Conference on Agricultural and Animal Science*; November 25–27, 2011; Maldives. Singapore: IACSIT Press, 2011; vol. 22: pp. 160–163.

Mohammed RS, Mujeeb K, Mufsir K, Abdulrahman Al-W, Hamad ZA, Mohammed RHS et al. Plant-extract-assisted green synthesis of silver nanoparticles using *Origanum vulgare* L. extract and their microbicidal activities. *Sustainability.* 2018; 10: 913–927.

Murray T, Miles C, Daniels C. Natural insecticides, A Pacific Northwest extension publication, PNW649; 2013; pp. 1–10.

Naderi MR, Danesh-Shahraki A. Nanofertilizers and their roles in sustainable agriculture. *Int. J. Agri. Crop. Sci.* 2013; 5, (19): 2229–2232.

Okuyama K, Lenggoro W, Iwaki T. Nanoparticle preparation and its application - A nanotechnology particle project in Japan. *Proceedings of 2004 International Conference on MEMS, NANO and Smart Systems (ICMENS'04)*, IEEE, 2004; pp. 1–4.

Pandey MM. Indian agriculture – an introduction. *Country report presented in the Fourth Session of the Technical Committee of APCAEM*, December 1–3 2008; Chiang Rai, Thailand, 2009; pp. 1–39.

Paul SK, Das GSK, *Pest Control.* Andhra Pradesh, India: International Crops Research Institute for the Semi-Arid Tropics (ICRISAT), Patancheru, 1994; pp. 1–56.

Pereira AES, Sandoval-Herrera IE, Zavala-Betancourt SA, Oliveira HC, Ledezma-Perez AS, Romero J et al. γ-Polyglutamic acid/chitosan nanoparticles for the plant growth regulator gibberellic acid: Characterization and evaluation of biological activity. *Carobohyd. Poly.* 2017; 157: 1862–1873.

Phanjom P, Sultana A, Sarma H, Ramchiary J, Goswami K, Baishya P. Plant-mediated synthesis of silver nanoparticles using *Elaeagnus latifolia* leaf extract. *Digest. J. Nano. Biostruct.* 2012; 7(3): 1117–1123.

Puji L, Indah P, Any J. Green synthesis of silver nanoparticle using rambutan (*Nephelium lappaceum* L.) peel extract and its antibacterial activity against *Salmonella parathypi* A. *MATEC Web Conference*, 2018; vol. 154: p. 01024.

Quinones JP, Bruggemann O, Covas CP, Ossipov DA. Self-assembled hyaluronic acid nanoparticles for controlled release of agrochemicals and diosgenin. *Carbohyd. Poly.* 2017a; 157–169.

Quinones JP, Mardare CC, Hassel AW, Bruggeman O. Self-assembled cellulose particles for agrochemical applications. *European Polym. J.* 2017b, 93: 706–716.

Ragaei M, Sabry AKH. Nanotechnology for insect pest control. *Int. J. Sci. Tech.* 2014; 3(2): 528–545.

Rai V, Acharya S, Dey N. Implications of nanobiosensors in agriculture. *J. Biomater. Nanobiot.* 2012; 3: 315–324.

Rouhani M, Samih MA, Kalantari S. Insecticidal effect of silica and silver nanoparticles on the cowpea seed beetle, *Callosobruchus maculatus* F. (Col.: *Bruchidae*). *J. Entomology. Res.* 2012; 4(4): 297–305.

Sankar MJ, Abideen S. Pesticidal effect of green synthesized silver and lead nanoparticles using *Avicennia marina* against grain storage pest *Sitophilus oryzae. Int. J. Nano. Biostruct.* 2015; 5(3): 32–39.

Sarada R, Jagannadharao V, Sunder BS. Plant mediated synthesis of silver nanoparticles by using dried stem powder of *Tinospora crispa* and study of its photoluminescence activity. *Int. J. Engg. Res. Appl.* 2015; 5(7): 82–89.

Sarah JS, Raji PK, Chandramohanakumar N, Balagopalan M. Larvicidal potential of biologically synthesised silver nanoparticles against *Aedes albopictus. Res. J. Rec. Sciences.* 2012; 1: 52–56.

Scalf J, West P. Part I: Introduction to nanoparticle characterization with AFM. *Pacific Nanotechnol. Santa Clara*, 2006; 1–8.

Shah S, Din SU, Khan A, Rehmanullah, Shah SA. Green synthesis and antioxidant study of silver nanoparticles of root extract of *Sageretia thea* and its role in oxidation protection technology. *J. Poly. Env.* 2018; 26: 2323–2332.

Shah V, Belozerova I. Influence of metal nanoparticles on the soil microbial community and germination of lettuce seeds. *Water Air Soil Pollut.* 2009; 197: 143–148.

Shao Y, Wu C, Wu T, Yuan C, Chen S, Ding T, Ye X, Hu Y. Green synthesis of sodium alginate-silver nanoparticles and their antibacterial activity. *Int. J. Biol. Macromol.* 2018; 111: 1281–1292.

Singh A, Singh NB, Hussain I, Singh H, Singh SC. Plant-nanoparticle interaction: An approach to improve agricultural practices and plant productivity. *Int. J. Pharma. Sci. Invention.* 2015; 4(8): 25–40.

Sivagnanam KJ. *Report on state agriculture profile of Tamilnadu – 2011.* Ministry of Agriculture, Government of India, Chennai, Tamilnadu, India, 2011: pp. 1–98.

Vedelago J, Gomez CG, Valente M, Mattea F. Green synthesis of silver nanoparticles aimed at improving theranostics. *Rad. Phy. Chem.* 2018; 146: 55–67.

Velmurugan P, Lee SM, Iydroose M, Lee KJ, Oh BK. Pine cone-mediated green synthesis of silver nanoparticles and their antibacterial activity against agricultural pathogens. *Appl. Microbiol. Biotechnol.* 2013; 97: 361–368.

Wang ZG, Zu XT, Xiang X, Lian J, Wang LM. Preparation and characterization of polymer/inorganic nanoparticle composites through electron irradiation. *J. Mater. Sci.* 2006; 7: 1973–1978.

Yao KS, Li SJ, Tzeng KC, Cheng TC, Chang CY, Chiu CY et al. Fluorescence silica nanoprobe as a biomarker for rapid detection of plant pathogens. *Adv. Mater. Res.* 2009; 79e82: 513–516.

Zabrieski Z, Morrell E, Hortin J, Dimkpa C, McLean J, Britt D et al. Pesticidal activity of metal oxide nanoparticles on plant pathogenic isolates of Pythium. *Ecotoxicol.* 2015; 24: 1305–1314.

Zahir AA, Bagavan A, Kamaraj C, Elango G, Rahuman AA. Efficacy of plant-mediated synthesized silver nanoparticles against *Sitophilus oryzae. J. Biopest.* 2012; 95–102.

4 Nanofertilizers

An Innovation towards New Generation Fertilizers for Improved Nutrient-Use Efficacy and Environmental Sustainability

Anu Kalia and Harleen Kaur*

CONTENTS

* Corresponding author.

4.1 INTRODUCTION

Fertilizers are chemical compounds of synthetic/natural origin, formulated in appropriate concentrations and combinations, which furnish the essential nutrient elements to plants on application to achieve improved growth and yield of crop plants (Behera and Panda, 2009). The prominent use of fertilizers is to achieve higher food production targets (more crop per unit arable land), in order to satiate the burgeoning human population, particularly witnessed in the developing world nations during the green-revolution phase. Thus, the soils of the intensively cultivated lands exhibit depleted capability to sustain the optimum nutrient contents for maintenance of appropriate plant growth due to voracious mining, uptake, and exhaustion of essential nutrients by the previous crop. Therefore, the nutrients have to be replenished back instantaneously as fertilizers. However, recurrent fertilization or, precisely, fertilizer misuse and abuse has led to multiplexed negative impacts, including decreased fertilizer response ratios and episodes of multi-micronutrient deficiencies in soils, besides weathering of soil minerals and development of ailing soils, incapable to sustain optimum fertility standards. Thus, strenuous research and development efforts in the next-generation innovation-driven technology, nanotechnology, are required to address the pitfalls of conventional fertilizers. The rising research interest, fund flow, and patents filed or published on novel nanofertilizers (NFs), showcase their tremendous potentiality. However, a similar or even greater amount of published literature has shown the possibilities of negative impacts on the rampant usage of these fertilizers. Nevertheless, the opportunities for the nano-interventions in plant nutrition and fertilizers have been growing and causing introgression into hitherto diverse disciplines of nanonutrient dynamics and nanotoxicity.

4.1.1 Nanofertilizers: Need and Possible Advantages over Conventional Fertilizers?

There are several factors which have expedited the research efforts on NFs, such as low fertilizer-response ratios, emergence of multi-micronutrient deficiencies in majority soils under agriculture due to imbalanced fertilization, and daunting environmental deterioration effects.

4.1.1.1 Controlled Release of Fertilizer Nutrients

The primary loss of nutrients from soils occurs due to an array of environmental abiotic and biotic factors: the physical volatilization of applied nitrogen (N) in gaseous ammonia and nitrous oxides, chemical precipitation of P-fertilizers such as calcium-, iron- and aluminum-phosphates, leaching of nitrates or phosphates to ground-water aquifers, and microbial transformation and fixation of the nutrients in organic form.

The high dissolution and mobility potentials of macronutrients (MNs) require adsorption, intercalation, sequestration, or encapsulation of the N, P (phosphorus), and K (potassium) nutrients. MNs, on or in nanovehicles, exhibit high nutrient-adsorption potential due to high surface area (zeolites, clay minerals), nanoporous materials (wood, biomass, ash), and protection by thin protective polymeric films or emulsions (de Rosa et al., 2010). The nitrogen release potential of a sustained, slow-release soft-wood impregnated nano-hydroxyapatite-urea formulation was evaluated, and it was observed to release more than 10 mg N in soil under rice cultivation (Kottegoda et al., 2011). Another report on nano-hydroxyapatite as a potential P-fertilizer to improve phosphorous availability in wheat has been evaluated and reported to be superior in comparison to the bulk hydroxyapatite and triple super phosphate for improving the growth of wheat (Montalvo et al., 2015).

Alternatively, the nutrient or its compound can be encapsulated in a protective polymeric film or in a hygroscopic, swelling-type nanocomposite material to ensure the controlled release. The latter type possesses the ability to absorb and retain large volume of water so that NH_4^+ nutrient release rate could be ensured to be more than 50% after ten days of application (Barati, 2010).

Apart from macronutrients, micronutrients are most benefitted by development of nanoformulations in terms of availability in plant uptake-friendly amounts and forms for longer durations, compared to their bulk counterparts that supply nutrients in ionic forms (Dimpka et al., 2017).

4.1.1.2 Enhanced Nutrient Use Efficiency (NUE)

Conventional fertilizers suffer from substantial low-use efficiency issues. Among the primary macronutrients, nitrogen, phosphorus, and potassium exhibit alarmingly low use efficiency. Approximately 40%–70% N, 80%–90% P, and 50%–70% K applied as conventional fertilizers are lost due to leaching as nitrates or phosphates contaminating the subsurface water aquifers, removed from soil as gaseous ammonia and nitrogen oxides, flushed to surface water bodies to cause eutrophication, or transformed into recalcitrant or mineralizable soil organic matter by activity of soil microflora, if not utilized by the plant in a timely manner (Monreal et al., 1986). Ironically, this leaves a meager amount of the applied fertilizers (~30%–60% N, 10%–20% P, and 30%–50% K) available for uptake by the plants. Thus, the primary loss of nutrients from soil and heavy usage of conventional fertilizers demonstrate many demerits which override their possible benefits of nutrient application.

Use of NFs is anticipated to circumvent, as well as effectively address, these serious low use efficiency issues of the conventional fertilizers so as to reduce vehement losses and ensure environment safety. Not just this, NUE stays very high for NFs equivalent or higher plant growth, and yield may occur at doses substantially lower than the same nutrient supplied in its bulk form (Herrera et al., 2016).

4.1.1.3 Additional Benefits

The additional benefits of using nanonutrient fertilizers include enhancement in the nutritive quality of the produce (i.e., biofortification), induction of stress tolerance to a variety of biotic and abiotic factors, and use of nano-based hydrogels and other polymeric substances as potent soil conditioners. The nanomaterials (NMs) of micronutrients such as zinc, iron, and selenium have been recognized by the World Health Organization (WHO) as the essential dietary nutrients for the better growth and development of animal and human life. Priming, coating, or fortification of these nutrients in nanoform will help in improving the nutritional quality of the produce (Subbaiah et al., 2016). Application of ferric oxide nanoparticles in ginger significantly increased the iron content in ginger plants (Siva and Benita, 2016). Similarly, application of nano-Fe improved the seed protein content by 2%, compared to Fe, in control black-eyed peas (Delfani et al., 2014). A plant with better nutritional status and metabolic activities has often been found to show greater capacity to adapt to stress conditions (Morales-Diaz et al., 2017). The NFs containing nutrient elements as transition metals (Zn, Cu, Fe, Mn and others) induce oxidative stress in microorganisms and aquatic and terrestrial plants. However, in the case of terrestrial plants, if the oxidative stress level does not exceed the toxic threshold level responsible for cell death, it induces a defense induction phenomenon that results in expression of resistance genes and accumulation of enzymatic and nonenzymatic antioxidants (Van Aken, 2015).

Application of NMs as soil-binding (Majeed and Taha, 2013) or land-reclamation agents (Liu and Lal, 2012) has many benefits. The applied NM-based composites or polymeric substances will not only ensure binding of the soil particles to curb erosion, but the swelling potential of the polymer also helps in absorbing and storing high contents of water, besides decontamination of the pesticide residues, heavy metals, and radionuclides toxic for the plants, animals, and humans.

4.1.1.4 Eco-Benefits

The ecological benefits of nano-inspired technology can be towards reuse and recycling of the agri-biowaste such as straw, husk, stovers, pod covers, and others, such that these can be utilized to fabricate nanoscale products for use as nanodelivery vehicles for the nutrient elements. The ecological problems posed by burning or dumping of the rice husk and straw are grave. These rice field biowastes can be converted into nanosilicate particles, which work quite well for nutrient adsorption or sequestration (Wanyika et al., 2012). The other eco-benefit is discussed in Section 4.1.1.3, explaining the dual role of applied NMs in decontamination of different chemicals and heavy metals. The nanosilicate, derived from rice husk, can be useful to decontaminate arsenic from ground and surface water bodies.

4.1.2 Types of Nanofertilizers

4.1.2.1 Micronutrient Nanofertilizers

Micronutrients are the mineral elements required in very low or trace quantities by plants for proper growth and development. These elements play a pivotal role in physiological and biochemical processes, as these function as cofactors or the conjugating metal central atom of several enzymes and primary/secondary biological macromolecules. Application of micronutrient NFs enhance seed germination, root and shoot growth (probably due to enhancement in the *in planta* indole acetic acid [IAA] levels), and increase micronutrient content in produce and grains. Pandey et al. (2010) found that zinc oxide (ZnO) nanoparticles (NPs) can enhance seed germination and root growth of the common gram *Cicer arietinum* (chick pea). They found that ZnO NPs increased the level of IAA, a phytohormone that regulates plant growth, in roots.

4.1.2.2 Macronutrient Nanofertilizers

Macronutrients (MNs) are mineral elements required in large quantities for proper growth, metabolism, and other processes/functions of a plant. The nine-element MNs can be categorized as those derived from air and water, that is, carbon, hydrogen, and oxygen, and those derived from soil, including nitrogen, phosphorus, and potassium (primary MNs) and calcium, magnesium, and sulfur (secondary/tertiary MNs). These elements contribute >95% of the entire plant biomass. The nanoscale macronutrient fertilizers can be obtained by preparing the emulsion of nutrients in synthetic or organic polymers as emulsions, encapsulation in polymeric films (Barati, 2010), and adsorption on nanovehicle assembly (Kottegoda et al., 2011, 2017).

4.1.2.3 Non-Nutrient Nanoparticulate Fertilizers

Non-nutrient nanofertilizers may include the nanoparticulate materials that have no nutritive relevance or contribution to the plant, such as the carbon-based NMs (e.g., C_{60} fullerenes, carbon nanotubes, graphene sheets, carbon Q-dots), transition metal/metal oxide NPs (e.g., TiO_2, CdSe, CeO_2, Li, Ag, Au), and the nanoclay mineral particle products. Corral-Diaz et al. (2014) have reported an increase in antioxidant level in plants on application of nano-CeO_2. Tripathi et al. (2017) have observed significantly higher growth and yield in gram by application of water-soluble carbon nano-onions. Further, the first generation seeds derived from of the plants treated with carbon nano-onions had higher protein and micronutrient contents.

4.1.3 Synthesis of Nanofertilizers

The NFs can be generated/fabricated through either of the two basic nanotechnological approaches, top-down or bottom-up. However, it is improbable that all types of NFs can be synthesized through one or both techniques, as an in-depth understanding to probe chemical nanothermodynamics of interacting reactants, products, and interfacial interactions of the constituent nanostructures is a prerequisite.

4.1.3.1 Top-Down Approaches

This involves fabrication of nanophase materials from bulk material using various physical, chemical, or biological methods, or a combination of any of these. The physical technique that can ensure large-scale synthesis in an economically feasible manner involves attrition or mechanical grinding. This technique has been utilized for generation of clay-mineral (Kottegoda et al., 2011) or apatite-based NF products (Kottegoda et al., 2017). The other top-down approaches—such as laser ablation, electro-explosion, and sputtering—have been utilized for the generation of NMs for industrial purposes only. However, the agri-nutrition requirement for the nano-nutrient materials is several orders higher, particularly for the macronutrients. Moreover, these techniques produce NMs without functionalization or surface modification, which is critical for enhancing the interaction and uptake by the plant cells on application.

4.1.3.2 Bottom-Up Approaches

The bottom-up techniques for the synthesis primarily involve chemical precipitation or phase separation techniques for the synthesis of various nanonutrient fertilizers. Biological or green synthesis is also gaining impetus due to the associated low-cost and environmental-safety benefits. The basic principle of biological synthesis of nanonutrients involves reduction of dissolved metal/nonmetal salts on interaction with the biomacromolecules such as proteins, sugars, polyphenols, and other compounds, which possess strong reducing potential.

4.1.4 Modes of Application of Nanofertilizers

NFs can be applied in myriad ways to provide the desired nutrients to the plants. The variability in the formulations of these fertilizers endear them to be applied as a soil dressing—powdered or nanoclay mineral-based NFs, foliar or electrospray, as primed/coated seeds, seedling root drenching/dipping, hydrogel pastes, films, and many other conjugate forms.

4.1.5 Nanofertilizers: Anticipated to Behave Differently

Matter in nanoscale exhibits unique properties, quite disparate from its bulk counterparts. NMs possess small size, greater surface area-to-mass ratios, higher surface energy, variable interaction potentials with electromagnetic radiations, increased dielectric constants, and improved mechanical properties (Dasgupta et al., 2015; Ranjan et al., 2016). Being an interstitial state of matter, NMs may either exhibit spontaneous dissolution of their surface to yield constituent molecules, atoms, or ions and may also agglomerate to form bulk aggregates. However, the fate of a NM in a complex, open, and dynamic system such as a soil system is governed by prevailing abiotic and biotic factors of the system (Kalia and Gosal, 2011). Thus, NM dissolution or aggregation as bulk due to complexation with the other components of the soil system depends on pH, electrical conductivity, organic carbon contents, and soil microbiota. The higher surface area and energy enable NMs to be chemically reactive, thereby imparting them with the ability to absorb or sequester onto other substrate materials at exceptionally higher rates compared to their bulk counterparts (Dasgupta et al., 2015; Ranjan et al., 2016).

The small size of the NMs enables these materials to be readily taken up by the plant/microbe from exposed root/leaf epidermal/cellular surfaces. Therefore, novel formulations of the nanoscale nutrient elements can be developed, which may be applied by alternate modes such as foliar/electrospray of the liquid suspensions/emulsions, polymer coatings/priming on seeds, granulated/pellet-forms, and gels or sticky emulsions. The high adsorption/sequestration potential of the nanoscale nutrient elements can be effectively harnessed to develop nanovehicle/carrier-based fertilizers. Such assemblies will ensure slow and continuous release of the adsorbed/sequestered nutrients in plant-available forms for a many-fold increase in durations, such as mesoporous silica-based urea NFs ensured slow release of nitrogen for a five-fold increase in the availability period (Wanyikaet al., 2012).

Another advantage of their small size-to-mass ratios and adsorption features can be utilized for the development of multinutrient fertilizer formulations, so as to achieve balanced fertilization of the crop plant (Liu and Lal, 2015). This objective is unique, as it will ensure balanced fertilization of the crop, that is, supply of several nutrients in appropriate proportions to ensure improved nutrient management.

4.1.6 Effect of Application of Nanofertilizers

4.1.6.1 Interactions of NFs with Soil and Its Components

Various studies have reported that metal/metal oxide NPs do not affect the soil properties such as soil organic matter and carbon content, pH, mineral concentration, and available nutrients

(Antisari et al., 2013; Ben-Moshe et al., 2013; Carbone et al., 2014). However, these can alter the bioavailability of nutrients and/or toxic compounds in the soil (Rashid et al., 2017). Thus, the contradictory interactions between NPs and soil limit the large-scale use of NFs in agriculture. Micrometric aggregate formation by certain nanofertilizers, such as TiO_2, CeO_2 and $Cu(OH)_2$, results in their decreased mobility, which leads to their decreased reactivity with soil components. Consequently, this may alter the pH and release of nutrients from the soil (Conway and Keller, 2016).

NPs can interact with the soil matrix in a number of ways. Soil components (i.e., humic acid and soil pore water) act as buffers to mitigate the antimicrobial effect of silver NPs towards the beneficial microorganism, *Pseudomonas chlororaphis* O6 (Calder et al., 2012). Similarly, application of ZnO NPs significantly affect reduced carbon mineralization (CO_2 emissions and dissolved organic carbon) and nitrogen mineralization in *Phoenix dactylifera* leaf litter-amended soil (Rashid et al., 2017). Application of potassium- and phosphorus-incorporated NFs, having zeolite as a carrier material, can improve the available phosphorus and potassium content of the soil compared to the conventional fertilizers, in the case of *Ipomoea aquatica* (Rajonee et al., 2017).

4.1.6.2 Probable Effects on Soil Microbiota

Soil acts as a reservoir for all the NFs, and the latter may enter the soil either by direct application or indirectly via run-offs or amendments of sewage sludge. Soil microorganisms are known to play a pivotal role in maintaining the soil fertility and are evaluated as indicators of soil health due to their involvement in key metabolic processes such as nutrient biogeochemical cycling. Changes in the soil biological profile can be critical for sustainable functioning of the soil. The impact of NFs on soil microbial diversity and its functioning is crucial to know as to how the functioning of the ecosystem gets affected by the application of NFs. The effect of various NPs on soil microflora is documented in Table 4.1.

4.1.6.3 Possible Impacts of NF Application in Crop Plants

The NFs, owing to their small size and larger surface area-to-volume ratio, are known to enhance plant growth and productivity. Application of NFs may lead to changes in morphology, physiology, and genetics of the plant. The effect of NPs on the plant depends not only on the type of NPs and their characteristics, but also on plant species and the type of soil in which the NPs are applied. The antimicrobial activities of certain metal/metal oxide NFs (Ag, TiO_2, ZnO) have rendered their usefulness for protection against plant pathogens (Duhana et al., 2017).

There are two ways by which NPs are taken up by plant: either by root uptake or by foliar uptake. Through roots, NPs can follow one of the two possible ways, the apoplastic route or the symplastic route (Deng et al., 2014). The apoplastic route is the nonliving pathway of uptake, in which the NPs travel through the cell walls and free spaces between the cells. The symplastic route, being the living pathway, employs cell-to-cell transport of NPs via plasmodesmata. In foliar uptake, the NPs enter the plant cells through stomatal pores (Hong et al., 2014).

Various studies have demonstrated the positive effects of NF application on crop plants (Table 4.2). Application of calcium phosphate NPs at lower doses (10 and 20 ppm) improves the growth and reduces the generation of reactive oxygen species in the rice plant (Upadhyaya et al., 2017).

4.1.7 Potential Bottlenecks

4.1.7.1 Mass Scale-Up Issues Seriously Affect Commercialization

Various studies have shown implicit application of NFs in agriculture. The number of research works being published in this area is increasing daily, alluding to the potential applications of NFs. However, when compared to other areas of application, nanotechnology based products have not yet attained tangible prospect in agriculture and are still under research and development (Cheng et al., 2016). The major issue is the cost of production of the NFs, which is not compensated by the profits

TABLE 4.1
Effect of Different Types of Nanoscale Nutrients as Fertilizers on Soil Microbial Communities

NF Used	Particle Size and Conc.	Mode of Application and Time of Exposure	Microorganism Affected	Effect	Mode of Action	Reference
TiO_2	(80% anatase and 20% rutile) 21 nm, 0.05–500 ppm	Soil microcosm, 90 days	Ammonia oxidizers and nitrate oxidizers (*Nitrospira* and *Nitrobacter*)	Reduction in abundance of ammonia oxidizers and *Nitrobacter*, no effect on *Nitrospira*	Increased aggregation at higher conc. resulting in their lower oxidative potential	Simonin et al. (2017)
	(anatase) 40–60 nm, 200 ppm	Soil application, 42 days (6 weeks)	Soil microbial communities	Alteration in arbuscular mycorrhizal fungal communities	Increased conc. of Ti in root tissues affecting colonization of root interior	Burke et al. (2014)
Fe_3O_4 (Magnetite)	18 nm, 100 and 200 ppm,	Soil application, 6 weeks	Soil microbial communities	Increase in arbuscular mycorrhizal colonization with increase in conc., higher nodule colonization	Essential micronutrient availability	Burke et al. (2015)
CNT (Carbon nanotube)	15 nm thickness, 50 and 200 ppm	Soil application, 9 weeks	Soil microbial communities	Increase in Bacteriodetes and Firmicutes, decrease in Proteobacteria and Verrucomicorbia	Shift in composition of minor phylotypes of microbial community in soil	Khodak-ovskaya et al. (2013)
CuO	<50 nm, 1% (w/w)	Soil application, 48 h	Soil bacterial communities	Decrease in bacterial community composition	Toxic effect of NPs at higher concentration	Ben-Moshe et al. (2013)
	40 nm, 1and100 ppm	Soil application, 30 days	Soil microbial communities	Increase in Alphaproteobacteria at 100 ppm, decrease in Firmicutes	Enhanced nutrient bioavailability favoring aerobic bacteria proliferation and destruction of less-competent cells	Asadishad et al. (2018)
ZnO	<50 nm, 0.5–2 mg/g	Soil application, 15 and 30 days	Soil bacterial communities	Decrease in total bacterial population in saline-alkali soil; increase in relative abundance of *Bacilli*; decrease in relative abundance of Actinobacteria, Gamma-proteobacteria, Alpha-proteobacteria, and Gemmatimonadetes	Enhancement in turnover of carbon and nitrogen is soil, decrease in bacterial population due to membrane disorganization, Zn ion production	You et al. (2017)

(Continued)

TABLE 4.1 (*Continued*)
Effect of Different Types of Nanoscale Nutrients as Fertilizers on Soil Microbial Communities

NF Used	Particle Size and Conc.	Mode of Application and Time of Exposure	Microorganism Affected	Effect	Mode of Action	Reference
	100–1600 μg/mL	Growth media supplementation	*Pseudomonas* and *Bacillus*	Decrease in bacterial growth with increase in NP concentration	Interaction with bacterial membrane leading to lysis of cell wall and disruption of proton motive force	Haris and Ahmad (2017)
	~90 nm; 1, 10, and 100 ppm	Soil application	Soil bacterial communities	Increase in abundance of Cyanobacteria (*Nostoc* and *Scenedesmus*), Xanthomonadales, Cytophagaceae at 10 ppm, Decrease in *Rhodanobacter, Chitnophagaceae, Rhizomicrobium, Nitrospira*	Growth promoting effect of Zn on Cyanobacteria, production of reactive oxygen species causing cellular damage	Xu et al. (2018)
CeO_2	<25 nm; 500, 1000 and 2000 ppm	Soil application	Soil bacterial communities	No significant change in overall bacterial community size	Insolubility of CeO_2 in nanoform as compared to bulk form	You et al. (2018)

Abbreviations: NPs, Nanoparticles; Ti, Titanium.

TABLE 4.2
Effect of Different Types of Nanonutrient Fertilizers on Crop Plants

Nutrient Provided	Particle Size, Conc., Mode of Exposure, Time of Exposure	Plant Species	Physiological Effect	Reference
Nitrogen (nano-urea)	0.002%, foliar and soil application	Borage (*Borago officinalis*)	Significant increase in oil production	Mahmoodi et al. (2017)
Nitrogen (nano-nitrogen)	50 nm, 250 and 500 ppm, foliar application	Pomegranate (*Punica granatum* cv. Ardestani)	Increase in yield and number of fruits at 500 ppm conc.	Davarpanah et al. (2017)
Potassium (zeolite)	Soil application, 30 days	Kalmi (*Ipomoea aquatica*)	Increase in growth and phytoavailability of potassium	Rajonee et al. (2017)
Phosphorus (hydroxyapatite)	10–40 nm, soil application	Soybean (*Glycine max*)	Increase in seed yield, shoot, and root biomass by 20.4%, 18.2%, and 41.2%, respectively	Liu and Lal (2014)
	20 mL, foliar application	Baobab (*Adansonia digitata*)	Increase in plant height, number of leaves, leaf area, root length, and plant weight	Soliman et al. (2016)
			Increase in NPK and crude protein content	
Zinc (ZnO)	25 nm, 50–2000ppm, foliar exposure, 90 days	Maize (*Zeamays*)	Improved germination rate and increase in root length, shoot length, and yield	Subbaiah et al. (2016)
	~90 nm; 1, 10, and 100 ppm; soil application; 7 weeks	Lettuce (*Lactuca sativa* L.)	Increase in net photosynthetic rate and fresh plant biomass at 10 ppm	Xu et al. (2018)
Iron (Fe_3O_4)	18 nm, 100 and 200 ppm, soil applicaton, 6 weeks	Soybean (*Glycine max*)	Significant effect on plant biomass	Burke et al. (2015)
Iron (Fe/SiO_2)	30–40 nm, 0–25 ppm, soil application	Barley (*Hordiumvulgarium*) and maize (*Zea mays*)	Increase in shoot length at 15 ppm: 8.25% for barley and 20.8% for maize	Disfani et al. (2016)
Iron (FeO)	<50 nm, 100–400 ppm, hydroponic solution, 10 days	Rice (*Oryza sativa*)	Increase in root and shoot length and weight, chlorophyll content, total protein content, and antioxidant activity	Mankad et al. (2017)
Titanium (TiO_2)	22–25 nm, 100 and 200 nm, soil application, 6 weeks	Soybean (*Glycine max*)	Increase in plant growth, leaf phosphorus, and carbon content	Burke et al. (2015)
	35 nm, 250–750 ppm, seed exposure, 7 days	Pea (*Pisum sativum*)	No effect of short term exposure on germination or root growth	Fan et al. (2014)
CNT (Carbon nanotube)	15 nm thickness, 50 and 200 ppm, soil application, 9 weeks	Tomato (*Lycopersicon esculentum*)	Slight increase in plant height and higher number of flowers	Khodakovskaya et al. (2013)

(Continued)

TABLE 4.2 (*Continued*)
Effect of Different Types of Nanonutrient Fertilizers on Crop Plants

Nutrient Provided	Particle Size, Conc., Mode of Exposure, Time of Exposure	Plant Species	Physiological Effect	Reference
	SWCNT: 1–2 nm diameter MWCNT: 150–200 nm diameter, 500 ppm, seed treatment with suspension in water and THF, 8 days	Wheatgrass	Increase in germination rate with pure-SWCNT and MWCNT dispersed in THF	Lee et al. (2018)
Zinc (ZnO)	<100 nm	Wheat (*Triticum aestivum*)	Reduction in biomass	Du et al. (2011)
	20 nm,10–2000 ppm, suspension in growth media, 60 h	Mung bean (*Vigna radiata*) Gram (*Cicer arietinum*)	Increase in root, shoot, and plant biomass	Mahajan et al. (2011)
Magnesium (MgO)	<50 nm, 50 and 100 ppm, suspension in growth medium, 28 days	Mung bean (*Vigna radiata*)	Concentration-dependent increase in root length, two-fold increase in shoot weight at 100 ppm	Mahawar et al. (2017)
Calcium ($CaPO_4$)	30 nm; 10, 20, and 50 ppm; suspension in growth medium; 48 h	Rice (*Oryza sativa*)	Increase in root and shoot length and biomass at 10 and 20 ppm but decrease at 50 ppm	Upadhyaya et al. (2017)
Silicon (SiO_2)	10–20 nm, 50 and 100 ppm, suspension in culture medium, 28 days	Mung bean (*Vigna radiata*)	Significant increase in chlorophyll content at 100 ppm	Mahawar et al. (2017)
Nickel (Ni)	20 nm; 20, 40, and 80 ppm; soil application; 22 days	Coriander (*Coriandrum sativum*)	Decrease in root and shoot elongation and antioxidant activity of the plant Increase in chlorophyll a content at 20 and 80 ppm Increase in chlorophyll b content at 40 ppm	Miri et al. (2017)
Copper (CuO)	40 nm, 500 ppm, soil application, 28 days	Wheat (*Triticum aestivum*)	Increase in root compactness for aged CuO NPs as compared to fresh CuO NPs	Gao et al. (2018)
Boron (Zn/B)	200–2000 nm, 10–40 ppm, foliar spray, 100 days	Coffee (*Coffea*)	Increase in nitrogen and phosphorus content of leaves, total chlorophyll content, transpiration rate, stomatal conductance, and net photosynthetic rate	Wang and Nguyen (2018)
Cerium (CeO_2)	10–30 nm, 100 ppm, soil application with varying moisture content, 3 weeks	Soybean (*Glycine max* (L.) Merr.)	Increase in plant biomass at 70%, 85%, and 100% moisture content, Nonsignificant increase in stomatal conductance and chlorophyll content	Cao et al. (2018)

Abbreviation: THF, Tetra hydro furan.

incurred. The uncertainties in public opinion, as well as the stigma associated with the use of NFs, act as the major hurdles in the large-scale manufacture.

4.1.7.2 Lack of Well-Delineated Protocols for Risk Assessment/Safety Standards

Whether NFs are applied as agricultural nutrient supplements or these are released into the environment as contaminants, it is necessary to understand the effect of these NPs on plants as well as on soil microflora (Figure 4.1). The increased use of NFs has raised concerns about their negative impacts on the environment. The safety aspect of NFs is another major issue that acts as a barrier against large-scale commercialization. Although NFs have shown promising results in relation to plant growth; yet their use in higher concentrations is shown to exhibit deleterious effects on plant growth through induction of oxidative stress and cytotoxic and genotoxic effects on plants as well as beneficial soil microbes, which results in impaired growth and low nutritional value of the crop (Peralta-Videa et al., 2014). Toxicity of NPs depends on the size and exposure doses, as well as the chemical nature of the NPs. Lower concentrations of NPs application have also been shown to result in retarded growth of the plant (Mahawar et al., 2017). Adsorption and uptake of NPs (smaller than the size of cell-wall pores) by cells via interaction with plant roots induces oxidative stress. Oxidative stress is a result of production of reactive oxygen species which imbalance the redox system of the cell. On the contrary, NPs larger than the pores cause blockage of the pores, thereby hindering water absorption and nutrient uptake. In plants, toxicity resulting in slower growth and oxidative stress leads to DNA damage and changes in genetic makeup, decreased leaf pigmentation, and decreased transpiration rate (Morales-Diaz et al., 2017). Foliar application of NFs may result in reduced rate of transpiration.

As NPs form relatively stable colloids in water, their effect on soil microflora remain for longer durations. Different organisms show different levels of sensitivity to the NPs, with algae being most

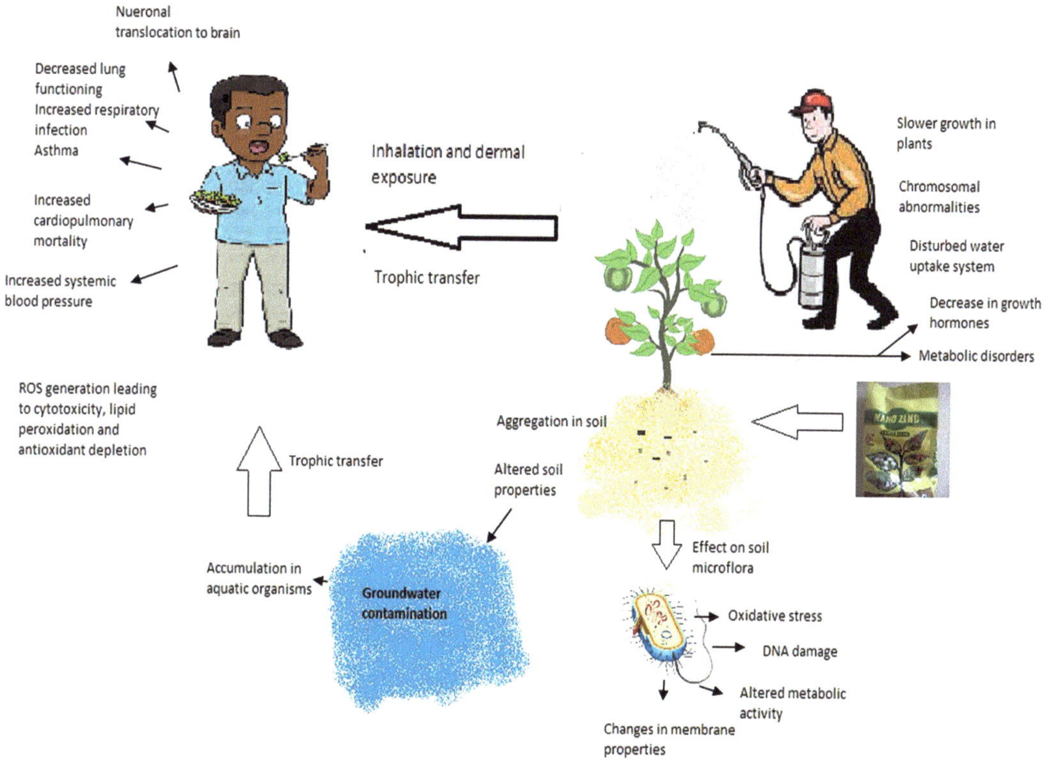

FIGURE 4.1 Toxicity issues arising due to contamination of soil, water, and air on agri-application of nano-Zinc formulations.

susceptible, followed by bacteria and protozoa (Aruoja et al., 2015). Toxicity is the result of changes in membrane properties (Huang et al., 2017), leading to absorption of NPs by the cell, and ultimately generation of reactive oxygen species. The fact that production of oxidative stress in plant, microbial, and animal cells occurs at only certain concentrations of NF application, it can be explained by the property of NPs to interact with proteins, membranes, nucleic acids, and other metabolites (Van Aken, 2015). The toxicity of NPs gets significantly affected by pH and the presence of complexing agents (i.e., citrate or ascorbate) in water or soil solution and bacterial or root exudates. These factors increase the stability of NPs, thereby decreasing aggregate formation, which increases the opportunity to contribute potentially toxic ions (Khan et al., 2011). Other environmental factors, such as temperature, can alter the rate of aggregation (Loria et al., 2011).

4.1.7.3 Technical Challenges for Overall Quantification in Dynamic Systems

Quantification of the NMs by *in vivo* and *in vitro* experiments is necessary to evaluate the impact of environmentally relevant doses. The physicochemical properties of NPs, such as size, shape, solubility, surface area, and aggregate formation, pose a challenge in quantification of the NPs. Before evaluation of responses of NPs in biological systems, the relationships among the various alterations in the properties of NPs need to be clarified (Hu et al., 2016). Quantification is necessary to reduce the release of NPs into the environment. This can be accomplished by matching the quantity of NFs with the stage of crop growth with higher responses.

4.1.8 Future Prospects of Nanofertilizers

4.1.8.1 Environmental Fate and Transformation in Soil

Various studies can be documented showing the size and concentration-dependent toxicity of the NPs, but these studies are limited only to growth in artificial media or hydroponics. When the NPs enter the biological environment, they may show completely different properties. The environmental fate of these NPs depends upon their physicochemical properties, which can undergo a variety of chemical and physical changes. The NPs can undergo various transformations, including aggregation, adsorption to cell surfaces (microbial and plant root), dissolution, and changes in surface chemistry. Dissolution of metal-oxide NPs may lead to release of metals to the environment, which may be toxic. After their release into the environment, NPs may undergo either homoaggregation or heteroaggregation, depending upon whether they are interacting with other NPs or mineral and organic compounds (Batley et al., 2013).

The impact of NFs on soil ecosystems is determined by their bioavailability, which is the actual concentration of the NPs to which the organisms are exposed (Karimi and Fard, 2017). However, the general rules and mechanisms for bioaccumulation and subsequent biotransformation are lacking. Although the NFs are applied in low concentrations, they may accumulate over time, resulting in much higher concentrations, which may eventually lead to their toxic effects.

Biomolecules present in the natural environments also influence the transformation of the NFs. NPs may interact with proteins, which are important constituents of any biological substrate, and form a protein corona on their surface (Hu et al., 2016), leading to surface changes in the NPs. Despite the involvement of proteins, the presence of electrolytes influences the dissolution and aggregation of NPs. The pH and the humic components of the soil are also known to affect the association of NPs with the solid phase. Humic substances present in the soil may act as stabilizers, but may also induce aggregation, depending upon their concentration and other abiotic and biotic conditions (Kralchevsky et al., 2016). Therefore, the type of soil and its conditions (i.e., moisture content, type and amount of organic materials, presence of electrolytes, acidity/alkalinity of soil, mineral composition, and aeration level) are expected to determine the stability and transformation of NPs.

The degradation of NMs is another phase of transformation. The soil organisms or the peroxidases enzymes can cause degradation of the NPs, which can either increase or decrease the toxicity of the by-products. In biological environments, the transformation of NPs is inevitable. The transformation of the NPs exhibits remarkable changes in the properties of NMs, which need to be assessed before their application.

4.1.8.2 Interactions between Nanoparticles and the Human Body

Although application of nanotechnology in agriculture has benefitted the crop production, it has also been reported to exhibit adverse effects like those of the conventional fertilizers. NFs applied to the crop can enter the human body through trophic transfer from plants, water, and soil, where it may linger through leachate and run-off. Foliar application of NFs may lead to aerosol formation of NPs. These NPs suspended in the atmosphere can be inhaled by humans and can deposit in their respiratory tracts, from where these may move to other organs such as the liver, spleen, and brain and may circulate via bloodstream. The distribution of C-13 NPs in rats showed the accumulation of C-13 in liver at 18 and 24 h after exposure (Oberdörster et al., 2002).

The impact of NPs on human health is immense. Hence, acute and chronic toxicity, DNA damage, breaching and damaging of tissue barriers, brain damage, and translocation of engineered NPs in the skin, gastrointestinal, or respiratory tract are some of the effects that have to be studied to assess the safety of NPs for human health (Som et al., 2011). Studies by Lucarelli et al., (2004) showed that silica NPs induced proinflammatory activity in human bone-marrow monocytes. The toxicity of the NFs can also be concentration-dependent: inhalation of low concentration (10 mg/m^3) of titanium dioxide NPs resulted in higher tumor incidence than exposure to high doses (250 mg/m^3) of the same (Lee et al., 1986; Heinrich et al., 1989). The main mechanisms responsible for the toxicity were reactive oxygen species (ROS) generation and subsequent oxidative stress.

4.1.8.3 Choosing Environmentally Relevant Doses

Most of the researches regarding NF application in crop plants are carried out either in hydroponic systems, in artificial media, or other nonsoil media with very high doses and relatively shorter exposure durations. The befitting applicability of the NFs can be achieved only when sustainable NMs can be generated for their full-scale release into the environment. The behavior of the NPs in artificial media is different than in agricultural soils (Dimkpa and Bindraban, 2017), considering the presence of various external factors in the soil. Moreover, high doses of NFs (>100 ppm) have been shown to induce toxic effects on the plants as well as on the soil ecosystem. Therefore, choosing the doses for maximizing the benefits by keeping check on the risks is crucial in NF application. Boxall et al., (2007) used algorithms to predict the concentration of NPs in soil (i.e., direct application) and showed that concentration of NPs is related to application rate, soil depth, and soil bulk density.

4.1.8.4 Adaptations to Toxic Effects

Studies regarding the application of NFs mainly involve high-dose, short-term experiments, and therefore, the adaptations to nanotoxicology are generally neglected. However, in biological environments, NPs are continuously entering from various nonagricultural sources, and their transformation may lead to either reversibility of or adaptation to the toxic effects. Continuous exposure of NPs results in adaptability of soil microflora to a short extent via exopolysacharride (EPS) secretion (Dhas et al., 2014). EPS secretions can play a major role in stabilizing the NPs or in inducing their aggregation; thus either increasing or reducing the toxic effects of NPs (Wilkinson and Reinhardt, 2005).

4.2 CONCLUSIONS

NFs are the products of the nanotechnology interventions at incipient stage of development. Scientific evidence and increased awareness in support of positive effects have been out-weighing the apprehensions related to environmental hazards associated with the usage of nanoparticulate nutrients at a large scale. Thus, enhanced NUE, targeted placement and release of the loaded

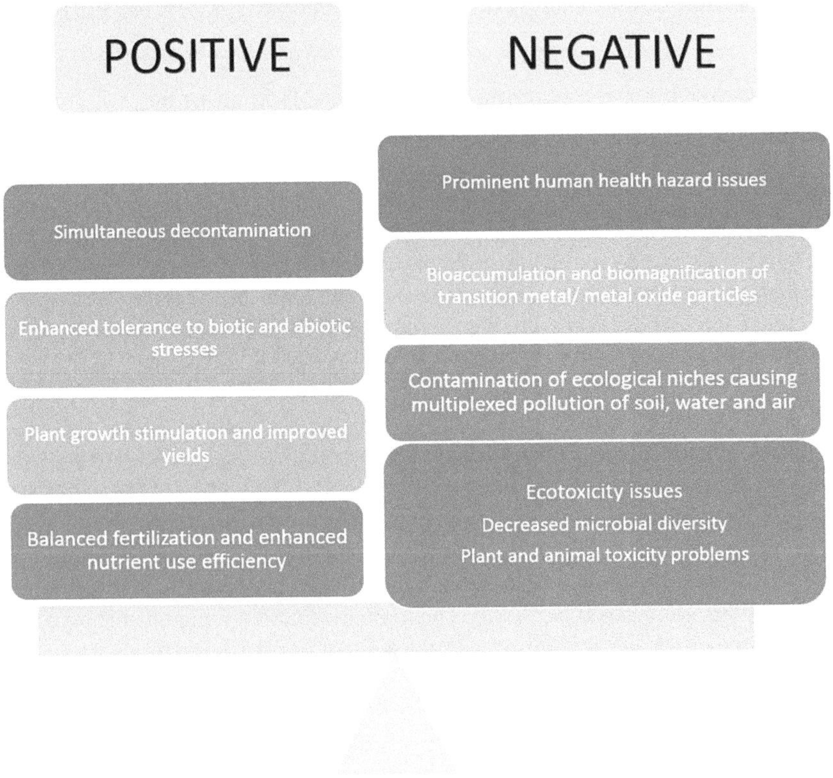

FIGURE 4.2 Assertive and grey dimensions of nanonutrient application.

nutrients, plant health with yield improvement, and environmental security features of the NFs all demonstrate a definitive promise to trigger a revolution in the fertilizer industry (Figure 4.2).

ACKNOWLEDGMENTS

The authors are thankful to the Head, Department of Soil Science and Dean, College of Agriculture, Punjab Agricultural University, Ludhiana, Punjab, India, for providing the necessary infrastructural facilities.

REFERENCES

Antisari LV, Carbone S, Gatti A, Vianello G, Nannipieri P. Toxicity of metal oxide (CeO_2, Fe_3O_4 SnO_2) engineered nanoparticles on soil microbial biomass and their distribution in soil. *Soil Biol Biochem.* 2013; 60: 87–94.

Aruoja V, Pokhrel S, Sihtmäe M, Mortimer M, Mädler L, Kahru A. Toxicity of 12 metal-based nanoparticles to algae, bacteria and protozoa. *Environ Sci Nano.* 2015; 2: 630–44.

Asadishad B, Chahal S, Akbari A, Cianciarelli V, Azodi M, Ghoshal S, Tufenkji N. Amendment of agricultural soil with metal nanoparticles: Effect on soil enzyme activity and microbial community composition. *Environ Sci Technol.* 2018; 52(4): 1908–18.

Barati, A. Nano-composite superabsorbent containing fertilizer nutrients used in agriculture US20100139347 A1; 2010.

Batley GE, Kirby JK, Mclaughlin MJ. Fate and risks of nanomaterials in aquatic and terrestrial environments. *Acc. Chem. Res.* 2013; 46: 854–62.

Behera SK, Panda RK. Integrated management of irrigation water and fertilizers for wheat crop using field experiments and simulation modeling. *Agric Water Manage.* 2009; 96: 1532–40.

Ben-Moshe T, Frenk S, Dror I, Minz D, Berkowitz B. Effects of metal oxide nanoparticles on soil properties. *Chemosphere*. 2013; 90: 640–6.

Boxall ABA, Chaudhry Q, Sinclair C, Jones A, Aitken R, Jefferson B, Watts C. Modelling exposure to engineered nanoparticles. *In*: *Current and Future Predicted Environmental Exposure to Engineered Nanoparticles*. Central Science Laboratory Sand Hutton York; 2007; pp. 37–65.

Burke DJ, Pietrasiak N, Situ SF, Abenojar EC, Porche M, Kraj P, Lakliang Y, Samia ACS. Iron oxide and titanium dioxide nanoparticle effects on plant performance and root associated microbes. *Int J Mol Sci*. 2015; 16: 23630–50.

Burke DJ, Zhu S, Pablico-Lansigan MP, Hewins CR, Samia ACS. Titanium oxide nanoparticle effects on composition of soil microbial communities and plant performance. *Biol Fertil Soils*. 2014; 50: 1169–73.

Calder AJ, Dimpka CO, McLean JE, Britt DW, Johnson W, Anderson AJ. Soil components mitigate the antimicrobial effects of silver nanoparticles towards a beneficial soil bacterium, *Pseudomonas chlororaphis* O6. *Sci Total Environ*. 2012; 429: 215–22.

Cao Z, Rossi L, Stowers C, Zhang W, Lombardini L, Ma X. The impact of cerium oxide nanoparticles on the physiology of soybean (*Glycine max* (L.) Merr.) under different soil moisture conditions. *Environ Sci Pollut Res*. 2018; 25: 930–9.

Carbone S, Antisari LV, Gaggia F, Baffoni L, Di Gioia D, Vianello G, Nannipieri P. Bioavailability and biological effect of engineered silver nanoparticles in a forest soil. *J Hazard Mater*. 2014; 280: 89–96.

Cheng HN, Klasson KT, Asakura T, Wu Q. Nanotechnology in agriculture. In: *Nanotechnology: Delivering on the Promise*. American Chemical Society; 2016; 232–42.

Conway JR, Keller AA. Gravity-driven transport of three engineered nanomaterials in unsaturated soils and their effects on soil pH and nutrient release. *Water Res*. 2016; 98: 250–60.

Corral-Diaz B, Peralta-Videa JR, Alvarez-Parrilla E, Rodrigo-García J, Morales MI, Osuna-Avila P, Niu G, Hernandez-Viezcas JA, Gardea-Torresdey JL. Cerium oxide nanoparticles alter the antioxidant capacity but do not impact tuber ionome in *Raphanus sativus* (L). *Plant Physiol Biochem*. 2014; 84: 277–85.

Dasgupta N, Shivendu R, Deepa M, Chidambaram R, Ashutosh K, Rishi S. Nanotechnology in agro-food: From the field to plate. *Food Res Int*. 2015; 69: 381–400.

Davarpanah S, Tehranifar A, Davarynejad G. Effects of foliar nano-nitrogen and urea fertilizers on the physical and chemical properties of pomegranate (*Punica granatum* cv. Ardestani) Fruits. *Hortsci*. 2017; 52: 288–94.

De Rosa MC, Monreal C, Schnitzer M, Walsh R, Sultan Y. Nanotechnology in fertilizers. *Nat Nanotechnol*. 2010; 5: 91.

Delfani M, Baradarn Firouzabadi M, Farrokhi N, Makarian H. Some physiological responses of black-eyed pea to iron and magnesium nanofertilizers. *Commun Soil Sci Plant Anal*. 2014; 45: 530–40.

Deng YQ, White JC, Xing BS. Interactions between engineered nanomaterials and agricultural crops: Implications for food safety. *J Zhejiang Univ-Sci A (Appl Phys & Eng)*. 2014; 15: 552–72.

Dhas SP, Shiny PJ, Khan S, Mukherjee A, Chandrasekaran N. Toxic behavior of silver and zinc oxide nanoparticles on environmental microorganisms. *J Basic Microbiol*. 2014; 54: 916–27.

Dimkpa CO, Bindraban PS. Nanofertilizers: New products for the Industry? *J Agric Food Chem*. 2017; DOI: 10.1021/acs.jafc.7b02150.

Disfani MN, Mikhak A, Kassaee MZ, Maghari A. Effects of nano Fe/SiO_2 fertilizers on germination and growth of barley and maize. *Archives Agron Soil Sci*. 2016; Doi: 10.1080/03650340.2016.1239016

Du WC, Sun YY, Ji R, Zhu JG, Wu JC, Guo HY. TiO_2 and ZnO nanoparticles negatively affect wheat growth and soil enzyme activities in agricultural soil. *J Environ Monit*. 2011; 13: 822–8.

Duhana JS, Kumar R, Kumar N, Kaur P, Nehra K, Duhan S. Nanotechnology: The new perspective in precision agriculture. *Biotechnol Rep*. 2017; 15: 11–23.

Fan R, Huang YC, Grusak MA, Huang CP, Sherrier DJ. Effects of nano-TiO_2 on the agronomically relevant *Rhizobium*-legume symbiosis. *Sci Total Environ*. 2014; 466:50–512.

Gao X, Avellan A, Laughton S, Vaidya R, Rodrigues SM, Casman EA, Lowry GV. CuO nanoparticles dissolution and toxicity to wheat (*Triticum aestivum*) in rhizospheric soil. *Environ Sci Technol*. 2018; 52(5): 2888–97.

Haris Z, Ahmad I. Impact of metal oxide nanoparticles on beneficial soil microorganisms and their secondary metabolites. *Int J Life Sci Scienti Res*. 2017; 3: 1020–30.

Heinrich U, Muhle H, Hoymann HG, Mermelstein R. Pulmonary function changes in rats after chronic and subchronic inhalation exposure to various particulate matter. *Exp Pathol*. 1989; 37: 248–52.

Herrera JM, Rubio G, Haner LL, Delgado JA, Lucho-Constantino CA, Islas-Valdez S, Pellet D. Emerging and established technologies to increase nitrogen use efficiency of cereals. *Agronomy*2016; 6: 25.

Hong J, Peralta-Videa JR, Rico C, Sahi S, Viveros MN, Bartonjo J, Zhao L, Gardea-Torresdey JL. Evidence of translocation and physiological impacts of foliar applied CeO_2 nanoparticles on cucumber (*Cucumis sativus*) plants. *Environ Sci Technol.* 2014; 48: 4376–85.

Hu X, Li D, Gao Y, Mu L, Zhou Q. Knowledge gaps between nanotoxicological research and nanomaterial safety. *Environ Int.* 2016; 94: 08–23.

Huang G, Ng TW, An T, Li G, Wang B, Wu D, Yip HY, Zhao H, Wong PK. Interaction between bacterial cell membranes and nano-TiO_2 revealed by two- dimensional FTIR correlation spectroscopy using bacterial ghost as a model cell envelope. *Water Res.* 2017; doi: 10.1016/j.watres.2017.04.023.

Kalia A, Gosal SK. Effect of pesticide application on soil microorganisms. *Arch Agron Soil Sci.* 2011; 57(6): 569–96.

Karimi E, Fard EM.Nanomaterial effects on soil microorganisms. *In*: *Nanoscience and Plant–Soil Systems.* Springer International Publishing AG; 2017; 137–200.

Khan SS, Mukherjee A, Chandrasekaran N. Impact of polysachharides on the stability of silver nanoparticles in water. *Water Res.* 2011; 45(16): 5184–90.

Khodakovskaya MV, Kim BS, Kim JN, Alimohammadi M, Dervishi E, Mustafa T, Cernigla CE. Carbon nanotubes as plant growth regulators: Effects on tomato growth, reproductive system, and soil microbial community. *Small.* 2013; 1: 115–23.

Kottegoda N, Munaweera I, Madusanka N, Karunaratne V. A green slow-release fertilizer composition based on urea-modified hydroxyapatite nanoparticles encapsulated wood. *Curr Sci (Bangalore).* 2011; 101: 73–8.

Kottegoda N, Sandaruwan C, Priyadarshana G, Siriwardhana A, Rathnayake UA, Arachchige DMB, Kumarasinghe AR, Dahanayake D, Karunaratne V, Amaratunga GAJ. Urea-hydroxyapatite nanohybrids for slow release of nitrogen. *ACS Nano.* 2017; 11(2): 1214–21.

Kralchevsky P, Miller R, Ravera F. Environmental impacts of nanomaterials. *In*: *Colloid and Interface Chemistry for Nanotechnology.* CRC Press; 2016; pp. 37–57.

Lee KP, Kelly DP, Schneider PW, Trochimowicz HJ. Inhalation toxicity study on rats exposed to titanium tetrachloride atmospheric hydrolysis products for two years. *Toxicol Appl Pharmacol.* 1986; 83:30–45.

Lee SM, Raja PMV, Esquenazi GL, Barron AR. Effect of raw and purified carbon nanotubes and iron oxide nanoparticles on the growth of wheatgrass prepared from the cotyledons of common wheat (*Triticum aestivum*). *Environ Sci Nano.* 2018; 5:103–14.

Liu R, Lal R. Nanoenhanced materials for reclamation of mine lands and other degraded soils: A review. *J Nanotechnol.* 2012; 1–18.

Liu R, Lal R. Synthetic apatite nanoparticles as a phosphorus fertilizer for soybean (*Glycine max*). *Sci Rep.* 2014; 4: 5686.

Liu R, Lal R. Potentials of engineered nanoparticles as fertilizers for increasing agronomic productions. *Sci Total Environ.* 2015; 514: 131–9.

Loria H, Pereira-Almao P, Scott CE. Determination of agglomeration kinetics in nanoparticles dispersions. *Ind Eng Chem Res.* 2011; 50(14): 8529–35.

Lucarelli M, Gatti AM, Savarino G, Quattroni P, Martinelli L, Monari E, Boraschi D, Woolley DE, Tetlow LC. Innate defence functions of macrophages can be biased by nano-sized ceramic and metallic particles. Mast cell activation and its relation to proinflammatory cytokine production in the rheumatoid lesion. *Eur Cytokine Netw.* 2004; 15: 339–46.

Mahajan P, Dhoke SK, Khanna AS. Effect of nano-ZnO particle suspension on growth of mung (*Vigna radiate*) and gram (*Cicer arietinum*) seedlings using plant agar method. *J Nanotechnol.* 2011; Article ID: 696535.

Mahawar H, Prasanna R, Simranjit K, Thapa S, Kanchan A, Singh R, Kaushik SC, Singh S, Nain L. Deciphering the mode of interactions of nanoparticles with mung bean (*Vigna radiata* L.). *Israel J Plant Sci.* 2017; doi: 10.1080/07929978.2017.1288516

Mahmoodi P, Yarnia M, Amirnia R, Tarinejad A, Mahmoodi H. Comparison of the effect of nanourea and nonoiron fertilizers with common chemical fertilizers on some growth traits and essential oil production of *Borago officinalis* L. *J Dairy Vet Sci.* 2017; J 2.

Majeed ZH, Taha MR. A review of stabilization of soils by using nanomaterials. *Australian J Basic Appl Scis.* 2013; 7: 576–81.

Mankad M, Fougat RS, Patel A, Mankad P, Patil G, Subhash N. Assessment of physiological and biochemical changes in rice seedlings exposed to bulk and nano iron particles. *Int J Pure App Sci.* 2017; 5(4): 150–9.

Miri AH, Shakib ES, Ebrahimi O, Shrifi-Rad J. Impacts of nickel nanoparticles on growth characteristics, photosynthetic pigment content and antioxidant activity of *Coriandrum sativum* L. *Orient J Chem.* 2017; 33(3): 1297–303.

Monreal CM, McGill WB, Nyborg M. Spatial heterogeneity of substrates: Effects on hydrolysis, immobilization and nitrification of urea-N. *Can J Soil Sci.* 1986; 66: 499–511.

Montalvo D, McLaughlin MJ, Degryse F. Efficacy of hydroxyapatite nanoparticles as phosphorus fertilizer in andisols and oxisols. *Soil Sci Soc Am J.* 2015; 79:551–8.

Morales-Diaz AB, Ortega-Ortíz H, Juárez-Maldonado A, Cadenas-Pliego G, González-Morales S, Benavides-Mendoza A. Application of nanoelements in plant nutrition and its impact in ecosystems. *Adv Nat Sci: Nanosci Nanotechnol.* 2017; 8: 013001.

Oberdörster G, Sharp Z, Atudorei V, Elder A, Gelein R, Lunts A, Kreyling W, Cox C. Extrapulmonary translocation of ultrafine carbon particles following whole-body inhalation exposure of rats. *J Toxicol Environ Health A.* 2002; 65: 1531–43.

Pandey AC, Sanjay SS, Yadav RS. Application of ZnO nanoparticles in influencing the growth rate of *Cicer arietinum. J Exp Nanosci.* 2010; 5: 488–97.

Peralta-Videa JR, Hernandez-Viezcas JA, Zhao L, Diaz BC, Ge Y, Priester JH, Holden PA, Gardea-Torresdey JL. Cerium dioxide and zinc oxide nanoparticles alter the nutritional value of soil cultivated soybean plants. *Plant Physiol Biochem.* 2014; 80: 128–35.

Rajonee AA, Zaman S, Huq SMI. Preparation, characterization and evaluation of efficacy of phosphorus and potassium incorporated nano fertilizer. *Advances Nanopart.* 2017; 6: 62–74.

Ranjan S, Nandita D, Bhavapriya R, Ganesh SA, Chidambaram R, Ashutosh K. Microwave irradiation-assisted hybrid chemical approach for titanium dioxide nanoparticle synthesis: Microbial and cytotoxicological evaluation. *Environ Sci Pollut Res.* 2016; 12:12287–302.

Rashid MI, Shahzad T, Shahid M, Ismail IMI, Shah GM, Almeelbi T. Zinc oxide nanoparticles affect carbon and nitrogen mineralization of *Phoenix dactylifera* leaf litter in a sandy soil. *J Hazard Mater.* 2017; 324:298–305.

Simonin M, Martins JMF, Roux XL, Uzu G, Calas A, Richaume A. Toxicity of TiO_2 nanoparticles on soil nitrification at environmentally relevant concentrations: Lack of classical dose-response relationships. *Nanotoxic.* 2017; 11:247–55.

Siva GV, Benita LFJ. Synthesis, characterization of Iron Oxide nanoparticles and their applications as nano-fertilizers on some quality characters of Ginger (*Zingiberofficinale*Rosc.)*Int J Sci Res Sci Technol.* 2016; 2: 11–8.

Soliman AS, Hassan M, Abou-Elella F, Ahmed AHH, El-Feky SA. Effect of nano and molecular phosphorus fertilizers on growth and chemical composition of Baobab (*Adansoniadigitata L.*)*J Plant Sci.* 2016; 11:52–60.

Som C, Wick P, Krug H, Nowack B. Environmental and health effects of nanomaterials in nanotextiles and facade coatings. *Environ Int.* 2011; 37: 1131–42.

Subbaiah LV, Prasad TN, Krishna TG, Sudhakar P, Reddy BR, Pradeep T. Novel effects of nanoparticulate delivery of zinc on growth, productivity, and zinc biofortification in maize (*Zea mays* L.). *J Agric Food Chem.* 2016; 64: 3778–88.

Tripathi KM, Bhati A, Singh A, Sonker AK, Sarkar S, Sonkar SK. Sustainable changes in the contents of metallic micronutrients in first generation gram seeds imposed by carbon nano-onions: Life cycle seed to seed study. *ACS Sustainable Chem Eng.* 2017; 5: 2906–16.

Upadhyaya H, Begum L, Dey B, Nath PK, Panda SK. Impact of calcium phosphate nanoparticles on rice plant. *J Plant Sci Phytopathol.* 2017; 1: 01–10.

Van Aken B. Gene expression changes in plants and microorganisms exposed to nanomaterials. *Curr Opin Biotechnol.*2015; 33: 206–19.

Wang SL, Nguyen AD. Effects of Zn/B nanofertilizer on biophysical characteristics and growth of coffee seedlings in a greenhouse. *Res Chem Intermed.* 2018; 1–13.

Wanyika H, Gatebe E, Kioni P, Tang Z, Gao Y. Mesoporous silica nanoparticles carrier for urea: Potential applications in agrochemical delivery systems. *J Nanosci Nanotechnol.* 2012; 12: 2221–8.

Wilkinson KJ, Reinhardt A. Contrasting roles of natural organic matter on colloidal stabilization and flocculation in freshwaters. *In*: IG Droppo, GG Leppard, SN Liss, TG Milligan, eds., *Flocculation in Natural and Engineered Environmental Systems*, CRC Press, Boca Raton, FL, 2005; pp. 143–70.

Xu J, Luo X, Wang Y, Feng Y. Evaluation of zinc oxide nanoparticles on lettuce (*Lactuca sativa* L.) growth and bacterial community. *Environ Sci Pollut Res.* 2018; 25: 6026–35.

You T, Liu D, Chen J, Yang Z, Dou R, Gao X, Wang L. Effects of metal oxide nanoparticles on soil enzyme activities and bacterial communities in two different soil types. *J Soils Sediments.* 2017; DOI: 10.1007/s11368-017-1716-2

You T, Liu D, Chen J, Yang Z, Dou R, Gao X, Wang L. Effects of metal oxide nanoparticles on soil enzyme activities and bacterial communities in two different soil types. *J Soils Sediments.* 2018; 18(1): 211–21.

5 Organophosphorus Compounds as Pesticides and Their Detection Using Nanomaterials

Divya Utreja, P. Sharma, S. Singh, and M. Kaur*

CONTENTS

5.1 INTRODUCTION

The insecticides having phosphorus as an active nucleus, commonly known as *organophosphates*, constitute the largest and most diverse group of insecticides known so far. To refer to these as "organophosphates" is not precisely correct, since one or more of the oxygen atoms in the phosphoric acid moiety may be replaced, and, in fact, the proper nomenclature reflects these substitutions. As pointed out by O'Brien (1967), these insecticides are associated with phosphoric acid esters of alcohols or anhydrides of phosphoric acid with different acids. Thus, they can be anticipated to be chemically reactive, subject to hydrolysis.

* Corresponding author.

It has been reported that the organophosphorus insecticides are the products of the research on "war-gas" during World War II. To the extent this is true, nerve gases such as toban, sarin, and soman developed during that period act in a manner similar to this class of insecticides. This research led to the discovery by Schrader of Bladan in the late 1930s of a compound that owed its insecticidal activity to tetraethylpyrophosphate (TEPP), one of the most potent aphicides ever developed (Schrader, 1965). TEPP was narrow in spectrum and unstable, especially under alkaline conditions, and because of this, it was not considered suitable for inclusion in sprays of Bordeaux mixture (i.e., copper sulfate and lime) used for control of fungal diseases on many crops. Further research by Schrader resulted in the discovery of parathion in 1944 and the systemic aphicide, schradan. After World War II, the chemistry behind these insecticides became known, and an estimated one hundred thousand organophosphorus compounds have been synthesized and tested against different insects since then.

The organophosphorus compounds are numerous, but form a well-recognized class. More is known about their mode of action, and the features of chemical structure and reactivity necessary for this action, than for any other class of poison. The field, therefore, has been concentrated on chemical research towards developing newer organophosphorus compounds. Many thousands of compounds have been synthesized, mainly in industrial laboratories. The objective of synthesis of organophosphorus compounds is, however, not to produce just another insecticide, but to find substantially improved selectivity, safety to mammals, and/or desirable level of persistence.

The structure of all active compounds in this class can be included in the general formula (**1**), as represented below in Figure 5.1:

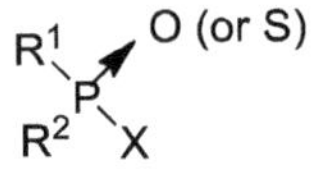

1

FIGURE 5.1 General formula of organophosphorus compound (**1**).

where, R^1 and R^2 are univalent groups and are stably linked to the central P atom, while X is an acidic group more easily split off by alkaline hydrolysis. A few compounds have been made in which a higher group VI element, Se or Te, replaces O or S as the recipient of the lone pair of electron of the central phosphorus, but this is not economical. One could expect that the NH group could also act as acceptor, but such compounds are subject to polymerization during attempted synthesis.

The stable groups, R^1 and R^2, usually for the convenience of manufacture, are the same and are most frequently EtO- or MeO- groups. Isopropyl esters occasionally appear, but esters of higher alcohols usually have greatly reduced activity. Alkylamino groups, especially $(Me)_2N$-, are used in some compounds, but much less frequently than ester groups. Wherever R^1 and R^2 are different, one is usually an alkyl group directly linked by R to P, that is, a phosphonic acid rather than phosphoric acid derivative.

The organophosphorus compounds were explored due to their stellar advantages over other classes of pesticides. They can be used as systemic insecticides, acaricides, fungicides, and herbicides, and possess a broad spectrum of activity against a number of pests, though these require special training for application. Due to high pesticidal toxicity, these are observed to be very economical and are used in small dosages. But some of the compounds possess an obnoxious odor and high acute toxicity. Particularly, in developing countries, poisoning with organophosphorus pesticides is one of the major issues causing serious health illness or even death. More than a decade ago, the number of intoxications with organophosphorus compounds was estimated at some three million per year (Eyer, 2003). Many organophosphorus compounds are easily decomposed and thus migrate from soil to water sources. Due to their high toxicity, persistence, and accumulation in the food chain, excessive use of these organophosphorus pesticides results in serious environmental pollution and other ecological problems. Their presence in water and food can lead to escalation of hazards to human health. Therefore, in order to overcome these problems, nanotechnology has emerged as a powerful tool for their detection. Thus, they require special training for application. This chapter

TABLE 5.1
General Classification of Organophosphorus Compounds on the Basis of Structure

Aliphatic	The organophosphorus compounds belonging to this group are carbon chain-like in structure and the compounds are monocrotophos, melathion, dimethoate, oxydemetonmethyl, dichlorvos, acephate, tetraethylpyrophosphate (TEPP), trichlorfon, dicrotophos, disulfoton, mevinphos, methamidophos, etc.
Phenyl derivatives	The organophosphorus compounds belonging to this group have a phenyl ring with one of the ring hydrogens displaced by attachment to the phosphorus moiety and other hydrogens frequently displaced by Cl, NO_2, Me, CN or S. These are generally more stable than the aliphatics, thus their residues are longer lasting. These compounds are methyl parathion, profenofos, fenitrothion, fenthion, etc.
Heterocyclic derivatives	The compounds of this group have ring structures composed of different or unlike atoms (O, N, S, etc.). The compounds are diazinon, chlorpyriphos, chlorpyriphos-methyl, etc.

includes different classification systems of organophosphorus compounds, their synthesis, modes of action, and different methods of their detection.

5.2 CLASSIFICATION OF ORGANOPHOSPHORUS COMPOUNDS

5.2.1 Classification on the Basis of Their Structure

The organophosphorus compounds are generally divided into three main groups: aliphatic derivatives, phenyl derivatives, and heterocyclic derivatives, as seen in Table 5.1 (Reddy and Latha, 1987).

5.2.2 Classification on the Basis of Their Activities

Organophosphorus compounds are classified and discussed on the basis of their biological activities in the following section (Handa, 2004).

5.2.2.1 Insecticides

Organophosphorus insecticides form an important class of pesticides, with about one hundred insecticides already registered for use on various crops in the world. Different organophosphorus compounds can be classified under the following heads.

1. Derivatives of phosphoric acid (**2**)
2. Derivatives of thiophosphoric acid (**3**)
3. Derivatives of dithiophosphoric acid (**4**)
4. Derivatives of phosphonic acid
5. Esteramides of phosphoric and phosphorothioic acids
6. Cyclic phosphates and phosphorothiolates
7. Derivatives of pyrophosphoric acid
8. Cyclic phosphates

The structures of derivatives of phosphoric acid (**2**), thiophosphoric acid (**3**), and dithiophosphoric acid (**4**) in this class are represented below in Figure 5.2:

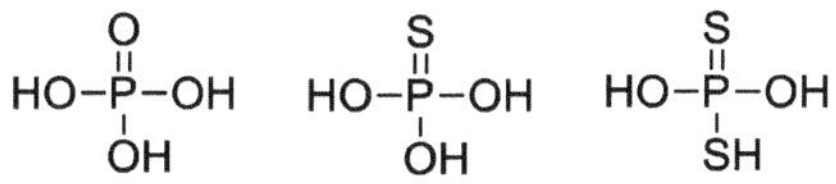

2 **3** **4**

FIGURE 5.2 Structures of organophosphorous insecticides (**2–4**).

5.2.2.1.1 Reactions Involved in the Preparation of Organophosphorus Insecticides

Michaelis-Arbuzov Reaction: This reaction is used for the preparation of phosphonate esters. Reaction between alkyl phosphites (**5**) and alkyl halides (**6**) yielded phosphonates (**7**) (Scheme 5.1). The intermediate is a quasi-phosphonium salt, which stabilizes by the loss of alkyl halide and formation of phosphonate (Scheme 5.2).

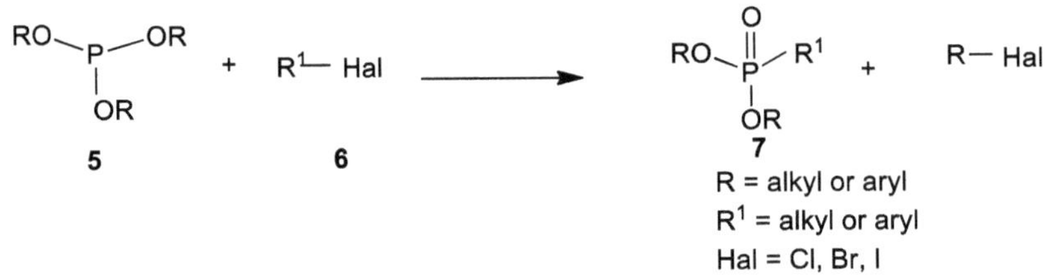

SCHEME 5.1 Synthesis of phosphonate esters (**7**).

Mechanism of the reaction

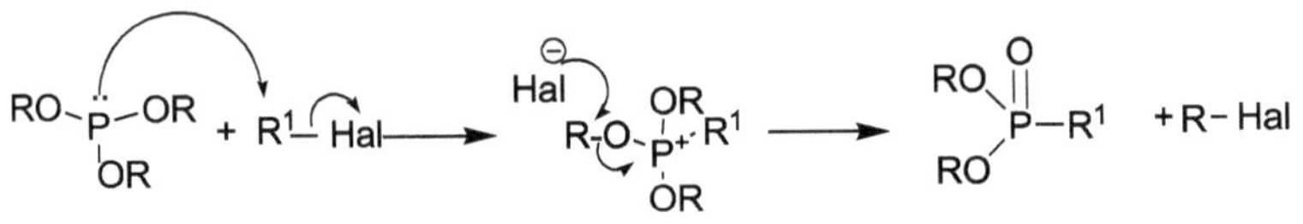

SCHEME 5.2 Plausible mechanism of formation of phosphonate esters (**7**).

Perkow reaction: In the Perkow reaction, when trialkylphosphites (**8**) are treated with α-halocarbonyl compounds (**9**), vinyl phosphates (**10**) are formed (Scheme 5.3).

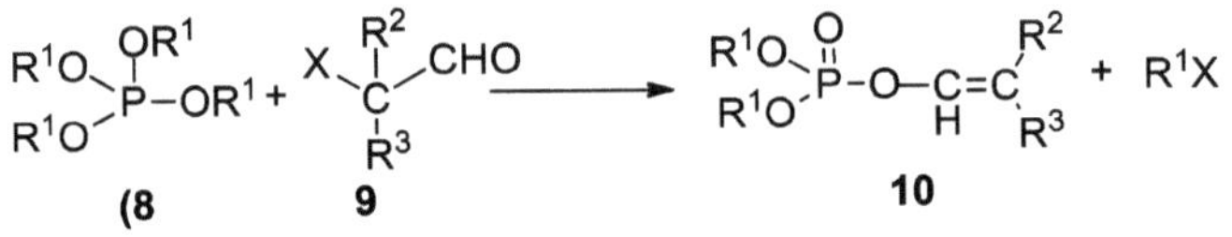

SCHEME 5.3 Synthesis of vinyl phosphates (**10**).

The proposed mechanism for the formation of vinyl phosphate (**10**) is represented in Scheme 5.4.

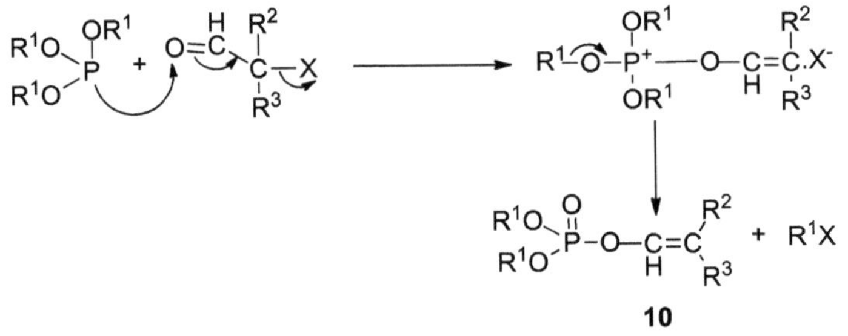

SCHEME 5.4 Plausible mechanism of formation of vinyl phosphates (**10**).

For instance, trimethyl phosphate (**11**) and chloral (**12**) tend to chemically synthesize dimethyl dichlorovinyl phosphate (**13**) (Scheme 5.5).

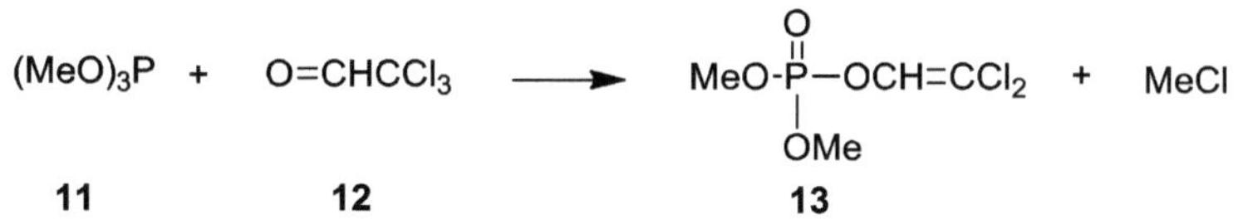

SCHEME 5.5 Synthesis of dimethyl dichlorovinyl phosphate (**13**).

*5.2.2.1.2 Derivatives of Phosphoric Acid (**2**)*

Dichlorvos (O,O′-dimethyl O-2,2-dichlorovinyl phosphate, DDVP, Nuvan, vaporoud, Vapona, **14**)

Dichlorvos (**14**) is prepared by treating trimethyl phosphate (**11**) with chloral (**12**) (Scheme 5.6).

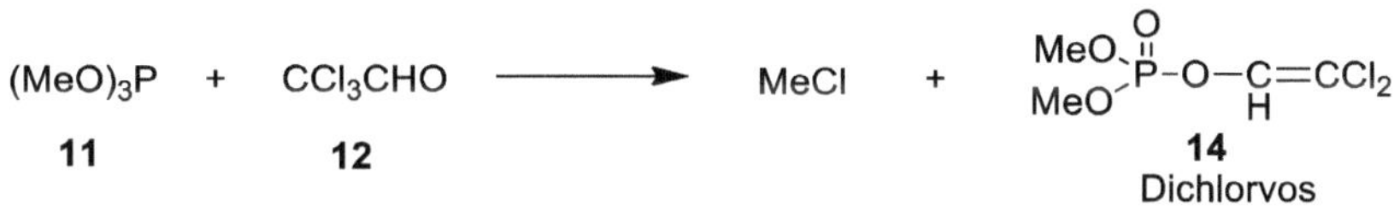

SCHEME 5.6 Synthesis of compound (**14**) from trimethyl phosphate (**11**) and chloral (**12**).

Dichlorvos (**14**) is also prepared by treating trichlorofon or dipterex (**15**) with alkali in aqueous solution at 40°C–50°C (Scheme 5.7).

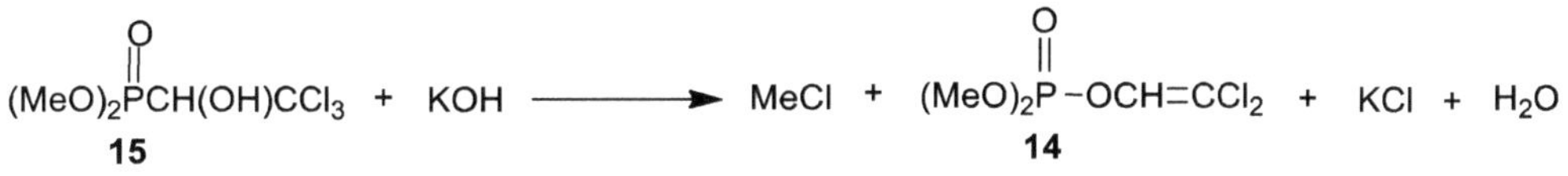

SCHEME 5.7 Synthesis of (**14**) from dipterex (**15**) under alkaline conditions.

It is a colorless liquid, with a boiling point (bp) of 120°C at 14 mm Hg and density 1.425 g/cc. It is highly soluble in the organic solvents. The lethal dose, 50% (LD_{50}) for rats is 80 mg/kg. It is hydrolyzed in water and forms dimethylphosphoric acid (**16**) and dichloroacetaldehyde (**17**) (Scheme 5.8).

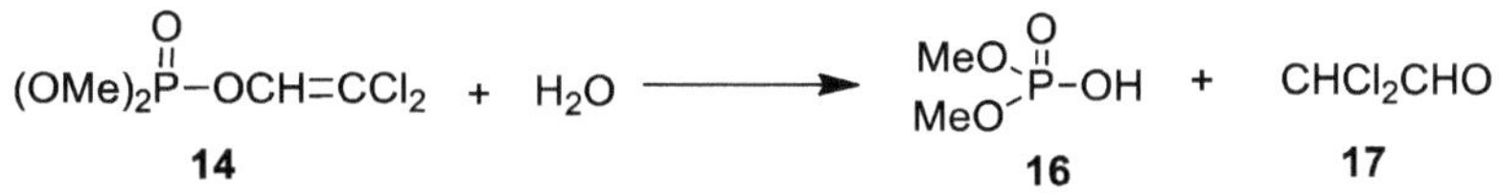

SCHEME 5.8 Hydrolysis of (**14**) to produce (**16**) and (**17**).

It is an interesting insecticide with a characteristic rapid knockdown of insects when used as a spray or bait. This compound is in the active form and thus does not require lethal synthesis within the insect. Dichlorvos is used extensively in Vapona strips, a preparation in which the insecticide is impregnated in resin and volatilizes at a fairly uniform rate over an extended period of time to control household pests, especially those which can fly. Although the LD_{50} values indicate that the insecticide is highly toxic, the rapid rate at which it is metabolized in the mammalian system permits its use in areas where human exposure occurs. Dichlorvos (**14**) is effective against ectoparasites and is used in flea collars for dogs and cats and a number of other veterinary applications. There is some question concerning the effect of dichlorvos (**14**) exposure on the ability of companion animals to withstand stress.

Naled (O,O′-dimethyl-1,2-dibromo-2,2,-dichloroethyl phosphate, Dibrom, **18**)

Dibrom (**18**) is brominated DDVP (**14**) and is prepared by brominating dichlorvos (**14**) (Scheme 5.9). It is a crystalline substance with a melting point (mp) of 25°C and a bp of 110°C. It is insoluble in water and soluble in organic solvents. Dibrom (**18**) is destructively hydrolyzed in water in two days and provides a short-lived insecticide for use in plants. The LD_{50} for rats is 430 mg/kg.

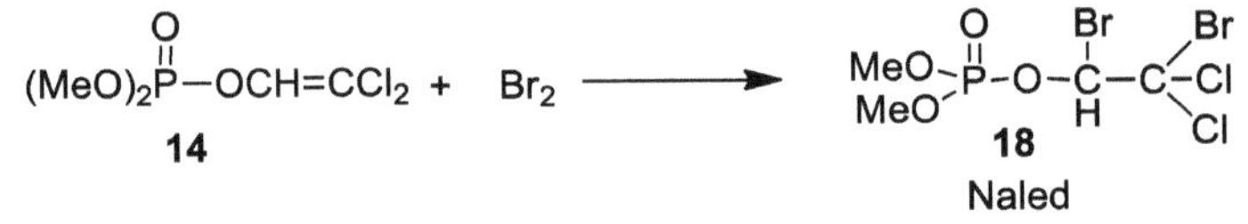

SCHEME 5.9 Synthesis of Dibrom (**18**).

Dibrom (**18**) is closely related chemically to trichlorfon (**15**) and has a similarly low mammalian toxicity. As it is quite volatile, this has been used as a "pour on" on heating pipes in greenhouses. This provides an effective fumigation for control of aphids and mites. In addition to its use in greenhouses, Naled is effective against adult Diptera and has been used to control these in agriculture and in mushroom houses.

Mevinphos (O,O′-dimethyl-O-[1-methyl-2-carbomethoxy vinyl]phosphate, Phosdrin, **20**)

Mevinphos (**20**) is prepared by treating trimethyl phosphite (**11**) with methyl α-chloroacetoacetate (**19**) (Scheme 5.10).

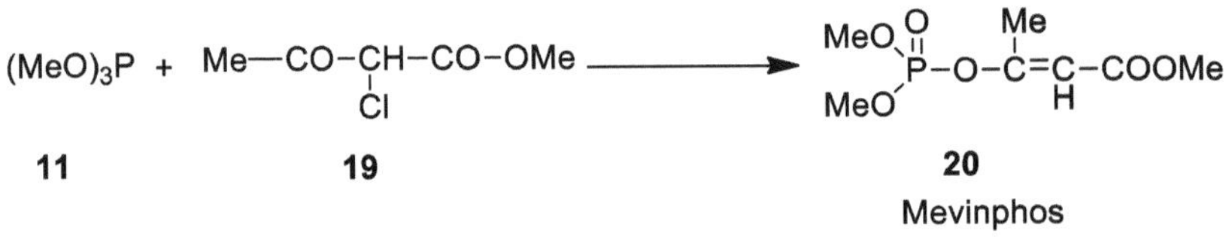

SCHEME 5.10 Synthesis of Mevinphos (**20**).

The commercial product is a mixture of *cis* and *trans* isomers containing 60% of the *cis* form and 40% *trans* form. The *cis* isomer is 100 times more toxic to both insects and mammals than the *trans* isomer. It is a colorless liquid, highly soluble in water, with a bp of 76°C at 0.2 mm Hg, vapor pressure of 2.2 × 10 mm Hg at 20°C, volatility of 27 mg/m^3 at 26°C. It is a short-lived systemic and contact poison useful on fruits and vegetables, where insect control is required near harvest. Mevinphos (**20**) is highly toxic by oral or dermal exposure (acute oral LD_{50} to the rat 3–7 mg/kg, dermal, 3–90 mg/kg).

Phosphamidon (O,O′-dimethyl-O-[2-chloro-N,N-diethyl carbamoyl] methylvinyl phosphate, Dimecron, **22**)

Phosphamidon (**22**) is prepared by Arbuzov reaction, accompanied by the Perkow rearrangement typical of α-halo carbonyls. Dichloroacetato N,N-diethyl acetamide (**21**) is reacted with trimethyl phosphate (**11**) to produce phosphamidon (**22**) (Scheme 5.11).

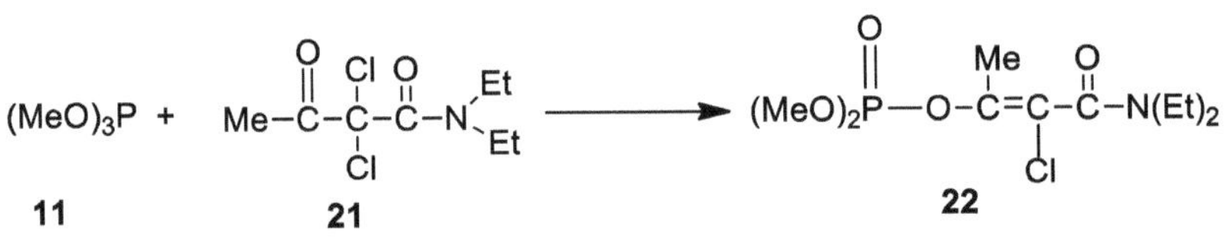

SCHEME 5.11 Synthesis of Phosphamidon (**22**).

When pure, it is a colorless and odorless oil. The bp is 1600°C at 1.5 mm Hg vapor pressure. It is soluble in water, alcohols, ketones, and most highly polar oxygenated and halogenated solvents. It is a highly toxic, general-purpose insecticide with systemic activity. It has high toxicity to birds, and when used in the spruce budworm program, caused some mortality in songbirds.

Dicrotophos (3-hydroxy-N, N-dimethyl-*cis*-crotonamide dimethyl phosphate, Bidrin, **24**)

Bidrin (**24**) is prepared by Perkow reaction from dimethylamide of monochloroacetoacetic acid (**23**) and trimethyl phosphate (**11**) (Scheme 5.12).

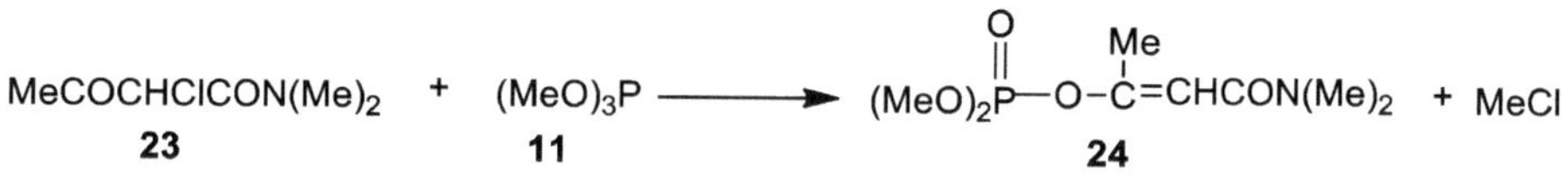

SCHEME 5.12 Synthesis of Bidrin (**24**).

Bidrin (**24**) is a brown liquid, with a bp of 115°C–120°C at 0.001 mm Hg and is soluble in water, ketones, and alcohols. The LD_{50} value to rats is 22 mg/kg with a dermal toxicity near 200 mg/kg. It is a systemic insecticide of interest owing to its unique method of application. When injected into tree trunks, it provides control of a wide range of sap borers and leaf miners.

Chlorfenvinphos (2-chloro-[2,4-dichlorophenyl]vinyl diethyl phosphate, Birlane, **27**)

Birlane (**27**) is prepared by the condensation of triethylphosphite (**25**) with 2,4,1′,1′-tetrachloroacetophenone (**26**) (Scheme 5.13).

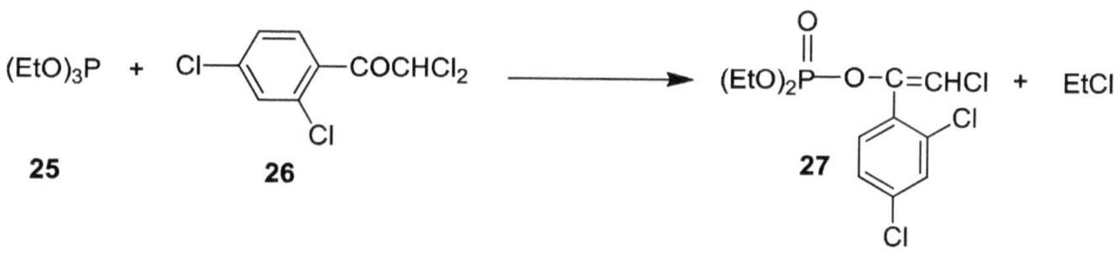

SCHEME 5.13 Synthesis of Birlane (**27**).

Birlane (**27**) is an amber-colored liquid. The technical grade product is a mixture of *cis* and *trans* isomers in the ratio 8.5:1, with the *trans* isomer being more active.

Stirofos (2-chloro-1-(2,4,5-trichlorophenyl)vinyl dimethyl phosphate, Gardona, **29**)

Gardona (**29**) is prepared by the condensation of trimethyl phosphite (**11**) with 2,4,5,1′,8-pentachloroacetophenone (**28**) (Scheme 5.14).

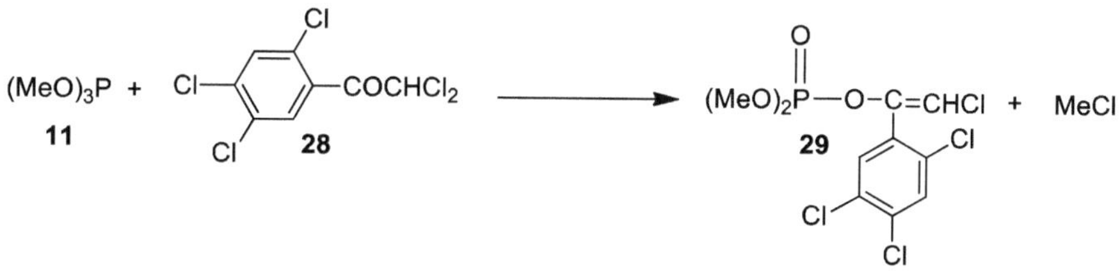

SCHEME 5.14 Synthesis of Gardona (**29**).

The crude product is a mixture of the geometric isomers; the *trans* isomer (with chlorine atoms on opposite sides of the double bond) is obtained (98%) by recrystallization. Gardona is a white crystalline substance, with a mp of 97°C –98°C. The LD_{50} for rats is 4000 mg/kg. Chlorfenvinphos (Birlane) (**27**) and stirofos (Gardona) (**29**) are closely related insecticides but with vastly different biological activity. Chlorfenvinphos (**24**) is highly toxic to most mammals (dog being an exception) and has a narrow spectrum of activity as a soil insecticide against root maggots and corn rootworms. By contrast, stirofos

(29) is less toxic to mammals and is selectively toxic to Diptera, Coleoptera, and Lepidoptera, whereas chlorfenvinphos **(27)** is a persistent (several months) soil insecticide, while stirofos **(29)** is not.

Monocrotophos (Dimethyl phosphate of 3-hydroxy N-methyl *cis*-crotonamide, Azodrin, **31)**

Monocrotophos **(31)** is prepared by treating trimethyl phosphite **(11)** with methyl amide of monochloroacetoacetic acid **(30)** (Scheme 5.15).

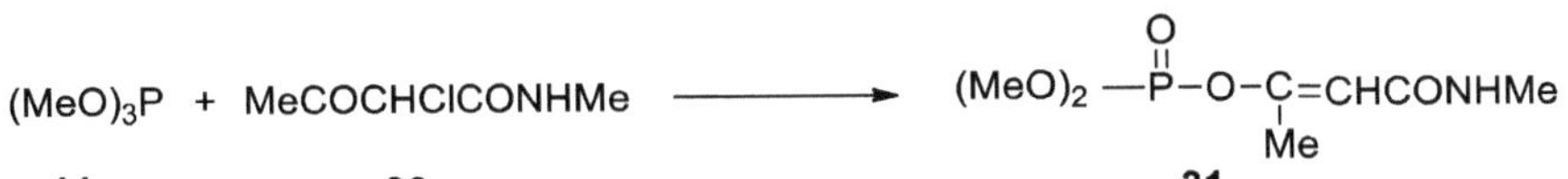

SCHEME 5.15 Synthesis of Monocrotophos **(31).**

Azodrin (**31**) is effective against economically important insects, which attack cotton. Pure azodrin **(31)** is a solid with a mp of 53°C–55°C. The acute oral LD_{50} for rats is 21 mg/kg. It is also a broad spectrum, highly toxic, and persistent (months) insecticide. Although it provides good control of many vegetable and cotton pests, its high toxicity and persistence on foliage have been associated in a few instances with illness among field workers.

Crufomate (4-*tert*-butyl-2-chlorophenyl methyl methyl phosphoramidate, Ruelene, **34**)

Ruelene **(34)** is synthesized by the reaction of 2-chloro-4-*tert*-butyl phenol **(32)** with O-methyl-N-methyl amidochlorophosphate **(33)** in the presence of base (Scheme 5.16). It is a white crystalline solid, insoluble in water, but soluble in organic solvents, with a mp of 61°C. The LD_{50} for rats is 770–950 mg/kg. Ruelene **(34)** is used to control intestinal worms and also for control of ectoparasites of cattle, as it has a high toxicity to cattle grubs but a low toxicity to the host.

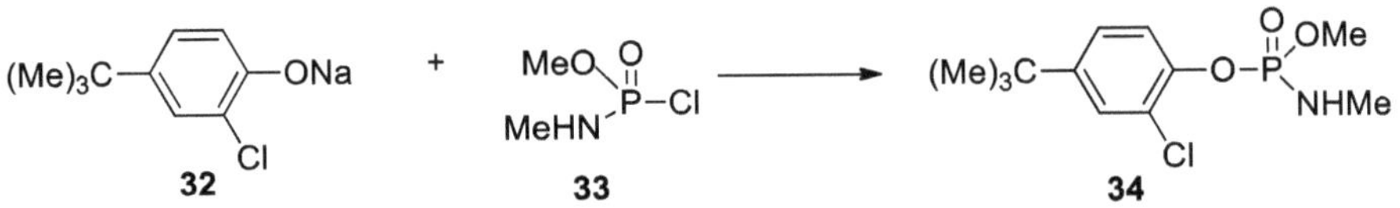

SCHEME 5.16 Synthesis of Ruelene **(34).**

5.2.2.1.3 Derivatives of Thiophosphoric Acid (3)

Parathion (O,O′-diethyl-O-nitrophenylphosphoro-thionate, thiophos, E-605, **37)**

Parathion **(37)** is prepared by reacting diethyl chlorothiophosphate **(35)** with *p*-nitrophenol **(36)** in the presence of potassium carbonate in acetone (Scheme 5.17).

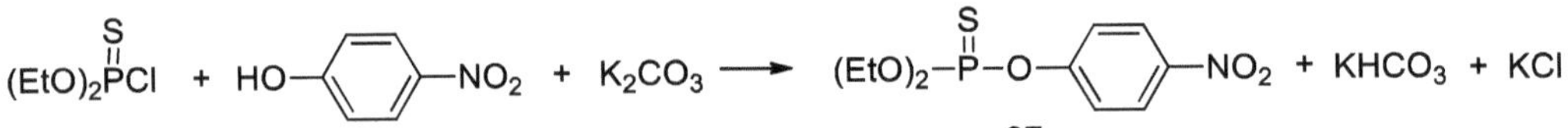

SCHEME 5.17 Synthesis of Parathion **(37).**

Parathion **(37)** is a clear liquid, with a bp of 157°C–162°C at 0.6 mm Hg, solubility of 24 ppm in water at 250°C, miscible with aromatic solvents, lower molecular-weight alcohols, acids, esters, and chlorinated hydrocarbons. It is resistant to acid hydrolysis but is hydrolyzed by alkali to form diethyl phosphorothioic acid and *p*-nitrophenol, both of which are of no significance as toxicants. Its acute oral LD_{50} to rats is approximately 4 mg/kg. Parathion is one of the most toxic and widely used insecticides in North America.

The conversion of parathion **(37)** to paraoxon **(38)** (Figure 5.3) activates the compound as an anticholinesterase, and this occurs in plants and on plant surfaces, as well as in animals. This conversion on plant surfaces is of special importance in hot, dry agricultural areas where the conversion to paraxon **(38)** is not accompanied by an equally rapid breakdown of this product. As a result, paraxon **(38)** accumulates on the dry, dusty foliage and presents a hazard to field workers, a hazard that may persist for many weeks after application and result in illness (McEwen and Stephenson, 1979).

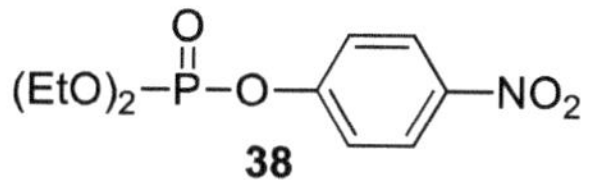

38

FIGURE 5.3 Structure of paraoxon **(38)**.

Although parathion **(37)** does not normally persist for long periods when applied to plant foliage, it can be quite persistent in soil, with measurable residues up to one year after application.

Methyl parathion (O,O′-dimethyl-O-*p*-nitrophenyl phosphorothionate, Folidol, metaphos, **41**)

Methyl parathion **(41)** is prepared by reacting dimethyl chlorothiophosphate **(39)** with sodium *p*-nitrophenolate **(40)** in the presence of potassium carbonate (Scheme 5.18).

39 **40** **41**

SCHEME 5.18 Synthesis of Methyl parathion **(41)**.

Methyl parathion **(41)** is a white crystalline solid powder, which is soluble in most common organic solvents and partially soluble in petroleum distillates. It has better solubility in water (60 ppm) than parathion **(37)**. It is used for the control of many insects and is effective against green leafhoppers, stem borers, and cutworms. The closely related methyl parathion **(41)** shares with parathion **(37)** many chemical properties, but it isomerizes and hydrolyzes more readily and is thus less persistent as a foliar residue and less stable in storage. Similarly, it is much less persistent in soil and is highly toxic (acute oral LD_{50} to rat near 20 mg/kg with dermal toxicity near 60 mg/kg). Although it has the same general spectrum of activity against insects as parathion has, it is more effective against aphids and beetles. Therefore, its greater effectiveness and slightly lower mammalian toxicity has resulted in its use instead of parathion in many agricultural programs, especially in cotton and citrus.

Oxidative stress is responsible for genetic damage in mouse spermatozoa (Guzman et al., 2006). It has been discovered that the oxidative stress is directly linked to methyl-parathion alterations on sperm DNA integrity. Also, the cells at meiosis and epididymal maturation are targets of methyl parathion. Methyl parathion deteriorates the sperm's functionality and capacity of fertilization after targeting spermatocytes and maturing spermatozoa (Guzman et al., 2009)

Methyl ethyl parathion (Thiophos, O-methyl O-ethyl-4-nitrophenyl thiophosphate, **43)**

Methyl ethyl parathion **(43)** is prepared by condensation of methyl ethyl chlorothiophosphate **(42)** with sodium salt of *p*-nitrophenol **(40)** and hydrochloric acid as acceptor in an organic solvent (Scheme 5.19).

42 **40** **43**

SCHEME 5.19 Synthesis of Methyl ethyl parathion **(43)**.

Methyl ethyl parathion (**43**) is a clear liquid, with bp of 116°C at 0.12 mm Hg vapor pressure at 20°C is 7.6 × 10^{-6} mm Hg. Its solubility in water is about 40 mg/L. It is soluble in most organic solvents except the paraffin solvents.

Fenitrothion (O, O′-dimethyl-O-3-methyl-4-nitrophenolphosphorothionate, Sumithion, Folithion, **45**)

Fenitrothion (**45**) is synthesized by condensation reaction of dimethyl chlorothiophosphate (**39**) and sodium 4-nitro-3-methylphenolate (**44**) (Scheme 5.20). Fenitrothion (**45**) is a liquid with an unpleasant odor, bp of 95°C at 0.01 mm Hg vapor pressure at 20°C. Fenitrothion (**45**) is effective to control rice stem borers and is also useful to control flies and mosquitoes in public health programs. The acute LD_{50} to rats is 1000 mg/kg.

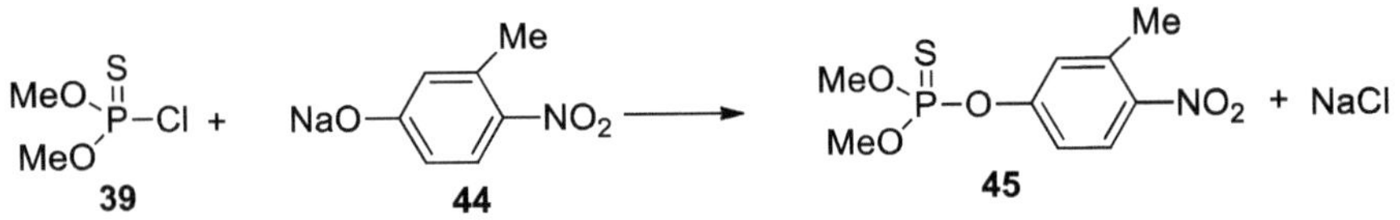

39 **44** **45**

SCHEME 5.20 Synthesis of Fenitrothion (**45**).

Fenitrothion (**45**) is a brownish liquid soluble in most organic solvents but with low solubility in water (20 ppm). It is converted by oxidases to the active fenitroxon in mammals, plants, and insects. Fenitrothion (**45**) is moderately toxic, with an acute oral LD_{50} for rat of 250–670 mg/kg and with dermal toxicity reported as 200–730 mg/kg.

Chlorothion (O,O′-dimethyl-O-4-nitro-3-chlorophenyl thiophosphate, **47**)

Chlorothion (**47**) is prepared by the reaction of dimethyl chlorothiophosphate (**39**) with 3-chloro-4-nitrophenol (**46**) in the presence of anhydrous potassium carbonate (Scheme 5.21).

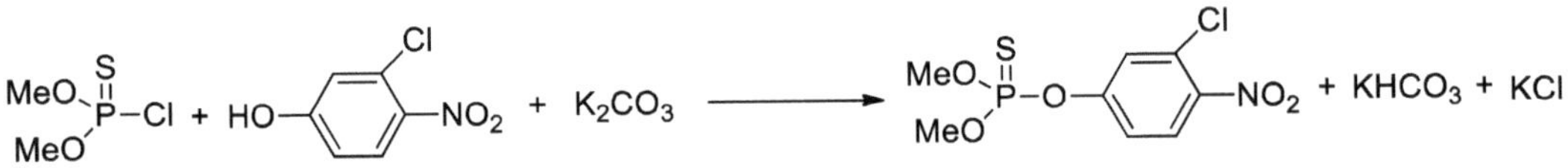

39 **46** **47**

SCHEME 5.21 Synthesis of Chlorothion (**47**).

The compound is a yellow crystalline solid, with a mp of 21°C. Its solubility in water at 20°C is 40 mg/L, while it is soluble in most organic solvents. The LD_{50} for rats is 850–980 mg/kg.

Ronnel (O,O′-dimethyl-O-2,4,5-trichlorophenyl phosphorothioate, Trolene, Korlan, **49**)

Ronnel (**49**) is prepared by treating sodium 2,4,5-trichlorophenolate (**48**) with dimethyl chlorothiophosphate (**39**) in aqueous medium in the presence of emulsifiers (Scheme 5.22).

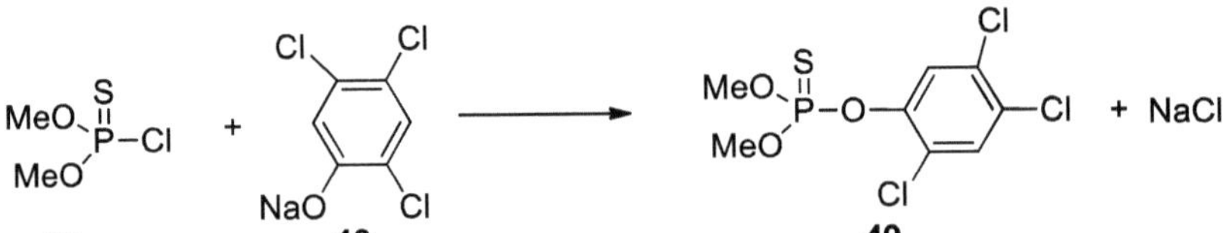

39 **48** **49**

SCHEME 5.22 Synthesis of Ronnel (**49**).

Ronnel (**49**) is a white nonhygroscopic powder with a mp of 36°C–37°C and soluble in organic solvents. Its solubility in water is 0.01 ppm. The LD_{50} for rats is 1250–1750 mg/kg. It has contact insecticidal activity against flies and cockroaches. Ronnel (**49**) is about equal in toxicity to crufomate (**34**) and provides control of cattle grubs when administered in the food at 100 mg/kg body weight or when given as a single 15 g bolus. In the animal, it is more persistent than crufomate (**34**) but is metabolized and excreted, the peak levels occurring in the urine 18–32 hours after administration.

Bromophos (O,O′-dimethyl-O-2,5-dichloro-4-bromophenyl thiophosphate, **51**)

Bromophos (**51**) is prepared by treating sodium salt of 2,5-dichloro-4-bromophenol (**50**) with dimethyl chlorothiophosphate (**39**) in aqueous medium in the presence of emulsifiers (Scheme 5.23).

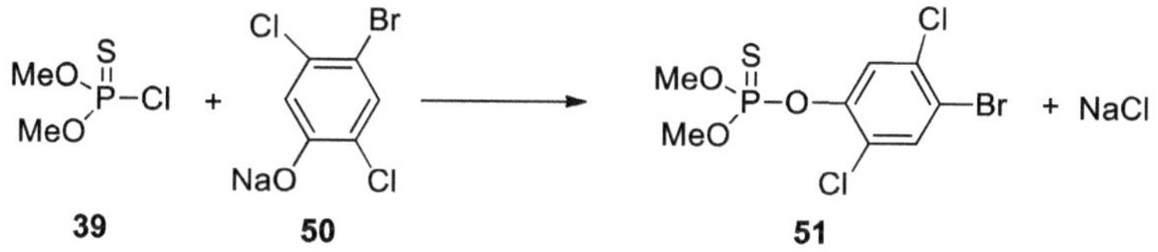

SCHEME 5.23 Synthesis of Bromophos (**51**).

Bromophos (**51**) is a white crystalline substance, with a mp of 54°C, solubility in water at 20°C as 40 mg/L, and is soluble in organic solvents. The LD_{50} for rats is 2000–4000 mg/kg. Bromophos is a nonsystemic insecticide and acaricide and is used on fruits and vegetables. It is useful as a public health insecticide.

Fenthion (O,O-dimethyl-O-[4-methyl mercapto-3-methyl phenyl]thiophosphate, Lebaycid, Baytex, **53**)

Fenthion (**53**) is prepared by reaction of 4-methyl mercapto-3-methylphenol (**52**) with dimethyl chlorothiophosphate (**39**) in methyl ethylketone solvent at 60°C in the presence of potassium carbonate (Scheme 5.24).

SCHEME 5.24 Synthesis of Fenthion (**53**).

Fenthion (**53**) is a colorless oil with a bp of 87°C and is a persistent (several months) insecticide of moderate toxicity with a fairly broad spectrum of activity, but is especially effective against Diptera. It finds use in controlling fruit flies, house flies, and mosquitoes, the last both as an adulticide and a larvicide. It is used against ectoparasites on livestock (Triguvon). It is highly toxic to birds and has been used in Africa against weave birds.

Fensulphothion (O,O′-diethyl-O-[4-methylsulfonyl phenyl] phosphorothiate, Dasanit, **56**)

Dasanit (**56**) is synthesized by the condensation of O,O′-diethyl chlorothiophosphate (**35**) with sodium 4-(methylthio)phenolate (**54**) and oxidation of the resulting ester (**55**) with hydrogen peroxide (Scheme 5.25).

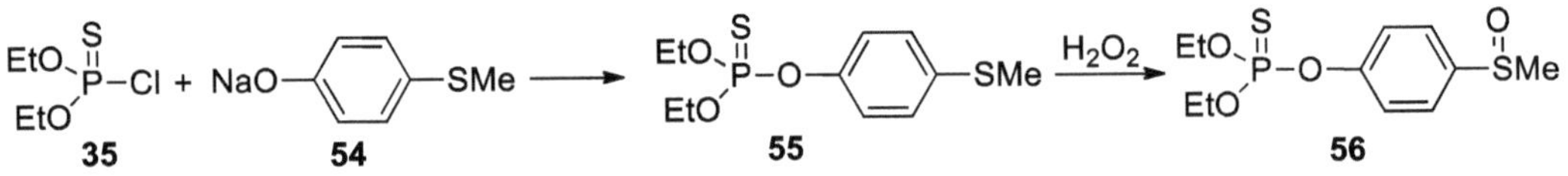

SCHEME 5.25 Synthesis of Dasanit (**56**).

Dasanit (**56**) is an oily liquid, with a bp of 138°C –141°C. The LD_{50} for rats is 4–10.5 mg/kg. It is effective as a soil insecticide with a long residual action. It displays both systemic and contact action. It is very similar in structure to fenthion (**53**), but is highly toxic. It is persistent (few months) and an effective soil insecticide against root maggots in vegetable crops and rootworms in corn. It is effective also against free-living cysts and root knot nematodes.

Diazinon (O,O′-diethyl O-2-isopropyl-4-methyl-6-pyrimidyl phosphorothionate, **58**)

Diazinon (**58**) is prepared by reacting 2-isopropyl-4-methyl-6-hydroxy pyrimidine (**57**) with diethyl chlorothiophosphate (**35**) in the presence of potassium carbonate (Scheme 5.26).

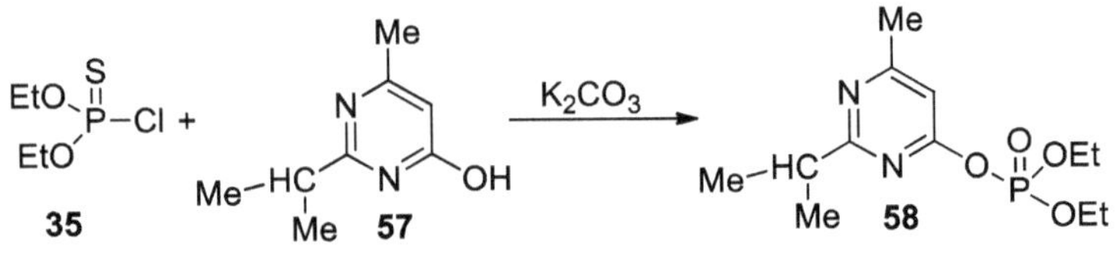

SCHEME 5.26 Synthesis of Diazinon (**58**).

Diazinon (**58**) is a reddish-brown liquid with a bp of 89°C at 0.1 mm Hg, miscible in most organic solvents, soluble in petroleum oils, and slightly soluble (40 ppm) in water. It is subject to oxidation and decomposes at high temperatures. It hydrolyzes rapidly in strong acid or alkaline conditions but only slowly where conditions are near neutral. In both animal and plant tissues, diazinon (**58**) is converted to the active diazoxon, a product much more susceptible to hydrolysis than the parent compound. Diazinon (**58**) is about as persistent in soil as parathion (**37**), being degraded by microorganisms and by hydrolysis. It is much less toxic than parathion (**37**), with acute oral LD_{50} values reported for rats as 66–600 mg/kg. Dermal toxicity is 379–1200 mg/kg.

Chlorpyriphos (O,O′-diethyl-O-3,5,6-trichloropyridyl thiophosphate, Dursban, **60**)

Dursban (**60**) is prepared by the reaction of sodium salt of 3,5,6-trichloro-2-hydroxypyridine (**59**) with diethyl chlorothiophosphate (**35**) in dimethyl formamide (Scheme 5.27).

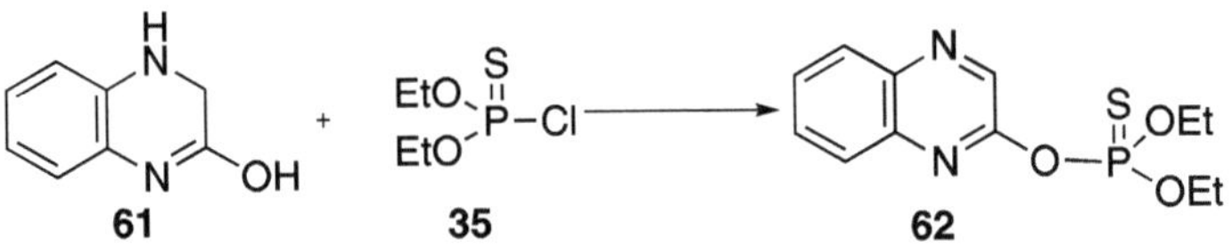

SCHEME 5.27 Synthesis of Dursban (**60**).

Dursban (**60**) is a white crystalline solid with a mp of 41.5°C–43°C. The LD_{50} value for rats is 150 mg/kg and for guinea pigs is 500 mg/kg. It is used for control of sucking and chewing pests. It is also used for control of termites. It is a moderately toxic broad-spectrum insecticide, of greatest interest from an environmental standpoint because of its use as a mosquito larvicide. Dursban (**60**) is relatively stable under acidic conditions but hydrolyzes when exposed to alkali. Many metabolites are formed in both animal and plant tissues, and when it is applied in an aquatic environment (e.g., as a mosquito larvicide), the residue is short-lived and found mostly in the phytoplankton. It has low solubility (2 ppm) in water. Its wide use in agriculture throughout the world has resulted in a series of toxicological and environmental problems, such as impacts on many nontarget aquatic species, including fish, as it is highly toxic to common carp. In addition, the results from the sub chronic toxicity test exhibited that its increasing concentration in the environment causes considerable stress for common carp, suggesting that chlorpyrifos (**60**) exposure causes immunotoxicity to common carp (Xing et al., 2015).

Quinalphos (O,O′-diethyl-O-[2-quinoxaline] phosphorothioate, Ekalux, **62**)

Quinalphos (**62**) is prepared by the reaction of 3,4-dihydroquinoxalin-2-ol (**61**) with diethyl chlorothiophosphate (**35**) (Scheme 5.28). Quinalphos (**62**) is a colorless and odorless crystalline solid with a mp of 31°C–32°C. It shows a high level of efficacy against a wide range of chewing, sucking, and leaf miner pests. The oral LD_{50} of the quinalphos (**62**) for rats is 71 mg/kg and the dermal LD_{50} is 1750 mg/kg.

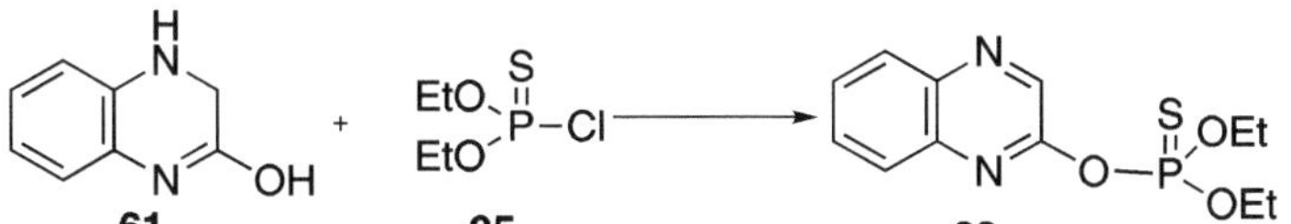

SCHEME 5.28 Synthesis of Quinalphos (**62**).

Pirimiphosmethyl (O,O′-dimethyl-O-[2-diethylamino-6-methyl-4-pyrimidinyl] phosphorothioic acid, **66**)

Pirimiphos methyl (**66**) is prepared by condensation of ethylacetoacetate (**63**) and N,N-diethyl guanidine (**64**) to afford 2-(diethylamino)-6-methylpyrimidin-4-ol (**65**), which on further reaction with dimethyl chlorothiophosphate (**39**), yielded pirimiphos methyl (**66**) (Scheme 5.29).

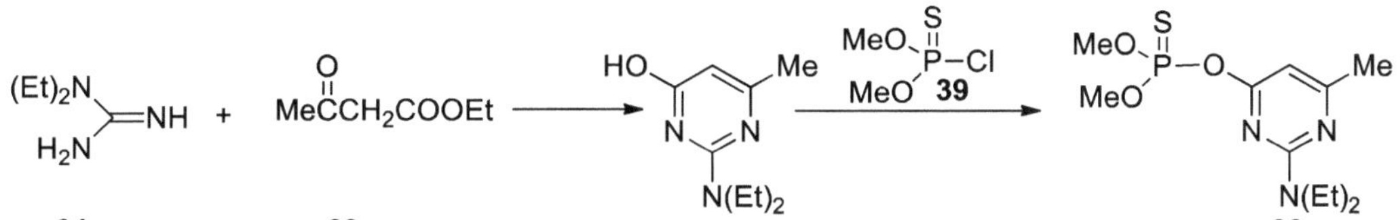

SCHEME 5.29 Synthesis of Pirimiphosmethyl (**66**).

It is highly effective against sucking pests. Its LD_{50} for rats is 2000 mg/kg. Pirimiphos methyl (**66**) acts as an insecticide and acaricide. It is mainly used in stored product protection.

Phoxim (O,O′-diethyl phenylglyoxylonitrile phosphorothioate, Volaton, **68**)

Phoxim (**68**) is prepared by treating α-cyanobenzaldoxime (**67**) with diethylchlorothiophosphate (**35**) (Scheme 5.30). Phoxim (**68**) is a light yellow liquid with a bp of 102°C at 0.01 mm Hg. It exhibits low toxicity to humans and warm-blooded animals. The LD_{50} for rats is 2000 mg/kg. Phoxim (**68**) has broad-spectrum insecticidal activity as a contact and stomach poison. It is effective to control leaf hoppers, aphids, cut worms, and insects of stored products, and thus is important in public health.

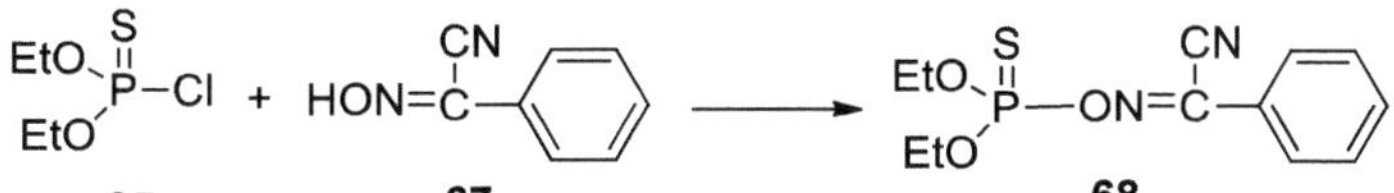

SCHEME 5.30 Synthesis of Phoxim (**68**).

Demeton (O,O′-diethyl-2-ethylmercepto ethyl thiophosphate, Systox, **72)**

Demeton **(72)** is prepared by the reaction of ethylmercapton **(69)** with ethylene oxide **(70)**, the resulting 2-ethylthio ethanol **(71)** is treated with diethyl phosphorothionochloridate **(35)**, in the presence of potassium carbonate (Scheme 5.31).

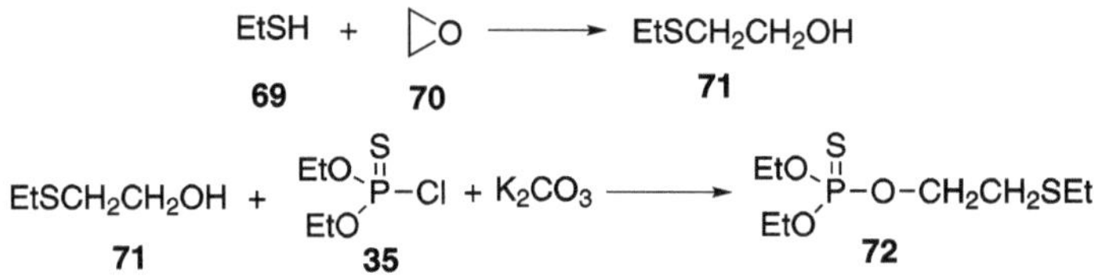

SCHEME 5.31 Synthesis of Demeton **(72).**

The compound is unstable in this form. It undergoes partial thiono-thiolo isomerization and the commercial product (demeton, systox) contains an equilibrium mixture of 30% of the thiolo isomer of (ethylthio) ethyl phosphorothiolate and 70% of the thiono isomer, O,O′-diethyl-2-ethylmercapto ethyl thiophosphate (**72a**).

Demeton **(72)** has a powerful systemic effect and is used for the control of many sucking pests. The oral LD_{50} value for rats is approximately 15 mg/kg. In plant tissues, both isomers are rapidly oxidized to the sulfoxide and sulfone derivatives. Oxydemeton methyl is less toxic to mammals (acute oral LD_{50} to rat about 75 mg/kg and dermal toxicity near 200 mg/kg). This compound has demonstrated some contact activity too.

Triazophos (O,O′-diethyl-O-[1-phenyl-1H-1,2,4-triazol-3-yl] phosphorothioate, **73)**

Triazophos **(73)** is a broad-spectrum insecticide effective against insects, mites, and nematodes. Its LD_{50} to rats is 82 mg/kg. Figure 5.4 represents the structure of triazophos.

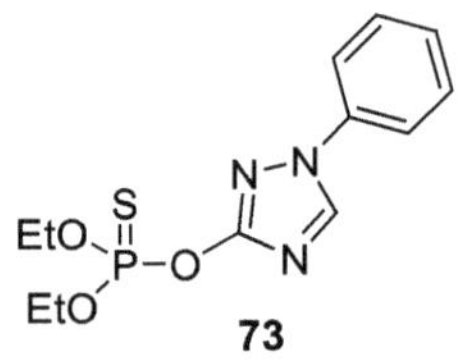

FIGURE 5.4 Structure of triazophos **(73)**.

Methyldemeton (O,O′-dimethyl-2-ethyl mercaptoethyl thiophosphate, Metasystox, **74)**

Methyldemeton **(74)** is the methyl derivative of demeton **(72)**. Technical grade methyl demeton is a 3:7 isomeric mixture of thiolo and thiono isomers. Methyl demeton **(74)** is synthesized by the reaction of 2-hydroxy diethyl sulfide **(71)** with dimethyl chlorothiophosphate **(39)** using sodium hydroxide (NaOH) as a base (Scheme 5.32).

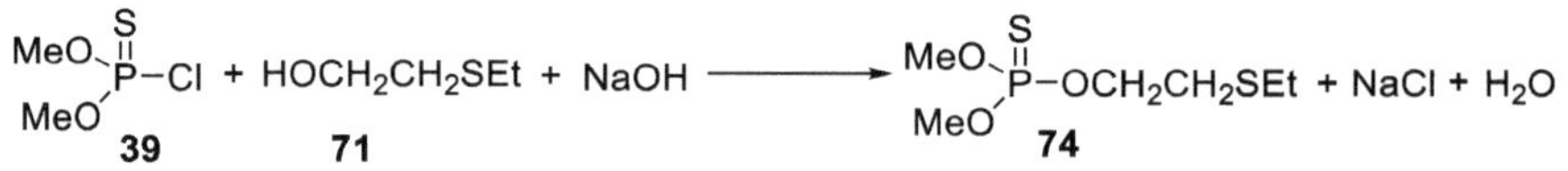

SCHEME 5.32 Synthesis of Methyldemeton **(74).**

The toxicity of methyl demeton (**74**) is quite less than that of Demeton (**72**). The LD_{50} values of thiono isomer and thiolo isomer of methyldemeton are 180 and 40 mg/kg, respectively.

5.2.2.1.4 Derivatives of Dithiophosphoric Acid (**4**)

Malathion (O,O′-dimethyl-S-[1,2-dicarbethoxyethyl] dithiophosphate, **77**)

Malathion (**77**) is synthesized by reacting dimethyldithiophosphoric acid (**75**) and maleic acid ester (**76**) in the presence of basic catalysts (Scheme 5.33).

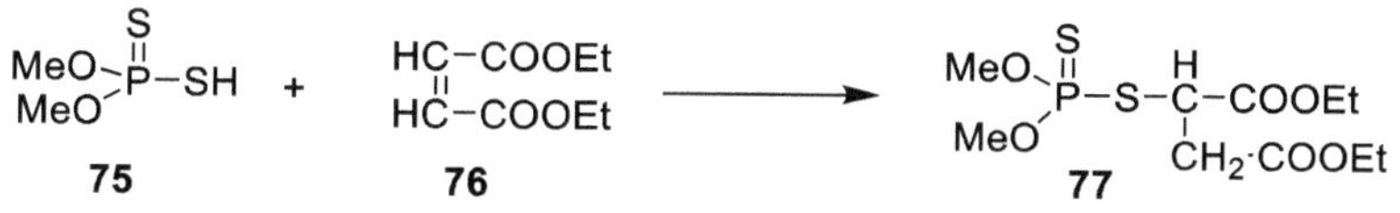

SCHEME 5.33 Synthesis of Malathion (**77**).

Malathion (**77**) has a broad-spectrum action and is effective against both sucking and chewing insects on fruits and vegetables. It is also used for the control of flies and mosquitoes. This product is toxic to insects, yet exhibits low toxicity to mammals. It is a clear, amber liquid, miscible in organic solvents, moderately soluble in light petroleum distillates, and slightly soluble in water (145 ppm). Malathion (**77**) is fairly stable in aqueous solutions in the pH range of 5.0–7.0, but hydrolyzes rapidly in more acidic or alkaline solutions. While its toxicity to insects approaches that of parathion (**37**), it has low mammalian toxicity, acute oral LD_{50} to the rate being about 2000 mg/kg with dermal toxicity in the 4000 mg/kg range.

Dimethoate (O,O′-dimethyl-S-[N-methyl carbamoylmethyl] phosphorodithioate, Roger, Cygon, **80**)

Dimethoate (**80**) is synthesized by reacting the potassium salt of dimethyldithiophosphoric acid (**78**) with N-methyl chloroacetamide (**79**) (Scheme 5.34).

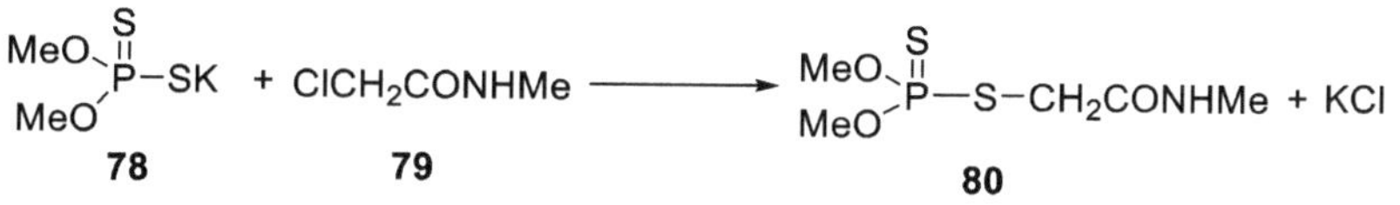

SCHEME 5.34 Synthesis of Dimethoate (**80**).

Dimethoate (**80**), a systemic insecticide, is used for control of insects on vegetables, cotton, tobacco, and sorghum, and decomposes slowly in plant tissues. Three processes occur simultaneously—namely, hydrolysis of the P-X bond, O-dimethylathion with the formation of products having low toxicity, and oxidation to the P = O analogue of dimethoate (**80**), a more powerful inhibitor of cholinesterase. In contrast to the high toxicity of most organophosphorus systemics, it exhibits moderate toxicity with an acute oral LD_{50} in rat as nearly 500 mg/kg by dermal or oral exposure. Moreover, cancer genes in breast cells are proposed to be induced by malathion (**77**) and parathion (**37**) in the presence of estrogen (Calaf and Roy, 2008).

Phorate (O-O′-Diethyl-S-[ethylthiomethyl] dithiophosphate, Thimet, **83**)

Thimet (**83**) is prepared by reacting diethyl dithiophosphoric acid (**81**) with formaldehyde (**82**) and ethyl mercaptan (**69**) (Scheme 5.35).

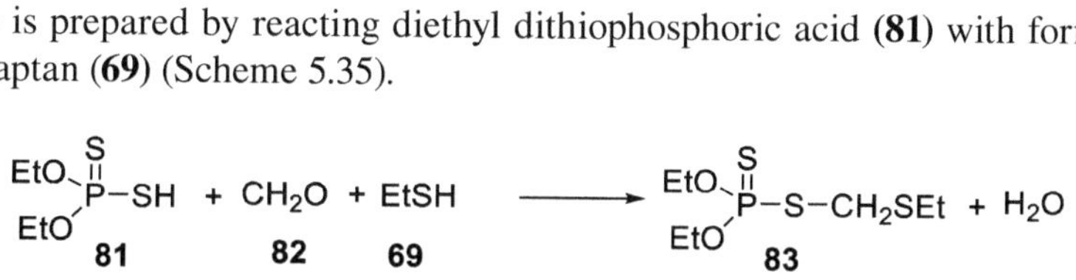

SCHEME 5.35 Synthesis of Thimet (**83**).

Phorate (**83**) is a systemic insecticide and is effective against sucking pests, larvae of the corn rootworms, and leaf-eating beetles. It is used as granules in soil and is relatively unstable to hydrolysis. However, it protects the plants for a relatively long time because of the greater persistency of the sulfoxide metabolite in plants and soil. It is oxidized to sulfoxide and then to sulfone in plants. It is highly toxic, with an acute oral LD_{50} to the rat of 1–5 mg/kg and a dermal toxicity reported as 2–3000 mg/kg. Because of its high mammalian toxicity, it is marketed only as a granular product. In common with other organophosphorus insecticides, phorate (**83**) does not accumulate in animal tissues; it is hydrolyzed and excreted in urine and feces.

Disulfoton (O,O′-diethyl-S-[2-ethylmercaptoethyl] dithiophosphate, Di-Syston, **86**)

Di-Syston (**86**) is prepared by treating salts of diethyldithiophosphoric acid (**84**) with 2-chlorodiethyl sulfide (**85**) (Scheme 5.36). Disyston (**86**) is a systemic insecticide and is used for control of sucking insects. Its principal disadvantage is its high toxicity to warm blooded animals, and its LD_{50} in rat is 2–12 mg/kg. It is oxidized to sulfoxide and then to sulfon. The sulfoxide of Di-Syston (**86**) is marketed as oxydesulfoton. Its biological properties are similar to disyston.

SCHEME 5.36 Synthesis of Disyston (**86**).

Phosalone (O,O′-diethyl-S-[6-chlorophenoxozolinyl-3-methyl] dithio phosphate, Zolone, **89**)

Phosalone (**89**) is prepared by the condensation of sodium diethyldithiophosphate (**87**) with 6-chloro-3-chloromethyl benzoxazolone (**88**) (Scheme 5.37).

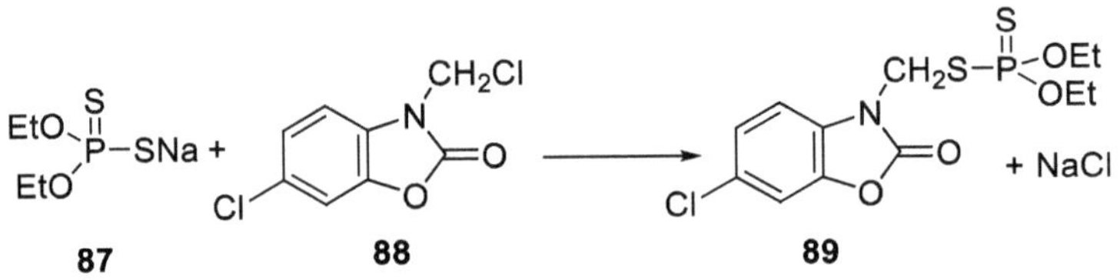

SCHEME 5.37 Synthesis of Phosalone (**89**).

Phosalone (**89**) is a broad-spectrum insecticide and acaricide useful for control of caterpillars, aphids, and mites on apples and field crops. The acute oral LD_{50} value for rats is 120 mg/kg.

Phosmet (Imidan) (**90**) and **phosalone** (Zolone) (**89**) are the general purpose insecticides, used to a significant extent on deciduous orchards. They exert moderate toxicity orally, while showing low toxicity to mammals on dermal exposure. Figure 5.5 represents the structure of Phosmet (**90**).

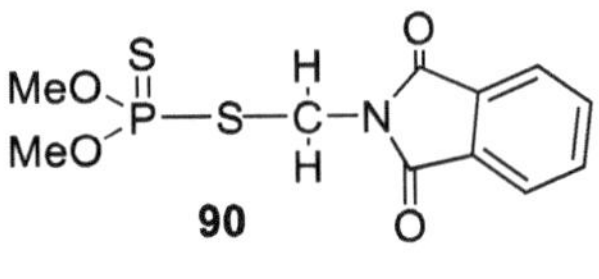

FIGURE 5.5 Structure of Phosmet (**90**).

Formothion (O,O′-dimethyl-S-[N-methyl-N-formylcarbamoylmethyl] phosphorodithioate, **92)**

Formothion (**92**) is prepared by the reaction of the potassium salt of dimethyl dithiophosphoric acid (**78**) with N-formyl-N-methyl chloroacetamide (**91**) (Scheme 5.38).

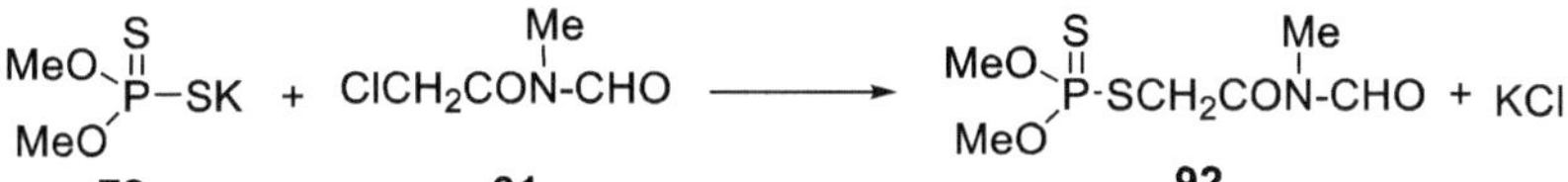

SCHEME 5.38 Synthesis of Formothion (**92**).

Formothion (**92**) is a yellowish liquid, which shows relatively low toxicity in mammals. Its LD_{50} for rats is 330–50 mg/kg and is effective against chewing and sucking pests by systemic routes and contact action.

Ethion (O,O,O′,O′-tetraethyl-S,S-methylene-*bis*-dithiophosphate, **95)**

Ethion is prepared by the condensation of salts of diethyldithiophosphoric acid (**93**) with dibromomethane (**94**) (Scheme 5.39).

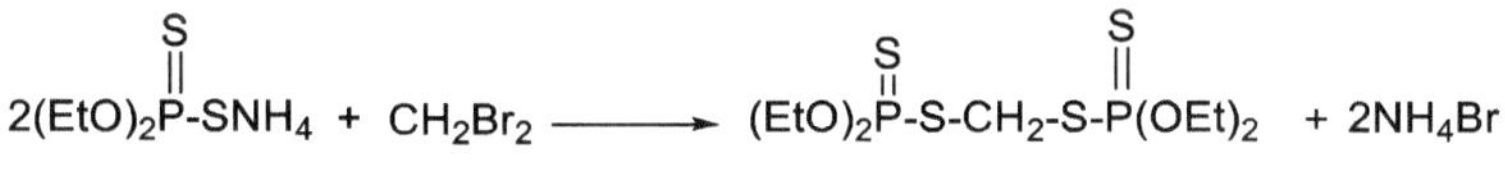

SCHEME 5.39 Synthesis of Ethion (**95**).

Ethion (**95**) is a yellow liquid with a bp of 164°C–165°C and is a persistent (several months in soil) insecticide and acaricide. It is practically insoluble in water and is used frequently in combination with oil as a dormant spray for control of aphids, mites, and scale insects in orchards. It is effective also when used without oil during the growing season. Ethion has been used for many years as a soil application in the seed furrow for maggot control on onions. In the soil, it is long lasting and has been shown to accumulate to significant levels (> 5 ppm) in organic soils. The LD_{50} for rats is 55 mg/kg.

Menazon (S-[4,6-diamino-1,3,5-triazin-2-methyl], O,O′- dimethyl phosphorodithioate, Sayfos, **98)**

Menazon (**98**) is synthesized by the reaction of the sodium salt of dimethyl dithiophosphate (**96**) with 2-chloromethyl-4,6-diamino-1,3,5-triazine (**97**) (Scheme 5.40).

SCHEME 5.40 Synthesis of Menazon (**98**).

Menazon (**98**) is a colorless crystalline solid with a mp of 60°C and is effective against aphids by foliar spray. The LD_{50} for rats is 900–1950 mg/kg.

Thiometon (O,O′-dimethyl-S-2[ethylmercaptoethyl] dithiophosphate, Ekatin, **100)**

Ekatin (**100**) is produced by reacting 2-hydroxyldiethyl sulfide (**71**) with dimethyldithiophosphoric acid (**75**) and *p*-toluene sulfonyl chloride (**99**) in the presence of sodium hydroxide (Scheme 5.41).

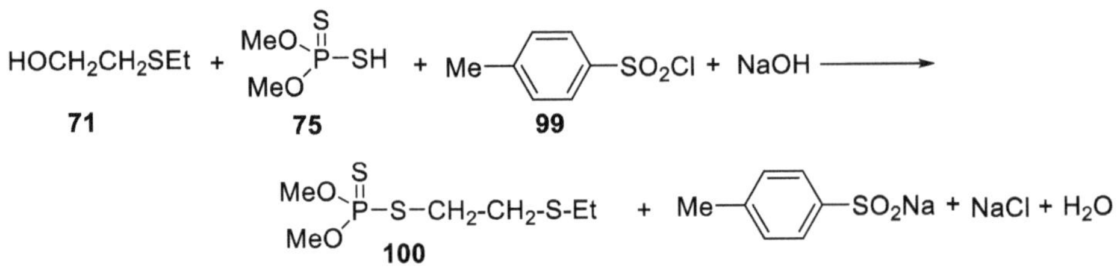

SCHEME 5.41 Synthesis of Ekatin (**100**).

Ekatin (**100**) is a colorless oil with a bp of 104°C,and is a systemic insecticide and acaricide with contact activity. It is effective against aphids, thirps, and mites. The LD_{50} for rats is 70–120 mg/kg.

Azinphosmethyl (O,O′-dimethyl-S-4-oxo-1,2,3-benzotriazin-3-[4H]-yl-methyl phosphoro dithioate, Azinphos, Guthion, **102**)

Guthion (**102**) is prepared by treating N-halo methyl derivative of azimidobenzoyl (**101**) with salts of dimethyl dithiophosphoric acid (**96**) (Scheme 5.42).

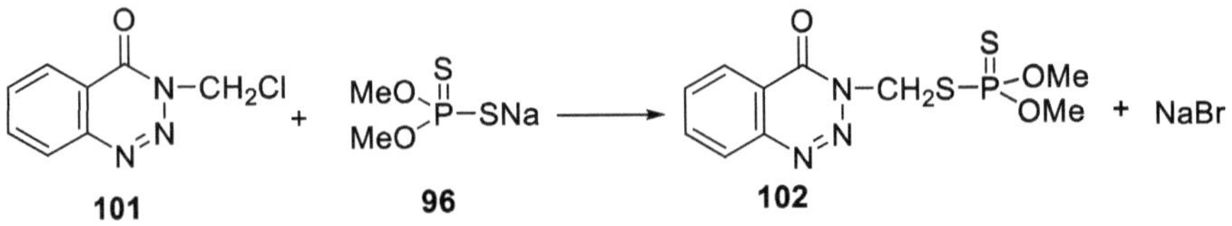

SCHEME 5.42 Synthesis of Guthion (**102**).

Guthion (**102**) is a white crystalline substance with a mp of 73°C–74°C. It is a nonsystemic insecticide and acaricide for use in fruit crops, vegetables, and cotton, as well as against foliage-feeding insects. It is soluble in organic solvents and slightly soluble (30 ppm) in water. It is readily hydrolyzed by acid or alkali. Azinphosmethyl (**102**) has high mammalian toxicity, the acute oral LD_{50} to the rat being near 15 mg/kg, with dermal toxicity near 225 mg/kg.

Phenthoate (O,O′-dimethyl-S-[1-carboethoxy benzyl] dithiophosphate, Cidial, **104**)

Cidial (**104**) is prepared by reacting the sodium salt of dimethyl dithiophosphate (**96**) with ethyl α-bromo phenylacetate (**103**) (Scheme 5.43).

SCHEME 5.43 Synthesis of Cidial (**104**).

Cidial (**104**) is a liquid with a bp of 70°C–80°C at 2×10^{-5} mm Hg. Cidial (**104**) has a broad spectrum of insecticidal and acaricidal activities and is particularly effective against scale insects. The oral LD_{50} value for rats is 4700 mg/kg.

5.2.2.1.5 Phosphonic Acid Derivatives

EPN (O-ethyl-O-[4 nitrophenyl] phosphonothioate, **111**)

Benzene (**105**) is reacted with phosphorus trichloride (**106**) in the presence of aluminum chloride to give benzene phosphonodichloridate (**107**). This is reacted with phosphorus thiochloride (**108**) or elemental sulfur to give benzene-phosphonodichlorothioate (**109**), which is reacted with sodium ethoxide to yield O-ethyl phenylphosphonochloridothioate (**110**), followed by the sodium salt of *p*-nitrophenol (**40**) to yield EPN (**111**) (Scheme 5.44).

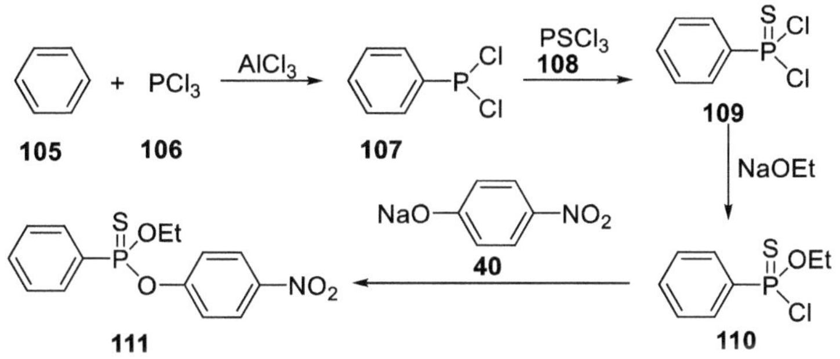

SCHEME 5.44 Synthesis of EPN (**111**).

The mp of EPN (**111**) is 36°C, and the technical material is a brown liquid. The commercial product has relatively low toxicity against spider mites, grasshoppers, and mosquito larvae. The LD_{50} for rats is 36 mg/kg.

Fonophos (O-ethyl-S-phenyl ethane phosphonodithioate, Dyfonate, **114**)

Fonophos (**114**) is prepared by treating O-ethyl-ethane phosphonothio-chloridate (**112**) and thiophenol (**113**) in the presence of base (Scheme 5.45). It is a highly toxic, persistent (months) general soil insecticide with a broad-spectrum of activity.

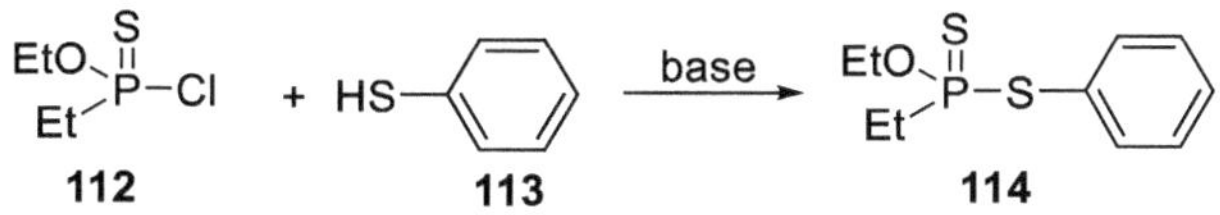

SCHEME 5.45 Synthesis of Fenophos (**114**).

Trichlorfon (O,O′-dimethyl-[1-hydroxy-2,2,2-trichloro] ethyl phosphonate, Dipterex, Dylox, **15**)

Dipterex (**15**) is prepared by addition of dimethylphosphonate (**115**) to chloral (**12**) (Scheme 5.46).

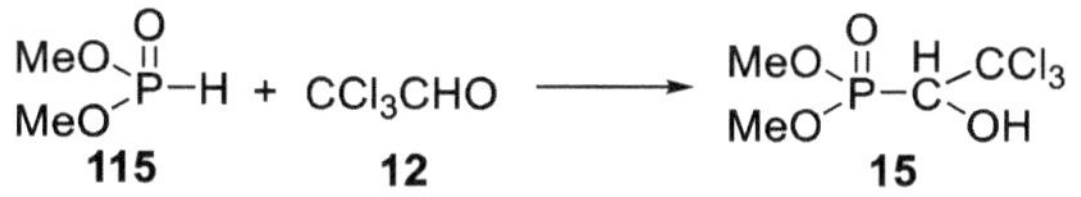

SCHEME 5.46 Synthesis of Dipterex (**15**).

Trichlorfon (**15**) is a broad-spectrum insecticide of low mammalian toxicity, and it is effective as a stomach and contact insecticide. It is used as a foliar spray against lepidopterous larvae and finds use in the dairy barn as bait for fly control. Trichlorfon (**15**) is converted readily to dichlorvos, a conversion believed responsible for much of its activity. The LD_{50} for rats is 630 mg/kg.

5.2.2.1.6 Esteramides of Phosphoric Acid and Phosphorothioic Acid

Phospholan (O,O′-diethyl-N-[1,3-dithiolan-2-ylidene] phosphoromidate, **120**)

Phospholan (**120**) is prepared by the reaction of ethylen-1,2-dithiol (**116**) with cyanogen chloride (**117**), followed by the reaction of 2-imino-1,3-dithiolane (**118**) with diethyl phosphorochloridate (**119**) (Scheme 5.47). Phospholan (**120**) has high activity against chewing and sucking insects.

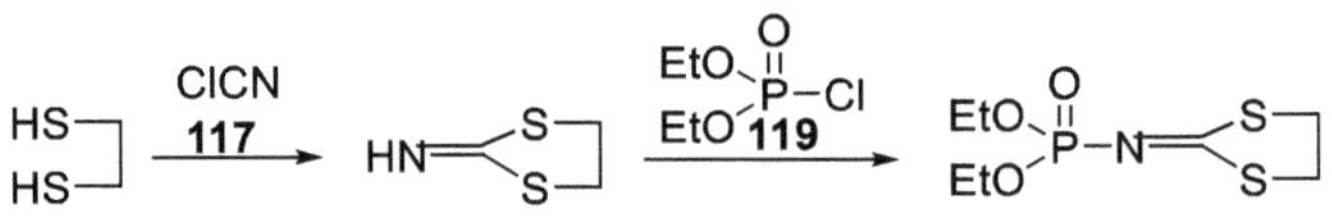

116 118 120

SCHEME 5.47 Synthesis of Phospholan (**120**).

Methamidophos (Tamaron, O,S-dimethyl phosphorothioamidate, Monitor, **122**)

Methamidophos (**122**) is prepared by converting dimethyl phosphorothiochloridate (**39**) with ammonia to O,O′-dimethyl phosphorothioamidate (**121**), which, in the presence of methyl iodide, undergoes thiono-thiolo isomerization and is converted to O,S-dimethyl analogue, methamidophos (**122**) (Scheme 5.48).

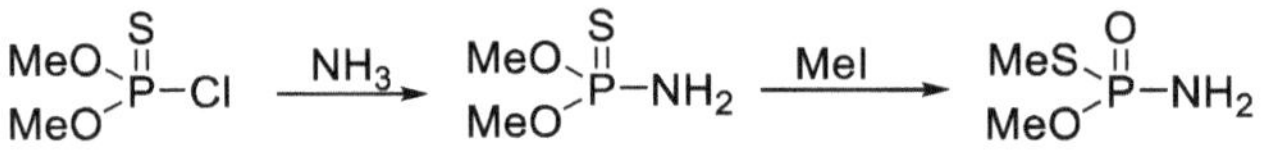

39 121 122

SCHEME 5.48 Synthesis of Methamidophos (**122**).

Methamidophos (**122**) is a colorless crystalline solid with a mp of 44.5°C. It is a broad-spectrum insecticide, persistent (weeks on foliage), particularly effective to control caterpillars and aphids. It is readily soluble in water and has some systemic activity. It has also systemic action against both biting and sucking pests. Its acute oral LD_{50} value for rats is 30 mg/kg.

Acephate (Orthene, O,S-dimethyl-N-acetyl phosphoramidothiolate, **124**)

Acephate (**124**) is N-acetyl derivative of the insecticide methamidophos (**122**), and is prepared by the acetylation of the latter with acetic anhydride (**123**) in the presence of $ZnCl_2$ or $FeCl_3$ (Scheme 5.49).

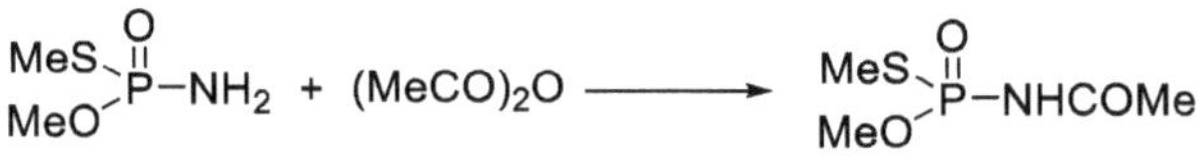

122 123 124

SCHEME 5.49 Synthesis of Acephate (**124**).

Acephate (**124**) is a white crystalline solid with a mp of 91°C–92°C. It is a systemic insecticide effective against chewing and sucking pests. The LD_{50} value for rats is 945 mg/kg.

5.2.2.1.7 Cyclic Phosphates and Phosphorothioates

Salithion (2-methoxy-4H-1,3,2-benzodioxophosphorin-2-sulfide, **127**) and **Salioxon** (2-methoxy-4H-1,3,2-benzodioxaphosphoran-2-one, **128**)

Among the cyclic phosphates, there are also insecticidal derivatives of Salioxon (**128**) and its thiono analogue salithion (**127**). Salithion (**127**) is prepared by treating saligenin (**125**) with methyl phosphorodichlorothioate (**126**) (Scheme 5.50).

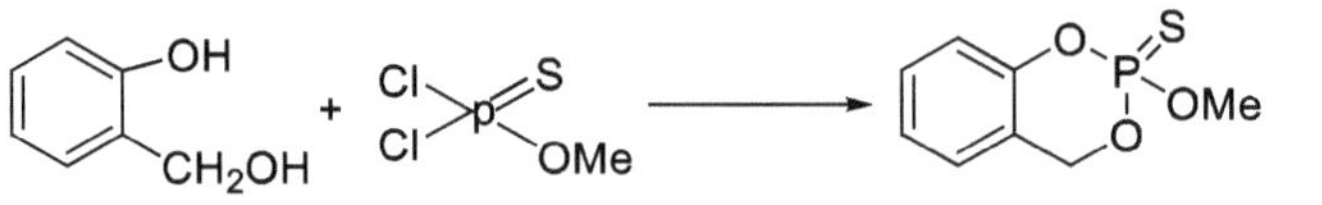

125 126 127

SCHEME 5.50 Synthesis of Salithion (**127**).

The LD_{50} value to rats for Salithion (**127**) is 102 mg/kg and for Salioxon (**128**) is 0.5–1 mg/kg. Figure 5.6 represents the structure of Salixon (**128**).

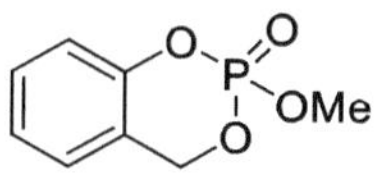

FIGURE 5.6 Structure of Salioxon (**128**).

5.2.2.1.8 Derivatives of Pyrophosphoric Acid

Tetraethylpyrophosphate (TEPP, 132)

TEPP was the first commercial organophosphate insecticide, discovered in 1938 by Schrader. It is miscible in water and hydrolyzes readily to nontoxic compounds. TEPP is colorless and odorless, with a bp of 124°C at 1 mm Hg. On plant foliage, it dissipates rapidly and controls mites and is used in aerosols in green houses. Because of its short persistence (less than 1 day) on plant foliage, it is useful for aphid control on food crops immediately prior to harvest. Tetraethylpyrophosphate (**132**) is synthesized by the reaction of the sodium salt of diethyl phosphite (**129**) with diethyl phosphorochloridate (**130**), followed by oxidation (**131**) (Scheme 5.51).

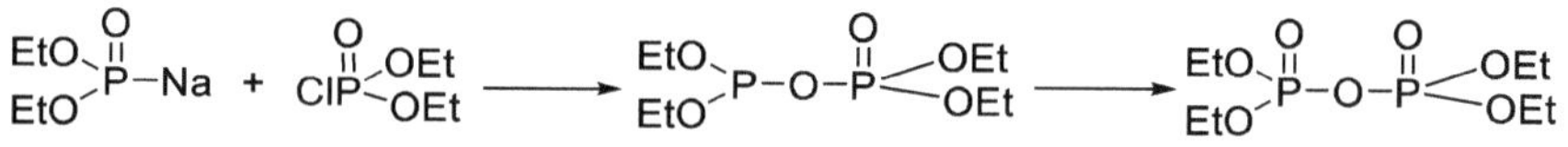

129 130 131 132

SCHEME 5.51 Synthesis of Tetraethylpyrophosphate (**132**).

The oral LD_{50} for rats is 2.0 mg/kg. Because of high toxicity, the use of TEPP is restricted.

Schradan (Octamethyl pyrophosphoamide, **136**)

Schradan (**136**) is prepared by treating dimethylamine (**133**) with phosphorus oxychloride (**134**), to afford N,N,N,N-tetramethylphosphorodiamidoyl chloride (**135**), which on treatment with tertiary amine furnished compound Schradan (**136**) (Scheme 5.52).

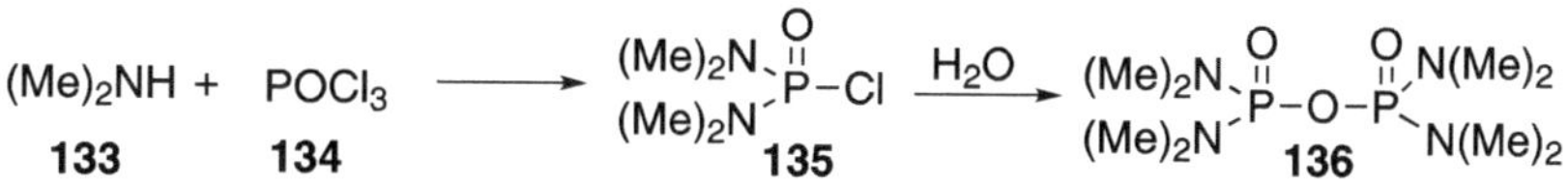

133 134 135 136

SCHEME 5.52 Synthesis of Schradan (**136**).

Schradan (**136**), developed by Schrader in the late 1940s, was the first of this group of plant systemics (Schrader, 1965). It is a mixture of octamethyl pyrophosphoramide and decamethyl triphosphoramide and is miscible in water. Schradan (**136**) is a colorless viscous liquid, having a bp of 118°C–122°C at 0.3 mm Hg. It is used to control aphids and mites on citrus, apples, and hops. It is highly toxic to mammals (acute oral LD_{50} to the rat of about 10 mg/kg and a dermal toxicity of about 15 mg/kg).

5.2.2.2 Fungicides

Organophosphorus fungicides also form an important class of pesticides. A large number of fungicides have been registered for use on various crops, worldwide. Different organophosphorus compounds can be classified as fungicides under the following heads:

5.2.2.2.1 Phosphoro-Thiolates

The fungicide registered in this group is Ediphenphos EC_{50} (50% Effective Concentration) pesticide used against paddy blast and brown leaf spot in the levels of 200–240 mL/acre. The fungicide, however, is not manufactured by most companies.

5.2.2.2.2 Phosphates

This group has many important systemic fungicides. The following are important fungicides used for controlling and inhibiting growth of fungi:

Wepsyn (P-[5-amino-3-phenyl-1H-1,2,4-triazol-1-yl]-N,N,N′,N′-tetramethyl phosphonic diamidate, **139**)

Wepsyn (**139**) is prepared by the condensation of the sodium salt of 3-amino-5-phenyl-1,2,4-triazole (**137**) with *bis* (dimethyl amino) phosphoryl chloride (**138**) (Scheme 5.53).

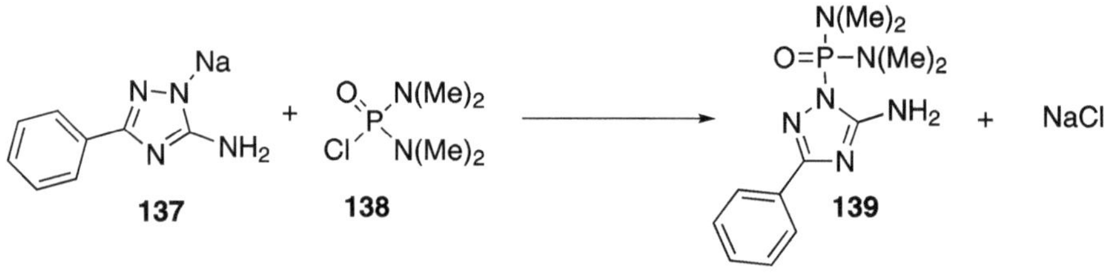

SCHEME 5.53 Synthesis of Wepsyn (**139**).

It is used for controlling powdery mildew on apples and ornamentals. The acute oral LD_{50} for rats is 20 mg/kg.

Kitazin P (S-benzyl O,O′-diisopropyl phosphorothiolate, **142**)

Kitazin P (**142**) is prepared by the condensation of O,O′-diisopropyl phosphorochloridate (**140**) and sodium benzylthiolate (**141**) (Scheme 5.54).

SCHEME 5.54 Synthesis of Kitazin P (**142**).

It is used as a fungicide to control rice blast, and works by inhibiting the mycelial growth in the tissue. The acute oral LD_{50} for rats is 660 mg/kg.

Conen (O-Butyl-S-ethyl-S-benzyl phosphorodithioate, **145**)

Conen (**145**) is prepared from sodium O-butyl-S-benzyl phosphorodithionate (**143**) and ethyl bromide (**144**) (Scheme 5.55).

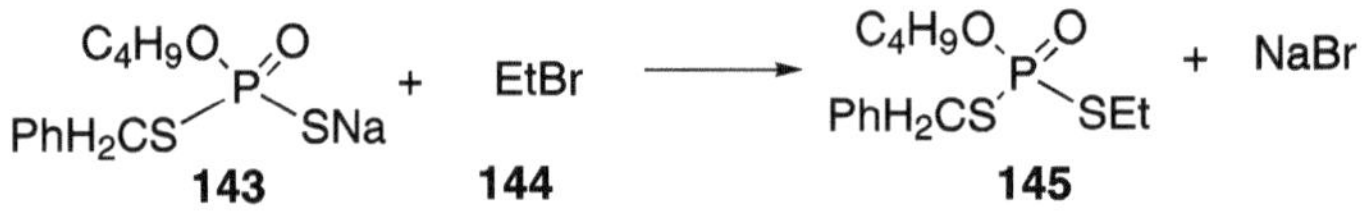

SCHEME 5.55 Synthesis of Conen (**145**).

It exerts a curative rather than a protective action. The oral LD_{50} for mice is 118 mg/kg.

Dowco (O,O′-diethylphathalimidophosphorothioate, **147**)

Dowco (**147**) is obtained by the condensation of potassium phathalimide (**146**) with O,O′-diethyl phosphorochloridothioate (**35**) (Scheme 5.56).

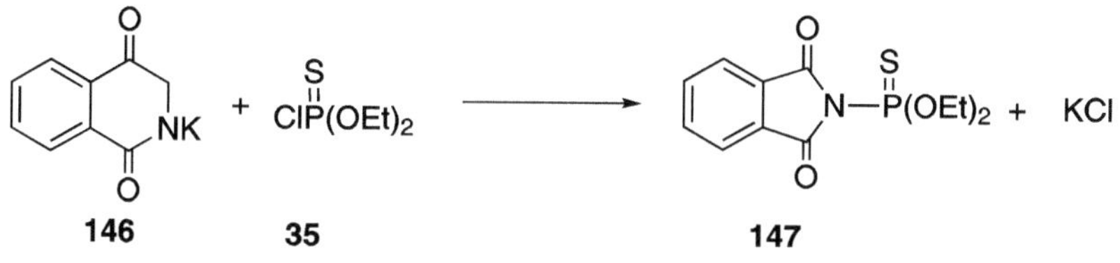

SCHEME 5.56 Synthesis of Dowco (**147**).

Dowco (**147**) is very effective against a wide range of powdery mildews and has low mammalian toxicity. The oral LD_{50} for rats is 5660 mg/kg.

Edifenphos (O-ethyl-S,S-diphenyl phosphorodithioate, **149**)

It is prepared by treating O-ethyl phosphorodichloridate (**148**) with thiophenol (**113**) (Scheme 5.57).

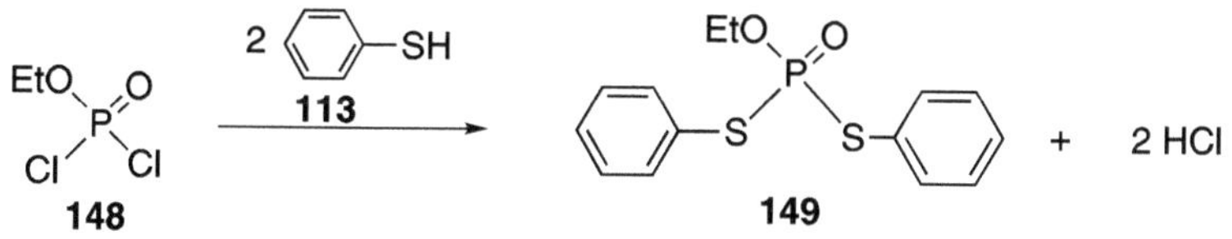

SCHEME 5.57 Synthesis of Edifenphos (**149**).

Edifenphos (**149**) is effective against *Pyricularia oryzae* in rice. The oral LD_{50} for rats is 218 mg/kg.

5.2.2.2.3 Phosphonates

The fungicide compound belonging to this group is Fosetyl-AL 80% wettable powder (WP). It is recommended to control grape downy mildew at 560–800 g/acre and cardamom Azhukal disease and damping off diseases at 900–1200 g/acre. The following are important phosphonate-based fungicides:

Inejin F 254 (O-methyl-S-benzyl benzene phosphonothioate, **150**)

Inejin is used as a fungicide to control rice blast and rice sheath blight. The oral LD_{50} for rats is 720 mg/kg.

Aliette (Aluminum *tris*[O-ethyl phosphonate], **151**)

Aliette (**151**) is mainly marketed in combination with a protective fungicide. For instance, Mikal (50% Aliette, 9.5% Folpel) is active against downy mildews in crops such as pineapple, citrus, avocado, lettuce, vegetables, grapes, and ornamentals.

Figure 5.7 represents the structures of (**150**) and (**151**).

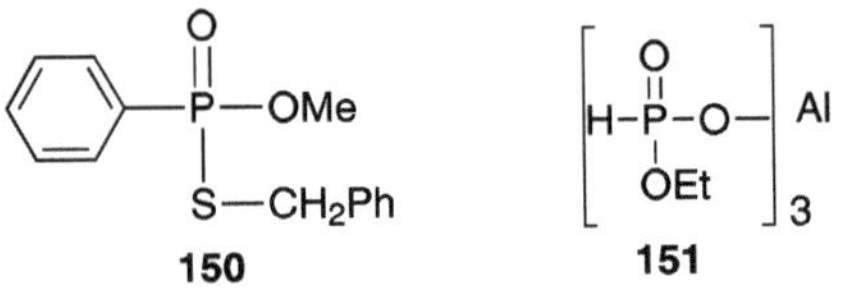

FIGURE 5.7 Structures of (**150**) and (**151**).

Cerezin (O-methyl-O-cyclohexyl-S-[4-chlorophenyl]phosphorothioate, **154**)

Cerezin (**154**) is prepared by reacting O-methyl-O-cyclohexylphosphate (**152**) in carbon tetrachloride with *p*-chlorophenylsulfenyl chloride (**153**) (Scheme 5.58). Cerezin (**153**) has a curative effect on rice blast diseases.

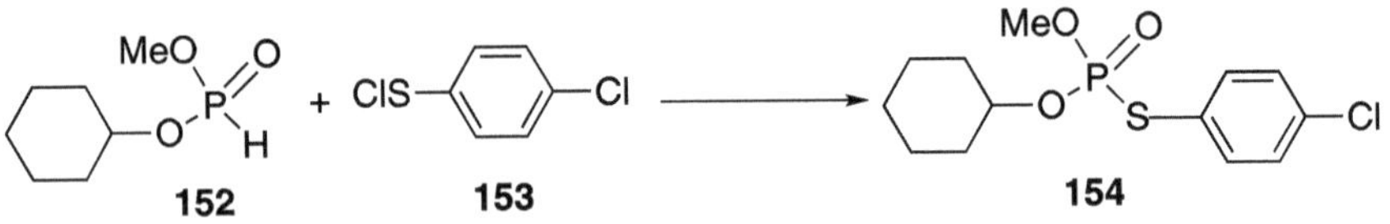

SCHEME 5.58 Synthesis of Cerezin (**154**).

5.2.2.2.4 *Mode of Action of Organophosphate Fungicides*

The activity of organophosphorus compounds is due to the reason that these compounds influence fungal metabolism by direct phosphorylation and thereby disturb the vital functions of the cell.

5.2.2.3 Nematicides

Nematodes have nervous systems similar to insects. All these organophosphates inhibit the nerve-transmitter enzyme, cholinesterase, resulting in paralysis and ultimately leading to death of the affected nematode. Figure 5.8 represents the structures of organophosphate nematicides used for killing nematodes: O,O diethyl-S-(ethylthiomethyl) phosphorodithioate, Phorate (**83**); O,O diethyl dichlorophenyl phosphorothioate, Dichlofenthion (**155**); ethyl-3-methyl-4-(methylthio) phenyl-(1-methyl ethyl) phosphoramidate, Fenamiphos (**156**).

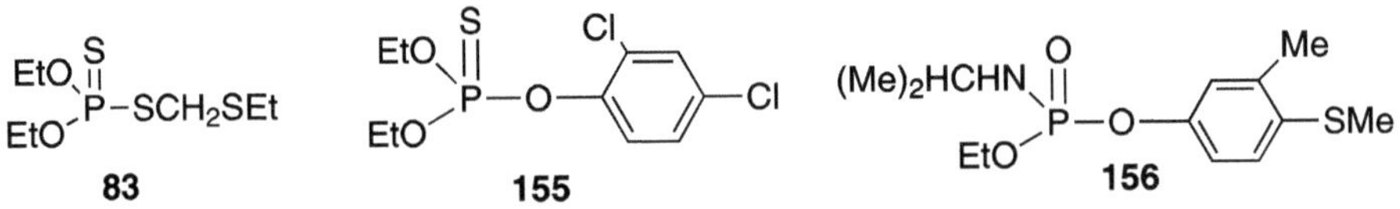

FIGURE 5.8 Structures of organophosphate nematicides (**83**), (**155**), and (**156**).

5.2.2.4 Rodenticides

Gophacide (O,O′-*bis*[*p*-chlorophenyl] acetimidoylphosphoramidothioate, **159**)

Gophacide (**159**) is prepared by condensation of O,O′-di-*p*-chlorophenyl phosphoro-chloridothioate (**157**) with acetamidine (**158**) (Scheme 5.59).

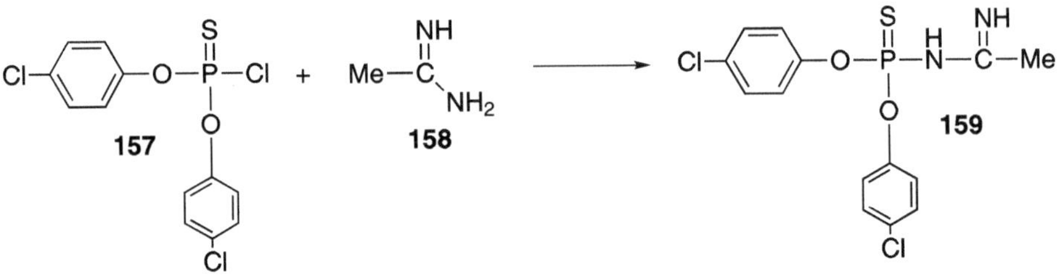

SCHEME 5.59 Synthesis of Gophacide (**159**).

Gophacide **(159)** is very effective for control of a range of rodent pests, but proves as hazardous to nontarget species.

5.2.2.5 Acaricides

Carbophenethion (S-[{4-chlorophenylthio} methyl] O,O′-diethyl phosphorodithioate, **162)**

Carbophenethion **(162)** is prepared by reaction of *p*-chloro thiophenol **(160)** with formaldehyde **(82)** and hydrochloric acid, followed by a reaction of the substituted chloromethyl phenyl sulfide **(161)** with sodium diethyl phosphorodithioate **(87)** (Scheme 5.60).

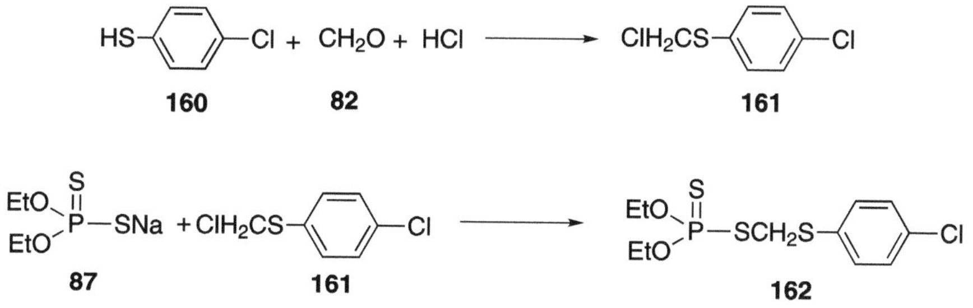

SCHEME 5.60 Synthesis of Carbophenethion **(162)**.

Its LD_{50} value for rats is 28–100 mg/kg.

5.2.2.6 Herbicides

A wide range of different types of organophosphorus compounds show some herbicide properties and the major groups are phosphorothiolates, phosphoroamidates, and phosphonates.

5.2.2.6.1 Phosphorothiolates

Bensulide (O,O′-disopropyl-S-[2-phenyl sulfonylamino] ethyl phosphorodithioate, **165)**

Bensulide **(165)** is prepared by the condensation of sodium O,O′-diisopropylphosphorodithioate **(163)** with N-(2-chloroethyl) benzene sulfonamide **(164)** (Scheme 5.61).

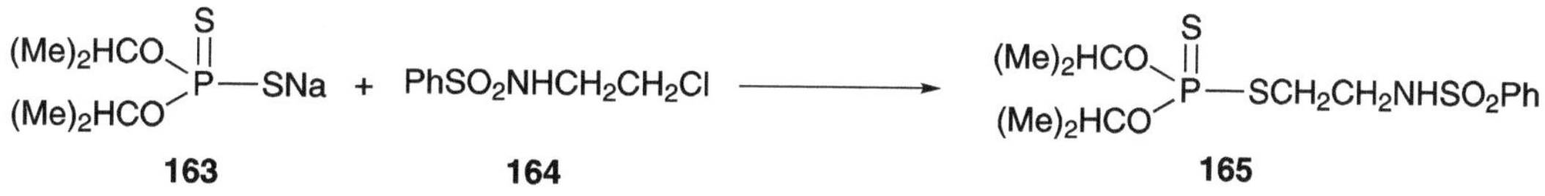

SCHEME 5.61 Synthesis of Bensulide **(165)**.

Bensulide **(165)** is used as a preplant and pre-emergence herbicide on cotton, brassica, and lettuce.

5.2.2.6.2 Phosphonates

Glyphosate (N-phosphonomethyl glycine, **168)**

Glyphosate **(168)** is obtained from chloromethyl phosphonic acid **(166)** and glycine **(167)** (Scheme 5.62). Glyphosate **(168)** is a nonselective pre-emergence herbicide. It is very effective

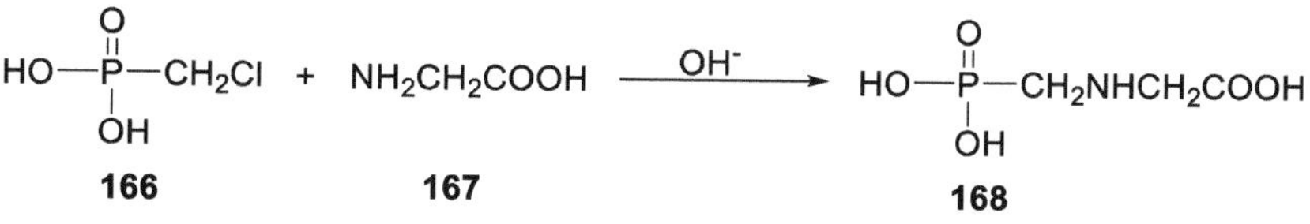

SCHEME 5.62 Synthesis of Glyphosate **(168)**.

against deep-rooted perennial species, as well as against some species of grasses and broad-leaved weeds. The acute oral LD_{50} value for rats is 4320 mg/kg.

5.2.2.7 Plant Growth Regulators (PGRs)

5.2.2.7.1 Ethylene Generators

Ethylene can inhibit growth and accelerate leaf and flower drop (abscission), the ripening of fruit, and flowering, depending upon the stage on which it is applied. Recently, chemicals have been developed which when sprayed onto plants, release ethylene and hence, control different phases of plant development.

Bis-benzoloxy-2-chloroethyl-methylsilane **(169)** produces ethylene **(170)** and is used as a chemical abscission agent for apples, peaches, oranges, and other fruits.

Ethephon (2-chloroethyl phosphonic acid) **(172)** is prepared from phosphorus trichloride **(106)** and ethylene oxide **(70)** (Scheme 5.63). Ethephon **(172)** generates ethylene **(170)** and benzyloxy methylsilane (171) in solutions with pH > 4. It is sprayed on fruit trees to promote ripening in rubber plantation to stimulate latex production and to improve stem strength in cereals.

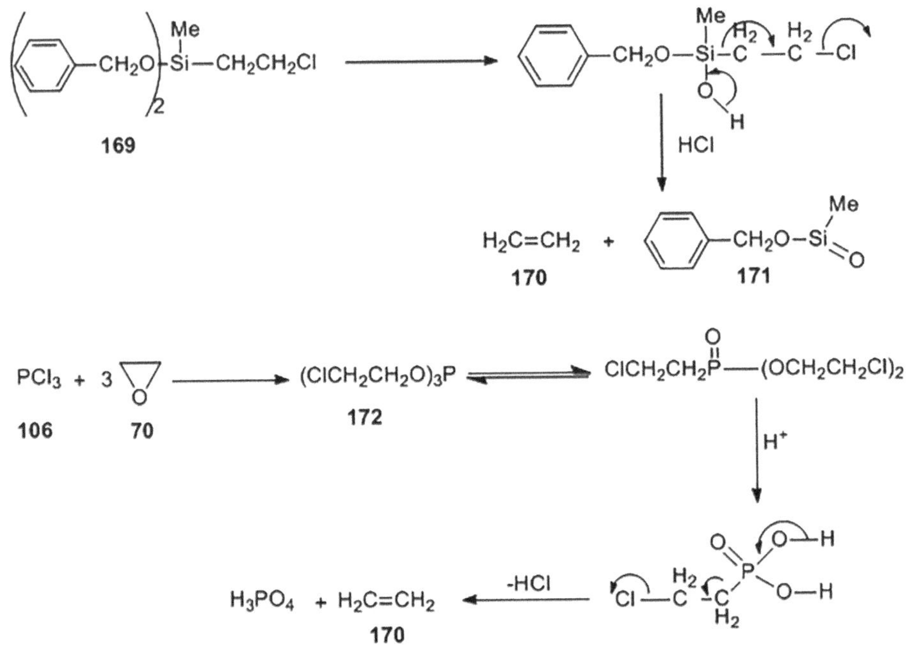

SCHEME 5.63 Synthesis of Ethephon **(172)** and **(170)**.

The plant growth regulator (PGR) market is currently dominated by six chemicals: chlormequat, diaminozide, maleic hydrazide, ethephon, gibberellic acid, and glyphosine. It is estimated that one million hectares of crops are annually treated with PGRs. Little is known about the mode of action of many PGRs, and a better understanding of the mechanism of the structural activity relationships should permit introduction of more effective compounds.

5.2.2.8 Proinsecticidal Organophosphates

In the pure state, phosphorothionate esters are poor inhibitors of cholinestrases. The insecticidal activity of the thiono compounds is the result of the oxidative desulfuration of the thionophosphoryl group *in vitro.* The conversion of thiono into oxo esters in animals is caused by the microsomal mixed function oxidase systems. In plants, peroxidases and/or light may play a role in the thiono-oxo-transformation, or oxidative desulfuration of the thiophosforyl sulfur has been demonstrated, as in parathion **(37)** and Malathion **(77)** to afford paraoxon (38) and Malaxon (173) respectively (Scheme 5.64).

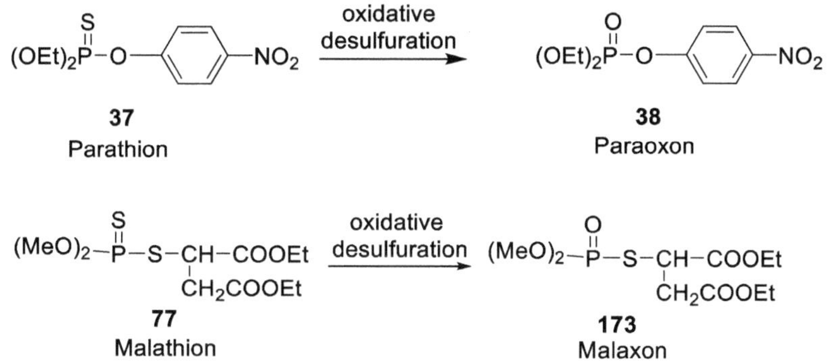

SCHEME 5.64 Oxidative desulfuration of (**37**) and (**77**) to give (**38**) and (**173**), respectively.

5.2.2.9 Chemosterilants

Chemosterilants are chemical substances that sterilize insects to prevent reproduction. Most chemosterilants relate to highly toxic compounds with a mutagenic, tetragenic and carcinogenic action. Alkylating compounds are a group of chemosterilants, in which hydrogen atom of molecule of a substance is replaced by an alkyl group. They readily enter into chemical alkylation reactions with various compounds of a cell, including proteins and nucleic acids. The most important compound of this group is ethyleneimine (**174**). Other chemosterilants are Tepa (1,1′,1″-phosfinylidyne-*tris*-aziridine, **175**) and Apholate (2,2,4,4,6,6-*Hexakis* [1-aziridinyl]-2,2,4,4,6,6-hexahydro-1,3,5,2,4,6, triazatriphos phorine, **176**) (Figure 5.9).

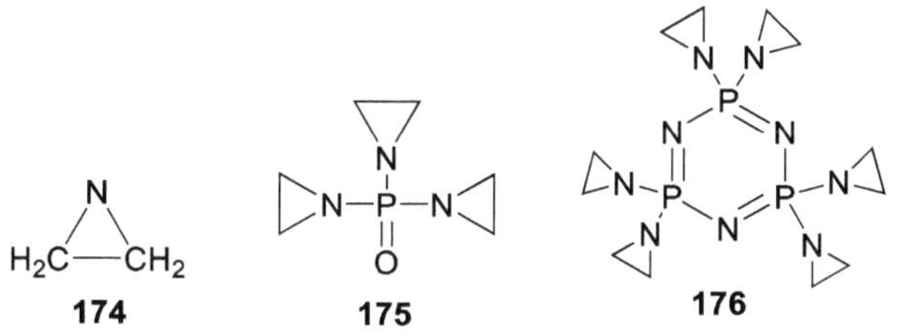

FIGURE 5.9 Structures of ethyleneimine (**174**), Tepa (**175**), and Apholate (**176**).

Antimetabolites are the substances whose structures are very close to that of the natural metabolites of an organism. When they enter the organism, they displace these metabolites in exchange reactions. Upon entering an insect organism, these substances violate the synthesis of nucleic acids (DNA and RNA) in the nuclei of the sexual cells.

5.3 MODE OF ACTION

The transmitter substance transmits the messages across the nerve synapse, and acetylcholine serves that role in cholinergic junctions. After passing the impulse, acetylcholinesterase acts on the acetylcholine to bring it into an active ionic state. The action can be visualized in Scheme 5.65 and as follows:

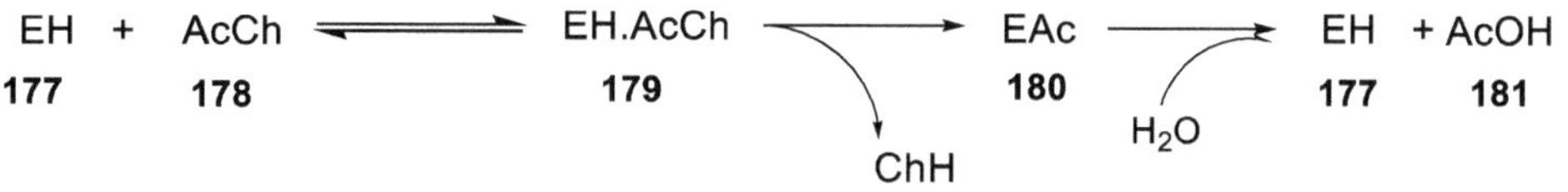

SCHEME 5.65 Mode of action of acetylcholinesterase

In the first step, the enzyme (EH) (**177**) combines with the acetylcholine (AcCh) (**178**) to form a reversible and transitory complex (**179**). This complex acetylates the enzyme (EAc, **180**) with the release of choline (ChH). Addition of water serves to deacetylate the enzyme to give free enzyme **(177)** and acetic acid (AcOH) **(181)**. In relation to the toxicity of organophosphorus insecticides, the important impetus for the initial steps in this reaction is the nucleophilic (attraction to a positive center) hydroxyl group of a serine component **(182)** in acetylcholinesterase attacking the positively charged nitrogen atom of acetylcholine **(178)** (Figure 5.10).

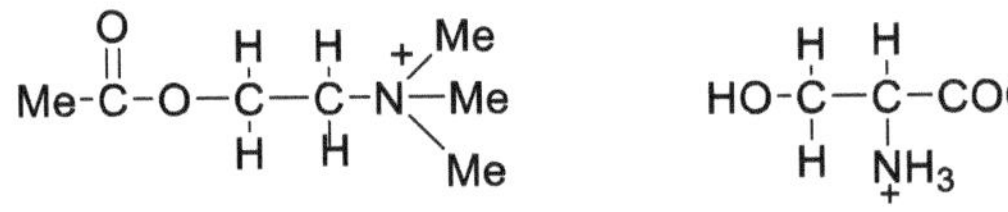

FIGURE 5.10 Structures of **(178)** and **(182)**.

Within the organophosphorus insecticides, the P = O nucleus has a strong electronic pull on the hydroxyl group of acetylcholinesterase, with the result that when one of these insecticides penetrates the synapse, it is attacked by the acetylcholinesterase, bonds with it, and thus ties up the enzyme so that it is not free to act on the transmitter substance. In the case of the organophosphorus insecticides, the bonding is strong and addition of water releases the enzyme very slowly. Thus, some literature on the subject refers to this as an irreversible step, in contrast to the situation with carbamate insecticides, wherein the bond is much less "tight." The continuous exposure being barred, concentration of the enzyme returns rapidly to the pre-exposure levels (McEwen and Stephenson, 1979).

The strength of the positive pull on the P = O group on the enzyme is altered when oxygen is replaced in this nucleus by sulfur (P = S), and the pull is also affected by other components of the insecticide molecule adjacent to the P = O center. In fact, P = S has little drawing power and organophosphorus insecticides of the thionate configuration must be oxidized to the corresponding P = O compound before they become active. This oxidation occurs readily in both animals and plants, a step referred frequently to as *lethal synthesis* (O'Brien, 1967).

Organophosphorus poisoning is thus brought about by phosphorylation of acetylcholinesterase and a resultant accumulation of acetylcholine in the nerve synapse. When this occurs, normal transmission of nerve impulses is prevented, expressed in insects as hyperactivity, tremors, convulsions, paralysis, and death. These symptoms are consistent with effects in the central nervous system, where cholinergic junctions occur, undoubtedly.

In higher animals, the susceptible cholinergic synapses occur throughout the nervous system. Symptoms of organophosphorus poisoning include nausea, salivation, giddiness, tremors, tearing, myosis, coma, convulsion, prostration, and death. Contrary to the situation in insects, where respiration through the spiracles is largely passive, respiration in higher animals requires controlled muscular activity. Disruption of this activity is controlled by poisoning the system, which results in asphyxiation.

The mechanism of inhibition of cholinesterase by organophosphorus compounds is fairly well-understood, but only by indirect evidence. More direct evidence is available on the much better characterized enzyme, chymotrypsin. It has been shown that, when paraoxon **(38)** is presented to this enzyme in dilute solution, the phosphorus atom of one molecule of the inhibitor becomes locked up in every molecule of enzyme inhibited. At the same time, one free nitrophenate ion appears. The active site of the enzyme evidently splits the molecule of the inhibitor, but is then unable to release the phosphoryl moiety. This blocks further action, whereas the unchanged enzyme can attack, split, and release some thousands of its normal substrate molecules per second.

Scheme 5.66 below roughly represents the course of normal and inhibition reactions, the active portion of the enzyme being represented as a vertical line with labile side groups. Actual behavior in the living organism, or even on enzymes *in vitro*, is of course more complex than this. Blockage of the enzyme by the substituted phosphate group is not necessarily quite permanent, and the phosphate ester may be hydrolyzed *in situ*. Thus, there may be reactivation of the attacked enzyme, although most recovery from partial intoxication must await replacement of the enzyme. The compound itself, in some cases, is not directly active and requires chemical alteration before it becomes so. The compound, or its derivatives, may be subject to other reactions, which compete with the toxic ones. Disposition of groups, other than those directly concerned in the inhibition reaction, may assist adsorption of the toxic molecule on to the enzyme in a suitable posture for reaction to occur (Hartley and West, 1969).

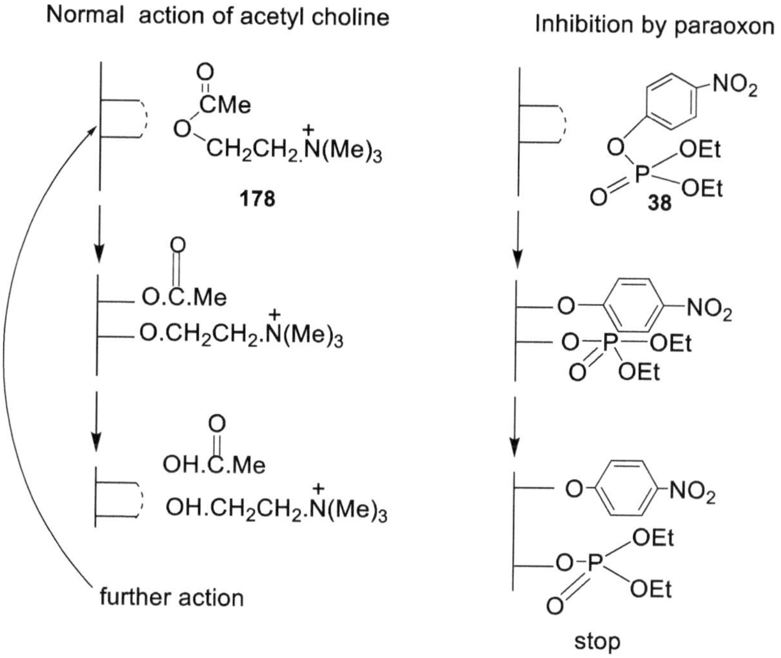

SCHEME 5.66 Representation of course of normal and inhibition reaction

5.4 SOME WIDELY USED ORGANOPHOSPHORUS INSECTICIDES

Dicapthon (183), although closely related to methyl parathion **(41)**, is much less toxic, especially by dermal exposure. It is useful for fly control and as a household insecticide.

Temephos (Abate) **(184)** is an important insecticide for control of larvae of mosquitoes, black flies, biting midges and sand flies of public-health or nuisance importance. Because of its widespread use in the aquatic environment, temephos (**184**) has been studied extensively with regard to its effects on nontarget organisms).

Leptophos (Phosvel) **(185)** is a broad-spectrum insecticide, especially effective against lepidopterous larvae. It is of environmental interest, because of its reported demyelinating effect, this being observed first in water buffalo that drank contaminated water. More recently, chronic poisoning from leptophos among workers exposed during its manufacture has been experienced. Leptophos was registered for cutworm control in Canada, but this registration has been withdrawn (Figure 5.11).

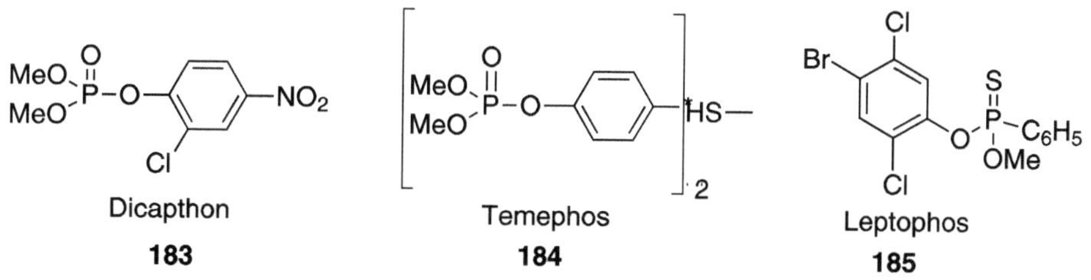

FIGURE 5.11 Structures of Dicapton **(183)**, Temephos **(184)** and Leptophos **(185)**.

5.5 SYNTHESIS OF ORGANOPHOSPHORUS COMPOUNDS

5.5.1 Main Reactions

In principle, there are two main types of reactions, by at least one of which, nearly all organophosphorus insecticides can be synthesized (Hartley and West, 1969). The first reaction is of phosphorus oxychloride **(134)** or thiophosphoryl chloride, $PSCl_2$, with an alcohol, mercaptan, or amine in the presence of an acid-binding agent, followed by reaction with the acidic component, as shown in Scheme 5.67 and described below:

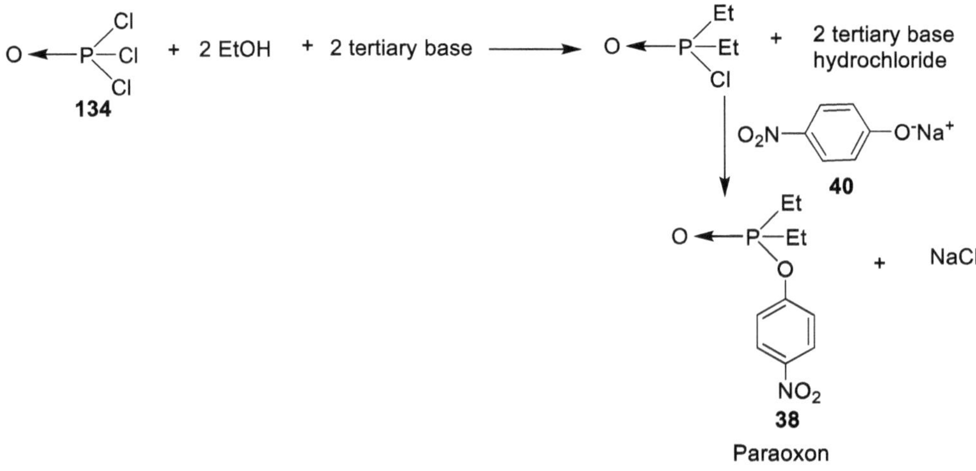

SCHEME 5.67 Synthesis of organophosphorus insecticides

When thiophosphoryl chloride is the starting phosphorus compound, parathion **(37)** can be made in this way. The second method is an alternative in the second stage only of this reaction. In this, a salt of the disubstituted phosphoric acid or thiophosphoric acid is formed and reacted with an acid halide. This is not a practicable method of synthesizing parathion **(37)**, since *p*-nitro-chlorobenzene is not sufficiently reactive. Reaction of potassium salt of dimethyldithiophosphoric acid **(78)** with N-methyl chloroacetamide **(79)** to form dimethoate **(80)** provides an example.

The second method for the second stage is much more efficient for preparing thiophosphates than phosphates, and in many cases, the reaction can proceed efficiently in the presence of water. When this reaction is carried out with an O,O′-disubstituted mono thiophosphate, the third group is attached through the S atom; that is, a thiolo rather than thiono compound is produced.

5.5.2 Addition Reactions

Dialkyl phosphonates can undergo important addition reactions, one of which gives the compound dipterex. The later is prepared by addition of dimethylphosphonate to chloral.

Reaction of an alcohol and phosphorus trichloride in the presence of a base can give the trialkyl phosphite the true phosphite structure, with the phosphorus trivalent and the alkoxy groups all equivalent. Trialkyl phosphites can undergo useful reactions, discovered by Michaelis and Arbuzov, with loss of an alkyl halide as shown in Scheme 5.68 below:

The second of these reactions gives DDVP **(14)** directly, whereas reaction of dimethyl phosphonate with chloral gives dipterex, from which DDVP is obtained, with loss of HCl by heating (Scheme 5.68).

The third of these reactions, an example of a general reaction with α-chloro-β-diketones, yields the rather toxic, but nonpersistent insecticide, Mevinphos **(20)** (Scheme 5.68). It will be noted that this compound can exist as two geometrical isomers. The α-isomer, which is some hundred times

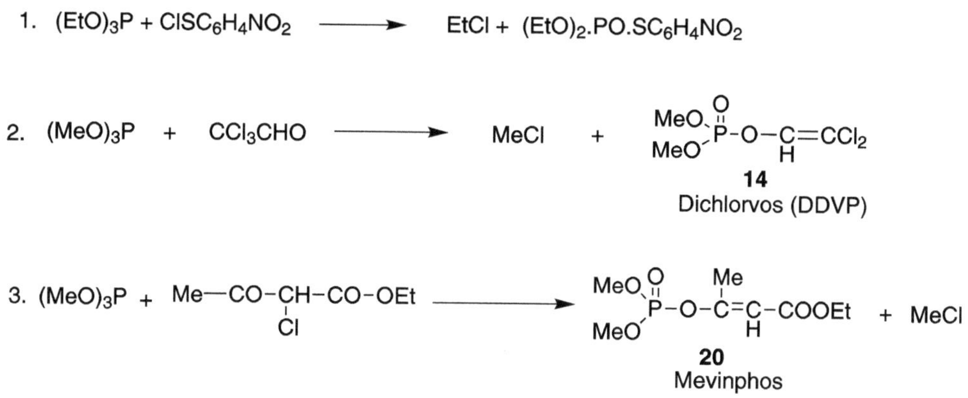

SCHEME 5.68 Reactions of trialkyl phosphites

more toxic to both insects and mammals, has the carboxylic and phosphoric esters on the same side of the plane of the double bond between the carbon atoms of the acidic group. The commercial mixture contains about 70% of this more active form, which is converted fairly cleanly to the less active form on heating.

Other types of addition reactions are important, particularly the addition to dialkyldithiophosphoric acids of substances with an easily opened ethylenic link. The important insecticide, Malathion (**77**), is prepared by addition of dimethyl-dithiophosphoric acid (**75**) to diethyl maleate (**76**).

5.5.3 Purification of Organophosphorus Insecticides and the Significance of Water during Preparation

Most organophosphorus insecticides are not industrially capable of purification by distillation and, indeed, even in the laboratory, a short-path, high-vacuum still is usually necessary. Many of the compounds are of such low melting point that the technical products are liquid so that purification by crystallization is not usually economical. But there is a tendency to develop more elaborate structures so that these compounds may have higher melting points. Recrystallization of the technical product, for example, in the case of dimethoate (**80**), mp 52°C, is sometimes undertaken during industrial manufacture (Hartley and West, 1969).

Purity of the technical product may be necessary not only to meet toxicity or other specification requirements, but to ensure adequate stability, since the decomposition products may catalyze the decomposition reactions. All the compounds generate acid on hydrolysis and most generate liquid products, which provide a solvent in which reaction of the active compound, if solid in the pure state is facilitated. Extensive storage tests, with physical, chemical, and biological assessment of the changes taking place, are a necessary part of the development of the industrial manufacturing process.

Most of the products partition favorably from water into good water-immiscible solvents, especially chloroform, so that it is usually practicable to wash out the acidic impurities (P-OH and P-SH compounds), retention of which is particularly undesirable for stability. Frequently, the last preparation step may be carried out in a two-phase agitated system, one phase being aqueous. Indeed, an important trend in industrial organophosphorus chemistry has been the working out of conditions under which water can desirably take part. Excess water, if it can be employed without undue loss of the main product, can convert some of the side products into more water-soluble forms, which are more easily removed and reduce the formation of pyrophosphate compounds.

As a good instance, the classical laboratory route to anhydrous dimefox (186) may be contrasted with an efficient industrial wet process for the synthesis of dimefox in solvent (187) (Scheme 5.69).

Laboratory anhydrous:

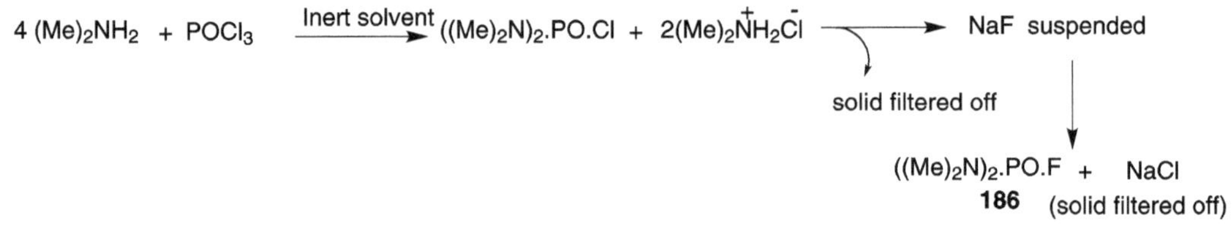

SCHEME 5.69 Laboratory anhydrous synthesis of dimefox **(186).**

While the anhydrous laboratory process may be made carefully to get a higher yield, the industrial wet process can give over 90% of the theoretical yield on both phosphorus and amine. The wet process has greater speed of operation, while filtration of highly toxic material is unnecessary. The dry process wastes half the amine, unless an extra stage is introduced for dry recovery. Overall, the wet process has great economic advantage (Scheme 5.69).

Industrial aqueous synthesis of dimefox (187):

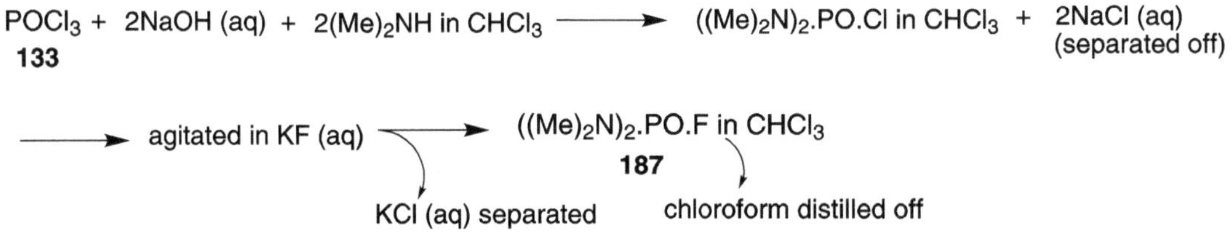

SCHEME 5.70 Industrial aqueous synthesis of dimefox in chloroform **(187).**

Contact with water is permissible in this reaction, as although phosphorus oxychloride **(133)** is very rapidly hydrolyzed in aqueous solution, it reacts even more rapidly with dimethyl amine (Scheme 5.70). Both reactants partition much towards chloroform from water. Dimethylamine is, therefore, greatly in excess of water in the immediate environment of the reactive oxychloride. The reaction of complex halides with dialkyldithiophosphates is another reaction preferably carried out in water, but in the simultaneous presence of an immiscible solvent such that the product is removed from the hydrolyzing environment.

5.6 DETECTION OF ORGANOPHOSPHORUS COMPOUNDS

Since soil is an integral component of the ecosystem, its survival is essential for humans. Human activities and use of pesticides lead to the contamination of soil, resulting in serious health risks to the humans. Organophosphorus compounds are one of the most commonly used pesticides in agriculture across the world (Du et al., 2009; Ding et al., 2012). Some of the organophosphorus compounds do not readily decompose and enter the soil ecosystem because of their frequent use on the soil surface (Naeeni et al., 2011). Analysis of organophosphorus residues is highly important because of their highly toxic nature and tendency to accumulate and interfere with human body functions. These pesticides could lead to diseases such as acute anemia, bone structure disorders, and teratogenic and embryologic diseases. Also, they can irreversibly inactivate acetyl cholinesterase, which is essential to nerve impulses in insects, animals, and human beings (Castilho et al., 2000; Mohammadi et al., 2015). Hence, search of highly specific analytical methods for quantification of organophosphorus compounds in trace levels is very important for public security and health protection (Karami et al., 2016; Dehghani et al., 2017).

The instrumental techniques for detection of organophosphorus compounds, such as gas chromatography (GC) (Cinelli et al., 2014), high performance liquid chromatography (HPLC), gas chromatography coupled with mass spectrometry (GC-MS), enzyme-linked immunosorbent assay (ELISA) (Shim et al., 2008), and electrochemical biosensing (Liu et al., 2012), have been used for their detection. These are highly sensitive, selective, and precise techniques for the detection of organophosphorus compounds, but are accompanied with some disadvantages such as being expensive, time-consuming, and often requiring well-trained technicians and experts to carry out the analysis. Apart from this, these techniques are not suitable for field use and also involve the crucial step of sample preparation. So far, different solid phase extraction (SPE) techniques have been employed to prepare samples in organophosphorus analysis. The most commonly used of such techniques are solid phase microextraction (SPME), matrix-solid phase dispersion (MSPD), head space-solid phase microextraction (HS-SPME), stir bar sorptive extraction (SBSE), and magnetic solid phase extraction (MSPE) (Sun et al., 2015). Most of these methods consume a lot of time during extraction of an analyte into the organic phase or onto the adsorbent surface due to limited interface between the absorbent and the aqueous samples.

Rodrigues et al. (2018) constructed a sensitive sensor for the detection of organophosphorus pesticides by immobilizing the acetylcholinesterase enzyme, via glutaraldehyde, on magnetic iron nanoparticles and functionalized with chitosan for determination of malathion. The sensor was found to be renewable, disposable, cost-effective, and very sensitive for real samples, such as tomato and pond-water samples.

Preparation of a composite film, consisting of ionic liquid, chitosan, and electrochemically synthesized gold nanoparticles on single-use pencil-graphite electrodes, was carried out by Bolat and Abaci (2018). It was found that the composite film had remarkable affinity for malathion, when tested in tomato and apple samples.

Also, in another report, an enzyme-based, miniaturized fluorescence biosensor was developed to detect paraoxon by Kim et al. (2018). The biosensor was fabricated with poly(ethylene glycol) (PEG) hydrogel microarrays that entrapped acetylcholine esterase and quantum dots (QDs) as fluorescence reporters. Metal-enhanced fluorescence (MEF) was utilized to amplify the fluorescence signal, which in turn was achieved by decorating QDs on the surface of silica-coated silver nanoparticles (Ag@ Silica). Paraoxon detection was found to be significantly detected using MEF-assisted fluorescence detection.

An electrochemical sensor was fabricated using pristine Ni and Cr-doped nano tin oxide (SnO_2) as a tool to detect organophosphorus compounds, such as chloropyriphos, by Khan et al. (2017). The nanomaterial was synthesized using the solution method, and nickel and chromium were used as the dopant during synthesis. These fabricated sensors were found to be suitable for easy detection of organophosphorus compounds to a level of 10 ppb.

An enzyme-based electrochemical transducer was made by Grace et al. (2017), immobilizing gold nanoparticles on enzyme acetylcholinesterase and further depositing it electrochemically over a screen-printed carbon working electrode. It was used for the detection of organophosphorus compounds. The developed biosensor gave optimum response within 25 s for a substrate acetylthiocholine at a concentration of 0.0699 mM. The developed biosensor exhibited good stability and reusability, thus making it a promising tool for on-field detection of organophosphorus compounds.

Magnetic solid-phase extraction based on $CoFe_2O_4$ magnetic nanoparticles was studied for the determination of diazinon and fenitrothion in aqueous samples. A gas chromatography flame ionization detector was used to investigate the preconcentration, extraction, and determination of pesticides from the aqueous samples (Mohammadi et al. 2017). The method has shown excellent selective cleanup of pesticides at low levels in real samples.

$CaWO_4$ nanoparticles were used for the removal of cyanophos pesticide from aqueous media by Yekta et al. (2016). The adsorption-degradation reaction of cyanophos as an organophosphorus pesticide on the $CaWO_4$ nanoparticles was monitored by nuclear magnetic resonance (NMR) spectroscopy and gas chromatography-mass spectrometry (GC-MS). The main product resulted from a degradation reaction between cyanophos and $CaWO_4$ was dimethyl phosphorothioic acid, which was less toxic than primary pesticide.

Sun et al. (2015) developed a two-dimensional carbon nonmaterial made from graphene and used it as a sorbent for sample pretreatment. The large specific surface area, chemical stability, flexibility, and π-electron rich structure, make it a useful candidate for use as a sorbent for sample pretreatment. The nanocomposites of graphene oxide were produced directly from graphite flakes using Hummers' method (Hummers and Offeman, 1958; Tung et al., 2009). Further reduction of graphene oxide and deposition of ZnO on graphene occur simultaneously. The synthesized reduced graphene oxide-ZnO composite was used to detect organophosphorus compounds in cucumbers, apples, and water. No residues of the pesticides were detected in the samples spiked at 1 and 10 ng/g levels of organophosphorus compounds.

Li et al. (2015) have developed a sensitive photoelectrochemical (PEC) biosensor for detection of organophosphorus pesticides using nanocomposites of CdSe@ZnS quantum dots, and graphene deposited in the indium doped tin oxide (ITO)-coated glass electrode acted as a photoactive electrode. Due to the narrower band gap and strong absorption in the visible-light region, semiconductor nanoparticles or quantum dots are important materials for solar energy conversion. Integration of the quantum dot-graphene hybrid system with AChE biomolecules yields a highly sensitive PEC assay for the detection of organophosphorus pesticides. The biosensor was advantageous, as it showed rapid response, excellent sensitivity, good reproducibility, ideal stability and fine applicability for the detection of organophosphorus compounds in real samples.

Zheng et al. (2015) have developed an acetylcholine esterase biosensor based on ionic liquid-functionalized graphene (IL-GR) and polyvinyl alcohol (PVA) to detect organophosphorus pesticides. PVA is synthetic polymer having excellent membrane-forming ability, good compatibility, and chemical stability, which provides a biocompatible microenvironment for enzyme immobilization and also gives sufficient accessibility to electrons to shuttle between the enzyme and the electrode. The IL-GR-PVA films, so developed, were used for the analysis of apple juices spiked with different concentrations of phorates (**83**), and it was found that these nanocomposites have excellent conductivity and biocompatibility and offered a hydrophilic surface for biomolecule adhesion.

Nanocomposites (AuDMBG/RGO) consisting of coral-like gold nanostructures on reduced graphene oxide (RGO), with the assistance of dimethyl biguanide (DMBG), were developed by Ju et al. (2015). The nanocomposite was further explored for the construction of AChE biosensor as a model system to detect the organophosphorus pesticide triazophos. Inhibition of triazophos to AchE at the biosensor, exhibits a linear relationship with its concentrations up to 200 ppb, with the detection limit of 0.35 ppb (S/N = 3). In addition, the biosensor exhibits great reproducibility and acceptable stability.

Ultra-preconcentration technique, composed of ultrasound-assisted leaching-solid phase extraction (USAL-SPE) and low density solvent-based dispersive liquid-liquid microextraction (LDS-DLLME) coupled with GC-MS, was developed for preconcentration and detection of organophosphorus pesticides in soil samples (Mohammadi et al., 2015). The salient advantage of this methodology was the low sample consumption, lesser use of toxic organic solvent, lesser extraction time, rejection of matrix constituent, and high enrichment factor.

Acetylcholine biosensor, based on a conducting polymer, poly(4-[2,5-di{thiophen-2-yl}-1H-pyrrol-1-yl] benzenamine) poly(SNS-NH_2), using a MWCNT-modified electrode was developed for the detection of organophosphorus pesticides by Kesik et al. (2014). The biosensor was used for the detection of pesticides in fortified tap-water samples and was found to be advantageous, as it offered a fast response time, a wide linear range, and a low detection limit, with a high sensitivity for acetylthiocholine.

In an another approach for the detection of pesticides, Feng et al. (2014) have also developed a HS-SPME using a stainless-steel fiber coated with 20 mm MWCNTs and gas chromatography with thermionic specific detector (GC-TSD). Determination of trace amounts of phorate (**83**), methyl parathion (**41**), chlorpyrifos (**60**), and parathion (**37**) in spiked soil samples was successfully carried out using this technique.

Yang et al. (2012) have prepared Au-ZrO_2-SiO_2 nanocomposite spheres by the hydrolysis and condensation of zirconium-butoxide on the surface of SiO_2 spheres and then the introduction of gold nanoparticles on the surface. These nanocomposites were used as selective sorbents for the solid-phase extraction of organophosphorus pesticides and proved as promising for the detection of organophosphorus pesticides.

5.7 CONCLUSIONS

In a nutshell, a wide variety of organophosphorus compounds have been employed for different uses in agriculture—as insecticides, fungicides, herbicides, and PGRs. The ease of their synthesis, mode of action, and low cost warrants their extensive use in the global market. However, in order to tackle the hazardous issues associated with their use, nanotechnology has emerged as a powerful tool for their detection. Nanomaterial-based detection methods are highly specific and selective *vis-à-vis* classic methods, such as GLC and HPLC-MS. We believe that this book chapter would be very useful to agrochemists working in the domain.

REFERENCES

Bolat G, Abaci S. Non-enzymatic electrochemical sensing of malathion pesticide in tomato and apple samples based on gold nanoparticles-chitosan-ionic liquid hybrid nanocomposite. *Sensors.* 2018; 18: 773–88.

Calaf GM, Roy D. Cancer genes induced by malathion and parathion in the presence of estrogen in breast cells. *Int J Mol Med.* 2008; 21: 261–68.

Castilho JAA, Fenzl N, Guillen SM, Nascimento FS. Organochlorine and organophosphorus pesticide residues in the Atoya river basin, Chinandega, Nicaragua: Environmental pollution. *Environ Pollut.* 2000; 110: 523–33.

Cinelli G, Avino P, Notardonato I, Russo MV. Ultrasound-vortex-assisted dispersive liquid–liquid microextraction coupled with gas chromatography with a nitrogen–phosphorus detector for simultaneous and rapid determination of organophosphorus pesticides and triazines in wine. *Anal Methods.* 2014; 6: 782–90.

Dehghani MH, Niasar ZS, Mehrnia MR, Shayeghi M, Al-Ghouti MA, Heibati B, McKay G, Yetilmezsoy K. Optimizing the removal of organophosphorus pesticide malathion from water using multi-walled carbon nanotubes. *Chem Eng J.* 2017; 310: 22–32.

Ding G, Wang P, Tian Y, Zhang J, Gao Y, Wang X, Shi R, Wang G, Shen X. Organophosphate pesticide exposure and neurodevelopment in young Shanghai children. *Environ Sci Technol.* 2012; 46: 2911–17.

Du D, Wang J, Smith JN, Tichalk C, Lin Y. Biomonitoring of organophosphorus agent exposure by reactivation of cholinesterase enzyme based on carbon nanotube-enhanced flow-injection amperometric detection. *Anal Chem.* 2009; 81: 9314–20.

Eyer P. The role of oximes in the management of organophosphorus pesticide poisoning. *Toxicol Rev.* 2003; 22: 165–90.

Feng X, Li Y, Jing R, Jiang X, Tian M. Detection of organophosphorus pesticides in soil samples with multiwalled carbon nanotubes coating SPME fibre. *Bull Environ Contam Toxicol.* 2014; 93: 769–74.

Grace R, Sundar K, Mohan N, Selvakumar D, Kumare NS. Fabrication of an enzymatic biosensor based on gold nanoparticles modified electrochemical transducer for the detection of organophosphorus compounds. *Nano Hybrid Compos.* 2017; 12: 67–73.

Guzman BP, Sanchez-Gutierrez M, Marchetti F, Hernandez-Ochoa I, Solis-Heredia MJ, Quintanilla-Vega B. Methyl-parathion decreases sperm function and fertilization capacity after targeting spermatocytes and maturing spermatozoa. *Toxicol App Pharm.* 2009; 238: 141–9.

Guzman BP, Solis-Heredia MJ, Rojas-Garcia AE, Uriostegui-Acosta M, Quintanilla-Vega B. Genetic damage caused by methyl–parathion in mouse spermatozoa is related to oxidative stress. *Toxicol App Pharm.* 2006; 216: 216–24.

Handa SK. *Principles of Pesticide Chemistry.* Published by Agrobios (India); 2004; 50–75.

Hartley GS, West TF. *Chemicals for Pest Control.* Pergamon Press: London; 1969; 64–89.

Hummers WS, Offeman RE. Preparation of graphitic oxide. *J Am Chem Soc.* 1958; 80: 1339–39.

Ju KJ, Feng JX, Feng JJ, Zhang QL, Xu T, Wei J, Wang AJ. Biosensor for pesticide triazophos based on its acetylcholinesterase and using a glassy carbon electrode modified with coral-like gold nanostructures supported on reduced grapheme oxide. *Microchim Acta.* 2015; 182: 1584–7.

Karami R, Mohsenifar A, Namini SMM, Kamelipour N, Rahmani-Cherati T, Shojaei TR, Tabatabaei M. A novel nanobiosensor for the detection of paraoxon using chitosan-embedded organophosphorus hydrolase immobilized on Au nanoparticles. *Prep Biochem Biotechnol.* 2016; 46: 559–66.

Kesik M, Kanik FE, Turan J, Kolb M, Timur S. An acetylcholinesterase biosensor based on a conducting polymer using multiwalled carbon nanotubes for amperometric detection of organophosphorus pesticides. *Sensors and Actuat B.* 2014; 205: 39–49.

Khan N, Athar T, Fouad H, Umar A, Ansari ZA, Ansari SG. Application of pristine and doped SnO_2 nanoparticles as a matrix for agro-hazardous material (organophosphate) detection. *Sci Rep.* 2017; 7: 1–11.

Kim M, Kwon EJ, Lee K, Koh W-G. Signal-amplifying nanoparticle/hydrogel hybrid microarray biosensor for metal-enhanced fluorescence detection of organophosphorus compounds. *Biofabrication.* 2018; 10: 035002, doi: 10.1088/1758-5090/aab004.

Li X, Zheng Z, Liu X, Zhao S, Liu S. Nanostructured photoelectrochemical biosensor for highly sensitive detection of organophosphorus insecticides. *Biosens Bioelectron.* 2015; 64: 1–5.

Liu D, Chen W, Wei J, Li X, Wang Z, Jiang X. A highly sensitive, dual-readout assay based on gold nanoparticles for organophosphorus and carbamate pesticides. *Anal Chem.* 2012; 84: 4185–91.

McEwen FL, Stephenson GR. *The Use and Significance of Pesticides in the Environment.* John Wiley & Sons New York; 1979; 179–99.

Mohammadi M, Tavakoli H, Abdollahzadeh Y, Khosravi A, Torkaman R, Mshayekhi A. Ultra-preconcentration and determination of organophosphorus pesticides in soil samples by a combination of ultrasound assisted leaching-solid phase extraction and low- density solvent based dispersive liquid-liquid microextraction. *RSC Adv.* 2015; 5: 75174–81.

Mohammadi S, Ebrahimi M, Khoshnood RS. Magnetic solid-phase extraction based on $CoFe_2O_4$ magnetic nanoparticles for the determination of diazinon and fenitrothion in aqueous samples. *Bulg Chem Commun.* 2017; 69–72.

Naeeni H, Yamini Y, Rezaee M. Combination of supercritical fluid extraction with dispersive liquid–liquid microextraction for extraction of organophosphorus pesticides from soil and marine sediment samples. *J Suoercrit Fluids.* 2011; 57: 219–26.

O'Brien RD. *Insecticides: Action and Metabolism*, Academic Press: New York; 1967.

Reddy DS, Latha MP. *Pesticides.* New Vishal Publications: New Delhi; 1987; 18–23.

Rodrigues NFM, Neto SY, de Cassia Silva Luz R, Damos FS, Yamanaka H. Ultrasensitive determination of malathion using acetylcholinesterase immobilized on chitosan-functionalized magnetic iron nanoparticles. *Biosensors.* 2018; 8: 16–27.

Schrader G. The modification of biological activity by structural changes in organophosphorus compounds. *World Rev Pest Contr.* 1965; 4: 140–9.

Shim JY, Kim YA, Lee EH, Lee YT, Lee H. Development of enzyme-linked immunosorbent assays for the organophosphorus insecticide EPN. *J Agric Food Chem.* 2008; 56: 11551–9.

Sun T, Jin Y, Yang J, Li L, Shi X, Li X. Dispersive solid-phase extraction of organophosphorus pesticides from apple, cucumber and water samples using reduced grapheme oxide coated with ZnO nanocomposites as a sorbent. *Anal Methods.* 2015; 7: 6095–102.

Tung VC, Chen LM, Allen MJ, Wassei JK, Nelson K, Kaner RB, Yang Y. Low-temperature solution processing of graphene – carbon nanotube hybrid materials for high-performance transparent conductors. *Nano Lett.* 2009; 9: 1949–55.

Xing H, Liu T, Zhang Z, Wang X, Xu S. Acute and subchronic toxic effects of atrazine and chlorpyrifos on common carp (*Cyprinus caprio L.*): Immunotoxicity assessments. *Fish Shellfish Immunol.* 2015; 45: 327–33.

Yang Y, Tu H, Zhang A, Du D, Lin Y. Preparation and characterization of Au-ZrO_2-SiO_2 nanocomposite spheres and their application in enrichment and detection of organophosphorus agents. *J Mater Chem.* 2012; 22: 4977–81.

Yekta S, Sadeghi M, Babanezhad E. Synthesis of $CaWO_4$ nanoparticles and its application for the adsorption-degradation of organophosphorus cyanophos. *J Water Process Eng.* 2016; 14: 19–27.

Zheng Y, Liu Z, Zhan H, Li J, Zhang C. Development of a sensitive acetylcholinesterase biosensor based on a functionalized grapheme-polyvinyl alcohol nanocomposite for organophosphorus pesticide detection. *Anal Methods.* 2015; 7: 9977–83.

6 Nanophytopharmaceuticals

Harnessing the Biopotential of Phytomolecules for Maximal Therapeutic Efficacy

Alka Mukne, Swapna Nair, and Misbah Momin*

CONTENTS

6.1 INTRODUCTION

The rich and diverse plant kingdom has been tapped since times immemorial, as a source of newer drugs and molecules against various diseases and disorders plaguing mankind. In fact, almost 50% of the drugs in the current usage—be it as anti-infectives, antioxidants, immune-modulating agents, or even anticancer agents—are derived either directly from natural sources or are as chemical analogues of the molecules of natural origin. Use of herbal medicines is gradually increasing day by day, with approximately 80% of the population worldwide principally relying on them for the treatment of various illnesses. Herbal medicines, due to the belief that they promote healthy living, have been widely accepted in the developed areas such as the UK, Europe, North America, and Australia, with complementary and alternative medicines taking a prominent place in the primary healthcare in these places. Availability of herbal remedies not only in the drugstores, but also in the supermarkets, is a testament to the fact that the public interest in herbal remedies has surged significantly over the past few years.

* Corresponding author.

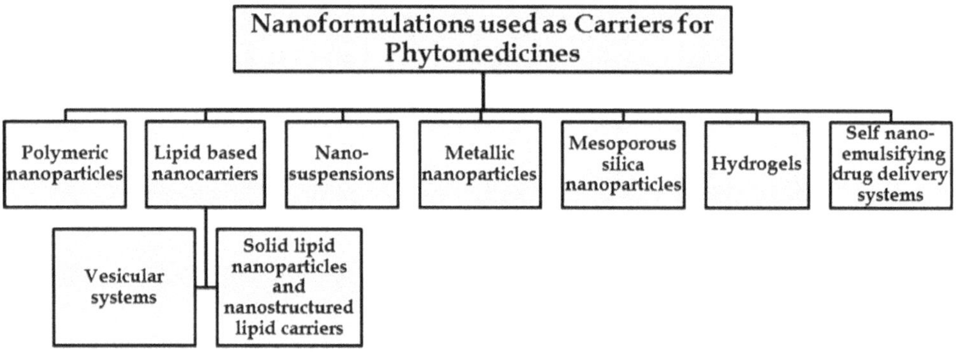

FIGURE 6.1 Types of nanoformulations used as carriers for phytomedicines.

The primary handicap in harnessing the potential of phytomedicines, hitherto unrealized, has often been the poor *in vivo* response of the phytoactives that have otherwise demonstrated excellent *in vitro* pharmacological activity. The major reasons that cap the *in vivo* efficacy of such phytomedicines are their large molecular size, hydrophobic nature, and poor solubility, which eventually leads to poor systemic absorption and low bioavailability. Many phytoconstituents/plant extracts are unstable at acidic pH levels, making it difficult to achieve their therapeutic concentrations in the blood stream. For attaining optimum bioavailability, it is important to strike the right balance between the hydrophilicity (for adequate drug dissolution in the physiological fluids) and lipophilicity (for complete absorption through the lipid-rich biomembranes) of the drug product. Various phytoconstituents, such as polyphenols, exhibit poor bioavailability either due to their highly hydrophilic nature, which restricts their entry into the lipid-rich membranes such as blood–brain barrier (BBB), or their large molecular size, which limits their ability to be absorbed by simple diffusion. Various other active phytoconstituents, such as resveratrol and quercetin, undergo rapid metabolism, which greatly reduces their bioavailability after oral administration.

Integrating the principles of novel drug-delivery systems (NDDS) with phytomedicines may prove to be the right step in circumventing the bioavailability issues of these versatile and therapeutically active moieties and maximizing their biopotential. Nanotechnology, in particular, has proven to be an effective strategy for improving solubility, dissolution, and absorption of such problematic molecules. The low particle size of nanosystems (in the range of 10–1000 nm) allows controlled and targeted release of the active constituent. This, in turn, channels the molecule to the desired site of action and thus increases its therapeutic efficacy while reducing its toxic effects. Various lipid, polymeric, and metallic nanocarriers have been utilized to load phytomolecules for targeted and improved drug delivery. A broad classification of the diverse types of nanoformulations used as carriers for phytomedicines is summarized as Figure 6.1.

Nano-based formulations, due to their obvious advantages over the conventional drug-delivery systems, have garnered a lot of interest among the scientific community. Accordingly, a multitude of research workers have lately been engaged in developing nano-based formulations for effective delivery of phytomedicines. This chapter provides an updated insight into various nanofomulations used for developing safe and efficacious phytopharmaceuticals.

6.2 NANOPARTICULATE SYSTEMS USED AS CARRIERS FOR PHYTOMEDICINES

Nanoparticles are drug-delivery systems, which can be applied for delivery of both hydrophilic and hydrophobic drugs by varying the materials used in formulating the matrix or shell of these systems. Usually, such nanoparticles are made up of biocompatible lipids or polymers. They can be broadly classified into two categories: nanocapsules and nanospheres. Nanocapsules are formulations that contain an inner core surrounded by a polymeric membrane. The drug to be encapsulated can either be incorporated in the core or in the polymeric membrane. On the other hand, nanospheres are made up of the polymeric matrix that contains the active constituent.

Different types of nanosystems that have been used for delivery of phytomolecules are discussed below.

6.2.1 Polymeric Nanoparticles

Polymeric nanoparticles are prepared from biodegradable and biocompatible polymers that are of either natural (chitosan, gelatin, albumin, sodium alginate) or synthetic (polylactide co-glycolic acid [PLGA], polyethylene glycol [PEG], poly ε-caprolactone, polyalkyl cyanoacrylates) in origin. Generally, polymers of natural origin are preferred, as they are bioabsorbable and/or biodegradable and have lower side effects. These nanoparticles facilitate controlled and targeted release of the bioactive payload by various mechanisms, such as swelling of the polymer due to hydration and enzymatic cleavage of the polymer.

6.2.1.1 Methods of Preparation of Polymeric Nanoparticles

Depending on the type of polymers used, various methods such as *in situ* polymerization and solvent precipitation are employed to prepare polymeric nanoparticles. Ionic gelation and nanoprecipitation are widely used for hydrophilic polymers, whereas solvent displacement followed by nanoprecipitation of the preformed polymer is the method of choice for hydrophobic polymers. An overview of the most commonly used methods for preparation of polymeric nanoparticles is presented in Figure 6.2.

6.2.1.1.1 Emulsion Polymerization

In emulsion polymerization, the monomer is dissolved in an aqueous phase containing the surfactant and drug. The initiation of polymerization occurs due to various mechanisms, such as collision of the continuous phase with the initiator molecule like an ion or a free radical. The suspension is then neutralized and filtered to obtain polymeric nanoparticles. Phase separation and formation of nanoparticles (nanospheres) may occur before or after the termination of the polymerization reaction. This is one of the fastest and most easily scalable methods.

6.2.1.1.2 Interfacial Polymerization

This method basically involves polymerization of a very fine oil-in-water emulsion at the oil/water interface. Generally, the monomer, oil, drug and surfactant are dissolved in the organic solvent and slowly added to the aqueous phase containing surfactant, which leads to initiation of polymerization. The organic solvent is then evaporated off, leading to the formation of polymeric nanoparticles (nanocapsules). Aprotic solvents such as acetone and acetonitrile selectively yield nanocapsules, whereas protic solvents such as ethanol and n-butanol lead to formation of a mix of nanocapsules and nanospheres. Interfacial polymerization is one of the most widely used methods for the formulation of polymeric nanoparticles.

6.2.1.1.3 Nanoprecipitation

Nanoprecipitation, also known as the solvent-displacement method, involves the precipitation of a preformed polymer from an organic solution. The polymer is dissolved in a medium-polarity,

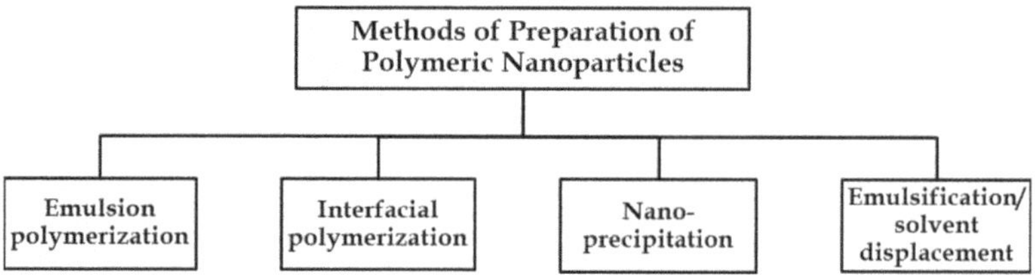

FIGURE 6.2 Commonly used methods for preparation of polymeric nanoparticles.

water-miscible solvent and is introduced quickly into the aqueous phase containing surfactant as the stabilizer. Deposition of polymer at the interface of the water and organic solvent is caused by fast diffusion of the solvent, followed by a phase separation process that leads to the formation of polymeric nanoparticles (nanospheres). The nanoprecipitation method also allows the preparation of nanocapsules by introducing a small amount of non-toxic oil in the organic phase. Due to the oil-based central cavities, high loading efficiencies are reported when lipophilic drugs are loaded into nanocapsules. Though nanoprecipitation is a simple method, it has some disadvantages, including difficulty in emulsification due to coalescence of solvents when mixed with water.

6.2.1.1.4 Emulsification/Solvent Displacement

In this method, the saturated organic solvent containing drug, polymer, and oil is added to an aqueous phase containing surfactant, followed by stirring at approximately 8000 rpm for 10 minutes. Solvent displacement is then carried out by addition of water, followed by evaporation of the organic solvent, to obtain polymeric nanoparticles (nanocapsules). There are various advantages of this method, such as ease of scale-up, high batch-to-batch reproducibility, and high entrapment efficiencies. The method suffers from some limitations, such as leakage of water-soluble drugs.

Achieving tumor-targeted delivery of the therapeutic agents and drug resistance are some of the major challenges in chemotherapy. Polymeric nanoparticles, due to their small size, can easily enter the leaky vasculature of tumors and facilitate targeted delivery of the drug to tumors, thereby reducing cytotoxicity to the peripheral healthy tissues. Formulating drugs as polymeric nanoparticles may help in improving their efficacy in drug-resistant cancers, especially if the resistance is due to the over-expression of efflux transporters. Paclitaxel, a diterpenoid pseudoalkaloid present in bark of the plant *Taxus brevifolia*, is reported to have potent antineoplastic activity. Paclitaxel induces programmed cell death and interferes with the process of microtubule growth. One of the major problems associated with the use of paclitaxel is its low bioavailability due to poor aqueous solubility. The drug is extensively pumped out by P-glycoprotein, which further reduces its oral bioavailability and causes drug resistance. Bhardwaj et al. (2010) evaluated the potential of PLGA nanoparticles in overcoming drug resistance in mice bearing tumor xenografts. The anticancer activity of paclitaxel-loaded nanoparticles was initially evaluated in a multidrug-resistant (MDR) human ovarian carcinoma cell line 2780 AD and sensitive cell line A2780 by 3-(4,5-dimethylthiazol-2-yl)-2,5-diphenyltetrazolium bromide (MTT) assay. The nanoparticulate formulation was found to be up to seven times more potent than free drug in both sensitive as well as resistant cell lines. Nanoparticulate formulation of paclitaxel, on oral administration, significantly delayed the growth of both sensitive and resistant tumors in an *in vivo* mouse xenograft model.

Stimuli-sensitive polymers, such as chitosan, that show phase transition in response to external stimuli, such as pH or temperature, have been explored as alternative polymers for developing nanoparticles for tumor targeting. Curcumin, a natural phenol obtained from *Curcuma longa*, is reported to have potent anticancer, antibacterial, antioxidant, hepatoprotective, and anti-inflammatory activities. Curcumin is extremely safe for clinical use in humans and is considered to be a "generally regarded as safe" (GRAS) compound by the United States Food and Drug Administration (USFDA). The clinical use of curcumin is limited due to its poor water solubility, poor absorption in the gastrointestinal tract, and rapid metabolism in the liver. Rejinold et al. (2011) prepared thermoresponsive polymeric nanoparticles loaded with curcumin for attaining tumor-targeted delivery of the drug. The nanoparticles were prepared by an ionic gelation method using chitosan-g-poly-(N-vinylcaprolactam). A hemolysis assay revealed that thermoresponsive nanoparticles loaded with curcumin were non-hemolytic at concentrations up to 10 mg/mL. Cytotoxicity in normal cells (L-929 mouse fibroblast cells), and cancer cells (MCF-7 breast cancer cell lines, KB oral cancer cells, and PC-3 prostate cancer cells) were evaluated by MTT assay. Concentration-dependent cytotoxicity was observed in both cancer and normal cells, and the apoptosis was mitochondria-dependent, as confirmed by JC-1 mitochondrial potential analysis.

Subsequently, Udompornmongkol and Chiang (2015) prepared polymeric nanoparticles loaded with curcumin for the treatment of colorectal cancer, this time by adding gum Arabic to chitosan by the emulsification-solvent diffusion method. *In vitro* release of curcumin from the nanoparticles in simulated gastric fluid was studied and it was observed that curcumin nanoparticles remained stable at acidic pH. *In vitro* anticancer activity and cell uptake of curcumin-loaded nanoparticles was evaluated in human colon carcinoma cell line HCT-116 and human colorectal adenocarcinoma cell line HT-29 by MTT assay and flow cytometry, respectively. The nanoparticles showed greater cytotoxic activity in both HT-29 and HCT-116 cells compared with free curcumin, due to increased uptake of nanoparticles by the cells.

Polymeric nanoparticles have also been developed for co-delivery of more than one drug to achieve synergistic effect and enhance the intracellular drug concentrations. Camptothecin, an alkaloid obtained from *Camptotheca acuminata*, is reported to have a broad spectrum of anticancer activity. The clinical use of camptothecin is restricted due to its hydrophobicity and toxicity and is often co-administered with other natural anticancer agents such as curcumin. Xiao et al. (2015) prepared chitosan-functionalized cationic polymeric nanoparticles loaded with camptothecin and curcumin, as a combination therapy for treatment of colon cancer. The nanoparticles exhibited an initial burst release, followed by simultaneous sustained release of both the drugs up to 72 h. *In vitro* studies revealed that the cellular uptake and anticancer activity of both drugs were significantly higher in colon-26 cells when the drugs were formulated as polymeric nanoparticles. A similar study has been reported by Katiyar et al. (2016), where PLGA nanoparticles were prepared by nanoprecipitation method for co-delivery of rapamycin and piperine for the treatment of breast cancer. Rapamycin, a macrolide obtained from the bacterium *Streptomyces hygroscopicus*, is a potent anticancer agent that is in phase II clinical trials for the treatment of breast cancer. Despite its potency, clinical use of rapamycin is limited due to its poor water solubility, low oral bioavailability, dose-dependent toxicity and low tumor specificity. Rapamycin is effluxed out by P-glycoprotein and MDR-associated proteins (MRP-1, -2, and -3) that further reduce its bioavailability. Piperine (an alkaloid obtained from *Piper longum*) was used as a chemosensitizer in the formulation to improve the efficacy of rapamycin. Piperine itself has mild anticancer activity and is reported to act as a natural bioenhancer by improving intestinal permeability. It also acts as a P-glycoprotein inhibitor and increases clinical bioavailability of drugs due to its ability to inhibit metabolizing enzymes. The formulation showed sustained *in vitro* release of both phytoconstituents for two weeks, with rapamycin getting released at a slower rate compared with piperine. Evaluation of effect of P-glycoprotein efflux on drug transport using an *in vitro* everted gut sac method revealed that uptake of rapamycin had increased in the presence of piperine. *In vivo* pharmacokinetic studies showed that bioavailability of rapamycin had increased 4.8 times in the presence of piperine. Rapamycin-piperine-PLGA nanoparticles demonstrated significantly greater *in vitro* anticancer activity in human-derived breast cancer cell lines (MDA-MB-231) compared with free rapamycin solution.

Surface-modified polymeric nanoparticles have been developed for oral delivery of drugs for treatment of ulcerative colitis. Chen et al. (2017) reported preparation of porous and nonporous polymeric nanoparticles of curcumin and evaluated their efficacy for the treatment of ulcerative colitis. In addition to the obvious advantages of higher drug loading and better control over release rates, porous nanoparticles can be particularly effective in colonic diseases by virtue of their ability to permeate the epithelial layer better and, therefore, greater drug retention at the target site. Porous and nonporous PLGA polymeric nanoparticles loaded with curcumin were prepared in the presence and absence of ammonium bicarbonate by an emulsion-solvent evaporation technique. Cell-uptake studies revealed that both porous and nonporous nanoparticles were taken up by RAW 264.7 murine macrophage cells. Porous PLGA nanoparticles loaded with curcumin were more effective in downregulating the expression of proinflammatory cytokines like tumor necrosis factor-α (TNF-α), interleukin-6 (IL-6), and interleukin-12 (IL-12). Efficacy of the formulations was further evaluated *in vivo* in a dextran sulfate sodium-induced ulcerative colitis model in mice. After completion of the treatment period, isolated colon tissues were evaluated histopathologically and myeloperoxidase activity (an important inflammatory marker in ulcerative colitis) was evaluated as an indicator of progression of the disease. Porous nanoparticles were found to demonstrate superior therapeutic efficacy in alleviating the symptoms of ulcerative colitis compared with the nonporous nanoparticles.

Introduction of ligands such as folic acid, peptides, and antibodies has been found to be an effective strategy for targeting polymeric nanoparticles to difficult-to-reach sites like tumors. Karra et al. (2013) prepared antibody-conjugated nanoparticles loaded with paclitaxel palmitate (prodrug of paclitaxel) and evaluated their efficacy in treatment of epidermal growth factor receptor (EGFR)-related tumors. EGFR is overexpressed in most cases of non-small cell lung cancer (NSCLC). PLGA nanoparticles loaded with paclitaxel palmitate were prepared and were conjugated with cetuximab. *In vivo* pharmacokinetic evaluation in mice revealed that the antibody-conjugated nanoparticles maintained persistent levels of paclitaxel palmitate with subsequent conversion and release of paclitaxel in a sustained manner. Intravenous administration of a paclitaxel-loaded, antibody-conjugated formulation in mice with metastatic lung tumors resulted in the increased survival rates and significantly higher inhibition of tumor as compared to the nonconjugated PLGA nanoparticles. A similar study has been reported by Han and Davis (2013), in which they prepared nanoparticles of camptothecin using a copolymer of mucic acid and PEG. The Herceptin (HER) antibody was then conjugated to the nanoparticles using boronic acid, which acted as a PEG spacer. The efficacy of the antibody-conjugated formulation was evaluated in nude mice bearing HER2-overexpressing BT-474 human breast cancer tumors. The antibody-conjugated formulation showed significantly greater *in vivo* anticancer activity as compared with both nonconjugated polymeric nanoparticles as well as irinotecan. S2.2-MUC1 aptamer-conjugated PLGA nanoparticles were reported by Liu et al. (2015), with curcumin as the payload. Aptamers are single-stranded DNA or RNA with higher stability and lower immunogenicity compared with other antibodies and can bind selectively to various target molecules. MUC1 is a large membrane glycoprotein that is overexpressed in most type of cancers and S2.2 is a modified version of the original aptamer that binds to MUC1 with high specificity. Cellular uptake and cytotoxicity studies in HepG2 and MCF-7 cells overexpressing MUC1 protein revealed that the uptake and cytotoxicity of aptamer-conjugated and nonconjugated formulations were identical in HepG2 cells, whereas the conjugated formulation was taken up to a larger extent compared with the nonconjugated nanoparticles by MCF-7 cells, with an increase in cytotoxicity. Salar and Kumar (2016) prepared folic acid-conjugated chitosan nanoparticles of vincristine to obtain tumor-targeted delivery of the drug. Conjugation with glutathione was attempted by Acharya and Reddy (2016). They developed the conjugated paclitaxel-PLGA nanoparticles for intranasal administration for delivery to the brain, using 1-ethyl-3-(3-dimethylaminopropyl) carbodiimide as a linker. The nanoparticles gave an initial burst release of paclitaxel, followed by sustained release for the next 24 h. *In vivo* biodistribution studies revealed that the uptake of paclitaxel-glutathione-conjugated PLGA nanoparticles in the brain was significantly higher than that of the unconjugated formulation.

6.2.2 Lipid-Based Nanoparticles

Lipid-based nanoparticles basically comprise vesicular systems, solid lipid nanoparticles (SLNs), and nanostructured lipid carriers (NLCs). The vesicular systems are basically composed of glycerophospholipids, whereas SLNs and NLCs are nano-assemblies, prepared using solid lipids like triglycerides and waxes. Lipid-based nanoparticles show the lowest *in vivo* toxicity and are reported to be suitable carriers for hydrophobic molecules (Table 6.1).

6.2.2.1 Vesicular Systems

Lipoidal vesicular systems are spherical, highly ordered assemblies of one or several concentric lipid bilayers that entrap an aqueous solvent core and separate the core from the surrounding

TABLE 6.1
Marketed Formulations of Lipid-Based Nanoparticles

Brand Names/Formulation Types	Phytoconstituents	Key Features/Therapeutic Indications
EndoTAG®-1 (Liposomes)	Paclitaxel	Pancreatic cancer
Onivyde® (Liposomal injection)	Irinotecan	Metastatic pancreatic cancer
Lipusu® (Liposomes)	Paclitaxel	NSCLC, ovarian and breast cancer
LEP-ETU (Liposomes)	Paclitaxel	Ovarian cancer
Marqibo®(Liposomes)	Vincristine sulfate	Philadelphia chromosome–negative (Ph−) acute lymphoblastic leukemia
DepoDur® (Liposomes)	Morphine sulfate	Management of postoperative pain
Relieva®(Novasome® technology-Liposomes)	Psorberine	Psoriasis and dandruff
MIAJ™ (Novasome® technology-Liposomes)	Vitamin C	Cosmeceutical (anti-wrinkle activity)
Bosexil™ (Phytosome®)	Boswellic acid	Cosmeceutical (anti-inflammatory and anti-photoaging activities)
Centella asiatica selected triterpenes (Phytosome®)	Selected triterpenes	Cosmeceutical (anti-wrinkle, collagen-restructuring effects)
Ginkgoselect® (Phytosome®)	Ginkgoflavonglucosides, ginkgoterpenes, bilobalide, ginkgolides, ginkgolic acids	Nutraceutical (vasokinetic, improvement of cognition and circulation)
Greenselect® (Phytosome®)	(−)-Epigallocatechin, (−)-epigallocatechin-3-O-gallate, caffeine	Nutraceutical (antioxidant, weight management)
Hawthorn (Phytosome®)	Vitexin-2″-O-rhamnoside, Hawthorn typical constituents	Nutraceutical (antioxidant and improvement of cardiovascular health)
Leucoselect® (Grape seed Phytosome®)	Proanthocyanidins	Nutraceutical (UV protection and improvement of cardiovascular heath)
Meriva® – (Turmeric Phytosomes®)	Curcuminoids	For joint and bone health
Proanthocyanidin A2 (Phytosome®)	Proanthocyanidin A2	Cosmeceutical (UV protection and skin-firming agent)
Rexatrol® – (Resveratrol Phytosome®)	Resveratrol	Nutraceutical (antioxidant and anti-aging)
Sericoside (Phytosome®)	Sericoside	Cosmeceutical (anti-wrinkle effect)
Silymarin (Phytosome®)	Silybin	Nutraceutical (antioxidant, hepatoprotective)
Visnadex® - (Visnadin Phytosome®)	Visnadin	Nutraceutical (Vasokinetic)
Lipoduction™ (Ethosomes)	Grape-seed extract and extract of roots of *Coleus forskohlii*	Anti-cellulite cream
Longvida® (Optimized curcumin SLNs)	Curcumin	Improved oral bioavailability
Tinoderm™ A (Nanotope™)	Vitamin A palmitate	Cosmeceutical (anti-aging and anti-wrinkle effects)
Tinoderm™ E (Nanotope™)	Vitamin E acetate	Cosmeceutical (prevention of premature ageing induced by UV irradiation and lipid peroxidation)
Tinoderm™ P (Nanotope™)	D-panthenol	Cosmeceutical (anti-inflammatory activity)

solvent. Of the numerous lipid-based systems, liposomes are considered as the first-generation vesicles. Similar to biological membranes, liposomes have low permeability to hydrophilic molecules and high permeability to hydrophobic chemicals. Retention of hydrophobic fractions in such systems, therefore, remains a challenge. The other issues associated with more effective use of liposomes as drug-delivery systems include rapid clearance, low drug entrapment, physical instability due to drug leakage and aggregation or fusion of vesicles to form larger particles and chemical instability caused by hydrolysis of ester bonds and/or peroxidation of unsaturated acyl chains of lipids. Various newer vesicular systems like niosomes, phytosomes, sphingosomes, ethosomes, transfersomes, archaeosomes, and pharmacosomes can be qualified as modified liposomes, albeit with manipulation of lipid composition and/or mode of drug entrapment or complexation, so as to overcome some of these inherent drawbacks associated with liposomes.

6.2.2.1.1 Liposomes

Liposomes are defined as microscopic, spherical vesicles consisting of an internal aqueous compartment entrapped by one or multiple concentric lipid bilayers composed of natural and/or synthetic lipids which are relatively biocompatible, biodegradable, and nonimmunogenic. Liposomes can be used as carriers for both lipophilic and water-soluble molecules. Lipophilic drugs are mainly entrapped within lipid bilayers, whereas the hydrophilic substances are encapsulated in the interior aqueous compartment.

Liposomes are further classified, based on their size and lamellarity, into multilamellar and unilamellar vesicles. Functionality of the liposomes can be modulated by varying the composition to form pH-sensitive, cationic, immuno-liposomes, and long-circulating liposomes. A schematic representation of different types of liposomes is presented in Figure 6.3.

Liposomal formulations of drugs show reduced toxicity and tissue irritation as the lipids used in the preparation of liposomes are biocompatible, biodegradable, and nonimmunogenic. Liposomes have high encapsulation efficiency, help in improving the drug solubility and protect the drug from getting degraded in acidic pH or in presence of light. The submicron size of liposomes allows relatively higher intracellular uptake, facilitates drug targeting, and provides sustained drug release, which subsequently leads to improved bioavailability of the drug.

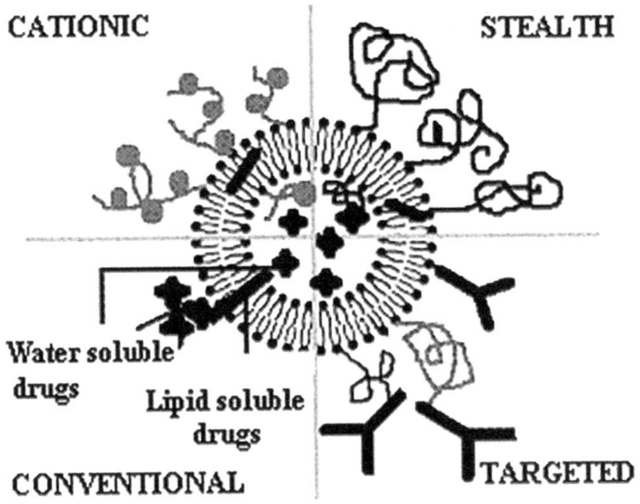

FIGURE 6.3 Schematic illustration of various types of liposomes.

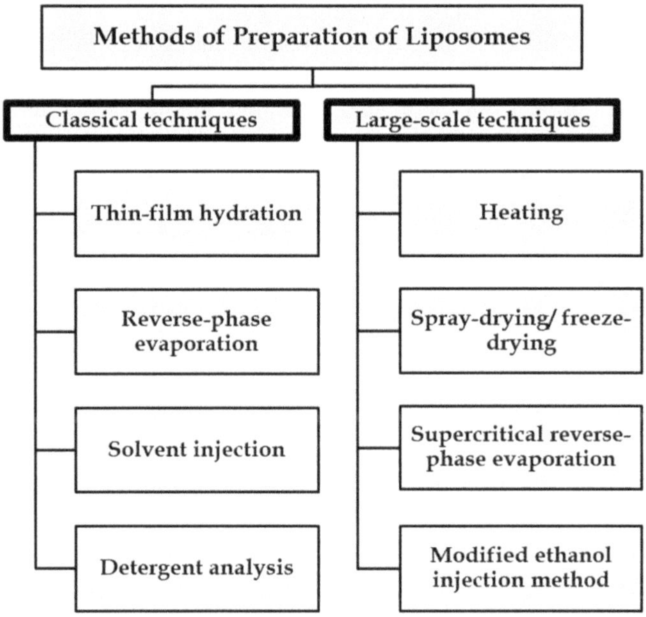

FIGURE 6.4 Methods of preparation of liposomes.

6.2.2.1.1.1 Methods of Preparation of Liposomes Liposomes are generally prepared using either the classical or large-scale techniques. Different methods used in the preparation of liposomes are summarized in Figure 6.4.

The difference between the various methods is the way in which the dry lipid film is formed, followed by hydration with aqueous media.

A. Classical techniques
   i. Thin-film hydration
      This technique involves the dispersion of phospholipid and cholesterol in an organic solvent, followed by evaporation of the organic solvent using a rotary evaporator to form a lipid film. The dry lipid film, thus formed, is hydrated by adding an aqueous buffer solution, under agitation at temperatures above the lipid phase transition temperature to get multilamellar liposomal vesicles (MLVs), about 1–5 μm in diameter. These MLVs are then subjected to sonication or extrusion through polycarbonate filters to get small unilamellar vesicles (SUVs) or large unilamellar vesicles (LUVs), respectively.
   ii. Reverse-phase evaporation (REV) technique
      Liposomes made by the REV method have aqueous volume-to-lipid ratios that are four times higher than hand-shaken liposomes or multilamellar liposomes. This method allows preparation of liposomes with a high aqueous space-to-lipid ratio and the ability to entrap high quantities of the aqueous agents.

      REV is based on the creation of inverted micelles, which are formed upon the sonication of a mixture of an aqueous phase containing water-soluble molecules and an organic phase in which the lipids are solubilized. The evaporation of the organic solvent under vacuum leads to the conversion of these inverted micelles into viscous gel. At a critical point in this process, the gel state collapses and the excess of phospholipids form a complete bilayer around the micelles, resulting in the creation of liposomes.

iii. Solvent (ether or ethanol) injection technique
The solvent injection method involves the injection of lipid-loaded organic phase (ethanol or ether) into aqueous media, forming liposomes. The ether injection method differs from the ethanol injection method in that heat is applied to the liposomal suspension to remove the immiscible ether. An advantage of the ether injection method compared with the ethanol injection method is that it allows removal of solvent from the product, enabling the formation of a concentrated liposomal product with higher entrapment efficiency.
iv. Detergent dialysis
In this method, lipids are solubilized with detergent, yielding defined mixed micelles. The detergent is removed by dialysis, which results in the formation of homogenous unilamellar liposomes in the size range of 40–180 nm with a large encapsulated volume.

B. Large-scale techniques
i. Heating method
This method involves the hydration of liposomal components in an aqueous medium, followed by the heating of this suspension, in the presence of glycerol (3% v/v), up to 120°C. It has been confirmed by thin-layer chromatography (TLC) that lipids used in the preparation of liposomes do not degrade at 120°C. Glycerol increases the stability of lipid vesicles and need not be removed from the final liposomal product because of its nontoxicity. The particle size is dictated by the phospholipid nature and charge, the speed of the stirring and the shape of the reaction vessel. No further sterilization procedures are needed since heat is employed in this method, thereby reducing the time and cost of liposome production.
ii. Spray-drying
The spray-drying process is considered to be the fastest single-step procedure for the preparation of liposomes. This technique involves suspending lipids and mannitol in organic solvent, followed by sonication of the suspension. This mixture is then subjected to spray-drying and the dried product is hydrated with different volumes of aqueous media with continuous stirring. Volume of the hydration medium governs the particle size of the liposomes. Mannitol plays a significant role in increasing the surface area of the lipid mixture, enabling complete hydration of the spray-dried product.
iii. Freeze-drying
This technique involves dissolving the liposome-forming lipids and water-soluble carrier materials, such as sucrose, in tert-butyl alcohol/water cosolvent systems in appropriate ratios to form a clear, isotropic, monophasic solution. The monophasic solution is then sterilized by filtration and filled into freeze-drying vials. After the completion of the freeze-drying cycle, the lyophilized product is hydrated to form a homogenous liposomal suspension. The lipid-carrier ratio used in the formulation is the key factor determining the size of the liposomes. This method allows the preparation of sterile and pyrogen-free submicron liposomes.
iv. Supercritical reverse-phase evaporation (SCRPE)
Emulsions are first formed by introducing a given amount of water into a homogeneous mixture of supercritical carbon dioxide, lipids, and/or ethanol, with sufficient stirring and subsequent pressure reduction. The entrapment efficiency by this method was found to be five times higher for the water-soluble drugs, when compared with multilamellar vesicles prepared by the thin-film hydration technique.
v. Modified ethanol injection method (MEIM)
Three novel methods for liposome production have been reported based on the principle of the ethanol injection technique, including the microfluidic channel

method, the crossflow-injection technique, and the membrane contactor method. These methods are readily amenable for scale-up. In the crossflow injection technique, an assembly with two tubes welded together, forming a cross, is fabricated, and a provision for an injection hole is made at the connection point. A homogenous batch can be prepared by keeping the buffer flow rate at minimum. Other important influencing factors are the lipid concentration and injection pressures. The microfluidization method makes use of a microfluidic hydrodynamic focusing (MHF) platform. The liposomes are formed by injecting the lipid and water phase into a microchannel. When multiple flow streams are injected in a microchannel, mixing occurs by interfacial diffusion. Microfluidic flow is usually laminar due to the small dimensions of the microchannels and low flow rates. The size of the liposomes is controlled by changing the flow rate. In the membrane contactor method, an organic phase composed of lipids (phospholipid and cholesterol) and dissolved in ethanol is pressed through a membrane with a specified pore size, using nitrogen gas at a pressure below 5 bars. Simultaneously, the aqueous phase is allowed to flow tangentially to the membrane surface, sweeping away the liposomes formed within the membrane device.

Various active phytoconstituents with a diverse group of pharmacological activities, such as antibacterials, antioxidants, and pain killers, have been developed as liposomes. The poor solubility of the active principles of *Orthosiphon stamineus* limits its pharmacological effects. To overcome the solubility issues of the active constituents, nanoliposomes of *O. stamineus* ethanolic extract with unpurified soybean phospholipids were prepared by Aisha et al. (2014). Liposomes containing *O. stamineus* extract showed improvement in 2,2-diphenyl-1-picrylhydrazyl (DPPH) scavenging effect (IC50 = 23.5 ± 1.1 μg/mL) compared with plain extract (32.4 ± 0.5 μg/mL).

Quercetin is a natural antioxidant, which has the ability to eliminate free radicals and is reported to have various important pharmacological activities, including amelioration of cognitive and memory dysfunction. The clinical use of quercetin is restricted due to its low solubility, extensive metabolism in intestine and liver, and reduced permeation across barriers such as the BBB. Jeon et al. (2015) prepared sodium hyaluronate-chitosan multilayered liposomes for improving skin permeation and stability of quercetin using layer-by-layer deposition technology. Layer-by-layer technology based on electrostatic interaction of polyelectrolytes is used to improve stability of active molecules. The multilayer liposomes were stabilized by dihexadecyl phosphate and were subsequently coated with cationic chitosan and anionic sodium hyaluronate. An increase in the number of layers deposited on the liposomal surface resulted in a sustained-release profile of quercetin that fit the Korsmeyer-Peppas model. *In vitro* skin permeation studies revealed that multilayered liposomes significantly improved the skin permeability of quercetin in comparison with uncoated liposomes.

The potential of liposome technology for improving permeation of molecules across the BBB has also been evaluated. Kuo and Tsao (2017) prepared modified liposomes of quercetin as an alternative approach to target the BBB to alleviate progression of Alzheimer's disease. The quercetin-encapsulated liposomes were grafted with RMP-7, a bradykinin analogue, and lactoferrin, a biomolecule that activates receptor-mediated transcytosis for BBB permeation. The grafted liposomal quercetin formulation showed significantly enhanced permeation across the BBB in comparison with free quercetin when evaluated *in vitro* in SK-N-MC human neuroblastoma cells. *In vitro* studies also demonstrated that the modified liposomes protected the neurons against apoptosis, probably due to the presence of RMP-7 and lactoferrin.

Eloy et al. (2016) prepared PEGylated liposomes co-loaded with paclitaxel-rapamycin and evaluated their efficacy in treatment of breast cancer. The formulation was prepared using soy phosphatidylcholine, cholesterol, and 1,2-distearoyl-sn-glycero-3-phosphoethanolamine- PEG_{2000} (DSPE-PEG_{2000}). The PEGylated liposomes showed high drug loading and retention within the liposomes, with sustained release of the drug. The liposomal formulation was found to be more cytotoxic to 4T1 breast cancer cell lines compared with the free-drug solution, when evaluated using MTT assay. The co-loaded paclitaxel-rapamycin liposomal formulation controlled the tumor growth to a greater extent as compared to the drug solution *in vivo* in 4T1 tumor-bearing mice.

Liposomes have been developed for protection of the phytoconstituents from degradation in gastric pH. The pomaces from red grapes (residue from winemaking) contain considerable amounts of phenolic antioxidants and have health-promoting effects. A large part of these active phytoconstituents are not absorbed by the human body due to their low bioavailability, which is majorly due to the instability of these phytoconstituents in the gastric environment. Manconi et al. (2016) developed polymer-associated liposomes loaded with grape pomaces. The grape pomace extract was incorporated in liposomes and stabilized by the addition of a natural polysaccharide, sodium alginate or Arabic gum. *In vitro* release studies revealed that polymers protected the vesicle from degradation in the acidic environment of the stomach, with consequent improvement of its bioavailability. Activity of a grape extract-loaded liposomal formulation against oxidative stress was evaluated in Caco-2 cell lines, and the cell viability was determined using the MTT assay. The liposomal formulation exerted significantly greater protective effect on Caco-2 cells against oxidative stress, compared with the plain extract.

Munich Biotech AG, Germany, developed a liposomal paclitaxel formulation, MBT-0206, by the thin-film hydration method. The formulation was prepared by incorporating paclitaxel in 1,2-dioleoyl-3-trimethylammonium-propane (DOTAP) and dioleylphosphatidylcholine (DOPC) (50:47 molar ratio). The drug and lipids were dissolved in chloroform and 5% w/v glucose solution was used as the hydrating medium. The multilamellar liposomes were prepared by ethanol injection method and were repeatedly extruded to give unilamellar vesicles. MBT-0206 demonstrated significantly reduced tumor growth compared with the conventional paclitaxel given as Taxol® ($p < 0.05$), at the same dose, when tested *in vivo* in Syrian Golden hamsters with syngenic A-Mel-3 melanomas. Significantly greater microvascular changes with no increase in vessel diameter and overall reduced blood flow ($p < 0.05$) were observed in MB-0206 treated animals (Streith et al. 2004). Since 2004, when Munich Biotech AG was acquired by Medigene MG, Germany, MBT-0206 has been renamed EndoTAG®-1. Later, in 2016, Syncore Biotechnology acquired the EndoTAG® technology from Medigene. The product is currently in phase III clinical studies for use in breast and pancreatic cancer.

NanoX™ is a PEGylated liposomal technology platform developed by Taiwan Liposome Company, Ltd. (TLCL) for fast and complete drug loading in liposomes. The technology makes use of saturated acyl chain phospholipids along with cholesterol. NanoX™ liposomes are composed of 1,2-distearoyl-sn-glycero-3-phosphocholine (DSPC), cholesterol, and DSPE-PEG$_{2000}$ (3:2:0.045 molar ratio). It is reported that NanoX™ efficiently encapsulates various drugs including vinca alkaloids and camptothecin, reduces the side effects of the drugs, protects drugs from degradation, helps in altering the pharmacokinetic profile of the active molecule and allows passive targeting to tumors or inflammatory sites (Lin et al., 2013).

NanoX™ technology-based vinorelbine liposomal formulation, NanoVNB® was developed using octasulfate triethylamine as the polyvalent counter-ion donor to prevent leakage of vinorelbine from the formulation before it reached the target site. NanoVNB® showed better anticancer activity in phase I clinical trials, but it also caused increased toxicity due to extended retention of vinorelbine in the formulation. Skin rash was the most common nonhematological and dose-limiting toxicity associated with the use of NanoVNB®. The formulation was suitably modified using other polyvalent counter-ion donors, such as dextran sulfate in combination with various concentrations of monovalent counter-ion donors such as ammonium sulfate (Kan et al., 2017 (US 9700511 B2)). Vinorelbine forms a salt with the counter-ions and precipitates in the intraliposomal aqueous core, leading to increased encapsulation efficiency. The release profile of the drug from the formulation into the plasma was manipulated by altering the proportions of monovalent and polyvalent counter-ions used, with consequent minimization of side effects. LV304 (batch of the formulation prepared using 300 mM ammonium sulfate and 2 mM dextran sulfate), with optimum *in vitro* release profile, was evaluated in HT-29 human colon cancer cell-bearing mice. The formulation showed reduced toxicity (including skin irritation and weight loss) in mice compared with NanoVNB® during the 60-day trial period, with the desired drug release profile.

TLC178 is a vinorelbine tartrate-encapsulated liposomal formulation prepared using NanoX™ technology. TLC178 is reported to improve the safety profile of vinorelbine. In October 2016, TLC178 was granted orphan-drug status for treatment of cutaneous T-cell lymphoma, the most common type of blood cancer, by the USFDA. Later, in April 2017, TLC178 was designated as a drug to treat rhabdomyosarcoma, a rare pediatric disease, by the USFDA. Subsequently, TLC178 received orphan-drug designation for the treatment of soft tissue sarcoma by the USFDA in July 2017. Further dose-escalation studies are being carried out to evaluate the maximum tolerated dose. TLC178 is under phase II clinical trials for treatment of non-Hodgkin lymphoma (NHL).

TLCL has also developed polymeric micelles loaded with lipotecan, which is a semisynthetic analogue of camptothecin. Lipotecan is reported to reduce drug resistance that is commonly encountered during chemotherapy. The lipotecan-polymeric micelles (TLC388) were prepared using polymeric surfactants and biocompatible lipid-based excipients. TLC388 has been granted orphan-drug status by the USFDA and the European Medicines Agency (EMA) for the treatment of hepatocellular carcinoma. The product has also been accepted in the "Green Path program" by the China State FDA. TLC388 is currently in phase II clinical trials for treatment of hepatocellular carcinoma and renal cell carcinoma.

DepoDur® is a single-dose, extended release, injectable liposomal formulation of morphine sulfate developed using Pacira Pharmaceutical Inc's DepoFoam® technology, which was approved by the USFDA in 2004 and is indicated for epidural administration for managing the postoperative pain in patients after major surgical procedures. The formulation is a sterile suspension of multivesicular liposomes encapsulating the water-soluble morphine sulfate and composed of DOPC, cholesterol, dipalmitoylphosphatidylglycerol (DPPG), tricaprylin, and triolein. Epidural administration of DepoDur® results in both systemic absorption and absorption through the meninges into the intrathecal space. DepoDur® was found to provide sustained release of the drug when tested in human patients, with its systemic AUC comparable to that of an equal dose of epidurally administered morphine sulfate injection, and C_{max} that was 30% of that of plain injection. A few opioid-related side effects, such as urinary retention, vomiting, and constipation, were reported as adverse events in the clinical trials. Some safety concerns of DepoDur® remain because of its unknown cerebrospinal fluid pharmacokinetics in humans.

Liposomal technology has been used extensively in cosmeceuticals. Blank liposomes—that is, liposomes with only water entrapped within—are also effective moisturizing agents and help in improving skin elasticity and suppleness. Topical liposomal cosmetic formulations are very popular, with inclusion of lipophilic compounds such as cholesterol and ceramides and polar components like antioxidants and vitamins. With herbal cosmetics becoming a rage the world over, an increasing number of studies focus on developing liposomes based on herbal agents.

Novasome® is a patented technology platform developed by Novavax, Inc., Rockville, Maryland, US, and exclusively licensed to IGI Laboratories, Inc., US. These paucilamellar vesicles are prepared using biocompatible, single-tailed amphiphiles (such as long-chain fatty acids, long-chain fatty alcohols, and glycerolipids), along with phospholipids and have up to seven bilayer membranes. The formulation is stable at a wide range of pH (2–13) and temperature (that of liquid nitrogen to above 100°C). Their encapsulation efficiency can be nearly 100% for lipid cargo and 85% for aqueous materials. Novasome® has vast range of applications and is widely being used as a dermal-delivery vehicle. Apollo Pharmaceutical, Inc., Canada, has developed Relieva® cream, lotion, and shampoo products for psoriasis and dandruff, by encapsulating psorberine in Novasome® . Psorberine is a refined extract from roots and bark of *Mahonia aquifolium*, an herb native to Canada, with alkaloids that relieve the effects of psoriasis, dermatitis, eczema, and other skin conditions. The presence of 5′-methoxyhydnocarpin in the extract further potentiates the antimicrobial action of the alkaloids. 5′-methoxyhydnocarpin is an MDR pump inhibitor that lowers bacterial resistance to antibacterial agents. IGI Laboratories, Inc. has recently developed its own range of anti-aging skin-care products (MIAJ™) using the Novasome® technology, and the products contain several active ingredients including 10% vitamin C serum. In 1995, IGI, Inc. provided Johnson and Johnson with a sublicense to produce and sell Novasome® technology-based products containing retinoid.

6.2.2.1.2 Sphingosomes

Scientists have been grappling with the challenge of maintaining adequate stability of liposomal formulations ever since these systems have gained prominence as viable and efficient drug-delivery options. Liposomes formulated using glycerophospholipids have been particularly vulnerable to chemical degradative changes, mainly due to the rapid hydrolysis of the labile ester linkage within the glycerophospholipids which causes loss of integrity of the vesicles. Sphingosomes that have been developed as improvised liposomes make use of sphingolipids instead of glycerophospholipids as the major lipid component. The partial double-bond character of the amide bond in sphingolipids makes them more resistant to thermal and acid hydrolysis compared with glycerophospholipids that have the ester bond. Absence of ester linkage makes sphingolipids resistant to action of phospholipases as well. Sphingosomes, therefore, have demonstrated better drug-retention characteristics than conventional liposomes. The presence of cholesterol further improves the drug retention, possibly due to hydrogen bonding between the amide nitrogen of sphingolipids and the hydroxyl group of cholesterol. Sphingosomes, however, score just as inadequately as the conventional liposomes in terms of physical instability aspects, including drug leakage due to aggregation and vesicle fusion.

Optisome™ is a sphingosomal drug-delivery platform developed for tumor targeting and extended drug release at tumor sites by Talon Therapeutics, Inc., a company acquired later by Spectrum™ Pharmaceuticals. The system uses sphingomyelin-based liposomes with an aqueous acidic core in which the chemotherapeutic payload is encapsulated. The Optisome™ platform is believed to offer various advantages, such as targeted delivery of medicines and improving their efficacy. Marqibo® is vincristine sulfate (0.16 mg/mL) injection based on Optisome™ technology that was approved by the USFDA in 2012 for Philadelphia chromosome negative (Ph−) acute lymphoblastic leukemia in adults. The product is currently being developed for NHL (under phase III studies) and multiple myeloma (preclinical phase). The company had developed Alocrest™, vinorelbine tartrate injection, again based on Optisome™ technology. A phase I dose-escalation trial of Alocrest™ has demonstrated promising anticancer activity, with acceptable toxicity in heavily pretreated patients with cancers for which no standard therapy was known to exist, including NSCLC and advanced stage breast cancer. Alocrest™ is scheduled for phase II trials for NSCLC and breast cancer. Lipid-based nanophytoformulations in various stages of clinical trials are summarized in Table 6.2.

6.2.2.1.3 Phytosomes®

Phytosomes® are colloidal, nanometric-sized lipoidal vesicular dispersions of drug-lipid complexes, wherein the drug molecules are covalently bound to lipids. A chemical bond is formed between the lipids and drug molecules with free and available –OH and $-NH_2$ groups (Kaur and Kanwar, 2002). Phytosome® is a patented technology developed by Indena®, Italy, a leading manufacturer of drugs and nutraceuticals. Water-soluble phytoconstituents such as flavonoids, terpenoids, and tannins or standardized plant extracts are incorporated into phospholipids to produce lipid compatible molecular complexes. Phytosomes® increase the bioavailability of herbal extracts due to their ability to permeate across the lipoidal biomembrane and reach the systemic circulation.

Phytosomes® hold some distinct advantages over liposomes and other vesicular carriers, particularly with respect to drug leakage. The release of drug from the phytosomes is brought about via hydrolytic cleavage of the covalent bond linking the drug to the lipid and is independent of alterations in bilayer fluidity due to changes in temperature, pH, or other physiological conditions. Other advantages of Phytosomes® include enhanced absorption of water-soluble polar phytoconstituents, remarkable oral as well as topical bioavailability, appreciable drug entrapment, reduced dose requirement of drug, and

TABLE 6.2
Lipid Vesicular Systems under Clinical Study

Brand Names/Formulation Types	Phytoconstituents	Therapeutic Indications
EndoTAG®-1 (Liposomes)	Paclitaxel	Phase III ongoing for metastatic adenocarcinoma of the pancreas (in combination with gemcitabine) Phase II completed for HER 2–negative breast cancer Phase II completed for triple receptor–negative breast cancer Phase II completed for metastatic hepatic cancer
LEP-ETU Liposomes	Paclitaxel	Phase II completed for metastatic hepatic cancer
TLC178 (Liposomes)	Vinorelbine tartrate	Phase II ongoing for NHL
Marqibo® (Liposomal injection)	Vincristine sulfate	Phase II completed for sarcoma, neuroblastoma, Wilms tumor, leukemia, lymphoma, brain tumors, acute lymphoblastic leukemia, and metastatic malignant uveal melanoma Phase II ongoing for recurrent adult acute myeloid leukemia Phase I completed for malignant melanoma
Nanoliposomal irinotecan	Irinotecan	Phase I ongoing for diffuse intrinsic pontine glioma and high-grade glioma
Nanoliposomal CPT-11	Irinotecan	Phase I completed for glioblastoma, gliosarcoma, anaplastic astrocytoma, and anaplastic oligodendroglioma
Onivyde® (Liposomal injection)	Irinotecan	Phase III ongoing for small cell lung cancer
CPX-1 (Liposomal injection)	Irinotecan HCl: Floxuridine	Phase II completed for colorectal neoplasms
Aerosolized liposomal camptothecin	Camptothecin	Phase II completed for corpus uteri lung cancer
Aerosolized liposomal 9-nitro-20 (S)-camptothecin	9-nitro-20 (S)-camptothecin	Phase II completed for lung cancer
Liposomal curcumin	Curcumin	Phase I completed for evaluation of safety
LipoCurc™ Liposomes	Curcumin	Phase II ongoing for glioblastoma patients who have failed first-line therapy
Alocrest™ (Sphingosomes)	Vinorelbine tartrate	Phase I completed for NSCLC and advanced stage breast cancer

Source: ClinicalTrials.gov. Available online at: https://clinicaltrials.gov (Accessed on 2 Aug 2017).

better stability owing to the chemical bond formed between the drug and the lipids. Phytosomes®, upon storage, are, however, susceptible to physical instability due to fusion and aggregation, as well as chemical degradation.

Phytosomes® have been developed as an alternative strategy to improve the therapeutic efficacy of an entire range of drugs on oral administration. Gymnemic acid, obtained from *Gymnema sylvestrae*, has a wide range of pharmacological activities, such as antioxidant and antidiabetic activities. Its use is restricted due to its poor bioavailability associated with its complex structure and low lipid solubility. A complex of gymnemic acid and phospholipid (GPC-phytosome) was prepared by Pathan et al. (2012) to evaluate the cardioprotective effects of phytosomal gymnemic acid on doxorubicin-induced cardiomyopathy in rats. Pretreatment with GPC-phytosome greatly reduced doxorubicin-induced cardiotoxicity and improved hemodynamic variables when compared with the untreated control group.

Phytosomes of milk thistle extract, a hepatoprotectant with low bioavailability (20%–50%) were prepared by El-Gazayerly et al. (2014), using lecithin from soybeans and egg yolks. Silymarin phytosomes were evaluated for their hepatoprotective effect on carbon tetrachloride-induced liver injury in rats compared with the plain milk thistle extract. The *in vivo* study revealed that silymarin-loaded phytosomes greatly decreased glutamic pyruvic transaminase and superoxide dismutase activities compared with milk thistle extract ($p < 0.05$).

6.2.2.1.4 Lipospheres

Lipospheres consist of water-dispersible microparticles with the drug and phospholipids embedded on the surface of a solid hydrophobic core of triglycerides. The internal core contains the drug, either dissolved or dispersed in solid fat matrix. Lipospheres have been developed for parenteral and topical delivery of anti-inflammatory compounds, local anesthetics, antibiotics, anticancer agents, insect repellent, vaccines, proteins, protein hydrolysates, and peptides. Some of the advantages of lipospheres listed in the literature include good stability, low cost of ingredients, ease of preparation and scale-up, high dispersibility in aqueous medium, and high entrapment of the hydrophobic compounds.

Lipospheres of methanolic extract of *Gongronema latifolium* leaves were formulated by melt homogenization using a lipid matrix consisting of a mixture of goat fat and Phospholipon® 90 H in the ratio 1:2. *In vivo* antidiabetic activity of the formulation was studied in alloxan-induced diabetic Wistar rats. The fasting blood glucose levels in the diabetic rats administered with lipospheres containing the extract (100 mg/kg) and the plain extract (100 mg/kg) were found to be 86.75 and 57 mg/dL and 117.25 and 97.5 mg/dL at 8 and 12 h, respectively. *G. latifolium*-loaded lipospheres had significantly higher antidiabetic effects than the plain extract at equivalent concentrations (Chime et al., 2013).

6.2.2.1.5 Archaeosomes

Archaeosomes are modified liposomes with higher stabilities compared with the conventional liposomes and are made using one or more of the polar lipids extracted from Archaea (a large and widely distributed group of organisms), which readily hydrate to form lipid vesicles. Archaeosomes are resistant to the action of lipolytic enzymes and demonstrate greater stability in acidic pH due to the ether linkages in the archaeol lipids. The saturated phytanyl chains give better protection under oxidative conditions and the bipolar lipids spanning the membrane make the vesicles more stable.

Due to their excellent stability under extreme physical and chemical conditions, archaeosomes have shown significant potential as hydrophilic and hydrophobic drug carriers for parenteral as well as oral use, with good tissue distribution profiles and considerably lower drug leakage. Archaeosomes are reported to induce a strong and sustained antigen-specific cytotoxic T lymphocyte (CTL) response to an appropriate acellular protein or peptide antigen that is encapsulated in and/or associated with the archaeosome, without needing the inclusion of another immune modulating adjuvant in the vaccine formulation. This makes them ideal vaccine adjuvants.

Alavi et al. (2014) formulated archaeosomes of paclitaxel as an alternative approach to overcome its side effects and improve its anticancer potential. Total polar lipids were extracted from methanogenic Archaea bacteria and archaeosomes were synthesized by hydrating the total lipids with phosphate-buffered saline containing paclitaxel. Cytotoxicity studies of archaeosomal paclitaxel and plain paclitaxel on MCF7 breast cancer cell lines were performed using an MTT assay. The IC_{50} value for archaeosomal paclitaxel (80.8 ± 1.37 μM) was found to be much lower than that of plain paclitaxel (166.9 ± 2.9 μM), showing that the anticancer efficacy of archaeosomal paclitaxel is much greater than that of plain paclitaxel.

6.2.2.1.6 Ethosomes

Classical liposomes are inefficient in exerting systemic action through the transdermal route as they form drug reservoirs only in the upper layer of the stratum corneum. This shortcoming of classical liposomes was overcome by development of various vesicle derivatives, with ethosomes being one of them. Ethosomes are soft, malleable lipid vesicles containing alcohol (ethanol and isopropyl alcohol) in relatively high concentrations, glycerophospholipids, and water and have been successfully used in various therapeutic applications, such as pilosebaceous targeting, delivery of antiarthritic, antifungal, and antiparkinsonian drugs. Drug delivery can be modulated by changing the alcohol-water or alcohol-polyol-water ratio. Lecithin and ethanol are known to have a significant effect on the size of ethosomes.

Though the exact mechanism of ethosome penetration through the skin remains unclear, it has been proposed that the increased drug penetration occurs due to a combination of processes involving the stratum corneum and the pilosebaceous pathways. The high concentration of ethanol in the ethosomes is thought to be responsible for causing disturbance in the highly organized skin lipid bilayer, thus providing the lipid vesicle more freedom to squeeze through the small openings created due to the disturbance in stratum corneum. Ethanol can also act as a blending agent for lipid vesicles, thus increasing their distribution in various skin layers.

Eucommia ulmoides extracts possess anti-inflammatory and anti-aging activities. Ethosomes, liposomes, and ethanolic solution of extracts of *E. ulmoides* were evaluated for their permeability across mice dorsal skin by Gu et al. (2013). Skin permeation was the highest in the case of ethosomes (77.16%), followed by liposomes (62.8%), and the ethanolic solution (41.59%).

Ethosomes of black-tea extracts containing gallic acid and caffeine were prepared by Yeh et al. (2013), to be used as hair dye and protective agent. The ethosomes were prepared using a nonionic surfactant (Brij®52), soya phosphatidyl choline, and cholesterol as excipients. When evaluated *in vitro*, the ethosomal formulation of black-tea extracts had enhanced adsorption of the extracts onto the hair surfaces compared with the hydroalcoholic solution of the extracts. Microscopic observations of hair pigmentation in hair treated with an ethosomal formulation of black-tea extracts or a solution of black-tea extracts revealed that the solution of black-tea extracts was retained with difficulty on the hair surface, but ethosomes containing the black-tea extracts coated the hair surface smoothly. The permeation of ethosomal formulation of caffeine, caffeine solution, and caffeine from black-tea extracts was studied on nude-mouse skin. The results showed that ethosomes increased the skin permeation of caffeine. Ethosomes also increased the transdermal delivery of gallic acid. Im et al. (2015) have reported development of ethosomes of liquiritrin and its aglycone liquiritrigenin, the flavonoids found in licorice-possessing antioxidant and anti-aging effects. Ethosomes loaded with liquiritigenin and liquiritin showed enhanced skin-permeation ability when compared with that of 20% ethanolic solution and conventional liposomes of the actives.

Lipoduction™ is a commercially available ethosomal anticellulite cream containing grape-seed extract and extract of roots of *Coleus forskohlii*, marketed in the US by Osmotics, Inc.

6.2.2.1.7 Transfersomes®

Transfersome® is a proprietary drug-delivery technology which is a trademark of IDEA AG, Germany. Transfersomes® are ultradeformable or elastic vesicles (EVs) that were developed for enhanced permeation and effective delivery of drugs through the intact skin. Transfersomes® mainly consist of glycerophospholipids and an edge activator (nonionic surfactant) and are more flexible and adaptable as compared with liposomes due to their self-optimized membrane properties and high surface hydrophilicity, which permit the transfersomes to squeeze through pores much smaller than their own diameter. Interaction between hydrophilic lipid residues and proximal water makes the polar lipids attract water and induce hydration, and hence, the lipid vesicles move to the site of higher water concentration. Penetration of transfersomes across skin occurs due to an osmotic gradient that forms due to a difference in water content across the stratum corneum and epidermis. The edge activators in transfersomes act as penetration enhancers that disrupt the stratum corneum and widen the pores, thus allowing easy penetration of transfersomes. Transfersomes® are also capable of squeezing themselves through the intact skin layer due to membrane flexibility and, hence, cause a gradual release of drugs.

EVs of curcumin were prepared to improve its cutaneous bioavailability and topical anti-inflammatory effect. *Ex vivo* skin permeation and retention studies were performed, and it was reported that improved permeation of curcumin-loaded EVs was observed across the mouse skin. Incorporation of curcumin-loaded EVs into a hydrophilic ointment base resulted in higher skin retention (51.66%) when compared with free-curcumin ointment (1.64%) and a marketed formulation (VICCO® turmeric skin cream). Acute and chronic anti-inflammatory effect of the EVs was evaluated using xylene-induced, acute ear edema in mice and cotton pellet-induced chronic inflammation in rats, respectively. The ointment containing EVs of curcumin showed an anti-inflammatory effect comparable to the marketed diclofenac sodium ointment (Omni-gel®) in reduction of acute inflammation in mice ($p < 0.05$). Remarkable inhibition of cotton pellet-induced chronic inflammation (28.8% for EVs of curcumin versus 3.91% for free curcumin) was also observed (Agrawal et al., 2015). Transfersomes of capsaicin were developed by Sarwa et al. (2015) to improve its antiarthritic potential. A transfersomal vesicular system was employed for the topical application of capsaicin in experimental rats. Results from *in vivo* antiarthritic activity study of capsaicin-loaded transfersomal nanoformulation were compared with that of conventional gel formulation available in the market (Thermagel). The transfersomal formulation possessed greater antiarthritic potential than the marketed Thermagel formulation at the same dose levels.

6.2.2.1.8 Niosomes

Niosomes are microscopic lamellar structures formed on an admixture of nonionic surfactant of the alkyl or dialkyl polyglycerol ether class and cholesterol with subsequent hydration in aqueous media. In niosomes, a nonionic surfactant such as Span 60 is the vesicle-forming agent, and it is usually stabilized by addition of cholesterol and a small amount of anionic surfactant such as dicetyl phosphate. Niosomes have architecture similar to liposomes as lipophilic drugs are entrapped within the lipid bilayers and hydrophilic drugs are entrapped in the hydrophilic tails. *In vivo*, niosomes act in a similar manner as liposomes, by prolonging the circulation of entrapped drugs and altering their organ distribution. Niosomal drug delivery has been used for various routes of administration, including intramuscular, intravenous, peroral, and transdermal routes. Niosomes also enhance absorption of some drugs across the cell membrane, to localize drugs in target tissues. Deformable and elastic niosomes have been prepared for improving transdermal penetration, by incorporating alcohol and edge activators, respectively, within the mix.

Niosomal preparations have been developed for targeting drugs to the brain. *Ginkgo biloba* is a plant reported to have numerous therapeutic effects, including anti-tumor and antioxidant activities. Various marketed formulations containing *G. biloba* extracts are available, though these formulations present several challenges, such as low bioavailability and short half-life. In an attempt to overcome these limitations, Jin et al. (2013) prepared a niosomal preparation of the extract for oral administration. Niosomes were prepared by a film dispersion homogenization method and were freeze-dried to improve the stability. An *in vitro* release study of the formulation in pH 6.8 phosphate buffer demonstrated a sustained release of the active constituents up to 48 h. *In vivo* biodistribution studies in rats showed that the content of active constituents in organs such as lungs, liver, and kidneys were significantly greater in animals treated with niosomal preparation when compared with those treated with *G. biloba* tablets. Interestingly, the active constituents were taken up by the brain tissues only in the group of animals that were administered with niosomal preparation of the extract, thus suggesting that niosomes may be promising carriers for targeting phytomedicines to the brain.

The role of niosomes in effectively targeting drugs to localized tumors, after oral administration, has been studied and reported by many research workers. Sezgin-Bayindir et al. (2013) prepared paclitaxel-loaded niosomes to improve the oral bioavailability of the drug. Extended release of paclitaxel was observed when the release profile of the niosomal preparation was evaluated *in vitro*. *In vivo* pharmacokinetic evaluation of the formulation in rats demonstrated that the bioavailability of paclitaxel, when administered as niosomes, was approximately three-fold higher compared with paclitaxel suspension. An elevated level of paclitaxel was found in the liver and intestine after administration of niosomes.

6.2.2.1.9 Nanotopes™

Nanotopes™, a trademark of Ciba Speciality Chemical Corporation (Tarrytown), are monolayered vesicles comprising a membrane with a well-defined ratio of a phospholipid and a cosurfactant.

Mixtures of phospholipids and cosurfactants are used to form Nanotopes™, at an optimum ratio that promotes intercalation of the cone-like cosurfactant between the cylindrical phospholipid molecules, creating a very densely packed membrane of small-sized, stable particles. Nanotopes™ are highly resistant against surface active molecules and tolerant to elevated ionic strength, compared with conventional liposomes. The mean size of Nanotopes™ depends on the encapsulated bioactive, but results in a very narrow particle size distribution, size ranging typically from 20 to 40 nm, with considerably smaller average particle size compared with the conventional liposomes. It is this difference in size that contributes to remarkably improved skin penetration and, consequently, greater efficacy of drugs, when formulated as Nanotopes™, in comparison with liposomes (Baschong et al., 2005).

6.2.2.1.10 Vesosomes

Almost all the modifications of liposomes have been an attempt to manipulate the chemical and physical attributes of the lipid bilayer, but with limited success. Vesosomes and capsosomes have been developed as novel systems that alter the structure of unilamellar liposomes, while retaining the lipid bilayer as the fundamental structural unit. Vesosomes are nested lipid-bilayer compartments or vesicles enclosed within a single, large bilayer vesicle. Vesosomes, thus, have distinct inner compartments separated from the outer limiting membrane that have the same or different bilayer compositions and can encapsulate different materials. Saturated phospholipids, including phosphatidylcholines, phosphatidylserines, and phosphatidylglycerols, which can be interdigitated by ethanol or other alcohols, glycerol, and propylene glycol, are used for forming the outer lipid bilayer. Polymer lipids such as 1,2-dipalmitoyl-sn-glycero-3-phosphoethanolamine-PEG_{2000} (DPPE-PEG_{2000}) can be incorporated into the interdigitated sheets to impart steric stability. The enclosed internal vesicles can be made from a variety of lipids, including 1,2-dipalmitoyl-sn-glycero-3-phosphocholine (DPPC), DSPC, ethyl-phosphatidylcholine (ePC), cholesterol, and cationic lipids. The nested compartment structure addresses one of the major problems with conventional liposomes—the overly fast release rates of most drugs in serum—possibly by shielding the internal compartments from the serum components responsible for degrading the bilayer and increasing permeability.

Vesosomes, therefore, have the most promising application in delivering drugs that have shown improved efficacy, when administered as liposomes, but are handicapped by their uncontrolled and fast drug release. Vincristine is one such example, in which increased duration of therapeutic alkaloid concentrations at the tumor site has shown to have a positive correlation with drug efficacy. Conventional liposomes, however, have not been able to sustain therapeutic vincristine levels for prolonged periods, though they have improved drug bioavailability.

6.2.2.1.11 Capsosomes

One of the more recent developments in liposome-type drug-delivery systems are capsosomes, cell mimics with a large, defined number of lipid compartments, encapsulated within a structurally stable polymer scaffold. Capsosomes are formed by sequentially layering liposomes and charged polymers onto particle substrates, followed by removal of the substrate cores as shown in Figure 6.5. In this context, degradable polymer capsules are employed as carriers, while the liposomes serve as entrapped drug deposits. Anionic silica particles, coated with single layers of cationic polylysine, have been used as substrate cores onto which small, unilamellar liposomes made of zwitterionic or anionic lipids are bound. These are then coated with cholesterol-modified polylysine, followed by sequential deposition of liposomes and polymer layers. The assembly is then sealed by sequential adsorption of poly(N-vinyl)pyrrolidone (PVP) and thiol-functionalized poly(methacrylic acid)-PVP5 (PVP-PMA_{SH}). The thiols in the outer layer can be cross-linked and the silica core etched to form capsosomes (Chandrawati et al., 2010). These polymer-hydrogel capsules with the subcompartmentalized liposomal vessels have demonstrated a cargo retention time of at least 2 weeks and can be positioned as drug depot-type delivery systems. The outer membrane of capsosomes have

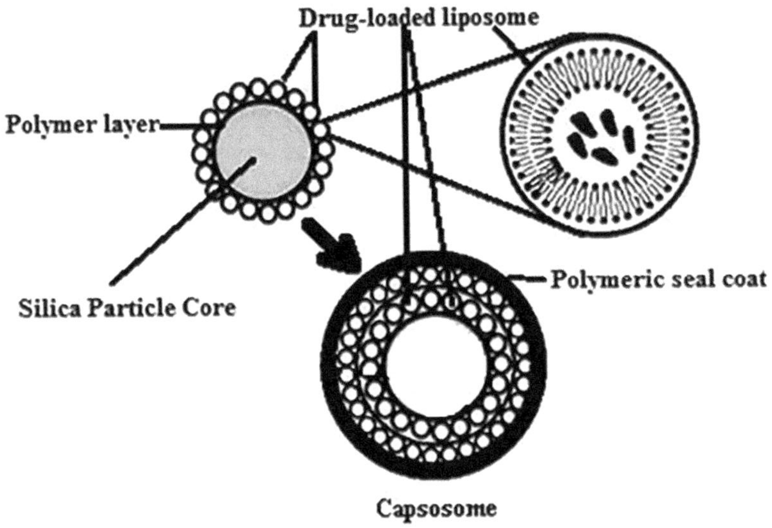

FIGURE 6.5 Schematic illustration of capsosome formation. (Adapted from Chandrawati et al. *ACS Nano.* 2010;4:1351–61.)

been PEGylated by immobilizing graft copolymers of PMA and PEG, with resultant improvement in stability of liposomal subcompartments, in the presence of extraneous phospholipases (Chandrawati et al., 2010). While the core strength of capsosomes is expected to be in cell mimicry by using enzyme-loaded liposomes and thus presenting greater opportunities in development of functional artificial cells, their potential for intracellular delivery of chemotherapeutic agents is something that ought to be worked on.

6.2.2.2 Solid Lipid Nanoparticles (SLNs) and Nanostructured Lipid Carriers (NLCs)

The first generation of SLNs were developed in the early 1990s as alternatives to other colloidal drug-delivery systems such as liposomes and polymeric nanoparticles. SLNs are colloidal carriers, mainly composed of high melting-point fats that are solid at room temperature as well as physiological temperature and water-soluble surfactants as stabilizers. SLNs provide a lipophilic matrix in which drugs can be incorporated. Judicious selection of the lipids and surfactants plays an influential role in determining the particle size, drug loading capacity, release profile and long-term stability of the SLN formulations. Greater biocompatibility, higher physicochemical stability, improved bioavailability of lipophilic molecules, protection of labile drugs from degradation, a preparation method without the use of organic solvents, and ease of scale-up are some of the advantages of SLNs over the conventional dosage forms. The solid lipid used in the preparation of SLNs, however, may crystallize into a perfect crystal lattice and may result in lower drug-entrapment efficiency and reduced drug release. SLNs may also undergo gelation or particle growth, which may lead to expulsion of drug from the colloidal vesicles during long-term storage.

NLCs, the newer generation of lipid based nanocarriers, have been developed to overcome the disadvantages of SLNs. NLCs make use of spatially different lipid molecules (both solid and liquid lipids) to create an imperfect lipid matrix, which increases the drug encapsulation capacity, as well as the stability of the formulation. The structures of SLNs and NLCs are presented in Figure 6.6.

Generally, lipids such as stearic acid, glyceryl behenate, and tristearin, which are solid at room temperature, are used as solid lipids for the preparation of NLCs, whereas digestible oils from natural sources such as Miglyol® 812 or Compritol® are used as liquid lipids. Emulsifiers such as Pluronic® F68, polyvinyl alcohol (PVA), and Tweens® are used as stabilizers to prevent particle aggregation. It has been reported that the addition of PEG during NLC preparation forms a coating

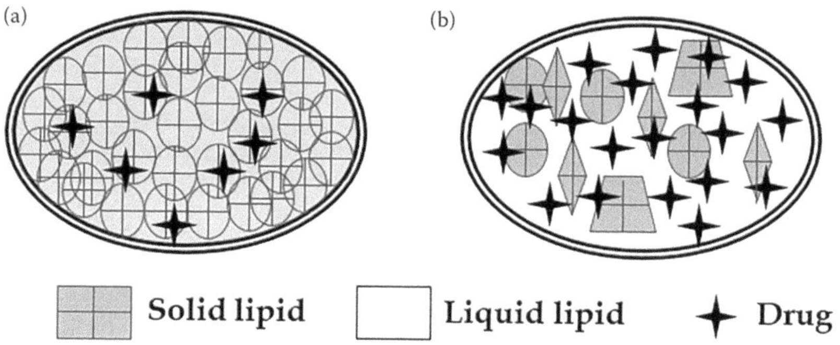

FIGURE 6.6 Structure of (a) SLNs and (b) NLCs.

on the nanoparticulate shell that prevents reticuloendothelial-system uptake and thus prolongs the residence time of the drug in the body.

Based on their structure, NLCs are classified as imperfect, formless, and multiple NLCs. Imperfect NLCs are formed when the solid and liquid lipids are mixed to form an extremely disordered matrix that improves the entrapment of the drug in the matrix. Formless NLCs have a noncrystalline structure that prevents the expulsion of drugs from the matrix. Multiple-type NLCs are similar to a w/o/w emulsion in which the solubility of the drug in the liquid lipid is greater than that in the solid lipid, thus reducing the possibility of degradation of drugs in solid lipids.

6.2.2.2.1 Methods for Preparation of SLNs and NLCs

Various methods such as high speed homogenization, solvent evaporation, solvent emulsification-diffusion, and microemulsion are used for the preparation of SLNs and NLCs. The lipid is dissolved in a water-immiscible organic solvent followed by emulsification in an aqueous solvent. Evaporation of the organic solvent leads to the formation of SLNs and NLCs and the particle size may further be reduced by homogenization or sonication.

High-pressure homogenization (HPH) is the most commonly used method for preparation of SLNs and NLCs. HPH is further classified into two types: hot homogenization and cold homogenization. In both methods, the active constituent is dissolved or dispersed into the melted lipid before the homogenization process. Liquid formulation in the HPH method is pushed through narrow gaps of the homogenizer at a high pressure of 100–2000 bars and the cavitation force and shear stress breakdown the particles into submicron size range. Hot homogenization may not be suitable for drugs that get degraded at high temperature. The procedures for hot and cold homogenization are summarized in Figures 6.7 and 6.8, respectively.

SLNs and NLCs may be administered by various routes, such as oral, dermal, and intravenous routes. SLNs and NLCs are also considered to be safe and effective for incorporation of poorly water-soluble molecules, to improve their oral absorption and penetration across cell membranes.

Triptolide, a component present in *Tripterygium wilfondil*, is a traditional Chinese medicine used as an anti-inflammatory, immunosuppressive, antifertility, and antineoplastic agent. The clinical use of triptolide is limited due to its poor water solubility and toxic effect on the gastrointestinal, urogenital, bone marrow, cardiovascular, and circulatory systems. Zhang et al. (2013) prepared triptolide-loaded SLNs by using different quantities of Compritol® 888 (solid lipid), Cremophor® RH40 (surfactant), and Lipoid® E dissolved in Transcutol® HP (cosurfactant). Gastric mucosa irritation caused by the SLNs was assessed by administering the formulation to rats. After oral administration of triptolide-loaded SLNs and triptolide solution in rats, it was observed that in contrast to triptolide solution, triptolide-loaded SLNs did not cause neutrophil infiltration, hyperemia, or any marked changes in the histology of gastric mucosa of rats. Tao et al. (2012) prepared SLNs of biochanin A. Biochanin A, a constituent present in various plants—including red clover, cabbage,

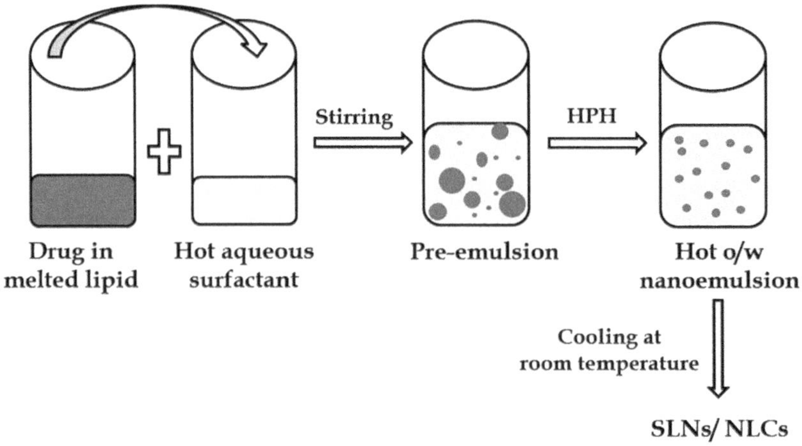

FIGURE 6.7 Methodology for the preparation of SLNs by hot homogenization.

and alfalfa—is a phytoestrogen and has various pharmacological activities, such as inhibition of melanogenesis, anticancer, anti-inflammatory, and protection of dopaminergic neurons. SLNs were prepared by emulsion-evaporation and low temperature-solidification techniques using glycerol monostearate, lecithin as the lipid phase, and Tween® 80/Pluronic® F-68 mixture as the aqueous phase. The formulation showed a sustained release of biochanin A, with approximately 95% drug release in 48 h. Ratcharin et al. (2012) prepared *Zingiber officinale* extract-loaded SLNs using stearic acid as the lipid, Cremophor® RH40 as surfactant, and ethanol as cosurfactant. *Zingiber officinale* (ginger) has reported anti-inflammatory activity, with gingerols as the major active constituent. Gingerols, however, are unstable in the presence of light and air, and SLNs of ginger extract were formulated to overcome the photostability issues. Reduction in particle size was observed when the surfactant and cosurfactant concentrations were increased with a concomitant increase in entrapment efficiency. Vitthal et al. (2013) prepared SLNs loaded with bacoside by a microemulsion probe sonication method. Bacoside, a triterpenoid saponin present in *Bacopa monniera* (Brahmi), is a dose-dependent inhibitor of acetylcholine esterase and is indicated in the treatment of brain- and memory-related disorders. Bacoside when administered orally is unable to reach the site of action due to its inability to cross the BBB. SLNs of bacoside were prepared using polysorbate 80 in water

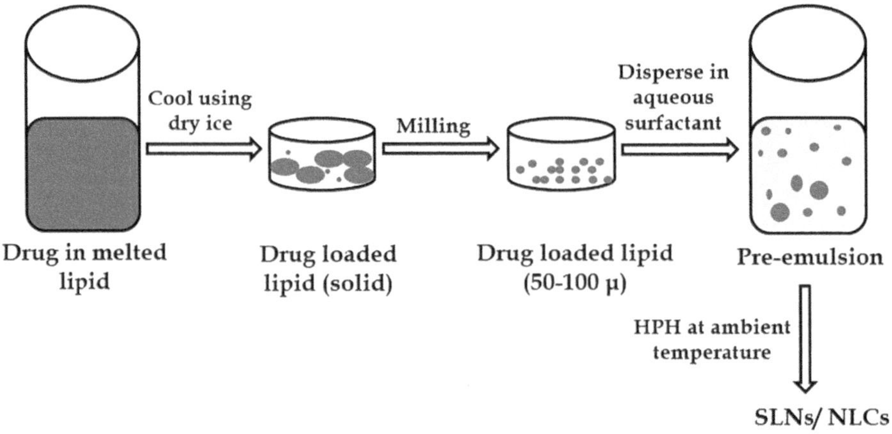

FIGURE 6.8 Methodology for the preparation of SLNs by cold homogenization.

as the aqueous phase and stearic acid as the lipid phase. The drug release from the SLNs was found to follow zero-order kinetics with approximately 85% of the drug released within 12 h from the SLNs.

Resveratrol, a stilbene compound present in high concentrations in grapes, mulberries, and peanuts, has attracted remarkable attention due to its potent cardioprotective, anti-inflammatory, anticarcinogenic, and neuroprotective effects. Resveratrol helps in maintaining cerebral homeostasis by increasing antioxidant defense and reducing proinflammatory cytokines. It is considered to be an effective molecule for treatment of neurological disorders such as Alzheimer's, Parkinson's, and Huntington's diseases. The clinical use of resveratrol is limited due to its pharmacokinetic parameters such as poor solubility, chemical instability, rapid metabolism, reduced bioavailability and restricted entry of the drug into the brain. Neves et al. (2013) have formulated SLNs and NLCs of resveratrol to enhance its oral bioavailability. SLNs loaded with resveratrol were prepared by a hot homogenization technique using cetyl palmitate and polysorbate 60, whereas, NLCs were prepared by using liquid lipid Miglyol® 812 in addition to the composition used for preparation of SLNs. *In vitro* release studies revealed that a more sustained and controlled release of resveratrol was obtained from NLCs as compared to SLNs, probably due to the less ordered lipid matrix of the NLCs. The same research group, in 2016, functionalized the resveratrol-loaded SLNs with apolipoprotein E, for targeting the LDL receptors that are overexpressed in the brain. The permeability through hCMEC/D3 monolayers, was 1.8-fold higher for the functionalized SLNs compared with the nonfunctionalized ones (Neves et al., 2016). Stearic acid–based SLNs of resveratrol, coated with poloxamer 188, were reported by Pandita et al. (2014). The nanoparticles were prepared by a solvent-diffusion-solvent evaporation method and showed controlled *in vitro* release for up to 120 h that followed Higuchi kinetics. Pharmacokinetics of the formulation was evaluated after oral administration in male Wistar rats, with around eight-fold improvement in bioavailability as compared to plain resveratrol solution.

Literature is replete with numerous reports about SLNs and NLCs of paclitaxel, developed for tackling an entire gamut of specialized applications, including improved tumor targeting, tackling MDR issues, and crossing the BBB. Chirio et al. (2014) prepared charged and noncharged SLNs of paclitaxel by a coacervation technique. The SLNs were found to be more effective in glioblastoma cells *in vitro* when compared to plain paclitaxel. Noncharged SLNs were prepared by dispersing sodium behenate and PVA_{9000} in deionized water followed by addition of an acidifying solution to adjust the pH to 4.0. Charged SLNs were prepared using the same method that was used for preparing noncharged SLNs, and a positive charge was introduced either by adding a fixed amount of 2-propanol stearylamine solution to the warm sodium behenate solution just above its Krafft point or a fixed amount of glycerol chitosan was added before formation of micellar solution. *In vitro* permeability across the BBB was evaluated using hCMEC/D3 cells, and it was observed that positively charged SLNs showed best permeability after 24 h. Cytotoxicity of both charged and noncharged paclitaxel-loaded SLNs was greater in NO3 glioblastoma cell line compared with the free drug. Co-culture experiments of different SLN formulations of paclitaxel with hCMEC/d3 and other glioblastoma cell lines demonstrated that paclitaxel-loaded SLNs increased the cytotoxicity of the drug towards glioblastoma cell lines due to improved permeation through the endothelial-cell monolayer. Encapsulation of paclitaxel in SLNs seemed to play an important role in improving the brain uptake of the drug, possibly by reducing the P-glycoprotein-associated drug efflux at the BBB. Xu and Lee (2015) developed SLNs and NLCs of paclitaxel and evaluated their efficacy in MDR cancer cells. SLNs were prepared using 5% w/w tristearin, 3.75% w/w egg phosphatidylcholine, and 1% w/w polysorbate 80 in water. In the case of NLCs, egg phosphatidylcholine was replaced with triolein. The cytotoxicity of SLNs and NLCs in human breast cancer cell line MCF-7 was evaluated, and the results were compared with that of the commercial micellar formulation of paclitaxel. Paclitaxel-loaded SLNs and NLCs showed activity similar to that of free paclitaxel and the commercial formulation. It was observed that SLNs and NLCs of paclitaxel showed greater anticancer activity than free paclitaxel in MDR cancer cell line MCF-7/ADR. Kim et al. (2015) prepared biocompatible SLNs of paclitaxel, mimicking the structure and composition of low-density lipoproteins, for effective tumor targeting. Paclitaxel was entrapped in the core of the SLNs and a tumor-targeting ligand, maleimide-poly(ethylene glycol)-succinimidyl carbonate, was covalently conjugated to the PEG surface. Evaluation of *in vivo* efficacy of the formulation in cancer xenograft mouse model revealed that the SLNs showed enhanced bioavailability, improved antitumor activity and better *in vivo* stability when compared with marketed anticancer formulations Taxol® and Genexol® PM. The biomimetic SLNs showed prolonged *in vivo* tumor-targeting effects, probably due to the increased half-life in comparison with free paclitaxel, that allow maintenance of therapeutic concentrations for a longer period.

6.2.3 Nanosuspensions

Nanosuspensions are submicron colloidal dispersions of drugs that are stabilized by surfactants. Unlike lipoidal systems, nanosuspensions can be used to formulate compounds that are insoluble in both water and oils. Nanosuspensions help in circumventing problems pertaining to delivery of compounds by maintaining the drug as small-sized crystals, eliminating the need for dissolving the drug. The advantages of nanosuspensions include enhanced solubility, stability, and bioavailability of drugs. It also allows higher drug loading, reduces therapeutic dose, and facilitates passive drug targeting.

6.2.3.1 Methods for Preparation of Nanosuspensions

Nanosuspensions can be prepared by two techniques—the top-down and bottom-up approaches—as described in Figure 6.9.

In top-down technology, the particle size of large molecules is reduced by techniques such as high-pressure homogenization and media milling, whereas in bottom-up technology, the drug is precipitated from the organic solvent using an antisolvent in the presence of a stabilizer.

6.2.3.1.1 High-Pressure Homogenization (HPH)

This method involves forcing a suspension of drug and stabilizers through a valve with a small orifice under high pressure. HPH is often classified into two groups: Dissocubes (homogenization in aqueous media) and Nanopure (homogenization in nonaqueous media or media with reduced water content).

In Dissocubes, the drug suspension is passed through a small gap at high pressure of up to 1500 bar, which causes an increase in the dynamic pressure and a decrease in the static pressure, resulting in the reduction of the boiling point of water. At room temperature, water starts boiling, creating gas bubbles, and as the suspension departs the orifice, pressure returns to atmospheric

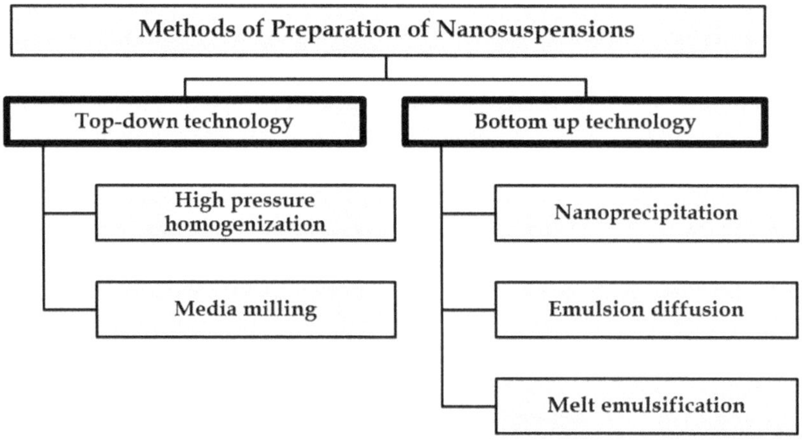

FIGURE 6.9 Methods of preparation of nanosuspensions.

level and the gas bubbles implode. This phenomenon, called cavitation, along with high shear and collisions, leads to fracture of the drug microparticles into nanosized particles. Particle size of the resultant nanosuspension is governed by homogenization pressure, number of homogenization cycles, hardness of drugs, and temperature. The major disadvantages of this method include the need for processing the drug to obtain microparticles before starting the homogenization process and the number of homogenization cycles required for nanosizing the drug particles.

Homogenization of drugs which are dispersed in low molecular weight PEG or in oil, that is, nonaqueous media or media with low water content, is independent of cavitation forces, as the high boiling point and low vapor pressure of oils fail to generate the pressure essential for cavitation to occur. In such cases, the Nanopure process is conducted at a reduced temperature of 0°C or below to achieve results comparable to Dissocubes.

6.2.3.1.2 Media Milling

In this method, high-shear media mills/pearl mills, comprising a milling chamber, a milling shaft, and a recirculation chamber, are used to produce nanosuspension. The milling media or balls are made up of glass (yttrium-stabilized), zirconium oxide, ceramics, or highly cross-linked polystyrene resins. The milling chamber is fed with an aqueous suspension of the drug and stabilizer. The milling media is then rotated at a very high shear rate. The high attrition rates among drug particles and the milling media generate nanoparticles. The advantage of this method is its ease of scale-up and little batch-to-batch variation, whereas its disadvantage is that the erosion of milling media may contaminate the final product.

6.2.3.1.3 Nanoprecipitation

In this method, drug that is insoluble with the aqueous phase is dissolved in a water-miscible organic solvent. This mixture is then rapidly introduced into an aqueous phase containing stabilizers, which causes precipitation of the drug.

6.2.3.1.4 Emulsion-Diffusion Method

Volatile organic solvents, such as butyl acetate, benzyl alcohol, triacetin, and ethyl acetate, which are partially water-miscible, are used as the dispersed phase in this technique. An emulsion of the organic phase containing the drug and water is prepared by HPH or another technique. Diffusion of the internal phase into the external phase converts the droplets into solid particles and leads to the formation of nanosuspensions. Use of organic solvents and the presence of residual solvents in the final products are major drawbacks of this method.

6.2.3.1.5 Melt Emulsification Method

In this method, the drug is initially dispersed in aqueous solution with a stabilizer, heated above the melting point of the drug, and subjected to HPH to produce an emulsion. The emulsion is cooled to a suitable temperature to produce a nanosuspension of drug. Factors affecting particle size include drug and stabilizer concentrations, type of stabilizer, and cooling conditions. The advantage of this method over the emulsion diffusion method is avoidance of organic solvents.

Nanosuspensions of active phytoconstituents have been developed as a novel strategy for improving their bioavailability. Aphidicolin, a tetracyclic diterpenoid, belonging to a class of diterpenes called labdanes, is an antimitotic and antiviral metabolite derived from the mold *Cephalosporium aphidicola*. Labdanes and their derivatives have been shown to possess *in vitro* antileishmanial activity in extracellular promastigotes and intracellular amastigotes of *Leishmania donovani* in murine macrophages. Bioavailability of aphidicolin is low due to its poor aqueous solubility. Nanosuspension of aphidicolin was prepared by Kayser (2000) using HPH. The *in vitro* leishmanicidal activity of aphidicolin nanosuspension against promastigotes of *L. donovani*, *L. infantum*, *L. enriettii*, and *L. major*, and intracellular amastigotes of *L. donovani*, was found to be 140-fold higher than plain aphidicolin.

The poor water solubility of flavonoids and lignans from *Cuscuta chinensis* is reported to limit their oral absorption. Nanosuspension of *C. chinensis* was prepared by Yen et al. (2008) by HPH, and its hepatoprotective and antioxidant effect was compared with that of *C. chinensis* ethanolic extract in acetaminophen-induced hepatotoxicity in murine models. The nanosuspension of *C. chinensis* (50 mg/kg) showed comparable hepatoprotective and antioxidant effects at much lower doses than the extract (125 mg/kg) ($p < 0.05$).

Yao et al. (2012) reported nanosuspension of camptothecin prepared by a supercritical antisolvent process and HPH. The cytotoxicity of camptothecin, topotecan, and the nanosuspension of camptothecin was evaluated by MTT assay in the human colorectal carcinoma cell line, HCT-8, the human prostate carcinoma cell line, PC-3, and the human breast adenocarcinoma cell line, MCF-7. Camptothecin nanosuspension showed less antitumor activity than camptothecin in each cell line, whereas it showed three, four, and six times higher activity than that of topotecan in HCT-8, PC-3, and MCF-7 cells, respectively. The *in vivo* antitumor activity of camptothecin nanosuspension was also evaluated in nude mice bearing the HCT-8 xenograft. It was observed that *in vivo* antitumor activity of camptothecin nanosuspension was comparable to that of topotecan, with tumor inhibitory rates of 95% exhibited by both, at equal concentrations.

Nanosuspensions of curcumin and sodium lauryl sulfate/β-cyclodextrin were prepared by Moorthi and Kathiresan (2013) using a nanoprecipitation technique to improve its bioavailability and wound-healing effects. The wound-healing capacity of the curcumin nanosuspension was evaluated *in vivo* in male Wistar albino rats, wherein curcumin nanosuspension showed greater wound contraction (34.5%) compared with plain ethanolic solution of curcumin (28%) after treatment for three days. Nanocrystals of curcumin didecanoate (CurDD), a decanoate ester prodrug of curcumin used to achieve prolonged drug delivery, were developed by Wei et al. (2013). The nanocrystals were prepared by wet ball milling, and the freeze-dried nanocrystals were readily dispersed in peanut oil to form stable nanosuspensions. The pharmacokinetics of oily nano- and microsuspensions after intramuscular administration to rats was compared. The nanosuspension exhibited slower clearance from the injection site after intramuscular injection compared with the microsuspension. A considerably higher maximum plasma curcumin concentration (69.0 ng/mL) was observed for the nanosuspension compared with the microsuspension (18.5 ng/mL), with levels of drug maintained in plasma and brain for about 15 days in the case of the nanosuspension.

Berberine, an isoquinoline alkaloid found in many plant species, including *Berberis aristata*, *Hydrastis canadensis*, and *Phellodendri chinensis*, is reported to have antidiabetic activity. The poor absorption and low bioavailability of berberine has restricted the full exploitation of its potential. Wang et al. (2015) formulated nanosuspension of berberine using d-α-tocopheryl polyethylene glycol 1000 succinate (TPGS) by HPH. Berberine nanosuspension (50 mg/kg) produced significantly greater reduction in blood glucose and total cholesterol levels, when compared with an equivalent dose of plain berberine and metformin (300 mg/kg) after oral administration in streptozotocin-induced diabetic C57BL/6 mice.

Annonaceous acetogenins have been reported to possess potent antitumor activity, but their clinical application has been limited owing to their poor solubility. Hong et al. (2016) prepared a nanosuspension of annonaceous acetogenins by a precipitation-ultrasonication method using monomethoxypoly (ethylene glycol)2000–poly (e-caprolactone)2000 ($mPEG_{2000}$–PCL_{2000}) as a stabilizer. Cytotoxicity studies of the nanosuspension of annonaceous acetogenins and the plain drug solution carried out in 4T1, HeLa, and MCF-7 cells showed that at same concentrations of the drug, the nanosuspension exhibited greater cytotoxic effects (1.70, 2.12, and 2.65 times, respectively), when compared with the free drug. An *in vivo* antitumor activity study performed in 4T1 tumor-bearing mice revealed that the nanosuspension of annonaceous acetogenins showed higher therapeutic efficacy than the conventional dosage form (oil solution) at one-tenth of the dose (74.83% and 45.53% inhibition of the tumor, respectively, $p < 0.05$) (Table 6.3).

One of the most recent studies reported the preparation of nanosuspensions of four poorly soluble plant compounds (quercetin, rutin, resveratrol, and hesperidin) by the use of a novel ultrasmall

TABLE 6.3
Clinical Trial Status and Marketed Preparations of Phytomolecules Developed as Nanosuspensions and Micellar Formulations

Brand Names/Formulation Types	Phytoconstituents	Key Features/Therapeutic Indications
	Clinical Trials	
TLC388 (Polymeric micelles)	Lipotecan	Phase II ongoing for hepatocellular carcinoma and renal cell carcinoma
SinaCurcumin® (Nanomicelles)	Curcumin	Phase II ongoing for prostate cancer, ankylosing spondylitis, and multiple sclerosis
Vineatrol®30 as native powder or liquid micelles	Resveratrol	Phase I completed for establishing safety after oral intake
	Marketed Preparations	
Juvedical® (Nanosuspension)	Rutin	Cosmeceutical (antioxidant activity, skin renewal)
SinaCurcumin® (Nanomicelle)	Curcumin	Dietary supplement

bead-milling method with a batch size of about 5 mg. The method was demonstrated to be a fast technique to screen many nanocrystal stabilizers simultaneously (Liu et al., 2017).

6.2.4 Metallic Nanoparticles

Metallic nanoparticles are of considerable interest to researchers due to the incredible antimicrobial activities of metals such as iron, palladium, zinc, copper, gold, and silver. Various physical and chemical methods are used for the synthesis of metallic nanoparticles. Physical methods for preparation of metal nanoparticles include processes such as plasma arcing, ball milling, spray pyrolysis, and the ultrathin film-formation method, whereas chemical methods include processes such as soft-gel process, chemical-solution deposition, and chemical-vapor deposition. These methods are hazardous, expensive, and labor-intensive. Constituents of plant origin, due to their inherent reducing effects, are increasingly being used to transform inorganic metal ions into metal nanoparticles. To prepare metallic nanoparticles using a plant extract, the extract is mixed with a solution containing metal salt at room temperature. Reaction between the extract and metal completes within minutes, leading to the formation of metallic nanoparticles that are generally in the size range of 18 to 50 nm. The procedure for preparation of metallic nanoparticles using plant extracts is depicted in Figure 6.10.

Different metallic nanoparticles, such as gold, silver, zinc, palladium, and copper nanoparticles, have been synthesized using plant resources. Size and shape of metallic nanoparticles can be influenced by varying amounts of metal and plant parts (for example, the stem, root and callus) in the reaction medium. Preparation of metallic nanoparticles using reducing agents of plant origin offers various advantages such as low cost of production, safety, and ease of scale-up. Using plant extracts for preparation of metal nanoparticles allows faster synthesis, facilitates easy purification, and provides more flexible control over the size and shape of the particles, which can be manipulated by changing the pH or temperature of the reaction medium (Table 6.4).

6.2.4.1 Silver Nanoparticles

Ponarulselvam et al. (2012) have reported green synthesis of silver nanoparticles using aqueous extract of leaves of *Catharanthus roseus* Linn. G. Don. *C. roseus* is a plant which contains various active constituents like vincristine, vinblastine, and ajmalicine, which are responsible for potent *in vitro* antiplasmodial activity. Concentration-dependent *in vitro* antiplasmodial activity of the synthesized metal nanoparticles of *C. roseus* was observed with 20%, 42%, 60%, and 75% growth inhibition at 25, 50, 75, and 100 μg/mL, respectively.

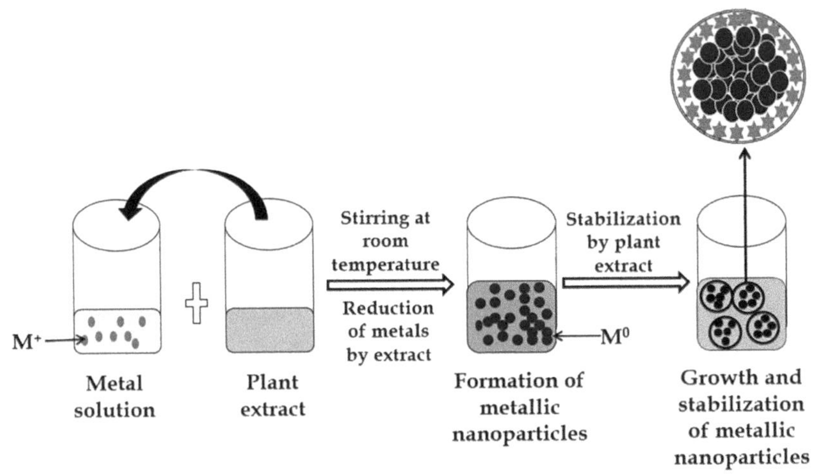

FIGURE 6.10 Procedure for preparation of metallic nanoparticles using plant extracts.

Ahmed et al. (2016) synthesized silver nanoparticles using aqueous extract of *Azadirachta indica* leaves. Formation of silver nanoparticles was confirmed by a change in color of the solution. They reported that the silver nanoparticles that were stabilized using *A. indica* extract showed appreciable *in vitro* antibacterial activity against both gram-positive and gram-negative organisms, whereas the extract alone did not exhibit any antibacterial activity against the tested organisms. It was reported that silver nanoparticles, due to their large surface area, provided better contact with the cell wall of the microorganism and showed increased antimicrobial activity. Hydroalcoholic extract of roasted *Coffea arabica* seeds was used to synthesize silver nanoparticles by Dhand et al. (2016). Formation of nanoparticles was visually confirmed by change in color from light to dark brown. The nanoparticles, thus formed, were spherical, highly crystalline in nature, had a cubic face-centered lattice, and exhibited remarkable antibacterial activity against *Escherichia coli* and *Staphylococcus aureus*, which was equal to the activity of ampicillin.

6.2.4.2 Gold Nanoparticles

Jin et al. (2015) leveraged the strong hydrophobic interactions of gelatin nanoparticles and developed a one-step method for synthesis of gold nanoparticles. Tannic acid, a polyphenol, was encapsulated into gelatin nanoparticles, and this was further used as a stabilizer for *in situ* synthesis of gold nanoparticles. Gold and silver nanoparticles, using root extract of Korean red ginseng, were prepared by Singh et al. (2016). The antimicrobial activities of gold and silver nanoparticles of red ginseng root were evaluated against pathogenic microorganisms such as *Vibrio cholera*, *S. aureus*, *Bacillus cereus*, and *Candida albicans*. The formulations showed appreciable *in vitro* antimicrobial activity against the organisms tested. The metal nanoparticles also exhibited biofilm degrading activity against *S. aureus* and *Pseudomonas aeruginosa*.

Gold nanoparticles have been reported to show synergistic cytotoxic activity in combination with plant extracts. Mishra et al. (2016) prepared gold nanoparticles using leaf and stem extracts of *Hibiscus sabdaiffa* and evaluated the cytotoxic potential in glioblastoma cells under hypoglycemic conditions. Hibiscus is reported to be effective as a chemopreventive agent, with substantial amounts of antioxidants. The authors studied the effects of hibiscus-stabilized gold nanoparticles in inhibiting the high glucose-stimulated migration of vascular smooth muscles, which signifies protection against diabetic vasculopathy. The cytotoxic potential in hypoglycemic conditions was studied in U87 glioblastoma cell lines and by molecular techniques such as 4′,6-diamidino-2-phenylindole (DAPI) staining, DNA degradation, and protein degradation analysis. The formulation was found to be nontoxic to 293 normal cell lines, whereas it was selectively toxic to the U87 cells. Various

TABLE 6.4
Metallic Nanoparticles Developed Using Plant Extracts

Plant Part/Extract	Mean Particle Size (nm)	Biological Activity Evaluated	Key Findings	Reference
		Silver Nanoparticles (AgNPs)		
Water extract of *Acorus calamus* rhizomes	31.83	Antioxidant activity, antibacterial activity against *B. subtilis*, *S. aureus*, and *B. cereus* and cytotoxic potential in HeLa cells and A549 cells	DPPH, superoxide scavenging and hydrogen peroxide scavenging assays revealed significant free-radical quenching ability of the AgNPs. The AgNPs showed significantly greater antibacterial activity than the plain extract against the tested organisms by disc-diffusion assay. IC50 for AgNPs by MTT assay at 48 h were 69.44 and 32.1 μg/mL in HeLa and A549 cells, respectively. It was confirmed by staining methods (such as Annexin staining) that the AgNPs induced apoptotic cell death.	Nakkala et al. (2014)
Ethanolic extract of *Onosma dichroantha* roots	5–65	Antimicrobial activity against *E. coli*, and *S. aureus*	The AgNPs showed higher antimicrobial activity against *E. coli* than against *S. aureus*.	Nezamdoost et al. (2014)
Water extract of banana peels	23.7	Antimicrobial activity against organisms such as *E. coli*, *S. aureus*, and *P. aeruginosa*	AgNPs demonstrated improved antibacterial activity compared with silver salt against the tested organisms by well-diffusion assay. The nanoparticles also showed synergistic antibacterial activity with levofloxacin against the tested organisms.	Ibrahim (2015)
Water extract of *Rosmarinus officinalis* L. leaves	33	Antimicrobial activity against *E. coli*, *B. subtilis*, and *S. aureus*	AgNPs inhibited growth of all tested organisms.	Ghaedi et al. (2015)
Water extract of *Cucurbita maxima* petals	76	Antimicrobial activity against *V. cholerae*, *P. aeruginosa*, *B. subtilis*, and *E. coli* and cytotoxicity in epidermoid A431 carcinoma cell lines	AgNPs showed significantly improved antimicrobial activity against all the tested organisms compared with the plain extract, by well-diffusion assay. It was observed that the formulations showed significant cytotoxicity in A431 cells, with the AgNPs synthesized using *M. oleifera* showing the highest activity followed by those prepared using *C. maxima* and *A. calamus*.	Nayak et al. (2015)
Water extract of *Moringa oleifera* leaves	94			
Water extract of *Acorus calamus* rhizomes	59			

(Continued)

TABLE 6.4 (*Continued*)
Metallic Nanoparticles Developed Using Plant Extracts

Plant Part/Extract	Mean Particle Size (nm)	Biological Activity Evaluated	Key Findings	Reference
Water extract of stem of *Tephrosia tinctoria*	<100	*In vitro* antidiabetic activity	Greater free radical scavenging ability, inhibition of digestive enzymes (α-glucosidase and α-amylase), and enhancement of *in vitro* glucose uptake rate in red blood cells (RBCs) was observed in the case of AgNPs compared with the plain extract.	Rajaram et al. (2015)
Methanol extract of *Leptadenia reticulate* leaves	50–70	Cytotoxic potential in HCT15 cancer cell lines	AgNPs demonstrated a dose-dependent cytotoxicity when evaluated by MTT assay and microscopic evaluation.	Swamy et al. (2015)
Water extract of *Fraxinus excelsior* leaves	25–40	Antioxidant activity	DPPH assay revealed that AgNPs showed greater antioxidant activity compared with plain extract.	Parveen et al. (2016)
Water extract of *Syzygium alternifolium* leaves	5–68	Antimicrobial activity against organisms like *Salmonella typhimurium*, *E. coli*, *Proteus vulgaris*, and *P. aeruginosa*	AgNPs caused significant growth inhibition in the case of all the microorganisms when evaluated by disc-diffusion method.	Yugandhar and Savithramma (2016)
Water extract of *Artocarpus altilis* leaves	34–38	Antioxidant activity and antibacterial activity against *E. coli* and *P. aeruginosa* and antifungal activity against *Aspergillus versicolor*	Significant antibacterial activity of AgNPs was observed when evaluated by zone of inhibition method. The antioxidant activity of AgNPs was comparable to that of ascorbic acid, when evaluated by DPPH assay.	Ravichandran et al. (2016)
Ethanolic and methanolic extracts of *Pelargonium endlicherianum* fenzl.	~35 for M1-AgNPs and ~17–50 for M2-AgNPs	Antimicrobial activity of AgNPs containing 11% ethanolic extract and 70% methanolic extract of *P. endlicherianum*	Antimicrobial activity of M1-AgNPs (11% ethanolic extract containing gallic acid and apocyanin) and M2-AgNPs (70% methanolic extract containing gallic acid, apocyanin, and quercetin) against *P. aeruginosa*, *E. coli*, and *Staphylococcus epidermis* was evaluated by microbroth-dilution assay, and penicillin was used as the standard drug. M2-AgNPs showed better activity against *S. epidermis* compared with M1-AgNPs as well as penicillin; whereas, both the formulations showed identical MIC values against *P. aeruginosa* and *E. coli*. The authors suggested that presence of three identified antimicrobial molecules in M2-AgNPs may be the probable reason for the improved antimicrobial potential against *S. epidermis* compared with M1-AgNPs.	Karatoprak et al. (2017)

(*Continued*)

TABLE 6.4 (*Continued*)
Metallic Nanoparticles Developed Using Plant Extracts

Plant Part/Extract	Mean Particle Size (nm)	Biological Activity Evaluated	Key Findings	Reference
Aqueous extract of leaves of *Tragia involucrate* (TIA)	47	Anti-urolithiatic potential of TIA and TIA-AgNPs.	*In vitro* anti-urolithiatic activity was evaluated by struvite crystal growth inhibitory assay, after growing crystals using sodium metasilicate. The results indicated that highest doses of TIA (2%) and TIA-AgNPs (200 μg/mL) showed significant inhibitory effect, with TIA-AgNPs showing better activity compared with TIA. *In vivo* anti-urolithiatic activity was evaluated in rats after inducing calcium oxalate stone formation by oral administration of ethylene glycol in water. The authors reported that treatment with TIA and TIA-AgNPs significantly reduced urinary calcium and phosphorus contents to a level which was comparable to the standard treatment group receiving cystone ($P < 0.001$), thus indicating that there was a reduction in stone formation in rats. The group of rats receiving TIA-AgNPs showed better results compared with the group treated with TIA alone. Treatment with both TIA and TIA-AgNPs restored magnesium levels to normal, with TIA-AgNPs showing better activity compared with TIA.	Velu et al. (2017)
Water extract of yellow bell pepper (YPE) and quercetin (Q)	16.25 for YPE-AgNPs and 13.9 for Q-AgNPs	Antibacterial and antibiofilm activity against ESβL(+) *E. coli*, ESβL(+) *P. aeruginosa* and methicillin sensitive and resistant *S. aureus*	Plain drugs *viz* YPE (1:8 v/v) and Q (50 μM) did not show any inhibition in growth of the tested strains of microorganisms. The zone of inhibition varied significantly between Q-AgNPs and YPE-AgNPs. The authors reported that gram-negative bacteria were more sensitive to both AgNPs compared with gram-positive strains. Increased production of ROS and membrane permeability was reported after staining isolates with fluorescent probes. AgNPs hampered the EPS production, endorsed DNA leakage, and generated superoxide radicals. A consistent decrease in bacterial growth was observed when time- and concentration-dependant growth inhibition assays were conducted. Both the AgNPs showed dose-dependent inhibition of biofilm formation and also demonstrated structural changes such as distortion, irregular margins, and shrinkage of cells when observed under SEM.	Ahmed et al. (2018)

(*Continued*)

TABLE 6.4 (*Continued*)
Metallic Nanoparticles Developed Using Plant Extracts

Plant Part/Extract	Mean Particle Size (nm)	Biological Activity Evaluated	Key Findings	Reference
Water extract of garlic, green tea, and turmeric	6.13–8.46	Evaluation of antioxidant and anticancer activity of chemically synthesized AgNPs (S1), AgNPs synthesized using garlic (S2), green tea (S3), and turmeric (S4) extracts	Green-synthesized AgNPs showed comparable antioxidant activity with respect to standard drugs ascorbic acid and rutin, whereas, the batch of AgNPs that was chemically synthesized exhibited lower antioxidant activity. The order of antioxidant activity by all the methods (DPPH assay and inhibition of p-NDA[a] bleaching activity) used for evaluation was as follows: S4>S2>S3>S1. The authors reported that S4-AgNPs showed most potent antioxidant activity due to presence of compounds such as cineole and borneol, in addition to phenolic compounds such as curcumin. Cytotoxicity of the AgNPs was further evaluated in human breast adenocarcinoma (MCF-7), cervical cancer (HeLa), epithelioma (Hep-2), and lung carcinoma (A549) cell lines along with noncancerous human dermal fibroblasts (NHDF) cell lines by MTT assay. All the AgNPs showed concentration-dependent cell killing in the cancerous cell lines and did not show any toxicity in the normal cell lines, with S4-AgNPs showing better activity compared with all other batches. The authors reported that all the green-synthesized AgNPs showed lower cytotoxicitycompared with the standard anticancer drug cisplatin. Apoptotic activity of synthesized AgNPs was evaluated in A549 cells by Hoechst staining assay in which cytological changes such as cell shrinkage and fragmentation was observed, indicating that green-synthesized AgNPs exhibited significant activity in killing the cells by apoptosis.	Selvan et al. (2018)

(*Continued*)

TABLE 6.4 (*Continued*)
Metallic Nanoparticles Developed Using Plant Extracts

Plant Part/Extract	Mean Particle Size (nm)	Biological Activity Evaluated	Key Findings	Reference
Water and methanol (1:1) extract of dried bark of *Eysenhardtia polystachya* (EP)	12 ± 0.15	Antioxidant and antidiabetic activity	Protective effects of EP-AgNPs against hydrogen peroxide-induced oxidative injury was evaluated in INS-I cells. It was reported that treatment of the cells with EP-AgNPs before exposure to hydrogen peroxide produced a marked increase in insulin secretion compared with the untreated cells. Further studies were carried out in hyperglycemia-induced zebrafish to confirm the antidiabetic activity of EP-AgNPs. The authors observed that treatment with 5 and 10 μg/mL EP-AgNPs in diabetes-induced zebrafish resulted in 38.1% and 44.5% reduction in blood glucose levels, respectively. Levels of total cholesterol (40%) and triglycerides (29.4%) in diabetes-induced zebrafish markedly decreased after treatment with EP-AgNPs compared with untreated controls, thus indicating that EP-AgNPs ameliorated dyslipidemia, reversed insulin resistance, and thus improved lipid metabolism.	Campoy et al. (2018)
Gold Nanoparticles (AuNPs)				
Water extract of aerial parts of *Salcorina brachiata*	25–35	Antibacterial activity against *P. aeruginosa*, *E. coli*, *Salmonella typhi*, and *S. aureus*	The AuNPs showed synergistic antibacterial activity with ofloxacin.	Ahmed et al. (2014)
Water extract of *Vetiveria zizanioides* roots	2.99	Antifungal activity against *Penicillium* spp., *Aspergillus flavus*, *A. fumigatus*, and *A. fusarium*	AuNPs prepared using the plant extract showed greater antimicrobial activity (by disc-diffusion method) compared with the AuNPs prepared using synthetic reducing agents such as sodium borohydride.	Swain et al. (2016)
Water extract of *Cannabis sativa* leaves	2.11			
Water extract of *Catharanthus roseus*	2–20	Cytotoxicity in MCF-7 and HepG2 cell lines and antimicrobial activity against *S. aureus*, *B. subtilis*, *E. coli*, and *P. vulgaris*	AuNPs prepared using a combination of both the extracts demonstrated greater antibacterial activity and cytotoxicity (dose-dependent) compared with those prepared using individual extracts.	Muthukumar et al. (2016)
Water extract of *Carica papaya*	3–9			
Combination of both extracts	6–18			

(*Continued*)

TABLE 6.4 (*Continued*)
Metallic Nanoparticles Developed Using Plant Extracts

Plant Part/Extract	Mean Particle Size (nm)	Biological Activity Evaluated	Key Findings	Reference
Water extract of *Commelina nudiflora* (both AuNPs and AgNPs were prepared)	24–150 for both AuNPs and AgNPs	Anticancer activity in human colon cancer HCT-116 cells	Both AuNPs and AgNPs showed significant cytotoxicity in HCT-116 cells. It was confirmed by RT-PCR that pro-apoptotic genes such as caspase-3 and caspase-8 were up-regulated in cells treated with the nanoparticulate formulations.	Kuppusamy et al. (2016)
Water extract of *Dendropanax morbifera* leaves (both AuNPs and AgNPs were prepared)	100–150 (AgNPs) and 10–20 (AuNPs)	Cytotoxic potential in A549 human lung cancer cell lines and HaCat human keratinocyte cells	AgNPs exhibited lower cytotoxic potential in HaCat cell lines, whereas, it showed potent cytotoxicity in A549 cells. AuNPs did not show cytotoxicity in both the cell lines. Synergistic activity between the synthesized nanoparticles and ginsenoside compound K was evaluated in A549 cells, and it was reported that both formulations enhanced the cytotoxicity of ginsenoside compound K.	Wang et al. (2016)
Water extract of *Cassia tora* leaves	57	Antibacterial activity and anticancer activity in Col320 colon cancer cell lines	The nanoparticulate formulation and plain extract did not show antibacterial activity. AuNPs showed reduction in cell viability by MTT assay and also suppressed the release of hydrogen peroxide, lipoperoxidase. and nitric oxide in a dose-dependent manner.	Abel et al. (2016)
Water extract of *Sargassum glaucescens*	3.65	Anticancer activity in cervical (HeLa), breast (MDA-MB-231), and leukemia (CEM-ss) cancer cell lines	AuNPs demonstrated significant anticancer activity in HeLa, HepG2, CEM-ss, and MDA-MB-231 cells, whereas, it did not show any cytotoxicity in normal human mammary epithelial cells (MCF-10A cells). The AuNPs also activated caspase-3 and caspase-9 activities and arrested HeLa cell cycle at G2/M phase.	Ajdari et al. (2016)
Water extract of *Spinacia oleracea* leaves (both AuNPs and AgNPs were prepared)	2–20 (AgNPs) 5–50 (AuNps)	Anticancer activity in C_2C_{12} mouse myoblast cancer cells and zebrafish embryo toxicity studies	Both the formulations demonstrated cytotoxicity in C_2C_{12} cells at concentration as low as 5 μg/mL. Significant up-regulation in caspase enzymes compared with plain extract in both AgNPs and AuNPs. AgNPs showed 100% mortality in zebrafish embryo toxicity studies at much lower concentrations compared with AuNPs.	Ramachandran et al. (2017)
Water extract of *Elettaria cardamomum* seeds	15.2 nm	Antioxidant activity, antibacterial activity against *S. aureus*, *E. coli*, and *P. aeruginosa* and cytotoxic activity inHeLa cell lines	Highest antibacterial activity was observed against *S. aureus*, followed by *E. coli* and *P. aeruginosa*. Significant cytotoxic activity of the formulation was observed with changes in morphology, such as rounding, shrinking, and granulation in the cytoplasm.	Rajan et al. (2017)

(*Continued*)

TABLE 6.4 (*Continued*)
Metallic Nanoparticles Developed Using Plant Extracts

Plant Part/Extract	Mean Particle Size (nm)	Biological Activity Evaluated	Key Findings	Reference
		Zinc Oxide Nanoparticles (ZnO-NPs)		
Water extracts of *Azadirachta indica* (AI), *Hibiscus rosa-sinensis* (HR), *Murraya koenigii* (MK), *Moringa oleifera* (MO), and *Tamarindus indica* (TI)	25–35	Antioxidant and antidiabetic activity of chemically and green-synthesized zinc oxide nanoparticles (ZnO-NPs)	Different batches of ZnO-NPs were prepared using different plant extracts, and one batch of ZnO-NPs was chemically synthesized. All the green synthesized ZnO-NPs showed significant antioxidant activity when compared with standard antioxidants ascorbic acid and rutin. The reported antioxidant activity for ZnO-NPs was in the following order: ZnO-NP (TI)>ZnO-NP (MO)>ZnO-NP (HR)>ZnO-NP (MK)>ZnO-NP (AI)>ZnO-NP- chemically synthesized. ZnO-NPs prepared using TI and MO showed higher antioxidant activity as compared to other samples due to presence of additional phytoconstituents such as proteins and amino acids. The antidiabetic activity of ZnO-NPs was evaluated by determining effects of the formulations on α-amylase and α-glucosidase. ZnO-NPs prepared using TI showed higher α-amylase (concentration-dependant activity) and α-glucosidase inhibition activity compared with other green-synthesized ZnO-NPs. Chemically synthesized ZnO-NPs showed the least antidiabetic activity.	Rehana et al. (2017)
Water extract of *Andrographis paniculata*	96–115 by SEM 57 ± 0.23 by TEM	Antioxidant, anti-inflammatory, and antidiabetic activities	Both the plant extract as well as green-synthesized AgNPs showed similar antioxidant activity when evaluated by DPPH assay. The AgNP formulation showed lower antidiabetic activity compared with the plain extract when α-amylase inhibition activity was evaluated.	Rajakumar et al. (2018)
Water extract of *Sargassum muticum* algae (SM)		Cytotoxicity in human liver cancer cell lines (HepG2)	The authors reported that at concentrations above 175 μg/mL, viability of HepG2 cells treated with SM-ZnO-NPs was less than 40%. The inhibition in growth of HepG2 cells followed a time- and concentration-dependent profile. The authors stained the cells with V-FITC/propidium iodide dyes and carried out flow cytometric assays to evaluate the apoptotic activity of the formulation. The results indicated that higher concentrations of SM-ZNO-NPs increased apoptotic cells. Angiogenesis assay revealed that the formulation showed no significant angiogenic effects at 150 μg/mL compared with the control group.	Sanaeimehr et al. (2018)

[a] p-NDA: p-nitrosodimethylaniline.

TABLE 6.5
Patents, Clinical Trial Status, and Marketed Preparations of Phytomolecules Developed as Polymeric Nanoparticles

Brand Names	Excipients	Plant Extracts/ Phytoconstituents	Key Features/Therapeutic Indications	Reference
		Patents		
—	Polymeric nanoparticles prepared using polymers (PLGA and PLGA-PEG), fatty alcohol (cetyl alcohol and octanol), amphiphilic compounds (phospholipids), and conjugated with targeting ligands (aptamer and transferrin)	Vincristine and vinorelbine	Prostate cancer, breast cancer, or NSCLC	Zale et al. (2016) (US20160338963 A1)
		Clinical Trials		
Abraxane® (nab-paclitaxel)	Human serum albumin and anti-vascular endothelial growth factor (anti-VEGF) antibody	Paclitaxel	Phase I completed for melanoma liver metastasis Phase II ongoing for prolonged therapy of metastatic breast cancer Phase III ongoing for tubular breast cancer stage II, mucinous breast cancer stage II, invasive ductal breast cancer, tubular breast cancer stage III, HER-2 positive breast cancer, and inflammatory breast cancer stage IV	ClinicalTrials.gov
		Marketed Preparations		
Genexol® PM injection	PEG-b-PLA	Paclitaxel	Breast cancer, NSCLC, and ovarian cancer	–
Abraxane® (nab-paclitaxel)	Human serum albumin and anti-VEGF antibody	Paclitaxel	Metastatic breast cancer (in combination with gemcitabine), metastatic pancreatic cancer, and advanced NSCLC (in combination with carboplatin)	

nanoformulations of phytoconstituents either in different stages of clinical trials or marketed are enlisted in Table 6.5.

6.2.5 Mesoporous Silica Nanoparticles (MSNPs)

Silica is of great interest in pharmaceutical and medical science due to its chemically inert, stable, optically transparent, and biocompatible nature. Surface functionalization of silica is a widely used procedure for coating. Mesoporous silica coated with amino acids is generally used for stabilization of phenolic and carboxylic group-containing compounds by hydrogen-bond interaction.

Ordered MSNPs, hollow/rattle-type MSNPs, and functionalized MSNPs have been reported for drug-delivery applications. The ordered MSNPs are prepared by two techniques. In the first technique, liquid crystal templating, high concentrations of surfactant form a lyotropic liquid-crystalline phase under the reaction conditions (temperature, pH) and in the absence of any precursor inorganic framework material, whereas the second technique involves self-assembly of inorganic silica species and supramolecular surfactants, leading to the formation of a lyotropic phase even at low surfactant concentrations. Hollow/rattle-type MSNPs having large voids inside shells, which possess empty channels (mesopores), are prepared by the soft-template method, selective etching strategy, and self-template method. The soft template method involves fabrication of a template using multisurfactants, micelles, microemulsion droplets, or vesicular structures to build a mesoporous shell and a hollow interior simultaneously. Rattle-type silica nanoparticles can be formed after the surfactants are removed. Disadvantages of this method include lack of good dispersity, less control over particle size, and difficulty in large-scale synthesis, as large amounts of surfactant(s) are required. Selective etching strategy involves use of varying compositions of silylated organic compounds or organosilane precursors to give MSNPs of different stabilities with different etching agents such as hydrofluoric acid and sodium borohydride at exceptional temperature and pH conditions.

Ease of scalability, controllable size range, and cost-effectiveness of this method make it advantageous. In the self-template method, surface dissolution of the precursor is followed by the nucleation of the second phase on the surface of the particles formed initially. A solid silica sphere is transformed to a rattle-type structure, which is then converted to a hollow structure by treatment with an etching agent or acid under different pH and temperature conditions, while using surface protectors such as PVP. Advantages of this method are that it is a template-free synthesis and no surfactant removal step is involved. The limitations associated with the conventional mesoporous nanoparticles—lack of targeted drug delivery and reduced tracking efficiency in diseased conditions (when used as an imaging agent)—can be overcome by preparation of functionalized MSNPs using organic and inorganic functional moieties by methods such as grafting, co-condensation, imprint coating and production of periodic mesoporous organosilicas. Functional groups such as chloropropyl, phenyl, benzyl, mercaptopropyl, cyanopropyl, and butyl can be introduced on either the exterior particle or interior pore surfaces for achieving controlled and targeted drug delivery.

Postsynthesis grafting is reported to be useful for functionalizing MSNPs. Grafting involves modification of inner surfaces of mesostructured silica phases by organosilanes or chlorosilanes, which requires a large number of free silanol groups after the removal of surfactant. It has the stellar advantage that the mesostructure of the silica phase is retained under the synthetic conditions used. In the co-condensation technique, the tetraalkoxysilanes and terminal trialkoxyorganosilanes are condensed in the presence of structure-directing agents, leading to the formation of mesospheres with organic residues anchored covalently to the pore walls. The organic moieties in co-condensation technique are more homogenously distributed than that in the grafting process. The disadvantages of this technique include small pore diameter and pore volume and reduction in surface area.

Arriagada et al. (2016) reported the preparation of MSNPs of morin, a phenolic compound present in several plants and vegetables. Morin is reported to have several therapeutic effects, though its use is restricted due to poor water solubility and sensitivity to environmental conditions such as light. Amino acid–modified MSNPs of morin were prepared, and antioxidant activity was evaluated by monitoring singlet oxygen deactivation and hydroxyl radical scavenging. The nanoparticles were found to have an appreciable antioxidant effect and morin behaved as an antioxidant against the hydroxyl radical generated *in situ*. Morin-loaded nanoparticles showed synergistic antioxidant activity with the silica nanoparticles, which increased with increases in concentration of morin adsorbed. Quercetin-loaded silica nanoparticles were prepared by Lee et al. (2016) using a oil-in-water microemulsion method. Spherical and monodispersed nanoparticles were formed that were highly stable in aqueous systems. The antioxidant and anti-inflammatory activities of quercetin-loaded nanoparticles were evaluated and the results were compared with those of quercetin. Superoxide scavenging effects of the nanoparticles were significantly higher than the untreated controls. The nanoparticles significantly down-regulated the levels of proinflammatory cytokines released by macrophage cells, such as IL-6 and TNF-α (Table 6.6).

6.2.6 Hydrogels

Hydrogels are cross-linked polymeric structures that are produced by reaction between one or two monomers due to interactions such as ionic forces or hydrogen bonds. Hydrogels have the ability to swell and retain a significant amount of water, but do not dissolve in water. Due to their high water content, they have a potential role in various biomedical applications. Various diblock, triblock, or pentablock copolymers are utilized in the formation of these hydrogels, and the formation is triggered by external factors such as temperature, pH, or ions. Hydrogels can be prepared by using either synthetic or natural polymers. Synthetic polymers are hydrophobic in nature and are chemically stronger compared with natural polymers. Techniques such as bulk, solution, and suspension polymerization have been reported for preparation of hydrogels. Monomer, initiator, and cross-linker form the three integral components in formulating these hydrogel formations.

Bulk polymerization is one of the simplest techniques used for the preparation of hydrogels and involves the polymerization of monomers to form a glassy, transparent hydrogel matrix, which is very hard. The hard matrix, after immersion in water, swells to become soft and flexible. In solution polymerization, ionic or neutral monomers are mixed with a multifunctional cross-linking agent, and the polymerization is initiated by ultraviolet irradiation or redox reaction. Hydrogel formation occurs due to phase separation, and the formed hydrogels are purified to remove traces of monomers, oligomers, and initiators. In suspension polymerization, monomers and initiators are dispersed in the hydrocarbon phase as a homogenous mixture, followed by agitation that leads to formation of hydrogels.

Liposome-in-hydrogel complex systems are one of the latest techniques attempted for improving skin permeation of water-insoluble compounds. Hydrogels temporarily disrupt the skin barrier by hydrating the skin and, at the same time, exert a shielding function on the epithelial cells, thereby preventing evaporation of water from the skin and protecting it from the external environment. Liposomes add on the beneficial effects of the hydrogel by reinforcing the intracellular lipid (which may disappear when the skin barrier is disrupted) and improving sequestration of drug in the stratum corneum. Park et al. (2013) developed a two-step delivery system to enhance transdermal permeation of quercetin and rutin by preparing a liposome-in-hydrogel complex. The complexes were prepared by incorporating ceramide liposomes into cellulose hydrogel. It was observed that rutin was encapsulated to a greater extent as compared to quercetin, but the skin permeability of quercetin from the complex was greater than that of rutin. Skin permeability of both rutin and quercetin, when loaded in the liposome-in-hydrogel complex, was significantly higher compared with permeability of the drugs from single systems of hydrogels or liposomes. A similar study has been reported by Kim et al. (2014) for improving the skin permeation of liquiritigenin and liquiritin. Greater skin permeation of liquiritigenin and liquiritin (56.55% and 66.99%, respectively) was observed from the liposome-in-hydrogel complex systems compared with that from single liposomal (43.34% for liquiritigenin and 48.97% for liquiritin) and hydrogel systems (38.21% for liquiritigenin and 55.07% for liquiritin).

Hydrogels have been evaluated as alternative vehicles for delivery of drugs with anticancer and anti-inflammatory activities. Teong et al. (2015) formulated a curcumin-loaded hydrogel using biopolymeric chitosan, gelatin, and hyaluronan and evaluated its anticancer activity. Stability of curcumin was significantly greater in hydrogels containing hyaluronan and gelatin. All the hydrogels showed greater *in vitro* apoptosis activity in A549 cells compared with pure curcumin. Zhang et al. (2015a) prepared a composite system of polymeric micelles in a thermosensitive hydrogel system for encapsulation of quercetin to treat colorectal peritoneal carcinomatosis. The nanoparticulate

TABLE 6.6
Phytomedicines Based on Mesoporous Silica Nanoparticles, Hydrogels, and SNEDDS

Plant Extract/ Phytoconstituent	Particle Size (nm)	Parameters Evaluated	Methods Used for Evaluation of the Parameter	Reference
Phytomedicines Based on MSNPs				
Paclitaxel	100 (pore size = 2.3 nm)	Anticancer activity	*In vivo* activity of the formulation, after intraperitoneal administration, was evaluated in mice bearing human pancreatic carcinoma MIA PaCa-2. Compared with a Cremophor®-containing formulation of paclitaxel, the MSNPs exhibited increased accumulation of the formulation at the tumor site and slower absorption of the drug into systemic circulation. Tissue-distribution imaging showed significantly greater accumulation of the MSNPs in the tumor tissues compared with other peritoneal tissues.	Fu et al. (2016)
Irinotecan	80 (pore volume = 0.7 cm^3/g)	Anticancer activity	Lipid-coated MSNPs were developed to reduce the toxicity (bone marrow and gastrointestinal tract) of irinotecan, and their efficacy was evaluated in orthotopically implanted, pancreatic ductal adenocarcinoma–bearing mice. MSNPs increased the drug concentrations at tumor sites compared with liposomal formulation of the drug, with subsequent reduction in toxicity of the drug and improvement in efficiency for treatment of tumor metastases.	Liu et al. (2016)
Curcumin	100–200 (pore size = 10 nm)	Oral bioavailability	The potential of amine-functionalized-MSNPs and amine-functionalized-mesoporous silica microparticles (MSMs) loaded with curcumin in improving bioavailability of the drug was investigated after oral administration in mice. Maximum plasma concentrations were observed in the group administered with MSNPs followed by MSMs and free curcumin.	Hartono et al. (2016)
Curcumin	150–200 (pore volume = 0.9 cm^3/g)	Anticancer activity	Encapsulating resveratrol in colloidal MSNPs enhanced its saturation solubility and *in vitro* release compared with pure resveratrol. The formulation showed increased cytotoxicity in HT-29 and LST147T colon cancer cell lines compared with resveratrol, mediated via the PARP and cIAP1 pathways.	Summerlin et al. (2016)
Quercetin	80–120	Anticancer activity	Folic acid-conjugated MSNPs loaded with quercetin were developed and their cytotoxicity in breast cancer cells was evaluated. Compared to the unconjugated MSNPs, the conjugated formulation showed higher uptake and cytotoxicity in folate receptor–overexpressing MDAMB-231 cells. Folic acid MSNPs caused cell-cycle arrest and apoptosis through regulation of AKT and Bax signalling pathways.	Sarkar et al. (2016)

(*Continued*)

TABLE 6.6 (*Continued*)
Phytomedicines Based on Mesoporous Silica Nanoparticles, Hydrogels, and SNEDDS

Plant Extract/ Phytoconstituent	Particle Size (nm)	Parameters Evaluated	Methods Used for Evaluation of the Parameter	Reference
Paclitaxel	200 (pore size = 5.79 nm)	Pharmacokinetics and anticancer activity	Core-shell MSNPs loaded with paclitaxel were developed for pulmonary delivery of the drug to treat lung cancer. *In vitro* and *in vivo* studies demonstrated that MSNPs improved the lung absorption of paclitaxel. The AUC after pulmonary delivery of the formulation in rabbits was 2.678-fold higher than pure paclitaxel, with no signs of lung inflammation. The MSNPs exhibited time-dependent cellular uptake and concentration-dependent antiproliferation effects in A549 cells.	Wang et al. (2017)
Curcumin	Mean pore diameter = 3.28 nm	Anticancer activity	pH responsive chitosan MSNPs containing curcumin were prepared to increase the solubility and improve the anticancer activity of curcumin. Anticancer activity of curcumin alone and the formulation was evaluated in U87MG glioblastoma cell lines. Results of cytotoxicity studies revealed that encapsulation of curcumin in chitosan MSNPs improved the cytotoxicity of curcumin, which may be due to enhanced accumulation of curcumin within the cancer cells.	Ahmadi et al. (2018)
Paclitaxel and curcumin	115 ± 15 (pore diameter = 2.754 nm)	Cytotoxic activity	PEGylated lipid bilayer coated MSNPs were prepared for co-delivery of paclitaxel and curcumin, and their cytotoxic effects in canine breast cancer cells was evaluated. The IC_{50} value for plain paclitaxel and curcumin were reported to be 0.19 and 53.16 μM, respectively. The authors reported that the PEGylated mesoporous silica nanoparticulate formulation of paclitaxel and curcumin exhibited the strongest cytotoxic effect in the breast cancer cell lines.	Lin et al. (2018)

(*Continued*)

TABLE 6.6 (*Continued*)
Phytomedicines Based on Mesoporous Silica Nanoparticles, Hydrogels, and SNEDDS

Plant Extract/ Phytoconstituent	Particle Size (nm)	Parameters Evaluated	Methods Used for Evaluation of the Parameter	Reference
Curcumin	Xanthan-galactomannan	*In vitro* cutaneous irritation analysis and anti-inflammatory activity	No cutaneous toxicity was observed for the hydrogel formulation as well as free curcumin. *In vivo* evaluation of anti-inflammatory activity in croton-oil induced mouse ear edema revealed that SNEDDS reduced the edema by 63.2%–76.8%, depending on the composition of the formulation.	Koop et al. (2015)
Ethanolic extract of Amaranth leaf extract, mixed with quartz sand and $CaCO_3$ powder	Reduced graphene oxide (rGO)-Amaranth extract (AE)-AuNPs composite	Antitumor activity	Amaranth extract was used in the formulation as a stabilizer and photosensitizer. Compared with rGo-AE, the rGO-AE-AuNPs composite showed improved *in vitro* antitumor activity in HeLa cells.	Chang et al. (2015)
Resveratrol	Niosome-in-Carbopol 934 hydrogel	Skin permeation, drug retention, and anti-inflammatory activity	When evaluated by Franz diffusion method, the niosomal gel showed greater permeation of resveratrol compared with the plain drug solution. The formulation showed reduction in edema and prolonged the therapeutic action in carrageenan-induced rat paw edema model.	Negi et al. (2017)
Water extract of *Hippophae rhamnoides* L. leaves	PVA-pectin	Wound-healing effect	Evaluation of wound healing properties in rats induced with excision injuries revealed that the recovery rate was higher in the animals treated with hydrogel formulation compared with those treated with the plain extract. The wound-healing effects were further confirmed by histological evaluation of the wound tissue.	Kim and Lee (2017)
Curcumin	PVA hydrogel	Antibacterial and wound-healing ability	Curcumin hydrogel with Curcumin to PVA volume ratio of 1:5 (20% Cur/PVA) exhibited potent antibacterial activity against *E. coli* and *S. aureus* when compared with other Curcumin-PVA hydrogel ratios and plain PVA hydrogel. Hydrogel with volume ratio of Cur to PVA 1:10 (10% Cur/PVA) demonstrated good wound healing in rats when compared with 20% Cur/PVA hydrogel and plain hydrogel.	He et al. (2018)

(*Continued*)

TABLE 6.6 (*Continued*)
Phytomedicines Based on Mesoporous Silica Nanoparticles, Hydrogels, and SNEDDS

Plant Extract/ Phytoconstituent	Excipients Used	Parameters Evaluated	Methods Used for Evaluation of the Parameter	Reference
		Phytomedicines Based on Hydrogels		
Curcumin	Hydrogel containing Curcumin-based supramolecular nanofiber (Cur-SNF)	Radiosensitization efficacy	Cur-SNFs exhibited enhanced performance as radiosensitizer to sensitize colorectal cancer cells to ionizing radiation when compared to free curcumin. Cur-SNFs in combination with radiation demonstrated significant reduction in tumor volume *in vivo*.	Xu et al. (2018)
Curcumin	Chitosan-β-glycerolphosphate disodium salt (CH/GP)-Curcumin hydrogel	Wound-healing and antibacterial activity	Wound appearance and wound closure rate in rats were enhanced in CH/GP-Curcumin hydrogel when compared with CH/GP and saline gauze. Also, the tissue bacterial count was much reduced in the case of CH/GP-Curcumin hydrogel when compared with CH/GP and saline gauze.	Zhao et al. (2018)
Curcumin	Curcumin nanoparticles (CNPs) enclosed in gelatin microspheres (GMs) loaded in thermosensitive hydrogels	Wound healing in diabetic animal models	The rate of wound contraction and re-epithelisation in STZ-induced diabetic mouse model was greater in CNPs-GMs/hydrogel treated animals when compared to Cur/hydrogel and hydrogel-treated animals.	Liu et al. (2018)
Curcumin	Injectable hydrogel-thiolated chitosan (TCS) and poly(ethylene glycol) diacrylate (PEGDA) with lysozyme	Enhanced drug loading and increased localized drug concentration with improved anticancer activity	Tumor volumes in male nude mice were significantly lower in animals treated with curcumin hydrogels with lysozyme compared with saline control group and those treated with free curcumin and curcumin hydrogel without lysozyme.	Ning et al. (2018)
Quercetin	Quercetin loaded liposomes-hydrogels	Wound healing	Quercetin-loaded liposomal hydrogels promoted accelerated wound healing when compared with plain liposome-hydrogel film.	Jangde et al. (2018)
Quercetin	Quercetin loaded NLCs-hydrogel composed of carboxymethyl chitosan and poloxamer	For ophthalmic delivery	Temperature and pH responsive hydrogel of quercetin exhibited sustained release with 80.52% of total quercetin being released within 3 days.	Yu et al. (2018)

(*Continued*)

TABLE 6.6 (*Continued*)
Phytomedicines Based on Mesoporous Silica Nanoparticles, Hydrogels, and SNEDDS

Plant Extract/ Phytoconstituent	Particle Size (nm)	Parameters Evaluated	Methods Used for Evaluation of the Parameter	Reference
Phytomedicines Based on Self Nanoemulsifying Drug-Delivery Systems (SNEDDS)				
Extract of *Tinospora cordifolia*	33.7–377	*In vitro* intestinal permeability	Formulating the plant extract in SNEDDS resulted in 3.95-fold enhancement of permeability coefficient compared with the control by everted gut sac method.	Kumar and Jaiswal (2014)
Naringenin	50	Oral bioavailability	*In vivo* studies in rats revealed that the absorption of naringenin from SNEDDS was significantly greater than that of plain naringenin ($p < .005$).	Khan et al. (2015)
Essential oils of *Piper cubeba*	7–52	Anti-inflammatory activity	When evaluated *in vivo* in rats, the pure *P. cubeba* essential oil (400 mg/kg), as well as SNEDDS formulation (40 and 80 mg/kg), showed significant reduction in pleural exudate volume and polymorphonuclear cells, with down-regulation in levels of myeloperoxidase, nitric oxide, and proinflammatory cytokines.	Shakeel et al. (2015)
Hydroalcoholic extract of *Berberis aristata*	145–147	*In vitro* antidiabetic activity	The formulation reduced *in vitro* glucose diffusion and the inhibition of hemoglobin glycosylation was found to be approximately 41% higher compared with the plain extract.	Kumar and Jaiswal (2016)
Curcumin (SNEDDS, NLCs and nanocapsules were prepared)	30–40	Efficacy in treatment of inflammatory bowel disease	*In vitro* permeability across Caco-2 monolayers revealed that nanocapsules showed highest (30-fold) increase in permeation of curcumin across the monolayers, followed by SNEDDS, NLCs, and curcumin suspension. SNEDDS and NLCs showed significant reduction in TNF-α in lipopolysaccharide-activated J774 macrophages. When evaluated *in vivo* in dextran sodium sulfate–induced colitis model in rats, only the NLCs were able to reduce the neutrophil infiltration, myeloperoxidase activity, and TNF-α secretion.	Beloqui et al. (2016)

(*Continued*)

TABLE 6.6 (*Continued*)
Phytomedicines Based on Mesoporous Silica Nanoparticles, Hydrogels, and SNEDDS

Plant Extract/ Phytoconstituent	Particle Size (nm)	Parameters Evaluated	Methods Used for Evaluation of the Parameter	Reference
Phytomedicines Based on Self Nanoemulsifying Drug-Delivery Systems (SNEDDS)				
Quercetin, resveratrol, and genistein	<200	Anticancer activity	SNEDDS demonstrated approximately 4.27-fold, 1.5-fold, and 2.8-fold increases in oral bioavailability of quercetin, resveratrol, and genistein, respectively. In comparison with quercetin-resveratrol-genistein solution, SNEDDS prolonged the growth of tumor in 7,12-dimethylbenzanthracene (DMBA)-induced breast cancer model in rats.	Tripathi et al. (2016)
Rutin and vitamin E	16.08	Amelioration of oxidative stress in Parkinson's disease model	Vitamin E is reported to have antioxidant activity, and it was used in the formulation to solubilize rutin. It was reported that the formulation improved the oral bioavailability of rutin by 2.3 times in rats. *In vivo* evaluation of efficacy in rats showing Parkinson's type symptoms revealed that the formulation containing rutin and vitamin E showed enhanced antioxidant activity and greater improvement in parameters such as locomotor activity and muscular coordination compared with rutin suspension, probably due to synergism between rutin and vitamin E.	Sharma et al. (2016)
Curcumin	11.2	Oral bioavailability	The SNEDDS formulation showed greater bioavailability of curcumin as compared to curcumin suspension in rats.	Nazari-Vanani et al. (2017)
Resveratrol and co-enzyme Q10	51–127	Antioxidant and anticancer activities	*In vitro* antioxidant activity was significantly greater in case of SNEDDS compared with solution containing resveratrol and coenzyme Q10. In DMBA-induced breast cancer model in rats, SNEDDS significantly reduced the tumor volume (~327 mm^3) compared with a solution containing the drugs (~1070 mm^3).	Jain et al. (2017)
Resveratrol and lignin	84.35 ± 2.32	Oral bioavailability	A 76.2-fold increase in bioavailability of resveratrol was observed in the case of SNEDDS when compared with resveratrol suspension in rats.	Dai et al. (2018)

formulation was a free-flowing solution at room temperature and got converted into a nonflowing gel at body temperature, serving as a drug depot. The formulation showed sustained *in vitro* release of curcumin and higher cytotoxicity compared with free curcumin and curcumin-polymeric micelles. The hydrogel showed higher AUC and prolonged biological $t_{1/2}$ compared with curcumin-polymeric micelles, after intraperitoneal administration in mice. It also reduced tumor growth, metastasis, and prolonged the survival time in colorectal peritoneal carcinomatosis-bearing mice.

Epigallocatechin galate–loaded *in situ* hydrogel implants were prepared using PLGA as the polymer (Zhang et al., 2015b). Antihyperlipidemic and antiobesity activities of the formulation, after subcutaneous administration for 30 days, were evaluated in mice fed a high-fat diet. After completion of the treatment period, the hydrogel formulation showed significant reduction in the body weight gain and also improved the high density lipoprotein-low density lipoprotein (HDL-LDL) ratio compared with the standard group treated with atorvastatin.

Fleck et al. (2016) prepared a hydrogel loaded with an ethanolic extract of *Punica granatum* peels, which is a potent anti-inflammatory agent. The hydrogel was prepared using a carbomer (i.e., Carbopol 980) and the stability of the bioactive components was found to be significantly increased, when the extract was loaded in hydrogels. The formulation was evaluated in a 76-year-old patient with a nonhealing ulcer on the leg, which was not reduced after following standard medical care, including the use of topical antimicrobial agents (e.g., neomycin, gentamycin, silver sulfadiazine, and clostebol), and topical corticosteroids (e.g., betamethasone dipropionate) for over a year. The *P. granatum*-hydrogel formulation was applied over the wounds and was then dressed with a cotton gauze. The size of the ulcer got reduced to one-quarter of its original size and was completely healed after six weeks (90 applications) of use of the hydrogel formulation.

6.2.7 Self Nanoemulsifying Drug-Delivery Systems

Self nanoemulsifying drug-delivery systems (SNEDDS) are isotropic, thermodynamically stable formulations consisting of surfactant, cosurfactant, oil, and drug which spontaneously form a nanoemulsion upon gentle stirring. Generally, a ternary phase diagram is used to determine the optimum concentration of the components required for preparation of SNEDDS. The required amounts of oil, surfactant, cosurfactant, and drug are mixed, and the mixture is solidified by various techniques such as spray-drying, freeze-drying, adsorption on carrier, and pelletization, for improving the stability of the formulation. SNEDDS have been identified as potential carriers for improving the bioavailability of poorly absorbed phytoconstituents by virtue of their ability to avoid first-pass metabolism owing to the nanosize of the globules (30–40 nm). Poorly water-soluble molecules with an intermediate partition coefficient are ideal candidates for formulation as SNEDDS).

Persimmon leaf extract obtained from the leaves of *Diospyros kaki*, belonging to the family Ebenaceae, is a mixture consisting of flavonoids, oligomers, tannins, and organic acids and has been used since early times in China to treat stroke and apoplexy. The extract has low solubility in water and poor bioavailability after oral administration. Avachat and Patel (2015) prepared SNEDDS loaded with ellagic acid, a polyphenol present in many dietary foods such as walnuts, strawberries, and pomegranates. Despite the numerous health benefits, ellagic acid continues to remain a poorly used molecule clinically, due to its low aqueous solubility, poor absorption extensive metabolism, and rapid elimination. The *in vitro* release of ellagic acid from SNEDDS was significantly greater as compared to ellagic acid suspension. It has been reported that the food interferes with the release and absorption of ellagic acid. It was established by *in vitro* studies that SNEDDS helped in overcoming the deleterious influence of food on release of ellagic acid.

6.3 CONCLUSIONS

Tapping the potential of phytomedicines for safe and efficacious treatment of various diseases has accumulated a lot of interest in the past few years. Albeit the phytomedicines possess a plethora of bioactivities—including oncolytic, anti-infective, antioxidant, and many other effects—their clinical use is restricted due to poor bioavailability owing to dissolution, absorption, and metabolism-related issues. Nanostructured systems, due to their nanosized, are able to transcend the biobarriers, with more efficient drug deposition at the target site and amenability to their design for manipulating release of the drug payload so as to obtain the desired pharmacokinetic profile. Judicious selection of the nanosystems from the wide options available, taking into consideration the physicochemical properties of the phytoactives may help in overcoming various limitations associated with the clinical use of phytomedicines, including large molecular size, hydrophobicity, and poor bioavailability. Application of nanotechnology for developing safe and effective nanophytopharmaceuticals continues to remain an open and dynamic area of research, which should open up newer vistas for modern day therapeutics.

REFERENCES

Abel EE, Poonga PR, Panicker SG. Characterization and *in vitro* studies on anticancer, antioxidant activity against colon cancer cell line of gold nanoparticles capped with *Cassia tora* SM leaf extract. *Appl Nanosci.* 2016;6(1):121–9.

Acharya SR, Reddy PR. Brain targeted delivery of paclitaxel using endogenous ligand. *Asian Journal of Pharmaceutical Sciences.* 2016;11(3):427–38.

Agrawal R, Sandhu SK, Sharma I, Kaur IP. Development and evaluation of curcumin-loaded elastic vesicles as an effective topical anti-inflammatory formulation. *AAPS PharmSciTech.* 2015;16(2):364–74.

Ahmadi Nasab N, Hassani Kumleh H, Beygzadeh M, Teimourian S, Kazemzad M. Delivery of curcumin by a pH-responsive chitosan mesoporous silica nanoparticles for cancer treatment. *Artif Cells Nanomed Biotechnol.* 2018;46(1):75–81.

Ahmed B, Hashmi A, Khan MS, Musarrat J. ROS mediated destruction of cell membrane, growth and biofilms of human bacterial pathogens by stable metallic AgNPs functionalized from bell pepper extract and quercetin. *Adv Powder Technol.* 2018;29(7):1601–16.

Ahmed KB, Subramanian S, Sivasubramanian A, Veerappan G, Veerappan A. Preparation of gold nanoparticles using *Salicornia brachiata* plant extract and evaluation of catalytic and antibacterial activity. *Spectrochim Acta Mol Biomol Spectrosc.* 2014;130:54–58.

Ahmed S, Ahmad M, Swami BL, Ikram S. Green synthesis of silver nanoparticles using *Azadirachta indica* aqueous leaf extract. *J Radiat Res Appl Sci.* 2016;9(1):1–7.

Aisha AF, Majid AM, Ismail Z. Preparation and characterization of nano liposomes of *Orthosiphon stamineus* ethanolic extract in soybean phospholipids. *BMC Biotechnol.* 2014;14(1):23–34.

Ajdari Z, Rahman H, Shameli K, Abdullah R, Abd Ghani M, Yeap S et al. Novel gold nanoparticles reduced by *Sargassum glaucescens*: Preparation, characterization and anticancer activity. *Molecules.* 2016;21(3):123–48.

Alavi SE, Mansouri H, Esfahani MK, Movahedi F, Akbarzadeh A, Chiani M. Archaeosome: As new drug carrier for delivery of paclitaxel to breast cancer. *Indian J Clin Biochem.* 2014;29(2):150–3.

Arriagada F, Correa O, Günther G, Nonell S, Mura F, Olea-Azar C et al. Morin flavonoid adsorbed on mesoporous silica, a novel antioxidant nanomaterial. *PloS One.* 2016;11(11):e0164507.

Avachat AM, Patel VG. Self nanoemulsifying drug delivery system of stabilized ellagic acid–phospholipid complex with improved dissolution and permeability. *Saudi Pharm J.* 2015;23(3):276–89.

Baschong W, Herzog B, Artmann CW, Mendrok C, Mongiat S, Lupia JA. Nanotopes™ A novel ultra-small unilamellar carrier system for cosmetic actives. In: Rosen MR, eds., *Delivery System Handbook for Personal Care and Cosmetic Products: Technology, Applications, and Formulations.* William Andrew Publishing, 2005, pp. 365–94. ISBN 0-8155-1504-9 (0-8155)

Beloqui A, Memvanga PB, Coco R, Reimondez-Troitiño S, Alhouayek M, Muccioli GG et al. A comparative study of curcumin-loaded lipid-based nanocarriers in the treatment of inflammatory bowel disease. *Colloids Surf B.* 2016;143:327–35.

Bhardwaj V, Plumb JA, Cassidy J, Kumar MR. Evaluating the potential of polymer nanoparticles for oral delivery of paclitaxel in drug-resistant cancer. *Cancer Nanotech.* 2010;1:29–34.

Campoy AH, Gutierrez RM, Manriquez-Alvirde G, Ramirez AM. Protection of silver nanoparticles using *Eysenhardtia polystachya* in peroxide-induced pancreatic β-cell damage and their antidiabetic properties in zebrafish. *Int J Nanomedicine.* 2018;13:2601–12.

Chandrawati R, Hosta-Rigau L, Vanderstraaten D, Lokuliyana SA, Städler B, Albericio F, Caruso F. Engineering advanced capsosomes: Maximizing the number of subcompartments, cargo retention, and temperature-triggered reaction. *ACS Nano.* 2010;4:1351–61.

Chang G, Wang Y, Gong B, Xiao Y, Chen Y, Wang S et al. Reduced graphene oxide/amaranth extract/AuNPs composite hydrogel on tumor cells as integrated platform for localized and multiple synergistic therapy. *ACS Appl Mater Interfaces.* 2015;7:11246–56.

Chen Q, Si X, Ma L, Ma P, Hou M, Bai S et al. Oral delivery of curcumin via porous polymeric nanoparticles for effective ulcerative colitis therapy. *J Mater Chem B.* 2017;5:5881–91.

Chime SA, Ikechukwu VO, Anthony AA, Onunkwo GC, Ajaraonye MC. Lipospheres: A potential delivery system of herbal extract for the treatment of diabetes mellitus. *Am J PharmTech Res.* 2013;3:480–91.

Chirio D, Gallarate M, Peira E, Battaglia L, Muntoni E, Riganti C et al. Positive-charged solid lipid nanoparticles as paclitaxel drug delivery system in glioblastoma treatment. *Eur J Pharm Biopharm.* 2014;88:746–58.

ClinicalTrials.gov. Available online at: https://clinicaltrials.gov (Accessed on 2 Aug 2017).

Dai L, Zhu W, Liu R, Si C. Lignin-containing self-nanoemulsifying drug delivery system for enhance stability and oral absorption of trans-resveratrol. *Part Part Syst Charact.* 2018;1700447.

Dhand V, Soumya L, Bharadwaj S, Chakra S, Bhatt D, Sreedhar B. Green synthesis of silver nanoparticles using *Coffea arabica* seed extract and its antibacterial activity. *Mater Sci Eng C.* 2016;58:36–43.

El-Gazayerly ON, Makhlouf AI, Soelm AM, Mohmoud MA. Antioxidant and hepatoprotective effects of silymarin phytosomes compared to milk thistle extract in CCl_4 induced hepatotoxicity in rats. *J Microencapsul.* 2014;31:23–30.

Eloy JO, Petrilli R, Topan JF, Antonio HM, Barcellos JP, Chesca DL et al. Co-loaded paclitaxel/rapamycin liposomes: Development, characterization and *in vitro* and *in vivo* evaluation for breast cancer therapy. *Colloids Surf B Biointerfaces.* 2016;141:74–82.

Fleck A, Cabral PF, Vieira FF, Pinheiro DA, Pereira CR, Santos WC, Machado TB. *Punica granatum* L. hydrogel for wound care treatment: From case study to phytomedicine standardization. *Molecules.* 2016;21:1059–62.

Fu Q, Hargrove D, Lu X. Improving paclitaxel pharmacokinetics by using tumor-specific mesoporous silica nanoparticles with intraperitoneal delivery. *Nanomed Nanotech Biol Med.* 2016;12:1951–9.

Ghaedi M, Yousefinejad M, Safarpoor M, Khafri HZ, Purkait MK. *Rosmarinus officinalis* leaf extract mediated green synthesis of silver nanoparticles and investigation of its antimicrobial properties. *J Ind Eng Chem.* 2015;31:167–72.

Gu HA, Kim HS, Kim MJ, Yu ER, Joe G, Jang J et al. Characterization and transdermal delivery of ethosomes loaded with *Eucommia ulmoides* extract. *Appl Chem Eng.* 2013;24:639–44.

Han H, Davis ME. Single-antibody, targeted nanoparticle delivery of camptothecin. *Mol Pharm.* 2013;10:2558–67.

Hartono SB, Hadisoewignyo L, Yang Y, Meka AK, Yu C. Amine functionalized cubic mesoporous silica nanoparticles as an oral delivery system for curcumin bioavailability enhancement. *Nanotechnology.* 2016;27:505605–13.

He F, Jiao H, Tian Y, Zhao L, Liao X, Fan Z, Liu B. Facile and large-scale synthesis of curcumin/PVA hydrogel: Effectively kill bacteria and accelerate cutaneous wound healing in the rat. *J Biomater Sci Polym Ed.* 2018;29(4):325–43.

Hong J, Li Y, Li Y, Xiao Y, Kuang H, Wang X. Annonaceous acetogenins nanosuspensions stabilized by PCL-PEG block polymer: Significantly improved antitumor efficacy. *Int J Nanomed.* 2016;11:3239–53.

Ibrahim HM. Green synthesis and characterization of silver nanoparticles using banana peel extract and their antimicrobial activity against representative microorganisms. *J Radiat Res Appl Sci.* 2015;8:265–75.

Im NR, Kim HS, Lim JW, Kim KJ, Noh GY, Park SN. Characterization and transdermal delivery of ethosomes loaded with liquiritigenin and liquiritin. *Appl Chem Eng.* 2015;26:563–8.

Jain S, Garg T, Kushwah V, Thanki K, Agrawal AK, Dora CP. α-Tocopherol as functional excipient for resveratrol and coenzyme Q10-loaded SNEDDS for improved bioavailability and prophylaxis of breast cancer. *J Drug Target.* 2017;25(6):554–65.

Jangde R, Srivastava S, Singh MR, Singh D. *In vitro* and *in vivo* characterization of quercetin loaded multiphase hydrogel for wound healing application. *Int J Biol Macromol.* 2018;115:1211–7.

Jeon S, Yoo CY, Park SN. Improved stability and skin permeability of sodium hyaluronate-chitosan multilayered liposomes by layer-by-layer electrostatic deposition for quercetin delivery. *Colloids Surf B: Biointerfaces.* 2015;129:7–14.

Jin W, Xu W, Ge H, Li Y, Li B. Polyphenol–gelatin nanoparticles as reductant and stabilizer for one-step synthesis of gold nanoparticles and their interfacial behavior. *RSC Advances.* 2015;5:26496–503.

Jin Y, Wen J, Garg S, Liu D, Zhou Y, Teng L et al. Development of a novel niosomal system for oral delivery of *Ginkgo biloba* extract. *Int J Nanomed.* 2013;8:421–30.

Kan P, Tseng Y-L, Ou HC. Controlled drug release liposome composition. US 9700511 B2; 2017.

Karatoprak GŞ, Aydin G, Altinsoy B, Altinkaynak C, Koşar M, Ocsoy I. The effect of *Pelargonium endlicherianum* fenzl. root extracts on formation of nanoparticles and their antimicrobial activities. *Enzyme Microb Technol.* 2017;97:21–6.

Karra N, Nassar T, Ripin AN, Schwob O, Borlak J, Benita S. Antibody conjugated PLGA nanoparticles for targeted delivery of paclitaxel palmitate: Efficacy and biofate in a lung cancer mouse model. *Small.* 2013;9:4221–36.

Katiyar SS, Muntimadugu E, Rafeeqi TA, Domb AJ, Khan W. Co-delivery of rapamycin-and piperine-loaded polymeric nanoparticles for breast cancer treatment. *Drug Deliv.* 2016;23:2608–16.

Kaur IP, Kanwar M. Ocular preparations: The formulation approach. *Drug Dev Ind Pharm.* 2002;28:473–93.

Kayser O. Nanosuspensions for the formulation of aphidicolin to improve drug targeting effects against Leishmania infected macrophages. *Int J Pharm.* 2000;196:253–6.

Khan AW, Kotta S, Ansari SH, Sharma RK, Ali J. Self-nanoemulsifying drug delivery system (SNEDDS) of the poorly water-soluble grapefruit flavonoid Naringenin: Design, characterization, *in vitro* and *in vivo* evaluation. *Drug Deliv.* 2015;22:552–61.

Kim J, Lee CM. Wound healing potential of a polyvinyl alcohol-blended pectin hydrogel containing *Hippophae rahmnoides* L. extract in a rat model. *Int J Biol Macromolec.* 2017;99:586–93.

Kim JH, Kim Y, Bae KH, Park TG, Lee JH, Park K. Tumor-targeted delivery of paclitaxel using low density lipoprotein-mimetic solid lipid nanoparticles. *Mol Pharm.* 2015;12:1230–41.

Kim SJ, Kwon SS, Jeon SH, Yu ER, Park SN. Enhanced skin delivery of liquiritigenin and liquiritin-loaded liposome-in-hydrogel complex system. *Int J Cosmet Sci.* 2014;36:553–60.

Koop HS, de Freitas RA, de Souza MM, Savi-Jr R, Silveira JL. Topical curcumin-loaded hydrogels obtained using galactomannan from *Schizolobium parahybae* and xanthan. *Carbohydr Polym.* 2015;116:229–36.

Kumar A, Jaiswal M. Enhanced intestinal permeability of *Tinospora cordifolia* extract through nanoemulsion formulation: *In-vitro* and *ex-vivo* studies. *J Nanopharm Drug Deliv.* 2014;2:209–18.

Kumar A, Jaiswal M. Effect of Self Nanoemulsifying Drug Delivery System (SNEDDS) on intestinal permeation and anti-diabetic activity of *Berberis aristata* extract: *In-vitro* and *ex-vivo* Studies. *J Nanopharm Drug Deliv.* 2016;3:51–62.

Kuo YC, Tsao CW. Neuroprotection against apoptosis of sK-N-Mc cells using rMP-7-and lactoferrin-grafted liposomes carrying quercetin. *Int J Nanomed.* 2017;12:2857–69.

Kuppusamy P, Ichwan SJ, Al-Zikri PN, Suriyah WH, Soundharrajan I, Govindan N et al. *In vitro* anticancer activity of Au, Ag nanoparticles synthesized using *Commelina nudiflora* L. aqueous extract against HCT-116 colon cancer cells. *Biol Trace Elem Res.* 2016;173:297–305.

Lee GH, Lee SJ, Jeong SW, Kim HC, Park GY, Lee SG et al. Antioxidative and antiinflammatory activities of quercetin-loaded silica nanoparticles. *Colloids Surf B: Biointerfaces.* 2016;143:511–7.

Lin J, Cai Q, Tang Y, Xu Y, Wang Q, Li T, Xu H, Wang S, Fan K, Liu Z, Jin Y. PEGylated Lipid bilayer coated mesoporous silica nanoparticles for co-delivery of paclitaxel and curcumin: Design, characterization and its cytotoxic effect. *Int J Pharm.* 2018;536(1):272–82.

Lin Y-Y, Kao H-W, Li J-J, Hwang J-J, Tseng Y-L, Li W-J, Lin M-H, Ting G, Wang H-E. Tumor burden talks in cancer treatment with PEGylated liposomal drugs. *PLoS One.* 2013;8:1–10.

Liu J, Chen Z, Wang J, Li R, Li T, Chang M, Yan F, Wang Y. Encapsulation of curcumin nanoparticles with MMP9-responsive and thermos-sensitive hydrogel improves diabetic wound healing. *ACS Appl Mater Interfaces.* 2018. (Accepted for publication)

Liu T, Yao G, Liu X, Yin H. Preparation nanocrystals of poorly soluble plant compounds using an ultra-small-scale approach. *AAPS Pharm Sci Tech.* 2017;1–8.

Liu X, Situ A, Kang Y, Villabroza KR, Liao Y, Chang CH et al. Irinotecan delivery by lipid-coated mesoporous silica nanoparticles shows improved efficacy and safety over liposomes for pancreatic cancer. *ACS Nano.* 2016;10:2702–15.

Liu Z, Zhao H, He L, Yao Y, Zhou Y, Wu J et al. Aptamer density dependent cellular uptake of lipid-capped polymer nanoparticles for polyvalent targeted delivery of vinorelbine to cancer cells. *RSC Advances.* 2015;5:16931–9.

Manconi M, Marongiu F, Castangia I, Manca ML, Caddeo C, Tuberoso CI et al. Polymer-associated liposomes for the oral delivery of grape pomace extract. *Colloids Surf B, Biointerfaces.* 2016;146:910–7.

Mishra P, Ray S, Sinha S, Das B, Khan MI, Behera SK et al. Facile bio-synthesis of gold nanoparticles by using extract of *Hibiscus sabdariffa* and evaluation of its cytotoxicity against U87 glioblastoma cells under hyperglycemic condition. *Biochem Eng J.* 2016;105:264–72.

Moorthi C, Kathiresan K. Fabrication of dual drug loaded polymeric nanosuspension: Incorporating analytical hierarchy process and data envelopment analysis in the selection of a suitable method. *Int J Pharm Pharm Sci.* 2013;5:499–504.

Muthukumar T, Sambandam B, Aravinthan A, Sastry TP, Kim JH. Green synthesis of gold nanoparticles and their enhanced synergistic antitumor activity using HepG2 and MCF7 cells and its antibacterial effects. *Process Biochem.* 2016;51:384–91.

Nakkala JR, Mata R, Gupta AK, Sadras SR. Biological activities of green silver nanoparticles synthesized with *Acorous calamus* rhizome extract. *Eur J Med Chem.* 2014;85:784–94.

Nayak D, Pradhan S, Ashe S, Rauta PR, Nayak B. Biologically synthesised silver nanoparticles from three diverse family of plant extracts and their anticancer activity against epidermoid A431 carcinoma. *J Colloid Interface Sci.* 2015;457:329–38.

Nazari-Vanani R, Moezi L, Heli H. *In vivo* evaluation of a self-nanoemulsifying drug delivery system for curcumin. *Biomed Pharmacother.* 2017;88:715–20.

Negi P, Aggarwal M, Sharma G, Rathore C, Sharma G, Singh B et al. Niosome-based hydrogel of resveratrol for topical applications: An effective therapy for pain related disorder(s). *Biomed Pharmacother.* 2017;88:480–7.

Neves AR, Lúcio M, Martins S, Lima JL, Reis S. Novel resveratrol nanodelivery systems based on lipid nanoparticles to enhance its oral bioavailability. *Int J Nanomed.* 2013;8:177–87.

Neves AR, Queiroz JF, Reis S. Brain-targeted delivery of resveratrol using solid lipid nanoparticles functionalized with apolipoprotein E. *J Nanobiotechnol.* 2016;14:27–38.

Nezamdoost T, Bagherieh-Najjar MB, Aghdasi M. Biogenic synthesis of stable bioactive silver chloride nanoparticles using *Onosma dichroantha* Boiss. root extract. *Mater Lett.* 2014;137:225–28.

Ning P, Lü S, Bai X, Wu X, Gao C, Wen N, Liu M. High encapsulation and localized delivery of curcumin from an injectable hydrogel. *Mater Sci Eng: C.* 2018;83:121–9.

Pandita D, Kumar S, Poonia N, Lather V. Solid lipid nanoparticles enhance oral bioavailability of resveratrol, a natural polyphenol. *Food Res Int.* 2014;62:1165–74.

Park SN, Lee MH, Kim SJ, Yu ER. Preparation of quercetin and rutin-loaded ceramide liposomes and drug-releasing effect in liposome-in-hydrogel complex system. *Biochem Biophys Res Commun.* 2013;435:361–6.

Parveen M, Ahmad F, Malla AM, Azaz S. Microwave-assisted green synthesis of silver nanoparticles from *Fraxinus excelsior* leaf extract and its antioxidant assay. *Appl Nanosci.* 2016;6:267–76.

Pathan RA, Bhandari U, Javed S, Nag TC. Anti-apoptotic potential of gymnemic acid phospholipid complex pretreatment in Wistar rats with experimental cardiomyopathy. *Indian J Exp Biol.* 2012;50:117–27.

Ponarulselvam S, Panneerselvam C, Murugan K, Aarthi N, Kalimuthu K, Thangamani S. Synthesis of silver nanoparticles using leaves of *Catharanthus roseus* Linn. G. Don and their antiplasmodial activities. *Asian Pacific J Trop Biomed.* 2012;2:574–80.

Rajakumar G, Thiruvengadam M, Mydhili G, Gomathi T, Chung IM. Green approach for synthesis of zinc oxide nanoparticles from *Andrographis paniculata* leaf extract and evaluation of their antioxidant, anti-diabetic, and anti-inflammatory activities. *Bioprocess Biosyst Eng.* 2018;41(1):21–30.

Rajan A, Rajan AR, Philip D. *Elettaria cardamomum* seed mediated rapid synthesis of gold nanoparticles and its biological activities. *Open Nano.* 2017;2:1–8.

Rajaram K, Aiswarya DC, Sureshkumar P. Green synthesis of silver nanoparticle using *Tephrosia tinctoria* and its antidiabetic activity. *Mater Lett.* 2015;138:251–4.

Ramachandran R, Krishnaraj C, Sivakumar AS, Prasannakumar P, Kumar VA, Shim KS et al. Anticancer activity of biologically synthesized silver and gold nanoparticles on mouse myoblast cancer cells and their toxicity against embryonic zebrafish. *Mater Sci Eng C.* 2017;73:674–83.

Ratcharin N, Wongtrakul P, Indranupakorn R. Preparation of *Zingiber officinale* extract loaded solid lipid nanoparticles. In: *Advanced Materials Research.* Trans Tech Publications, 2012, pp. 389–92.

Ravichandran V, Vasanthi S, Shalini S, Shah SA, Harish R. Green synthesis of silver nanoparticles using *Atrocarpus altilis* leaf extract and the study of their antimicrobial and antioxidant activity. *Mater Lett.* 2016;180:264–7.

Rehana D, Mahendiran D, Kumar RS, Rahiman AK. *In vitro* antioxidant and antidiabetic activities of zinc oxide nanoparticles synthesized using different plant extracts. *Bioprocess Biosyst Eng.* 2017;40(6):943–57.

Rejinold NS, Muthunarayanan M, Divyarani VV, Sreerekha PR, Chennazhi KP, Nair SV et al. Curcumin-loaded biocompatible thermoresponsive polymeric nanoparticles for cancer drug delivery. *J Colloid Interface Sci.* 2011;360:39–51.

Salar RK, Kumar N. Synthesis and characterization of vincristine loaded folic acid–chitosan conjugated nanoparticles. *Resour Efficient Technol.* 2016;2:199–214.

Sanaeimehr Z, Javadi I, Namvar F. Antiangiogenic and antiapoptotic effects of green-synthesized zinc oxide nanoparticles using *Sargassum muticum* algae extraction. *Cancer Nanotechnol.* 2018;9(1):3.
Sarkar A, Ghosh S, Chowdhury S, Pandey B, Sil PC. Targeted delivery of quercetin loaded mesoporous silica nanoparticles to the breast cancer cells. *Biochin Biophys Acta-General Subjects.* 2016;1860:2065–75.
Sarwa KK, Mazumder B, Rudrapal M, Verma VK. Potential of capsaicin-loaded transfersomes in arthritic rats. *Drug Deliv.* 2015;22:638–46.
Selvan DA, Mahendiran D, Kumar RS, Rahiman AK. Garlic, green tea and turmeric extracts-mediated green synthesis of silver nanoparticles: Phytochemical, antioxidant and *in vitro* cytotoxicity studies. *J Photochem Photobiol B.* 2018;180:243–52.
Sezgin-Bayindir Z, Onay-Besikci A, Vural N, Yuksel N. Niosomes encapsulating paclitaxel for oral bioavailability enhancement: Preparation, characterization, pharmacokinetics and biodistribution. *J Microencapsulation.* 2013;30:796–804.
Shakeel F, Shazly GA, Raish M, Ahmad A, Kalam MA, Ali N et al. Biological investigation of a supersaturated self-nanoemulsifying drug delivery system of *Piper cubeba* essential oil. *RSC Adv.* 2015;5:105206–17.
Sharma S, Narang JK, Ali J, Baboota S. Synergistic antioxidant action of vitamin E and rutin SNEDDS in ameliorating oxidative stress in a Parkinson's disease model. *Nanotechnology.* 2016;27:375101–21.
Singh P, Kim YJ, Wang C, Mathiyalagan R, El-Agamy Farh M, Yang DC. Biogenic silver and gold nanoparticles synthesized using red ginseng root extract, and their applications. *Artificial Cells Nanomed Biotechnol.* 2016;44:811–6.
Streith S, Eichhorn ME, Sauer B, Schulze B, Teifel M, Michealis U et al. Neovascular targeting chemotherapy: Encapsulation of paclitaxel in cationic liposomes impairs functional tumor microvasculature. *Int J Cancer.* 2004;110:117–24.
Summerlin N, Qu Z, Pujara N, Sheng Y, Jambhrunkar S, McGuckin M, Popat A. Colloidal mesoporous silica nanoparticles enhance the biological activity of resveratrol. *Colloids Surf B Biointerfaces.* 2016;144:1–7.
Swain S, Barik SK, Behera T, Nayak SK, Sahoo SK, Mishra SS et al. Green synthesis of gold nanoparticles using root and leaf extracts of *Vetiveria zizanioides* and *Cannabis sativa* and its antifungal activities. *BioNanoScience.* 2016;6:205–13.
Swamy MK, Sudipta KM, Jayanta K, Balasubramanya S. The green synthesis, characterization, and evaluation of the biological activities of silver nanoparticles synthesized from *Leptadenia reticulata* leaf extract. *Appl Nanosci.* 2015;5:73–81.
Tao C, Cheng H, Zhou K, Luo Q, Guo L, Chen W. Preparation and characterization of Biochanin A loaded solid lipid nanoparticles. *Asian J Pharm.* 2012;6:275–81.
Teong B, Lin CY, Chang SJ, Niu GC, Yao CH, Chen IF et al. Enhanced anti-cancer activity by curcumin-loaded hydrogel nanoparticle derived aggregates on A549 lung adenocarcinoma cells. *J Mater Sci Mater Med.* 2015;26:49–64.
Tripathi S, Kushwah V, Thanki K, Jain S. Triple antioxidant SNEDDS formulation with enhanced oral bioavailability: Implication of chemoprevention of breast cancer. *Nanomed Nanotech Biol Med.* 2016;12:1431–43.
Udompornmongkol P, Chiang BH. Curcumin-loaded polymeric nanoparticles for enhanced anti-colorectal cancer applications. *J Biomater Appl.* 2015;30:537–46.
Velu V, Das M, Dua K, Malipeddi H. Evaluation of *in vitro* and *in vivo* anti-urolithiatic activity of silver nanoparticles containing aqueous leaf extract of *Tragia involucrata. Drug Deliv Transl Res.* 2017;7(3):439–49.
Vitthal KU, Pillai MM, Kininge P. Study of solid lipid nanoparticles as a carrier for bacoside. *Int J Pharma BioSci.* 2013;3:414–26.
Wang C, Mathiyalagan R, Kim YJ, Castro-Aceituno V, Singh P, Ahn S et al. Rapid green synthesis of silver and gold nanoparticles using *Dendropanax morbifera* leaf extract and their anticancer activities. *Int J Nanomed.* 2016;11:3691–701.
Wang T, Liu Y, Wu C. Effect of paclitaxel-mesoporous silica nanoparticles with a core-shell structure on the human lung cancer cell line A549. *Nanoscale Res Lett.* 2017;12:66–74.
Wang Z, Wu J, Zhou Q, Wang Y, Chen T. Berberine nanosuspension enhances hypoglycemic efficacy on streptozotocin induced diabetic C57BL/6 mice. *J Evid Based Complementary Altern Med.* 2015;2015:1–5.
Wei XL, Han YR, Quan LH, Liu CY, Liao YH. Oily nanosuspension for long-acting intramuscular delivery of curcumin didecanoate prodrug: Preparation, characterization and *in vivo* evaluation. *Eur J Pharm Sci.* 2013;49:286–93.
Xiao B, Si X, Han MK, Viennois E, Zhang M, Merlin D. Co-delivery of camptothecin and curcumin by cationic polymeric nanoparticles for synergistic colon cancer combination chemotherapy. *J Mater Chem B.* 2015;3:7724–33.

Xu H, Wang T, Yang C, Li X, Liu G, Yang Z, Singh PK, Krishnan S, Ding D. Supramolecular nanofibers of curcumin for highly amplified radiosensitization of colorectal cancers to ionizing radiation. *Adv Funct Mater.* 2018;28(14):1707140.

Xu W, Lee MK. Development and evaluation of lipid nanoparticles for paclitaxel delivery: A comparison between solid lipid nanoparticles and nanostructured lipid carriers. *J Pharm Inves.* 2015;7:675–80.

Yao L, Zhao X, Li Q, Zu Y, Fu Y, Zu B et al. *In vitro* and *in vivo* evaluation of camptothecin nanosuspension: A novel formulation with high antitumor efficacy and low toxicity. *Int J Pharm.* 2012;423:586–8.

Yeh MI, Huang HC, Liaw JH, Huang MC, Wu TH, Huang KF, Hsu FL. Ethosomes in hair dye products as carriers of the major compounds of black tea extracts. *Int J Dermatol.* 2013;52(7):868–75.

Yen FL, Wu TH, Lin LT, Cham TM, Lin CC. Nanoparticles formulation of *Cuscuta chinensis* prevents acetaminophen-induced hepatotoxicity in rats. *Food Chem Toxicol.* 2008;46:1771–7.

Yu Y, Feng R, Yu S, Li J, Wang Y, Song Y, Yang X, Pan W, Li S. Nanostructured lipid carrier-based pH and temperature dual-responsive hydrogel composed of carboxymethyl chitosan and poloxamer for drug delivery. *Int J Biol Macromol.* 2018;114:462–9.

Yugandhar P, Savithramma N. Biosynthesis, characterization and antimicrobial studies of green synthesized silver nanoparticles from fruit extract of *Syzygium alternifolium* (Wt.) Walp. an endemic, endangered medicinal tree taxon. *Appl Nanosci.* 2016;6:223–33.

Zale SE, Troiano G, Ali MM, Hrkach J, Wright J. Therapeutic polymeric nanoparticles comprising vinca alkaloids and methods of making and using same. US20160338963 A1; 2016.

Zhang C, Gu C, Peng F, Liu W, Wan J, Xu H et al. Preparation and optimization of triptolide-loaded solid lipid nanoparticles for oral delivery with reduced gastric irritation. *Molecules.* 2013;18:13340–56.

Zhang W, Cui T, Liu L, Wu Q, Sun L, Li L et al. Improving anti-tumor activity of curcumin by polymeric micelles in thermosensitive hydrogel system in colorectal peritoneal carcinomatosis model. *J Biomed Nanotechnol.* 2015a;11:1173–82.

Zhang XZ, Guan J, Cai SL, Du Q, Guo ML. Polymeric *in situ* hydrogel implant of Epigallocatechin Gallate (EGCG) for prolonged and improved antihyperlipidemic and anti-obesity activity: Preparation and characterization. *J Biomater Tissue Eng.* 2015b;5:813–7.

Zhao Y, Liu JG, Chen WM, Yu AX. Efficacy of thermosensitive chitosan/β-glycerophosphate hydrogel loaded with β-cyclodextrin-curcumin for the treatment of cutaneous wound infection in rats. *Exp Ther Med.* 2018;15(2):1304–13.

7 Role of Nanoconstructs in Ameliorating Chemopreventive and Chemotherapeutic Potential of Plant Bioactives

*Rajneet Kaur Khurana, Teenu Sharma, Satvinder Kaur, Ashay Jain, and Bhupinder Singh**

CONTENTS

7.1 INTRODUCTION

Plant bioactives represent either the phytoconstituents or herbal products capable of being absorbed when formulated for producing therapeutic results (Jain et al., 2016). These have been used for medicinal purposes since history has started being recorded (Aqil et al., 2005). The potential of Ayurveda is also being reexplored in India, and there has been rising attention on the significance of medicinal plants in the traditional healthcare system, namely, Ayurveda, Unani, and homeopathy for handling healthcare problems (Shrivastava et al., 2015). Plant bioactives are being looked upon with attractive sources by the researchers as well as by patients.

Phytomedicines, the complex chemical mixtures prepared from plants, have been employed in medicine since primeval times and theses enjoy status to date (Landete, 2011). Being derived from

* Corresponding author.

botanicals, unlike pharmaceuticals, such phytomedicines comprise natural ingredients, which are better absorbed in the bloodstream, resulting in holistic and effective treatment, which is more easily and readily metabolized by the body, thus producing fewer and milder side effects, if any (Romero-Benavides et al., 2017). Most phytocontstituents exert their action primarily as antioxidants, that is, agents capable of preventing, retarding, or exterminating the oxidative stress of a molecule. Natural antioxidants obtained from plants mostly include phenolic and polyphenolic compounds such as flavonoids, cartenoids and carnosine, enzymes, vitamins, and chelators. Recent findings have demonstrated the antibacterial, anti-inflammatory, anticancer, and antimutagenic effects of various plant-derived antioxidants in cells.

Several phytomedicines are also currently available as tablets, pressed juices, capsules, infusions, tinctures, solvent-extracted preparations, or combinations of such product forms. However, these traditional delivery systems of herbal medicines suffer from various limitations, including low patient acceptability and compliance (Kumar et al., 2017). Understanding the potential of newer technologies, research efforts have been focused on developing various drug-delivery systems (such as rapid-dissolving tablets, controlled extended-release formulations, mucoadhesive systems, transdermal dosage forms, microparticles, microcapsules, and implants) of herbal dugs, but with only quite-limited fruition.

Nanostructured systems, in this regard, have been used, quite successfully, to tackle the limitations of the traditional drug-delivery systems. Currently, researchers are, accordingly, more inclined towards exploring the application of nanotech-enabled systems of herbal actives for obtaining useful therapeutic outcomes. Some of the nanocarrier systems that have been investigated for herbal actives and phytomedicines are portrayed in Figure 7.1. Not only do such nanoconstructs facilitate the delivery of the drug at a rate and over a period of treatment dictated as per the needs of the body, but they also channelize the active entity solely at the site of action in an optimally controlled manner. Such cutting-edge strategies are developed as a consequence of synergistic amalgamation of interdisciplinary fields such as material science, drug delivery, conjugate chemistry, and molecular biology (Garg et al., 2017). Application of novel and nanostructured drug-delivery technologies to herbal medicine have demonstrated remarkable potential to achieve improved therapeutic outcomes of various herbal compounds with minimal associated off-site side effects (Jain et al., 2017a,b).

Novel nanoconstructs are the nanoscale carriers, designed such that on administration to the patient, an optimum amount of the concerned bioactive molecule reaches the "action site" and starts off its action with minimal side effects (Jain et al., 2015). These systems circumvent the associated limitations by overcoming the hiccups such as low gastric pH, hepatic enzymes, along with increased prolonged circulation into the blood (Tripathi et al., 2016). Nanonization, when applied to herbal medicines, offers stellar advantages of increased solubility, stability, and dose reduction and improved

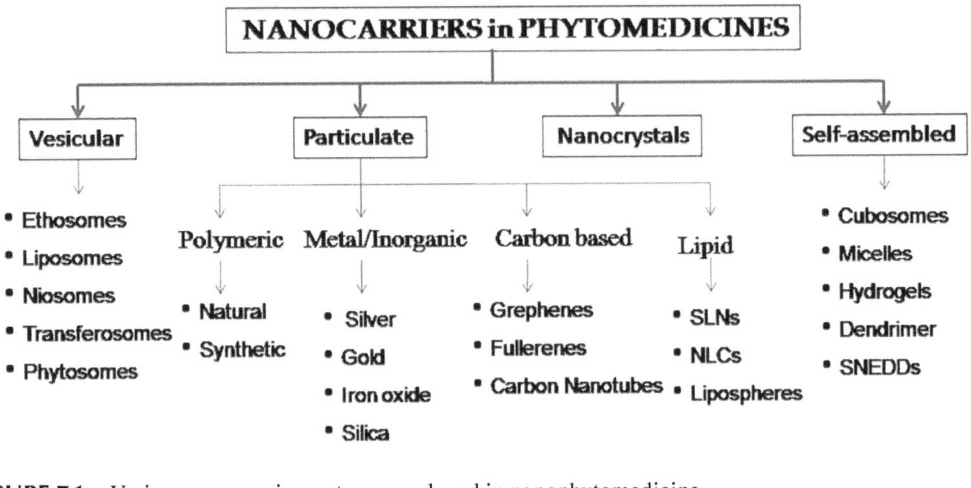

FIGURE 7.1 Various nanocarrier systems employed in nanophytomedicine.

absorption and/or distribution *vis-à-vis* the respective crude preparations (Namdari et al., 2017). Nevertheless, despite the documented phenomenal potential of these novel nanocarriers—such as lipidic, polymeric, lipid-polymer hybrids; metallic or inorganic nanoparticles; self-emulsifying nano- or microemulsions; dendrimers; and vesicular systems, such as liposomes and phytosomes—these still have not been explored much for various plant bioactives (Sandhu et al., 2017).

The current chapter presents various nanoconstructs, which have been utilized for efficient delivery of phytobioactives, including an account on various nanocarrier systems being popularly employed to deliver various phytobioactives.

7.2 NANOSTRUCTURED DELIVERY SYSTEMS

7.2.1 Nanoparticles

Nanoparticles encompass both nanocapsules and nanospheres, though the latter have a matrix-type structure, wherein the active ingredient is dispersed throughout the particles, and the former have a polymeric membrane and an active ingredient core. Nanoparticles offer advantages such as higher stability and drug-loading capacities and protection by the solid matrix of the incorporated drug against degradation, when compared to other colloidal carriers. Nanoparticles can be modified and/or functionalized to alter their biodistribution for effective drug targeting (Kesharwani et al., 2011). New polymers are being investigated to develop nanoparticles for their application as drug carriers. Various nanoparticulate delivery systems utilized for phytomedicines have been compiled in Table 7.1, and are detailed in the following sections.

7.2.1.1 Biodegradable Polymeric Nanoparticles

Biodegradable polymeric nanoparticles offer higher stability, even on contact with biological fluids (Kumar et al., 2008). These have attracted substantial consideration as potential carriers, owing to their drug targeting ability, and as carriers of DNA, proteins, peptides, and genes, through the oral route. A polymer used in nanoparticle fabrication must be chemically inert, nontoxic, and free of leachable impurities. For fabrication, the drug is dissolved, entrapped, encapsulated, or attached to nanoparticles, nanospheres, or nanocapsules, obtained depending upon the method of preparation. In nanocapsules, the drug is restrained to a void surrounded by a polymer membrane, while nanospheres are matrix systems in which the drug is physically and uniformly dispersed. Examples of synthetic biodegradable polymers widely employed in nanoparticle fabrication are poly(lactic acid) (PLA), poly(glycolic acid) (PGA), and poly(D, L-lactide-co-glycolide acid) (PLGA).

Aescin, an antileishmanial saponin, was fabricated into PLGA nanoparticles by Ven et al. (2011). β-Aescin, a major saponin of seeds of *Aesculus hippocastanun*, has shown activity against the intracellular amastigote stage of *L. Infatum*. Tsai et al. (2011) developed curcumin-loaded PLGA nanoparticles (C-NPs) by a high-pressure emulsification-solvent evaporation method. Curcumin, though extensively studied for its anticarcinogenic, anti-inflammatory, antimicrobial, and antimalarial properties, is associated with the limitation of low bioavailability. This is because of its limited absorption from the intestine, together with high hepatic metabolism and rapid biliary elimination. Therefore, the penetration ability of curcumin across the organs and its circulation time are increased owing to nanoparticulation. The mean particle size and entrapment efficiency were found to be 163 nm and 46.9%, respectively. Nanoformulations significantly enhanced the concentration of curcumin, along with increase in pharmacokinetic parameters such as area under curve (AUC), biological half-life ($t_{1/2}$), and mean residence time (MRT). Moreover, curcumin was able to transverse the blood–brain barrier (BBB) and was found to be concentrated in the hippocampus, chiefly. Nanoparticulation prolongs the retention time of curcumin in the cerebral cortex (increased by 96%) and hippocampus (increased by 83%). These observations provided better understanding for the probable beneficial effects of curcumin and its further translation to preclinical and clinical research.

TABLE 7.1
Nanoparticulate Drug Carrier Systems Used for Fabrication of Nanophytobioactives

Method of Preparation	Herbal Active/Extract	Role in Drug Delivery	Reference
		Silver Nanoparticles	
Addition of silver nitrate solution to leaf extracts	Aqueous leaf extract of *Euphorbia prostrata* Ait.	Eco-friendly treatment against adult cattle tick with no toxicity to cattle	Zahir and Rahuman (2012)
Precipitation of lactic acid in chitosan solution by dropping method	Anthraquinone	Enhanced *in vitro* anticancer activity with reduction in cell viability	Jeevitha et al. (2013)
		Gold Nanoparticles	
Addition of chloroauric acid solution to leaf extracts	*Crocus sativus* (saffron)	Increased antitumor potential	Vijaykumar et al. (2011)
Incubation method	Ginkgolide A	Significant inhibition of smooth muscle cell proliferation and migration *in vitro* in comparison with ginkgolide A treatment alone	Weakley et al. (2011)
		Iron Oxide Nanoparticles	
Chemical synthesis method	*Argemone mexicana* L. leaf extract	Enhanced inhibitory effects on *P. mirabilis* and *E. coli*	Arokiyaraj et al. (2013)
		Silica Nanoparticles	
Vacuum nanocasting technique	Rhodamine B (RhB) covalently grafted salvianolic acid B-15 SAB	Effective protection of salvianolic acid B-loaded nanoparticles from decomposition and outstanding sustained-release and relatively high release rates	Qianjun et al. (2010)
		Lipid Nanoparticles	
Solvent injection method	Thymoquinone	Significant decrease in the serum liver enzymes following oral administration *vis-à-vis* control and marketed formulations	Singh et al. (2013)
High temperature melt-cool solidification method	Icariin	Increased icariin oral bioavailability	Ke-pan et al. (2012)
		Polymeric Nanoparticles	
Methoxy PEG-PLA and melamide PEG-PLA via ring opening polymerization of L-Lactide	Tanshinone IIA (TIIA), a major active ingredient of a traditional herbal medicine *Salvia miltiorrhiza*	Increased permeability across blood–brain barrier (BBB) for treating cerebral ischemia	Liu et al. (2013)
W/O/W emulsification solvent evaporation technique	β-Aescin, a major saponin of seeds *Aesculus hippocastanun*	Increased antileishmanial activity and reduction in cytotoxicity	Ven et al. (2011)

(Continued)

TABLE 7.1 (*Continued*)
Nanoparticulate Drug Carrier Systems Used for Fabrication of Nanophytobioactives

Method of Preparation	Herbal Active/Extract	Role in Drug Delivery	Reference
		Chitosan Nanoparticles	
Ionic gelation method	Curcumin	Increased absorption of nanoparticulate curcumin from intestine, resulting in increased chelation of arsenic for treatment of arsenic poisoning, with high antioxidant potential and neuroprotective efficacy	Yadav et al. (2012)
		Chitosan and Poly(Lactic Acid) (PLA)	
Addition of silver nitrate solution to leaf extracts	Aqueous leaf extract of *Catharanthus roseus*	75% inhibition of *P. falciparum* parasites at just 100 μg/mL	Ponarulselvam et al. (2012)
		Nanocapsules	
Addition of silver nitrate solution to leaf extracts	Aqueous extract from *Eclipta prostrate*	Eco-friendly approach for the control of the *Culex tritaeniorhynchus* and *A. subpictus*	Rajakumar et al. (2011)

PLGA, though biocompatible and biodegradable, has restricted parenteral applications because of its hydrophobicity. Copolymerization with PEG can help in encapsulating hydrophilic drugs. Han et al. (2012) developed an invasive nanocarrier system based on copolymers, monomethyl polyethylene glycol–poly(lactide-co-glycolide)–monomethyl-polyethylene-glycol (MeO–PEG–PLGA–PEG–OMe, PELGE), which was used as a platform for coencapsulation and sustained release of four active components: ginkgolides A, B, and C and bilobalide in *Ginkgo biloba* extract. Ginkgo-leaf extract has been widely used for cardiovascular and cognitive disorders, including dementia and peripheral arterial occlusive diseases. Drug-loaded nanoparticles, prepared by a coprecipitation method, resulted in high encapsulation efficiency of 78.84 ± 2.06% of the total ginkgo terpenes (GT) with a loading dose of 11.90 ± 0.31 mg/150 mg PELGE. The particles presented a spherical shape with a mean diameter of 123.3 ± 44.0 nm and a zeta potential of –30.86 ± 2.49 mV. Differential scanning calorimetry (DSC) analysis revealed uniform dispersion and interaction of GT in nanoparticles. Thus, the AUCs of tested compounds were found to be enhanced significantly after incorporation into PELGE-NPs. The $t_{1/2\beta}$ of each terpene lactone was prolonged markedly. *In vitro* and *in vivo* studies exhibited sustained, as well as synchronized, release of the four components from PELGE nanoparticles, which was the main contributor to the long circulation of PEGylated nanoparticles and the slow degradation of PLGA. The results indicated that the release of ginkgo terpenes was mainly dependent on the rate of degradation of PELGE. Thus, PELGE-based nanoparticles could serve as promising carrier systems for herbal medicines containing multiple components.

Although the usage of various synthetic and semisynthetic polymers has lately been augmented in drug delivery, useful natural polymers—including gum acacia, guar, chitosan, gelatin, sodium alginate, and others—still enjoy their popularity in drug delivery.

Chitosan nanoparticles have been established as drug carriers with enormous potential, as these offer several key advantages (Liang et al., 2017):

- Sustained and prolonged drug release, with improved drug solubility and stability, thus enhancement of efficacy, thereby reduction in toxicity

- Ability to transverse through biological barriers and deliver drugs to the target site in order to enhance efficacy
- High biocompatibility and biodegradability, as well as ease of rapid modification
- Wide-ranging list of materials such asproteins, drugs, gene drugs, and anticancer chemical drugs, which can be loaded and administered via various routes, incluiding oral, nasal, intravenous and ocular routes
- Enhancement of transport of drugs across the nasal and intestinal mucosal barriers, and exhibition of very promising behavior as topical ocular drug-delivery systems

The aforementioned desirable properties of chitosan, when combined with the superior mechanical strength of PLA, create a novel kind of biohybrid amphiphile. The engineered biodegradable (chitosan [CS] and poly[lactic acid] [PLA]) nanoparticles encapsulate anthraquinone (AQ), an antioxidant useful in treating oxidative stress-related diseases such as atherosclerosis and cancer (Jeevitha et al., 2013).

7.2.1.2 Metallic Nanoparticles

Metallic nanoparticles exhibit size- and shape-dependent properties that are of interest for their varied applications in catalysis, sensing, optics, antibacterial activity, and data storage (Rajkumar et al., 2017). For instance, the antibacterial activity of silver nanoparticles is closely related to their size—that is, the smaller the silver nuclei, the higher are the antibacterial and catalytic activities. Thus, size and distribution control becomes highly crucial (Zhang et al., 2004).

7.2.1.2.1 Silver Nanoparticles

Silver phytonanoparticles have been synthesized by Prabhu et al. (2010) with the leaves of *Ocimum sanctum* and *Vitex negundo.* The composite release of silver showed antimicrobial efficacy against various microbes. Silver phytonanoparticles collected from each herb were tested for antibacterial activity against *Staphylococcus aureus, Vibrio cholerae, Proteus vulgaris*, and *Pseudomonas aeruginosa*. Silver nanoparticles of herbal-leaf extract showed a superior antibacterial activity than naïve herbal-leaf extract. Maximum inhibitory rate was observed using the *Ocimum sanctum* plant leaf nanoparticles. The leaf extract nanoparticle of *Vitex negundo* showed inhibitory action against bacteria. The antimicrobial efficacy of *Ocimum sanctum* was more pronounced than *Vitex negundo*, especially against *Proteus vulgaris* and *Vibrio cholerae.* Silver nanoparticles (AgNPs encapsulating aqueous extract from *Eclipta prostrata*, a member of the Asteraceae family, exhibited larvicidal activity against the filariasis vector.

7.2.1.2.2 Gold Nanoparticles

Controlled green synthesis of stable and crystalline gold nanoparticles (AuNPs) with well-defined dimensions encapsulating *Crocus sativus* (saffron) was performed and evaluated for antitumor activity (Vijaykumar et al., 2011). Proteins coat the AuNPs to prevent aggregation of the particles, thus providing stability to the particles in the medium. This further suggests that the biomolecules could possibly be employed to enhance the stability of the AuNPs in aqueous medium.

Weakley et al. (2011) conjugated ginkgolide A (GA) to 80 nm AuNPs in an overnight incubation. GA-AuNP treatment significantly inhibited smooth muscle cell (SMC) proliferation and migration *in vitro vis-à-vis* AuNP treatment alone, probably through reduced activation of ERK1/2, and the effect persisted up to 72 h after treatment.

7.2.1.2.3 Iron-Oxide Nanoparticles

Synthesized iron-oxide nanoparticles (IO-NPs) and ethanolic extract of *A. Mexicana* leaves, when taken in a 1:1 ratio after dissolving in ethanol, were incubated at 45°C for 12 h to find the bactericidal activity (Arokiyaraj et al., 2013). Furthermore, the antibacterial efficacy of IO-NPs, leaf extract, and IO-NPs treated with plant extract was screened against *Escherichia coli* MTCC 443, *Proteus mirabilis* MTCC 425, and *Bacillus subtilis* MTCC 441. IO-NPs treated with plant extract showed a significant inhibitory activity on *P. mirabilis* and *E. coli.* This study has demonstrated the use of IO-NPs as magnetic nanoparticles as efficient drug carriers to cure bacterial diseases. The study proved that the immobilized magnetite nanomaterials effectively improve the drug loading and the antibacterial efficiency against the microbes. Thus, this establishes enormous importance of such studies in the field of developing eco-friendly nanobiomaterials.

7.2.1.3 Silica Nanoparticles

Application studies of mesoporous silica nanoparticles (MSNs) in the biomedical field have attained enormous interest (Zhu et al., 2005). MSNs have been employed as effective carriers for controlled drug delivery, biosignal probing, gene transport and expression (Liong et al., 2008), biomarking (Lin et al., 2005), and many other important bioapplications (Trewyn et al., 2007). Sustained drug release of MSNs, due to their unique pore structures and simplistic surface alteration, are drawing growing consideration in the biomedical field. MSNs often need to be modified for providing special charge properties on the surface. The schematic representation of mesoporous structure is shown in Figure 7.2. Qianjun et al. (2010) have reported rhodamine B (RhB) covalently grafted salvianolic acid B-15-structured MSNs (SAB@MSNs-RhB), having high surface area (750 m^2/g), large pore volume (0.7 cm^3/g), uniform particle size (about 400 nm), and a positively charged surface (29.6 $\pm$ 5.0 mV), for anti-reactive oxygen species (ROS) or hepatic fibrosis.

Using porous silica nanoparticles (PSNs), peroral bioavailability of a poorly water-soluble drug, silymarin, has been reported to be improved by Cao et al. (2012). Synthesized by reverse microemulsion and ultrasonic corrosion, PSNs feature a large specific surface area and narrow pore-size distribution of approximately 10 nm. Silymarin-loaded PSNs showed a typical sustained-release pattern over a long period of about 72 h.

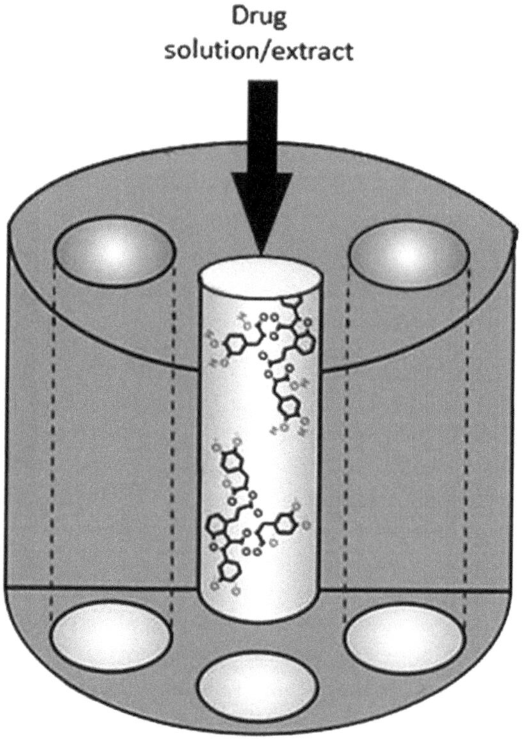

FIGURE 7.2 Schematic cutaway diagram of structure of mesoporous drug-delivery system.

7.2.1.4 Lipid Nanoparticles (LN)

Lipid nanoparticles with solid particle matrix are o/w emulsions with replacement of the liquid lipid (oil) by a solid lipid, to keep these solids at body temperature. The first generation of solid lipid nanoparticles (SLNs) was fabricated in the early 1990s, and these combined the advantages of polymeric nanoparticles, lipidic emulsions, and liposomes, while simultaneously avoiding their disadvantages, in particular. Thus, owing to their physiological nature and lower risk of toxicity and production feasibility and scalability, use of lipid nanoparticles has widely been investigated (Jain et al., 2014).

Singh et al. (2013) have reported SLNs of thymoquinone (THQ-SLNs) for efficient treatment of liver cirrhosis. Pharmacodynamic data following oral administration exhibited a considerable decline in the serum enzymes (SGOT, SGPT, and ALP) of THQ-SLNs *vis-à-vis* control and marketed (SILYBON) formulations for the paracetamol-induced liver cirrhosis. The treatment efficiency was confirmed by the histopathology of the liver.

Further, despite its several beneficial pharmacological effects reported in the literature—such as anticancer, antibacterial, and antiviral activities—the efficiency of widely used oral dosage forms of baicalin is not considered satisfactory. Thus, in order to improve the therapeutic effect of baicalin, its SLNs were prepared (<100 nm), and the *in vitro* antioxidant results corroborated that SLNs have a strong reducing capacity, as well as an ability to remove hydroxide radicals, with 6.931 as the targeting rate for liver (Yang et al., 2012).

7.2.2 Nanocapsules

Nanocapsules differ from nanoparticles, as these consist of a liquid/solid core in which the drug is placed, surrounded by a distinctive natural or synthetic polymer membrane. These have been gaining enormous importance because of their easily oxidizable protective coating, usual pyrophoric nature, and role in prolonging the release of encapsulated active ingredients. Lertsutthiwong et al. (2008) encapsulated turmeric oil, a model oily compound, in an alginate biopolymer coating, forming nanocapsules. The three-step procedure for alginate nanocapsules preparation uses emulsification and cross-linking with calcium chloride, followed by solvent removal. Average size, zeta potential, morphology, loading capacity, and stability at 4°C and 25°C of the nanocapsules were all affected by the solvent type, oil concentration, sonication, and oil-to-alginate mass ratio. Dissolution of turmeric oil in ethanol and Tween 80 was found to be crucial in the preparation process. An increase in the oil concentration or oil-alginate mass ratio resulted in increasing the average size of the nanocapsules. However, alginate nanocapsules showed good physical stability during long-term storage at 4°C.

7.2.3 Nanosuspensions and Nanocrystals

Nanosuspension strategy is most suitable for the molecules with a high partition-coefficient value, high melting point, and high administration dose. To overcome the poor solubility of apigenin, nanosuspensions were prepared by Loaye et al. (2011) using combination technology (CT)—that is, bead milling, followed by high pressure homogenization. After each bead-milling passage, a discrete decrease in particle size was noted, resulting in average size (Photon Correlation Spectroscopy, PCS) of 413 nm and polydispersity index of 0.202. The laser diffraction data showed a similar pattern of particle size reduction, reaching a diameter 99% of 0.515 μm. The antioxidant capacity of apigenin nanocrystals was found to be two-fold relative to the original coarse suspension. The development of smart crystals after incorporation into gels on dermal application made apigenin available as an efficient antioxidant.

Zhang et al. (2011) investigated the oral and pulmonary delivery of nanocrystals to augment the bioavailability of baicalein, a bioactive flavonoid, isolated from the root of *Scutellaria baicalensis* Georgi. Baicalein nanocrystals were prepared by antisolvent recrystallization, followed subsequently by high-pressure homogenization. No change in crystalline structure was deciphered after fabrication. Baicalein nanocrystals containing small quantities of stabilizer significantly enhanced its dissolution rate. The mean relative bioavailability of baicalein nanocrystal administered perorally was found to be 1.67-fold that of oral baicalein. Pulmonary delivery of baicalein nanocrystal exhibited rapid and extensive absorption with pharmacokinetic parameters analogous to intravenous baicalein injection.

7.2.4 Self-Assembled Nanosystems

Self-assembled nanosystems are self-aggregates of amphiphilic prodrugs at the molecular level. However, the knowledge regarding amphiphilic prodrugs and their formation of self-assembled nanosystems is very limited (Jin et al., 2008). Poor aqueous solubility of various entities has led to the use of a delivery system for achieving sufficient drug bioavailability and/or to facilitate its translation to clinical or even preclinical research. Self-assembled drug-delivery systems (SEDDS), apart from allowing improved solubilization, offer some other advantages, including controlled drug release, protection from drug hydrolysis against chemical or enzymatic degradation, reduction of toxicity, and improvement of drug bioavailability (Verma et al., 2012). Some of the self-assembled systems are discussed in the following sections.

7.2.4.1 Cubosomes

Cubosomes are emerging as active self-assembled drug-delivery systems for efficient delivery of bioactives. These are honeycombed (cavernous) structures with two internal aqueous channels separated from a large interfacial area. These demonstrate a diverse internal cubic configuration and composition with different drug-loading modalities. Cubosomes offer various advantages, including a relatively simple preparation method and high drug-loading efficiency owing to high internal surface area and cubic crystalline structures, use of biodegradable lipids, ability of encapsulating hydrophobic, hydrophilic, and amphiphilic substances, site-specific delivery and controlled release of bioactive agents (Thadanki et al., 2011). The preparation of cubosomes involves simple emulsification of a monoglyceride and a polymer, accompanied by sonication and homogenization. The preparation methods generally include both top-down and bottom-up techniques.

Seung et al. (2013) have reported monoolein cubosomes of six herbal extracts, formed voluntarily by hydration. Incorporation of herbal extracts had no effect on the tortuous striped structure of the cubosome, comprising two intercrossing water channels spanning throughout monoolein cubic phase. Among these, *Thuja orientalis* extract, *Polygonum multiflorum* Thunberg extract, and *Espinosilla* extract showed hair growth–promoting efficacy. Cubosomes tend to significantly enhance the efficacy of the herbal extracts, and their potency was equivalent to that of minoxidil solution (2.4%) in propylene glycol/water/ethanol (20/30/50, v/v/v).

7.2.4.2 Reverse Micellar System

To enhance oral bioavailability and antidiabetic efficacy of berberine (BER), an anhydrous reverse micelle (ARM) delivery system was prepared through lyophilization of water-in-oil (w/o) emulsions by Wang et al. (2011). BER-containing w/o emulsions were prepared using soy phosphatidylcholine as emulsifiers and then subsequently lyophilized to form dry products, which, upon addition of oil, formed clear ARMs containing amorphous BER nanoparticles. A schematic diagram of release of the herbal active from an anhydrous reverse micelle is shown in Figure 7.3. *In vivo* studies demonstrated that the 22% reduction of the initial blood glucose levels (BGLs) in diabetic mice after 4 h of BER solution intravenous injection (2.5 mg/kg body weight), while with ARMs (100 mg/kg body weight), the average BGL reduction was about 57%.

7.2.4.3 Self-Nanoemulsifying Drug-Delivery System (SNEDDS)

Lipid-based suspension and emulsions have been employed to enhance the oral bioavailability, though, recently there has been increasing focus on the utility of self-nanoemulsifying drug-delivery systems

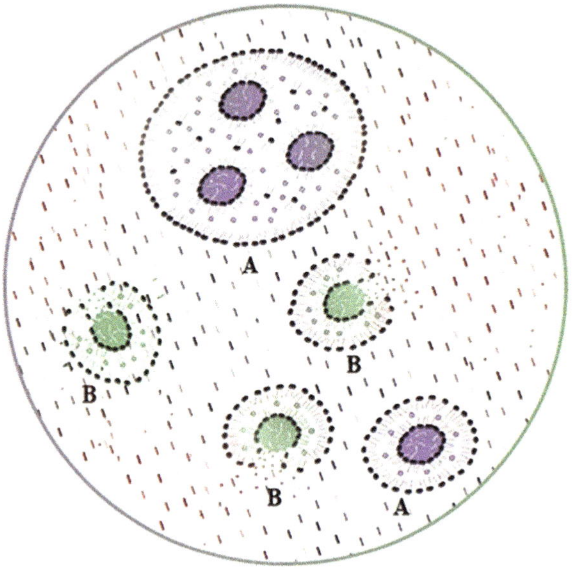

FIGURE 7.3 A schematic diagram of berberine release from the freeze-drying of water-in-oil. Emulsions (FWE) anhydrous reverse micelles (ARMs). (A) Oil droplets containing berberine-loaded ARMs, which are formed when the oily formulation is dispersed in an aqueous solution. (B) Broken berberine-loaded ARMs from which berberine is released. Small hollow dots represent oil phase, while short black bars represent water phase.

(SNEDDS). Being lipophilic, the SNEDDS ideally work for the oral bioavailability enhancement of a poorly water-soluble drug (Chavla and Bansal, 2008).

SNEDDS for zedoary turmeric oil (ZTO), an essential oil extracted from the dry rhizome of *Curcuma zedoaria,* for oral delivery, have been reported by Zhao et al. (2010). Pseudoternary phase diagrams facilitated the embarking of efficient emulsification regions. ZTO as a partial oil phase-enhanced drug loading with the aid of a secondary oil phase. With the increase in surfactant concentration, droplet size was reduced, whereas the emulsification time increased, while the reverse effect was observed by increasing the cosurfactant concentration. Thus, an optimized formulation consisting of ZTO, ethyl oleate, Tween 80, Transcutol P, (30.8:7.7:40.5:21, w/w) and with 30% drug was obtained.

Iosio et al. (2011) developed solid self-emulsifying pellets to deliver silymarin extract. These pellets were developed employing extrusion/spheronization to a self-emulsifying system of Akoline MCM, Miglyol, Tween 80, soy lecithin and propylene glycol, microcrystalline cellulose, and lactose monohydrate. The data demonstrated the application of extrusion/spheronization as a feasible technology for production of self-emulsifying pellets of fine quality. Further, this was able to augment *in vivo* oral bioavailability by more than 100-fold of the main components of phytotherapeutic extract, owing to enhanced lymphatic absorption.

7.2.5 Hydrogels

Hydrogels are the three-dimensional and cross-linked networks of water-soluble polymers, varying in chemical composition and bulk physical properties. Furthermore, hydrogels can be prepared in a diversity of physical forms: slabs, microparticles, nanoparticles, coatings, and films. As a result, hydrogels can be used for diverse applications in clinical practice and experimental medicine, including tissue engineering and regenerative medicine (Lee and Mooney, 2001), diagnostics (Van der Linden et al., 2003), cellular immobilization, separation of biomolecules or cells (Wang et al., 1993), and barrier materials to regulate biological adhesions (Bennett et al., 2003).

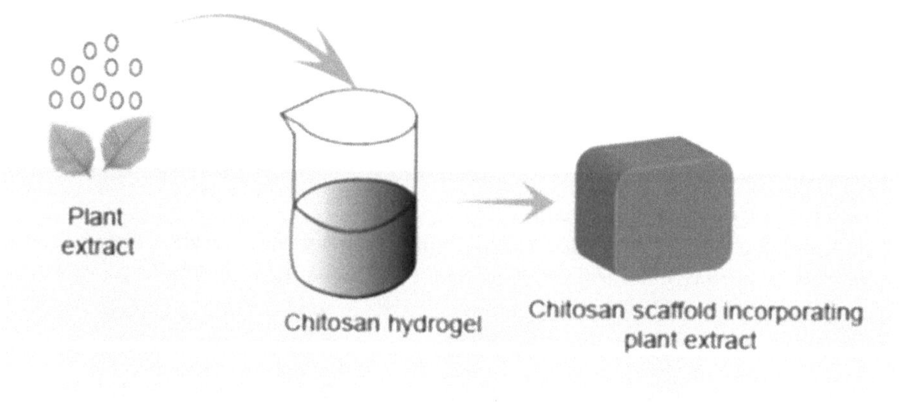

FIGURE 7.4 Schematic representation of the preparation of the plant extract-incorporated chitosan scaffold.

The highly porous structure of hydrogels can be maintained by keeping a check on the density of cross-links in the matrix and the affinity of the hydrogels for the aqueous environment in which they are swollen. Porosity is responsible for loading of drugs into the gel matrix, resulting subsequently in a rate-dependent manner of drug release through the gel network. Besides their use in systemic delivery of actives, the benefits of hydrogels are largely pharmacokinetic in nature such that a depot formulation is created from which drugs slowly release, maintaining a high local concentration of drug in the surrounding tissues over an extended period. Annapoorna et al. (2013) synthesized a *Hemigraphis alternata* extract–based chitosan scaffold for wound healing. Reduction in total colony forming units by 45-fold towards *Escherichia coli* and 25-fold against *Staphylococcus aureus* was observed employing an *H. alternata*–incorporated chitosan (HIC) scaffold. Human dermal fibroblast (HDF) cell cultures showed 90% viability after 48 h too, compared with the chitosan control. The herbal scaffold of chitosan demonstrated hemostatic and antibacterial characteristics. The obtained results ratified the fact that the herbal scaffold can be effectively applied for infectious wounds. The hydrogel yield was 125.0% and the hydrogel pellet obtained was freeze-dried to obtain the scaffold. Figure 7.4 shows the schematic representation of the preparation of a chitosan scaffold.

7.2.6 Carbon-Based Delivery Systems

These novel materials comprising carbon nanotubes (CNTs), fullerenes, and graphenes have been gaining significant attention. These materials exhibit a wide range of outstanding and novel features which makes these materials promising for numerous applications. The pioneer application for drug delivery was conceptualized by Dai et al. (2012). The authors reported that the physisorption via π-stacking can be applied for loading anticancer drugs. It is noteworthy that graphene and other graphene-derived materials have now been extensively researched in the area of herbal biomedicine, demonstrating an obvious future in this field. Dolatabadi et al. (2011) synthesized quercetin-based single-walled carbon nanotubes (SWCNTs) delivery system for anticancer activity. This introduced the concept of "functionalization partitioning" of SWCNTs, that is, imparting various chemical species, such as PEG, for controlling quercetin release by varying the pH.

7.3 ANTIOXIDANTS AS POTENTIAL ANTICANCER AGENTS

Lately, antioxidants have caught the vision of researchers for their promise in chemoprevention and management of cancer. Several bioactives with potential antioxidant activity have been explored and identified by the scientists and industrial researchers. Fundamentally, these act by scavenging free radicals, thereby disrupting the oxidative reactions chain. Food rich in antioxidants have been known to possess cancer-protective effects. These exert strong free-radical scavenging and chelating activities (De Marino et al., 2014). Antineoplastic activities of various phytoconstituents—such as lycopene (Jain et al., 2017a,b), resveratrol (Tatlidede et al., 2009), ascorbic acid (Pathak et al., 2005), α-tocopherol (Benzie, 2000), capsaicin (Shimeda et al., 2005), curcumin (Kuhad et al., 2007), garlic (Razo-Rodríguez et al., 2008) and quercetin (Jain et al., 2013)—have been well documented. The burgeoning interest in natural bioactives as a potential source of new drugs can be attributed to

- Need for relatively safer therapeutics
- Wide-ranging chemical structures and biological activities of natural secondary metabolites
- Capability of bioactive natural products to act as biochemical and molecular probes
- Recent development of efficient techniques to accurately detect, isolate, and structurally characterize the bioactive natural products
- Advances in solving the demand for supply of complex natural products

7.3.1 Proposed Mechanism of Action

The mechanisms of action for scavenging the free radicals of antioxidant molecules from plants sources (to prevent rancidity and control autooxidation of foodstuffs) varies greatly. The potential antioxidants consist of polyphenols which scavenge free radicals, and these possibly exert their action through inhibition of lipooxygenases by the formation of iron chelates and scavenging free radicals (Andreu et al., 2005). Iron, being an essential micronutrient, plays a vital role in cellular physiology. Iron concentration, on attaining certain levels, initiates a catalytic response towards oxyradicals. The free radicals have the ability to interrupt the membrane lipids, nucleic acid chains, carbohydrates, and proteins, eventually impairing the normal cellular physiology and functioning. Antioxidants, thus, may decrease the oxidative damage by formation of iron chelates (Fibach and Rachmilewitz, 2010). For eons, phytopolyphenols have been known to change multiple molecular targets and signalling pathways. Moreover, it is well-documented that the anticancer activity of antioxidants is owing to cell-cycle arrest and apoptosis, while its chemoprotective action is essentially attributed to its antioxidant and anti-inflammatory actions (Rajendran et al., 2015). Figure 7.5 portrays various such potential antioxidant agents categorized as per their source. Table 7.2 incorporates various targets aimed to curb cancer-using antioxidants. Table 7.3 highlights the selected plants providing selective antioxidant activity.

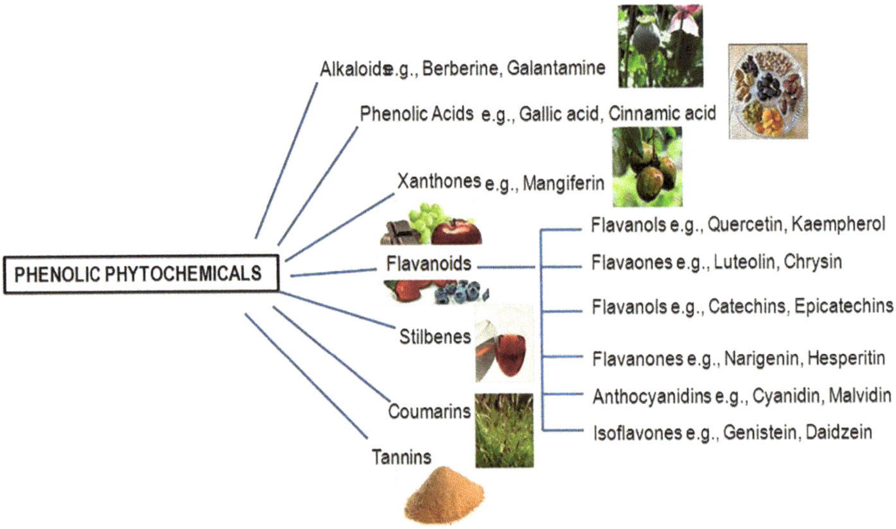

FIGURE 7.5 Classification of potential antioxidants.

TABLE 7.2
Various Molecular Targets of Antioxidants to Curb Cancer

Target	Biochemical Markers		References
Reactive oxygen species (ROS): Antioxidants scavenge ROS generated in the, cells which further inhibits xanthine oxidase, the enzyme responsible for oxidation	ROS	↓	Alfadda and Sallam (2012)
Anti-inflammation mediators: Inhibition of NFκB, IL-6, IL-8, CXCR4, XIAP, TNF, ICAM1, and COX2	NF-κB	↓	Garcia-Rivera et al. (2011)
	IκB-α	↓	
	JAK1/2	↓	
	STAT3	↓	
	AKT	↓	
	MAPK	↓	
	IKK	↓	
Anti-angiogenesis: Inhibition of tumor growth	VEGF	↓	Daud et al. (2010)
	TNFα	↓	
	B16F10	↓	
	FGF	↓	
Apoptosis: Supression of NF-κB activation and expressions of Bcl-xL and XAIP	Caspase 9,7,3	↑	Pan et al. (2014), Padma et al. (2015)
	XIAP	↓	
	Bcl2	↓	
	Bcl-xL	↓	
	PARP	↑	
Metastasis: Inhibition of β-catenin pathway	MMP 7-9	↓	Li et al. (2013)
	COX2	↓	
	VEGF	↓	
	ICAM	↓	
	E-cadherin	↑	
Cell cycle arrest: Down-regulating cdc2-cyclinB1 signalling pathway delays S-phase, arrest G_2/M phase, leading to apoptosis			Lv et al. (2013)
Mitochondrial membrane potential: Induction of loss of mitochondrial membrane potential and activate apoptotic proteins			Kavitha et al. (2014)

7.4 CONCLUSIONS AND FUTURE PROSPECTS

Exploring the herbal wealth for forbearance and treatment of ailments of the modern era has always been an attraction for the scientific community. A plethora of herbal drugs provides unlimited opportunities for further research to translate traditional herbal medicine into novel drug-delivery systems, thereby achieving controlled and target-specific delivery. Principles of nanotechnology can be further applied to increase the efficacy, site-specific delivery, and toxicity reduction of the herbal drugs. Such nanoscale transformations in the methods of delivery and administration demonstrate tremendous potential for increased outreach of herbal drugs with high patient compliance. These offer unlimited opportunities and challenges for future development. Thus, in a nutshell, novel nanophytomedicines, explored through a rational blend of the knowledge of traditional herbal

TABLE 7.3
Selected Plants Providing Selective Antioxidant Activity (in Alphabetical Order)

Plant(s)	Active Moiety(ies)	Details	Reference
Achyranthes aspera	Alkaloid, saponin	The methanol extract has been shown to exhibit significant inhibitory effects on the Epstein-Barr virus early antigen activation induced by the carcinogen 12-O-tetradecanoylphorbol-13-acetate in Raji cells.	Khan et al. (2015)
Allium sativum	Allicin, ajoene	Cytotoxicity was observed in higher doses for the primary cell fibroblasts than the permanent line baby hamster kidney cells, whereas the tumorgenic Burkitt lymphoma cells were quite sensitive.	Lu et al. (2011)
Andrographis paniculata	Flavonoids and labdane diterpenoids	These phytoconstituents exhibited potent tumor inhibition against the tested cell lines.	Qiao et al. (2017)
Annona muricata	Acetogenins	This blocks the production of adenosine triphosphate, inhibiting the efflux pump and thus acts against multidrug-resistant cancer.	Kim et al. (2016)
Apis mellifera	Protein	Enhances the proliferation of primary-cultured rat hepatocytes and in normal human lymphocytes, HL-60 cells.	Lee et al. (2007)
Astralagus hedysarum	Polysaccharide	Exhibits immuno-potentiating action in cancer chemotherapy.	Huang et al. (2013)
Bidens pilosa	Polyacetylenes, flavonoids, terpenoids, phenylpropanoids	Phenyl-1, 3, 5-heptatriyne revealed antioxidant activity on normal blood cell cultures.	Orhan et al. (2016)
Bolbostemma paniculatum	Tubeimoside-V	Exhibits apoptotic activity on glioblastoma cells.	Yang et al. (2013)
Camellia sinensis	Epigallocatechin-3-gallate	Inhibits invasion and migration of human colon and oral cancer cells.	Zhao et al. (2017)
Cannabis sativa	Terpenes and sesquiterpenes	Inhibits human breast cancer cell cultures	Frankova et al. (2014)
Centaurea ainetensis	Flavanoids	Cytotoxic activity has been reported in human colon carcinoma cells.	Ghantous et al. (2008)
Curcuma longa (Turmeric)	AR-turmerone and turmerone	Induces apoptosis by hindering cell-cycle progression, through inhibition of protein tyrosine kinase, protein kinase C, and bcl-2 mRNA expression and exhibits antibacterial and anticoagulant activities by inhibiting collagen and adrenaline-induced platelet aggregation.	Chattopadhyay et al. (2004)
Cymbopogon (Lemon grass)	Limonene, borneol, geraniol, geranyl acetate	Activates the apoptotic process by inhibiting tumor-cell viability	Sharma et al. (2009)
Eucalyptus	Eugenol,1,8-cineole,	Extract of eucalyptus leaves activates macrophages producing cytokine products such as ILs, and tumor necrosis factor (TNF), which causes killing of tumor cells.	Prabuseenivasan et al. (2006)

(*Continued*)

TABLE 7.3 (*Continued*)
Selected Plants Providing Selective Antioxidant Activity (in Alphabetical Order)

Plant(s)	Active Moiety(ies)	Details	Reference
Ferula asafoetida (Hing)	Asafoetida (oleo-gum-resin)	Presence of phenols induces phase II enzymes such as glutathione S-transferase and quinonereductase, and inhibits polyamine biosynthesis and DNA synthesis.	Iranshahy and Iranshahi (2011)
Gossypium hirsutum	Gossypol	Commonly known as cottonseed oil, is used in the treatment of metastatic carcinoma of endometrium or ovary and also used in HIV	Venkatachalam et al. (2017)
Hydrocotyle asiatica	Asiatic acid	Commonly known as brahmi, exhibited cytotoxic and antitumor properties.	Wu et al. (2017)
Momordica Charantia (Bitter Melon)	Momordin	Induces apoptosis by increasing caspase 3 activity. Extract of *M. charantia* augments Bax/Bcl-2, Bak/Bcl-2, or Bad/Bcl-2 and thus inhibits the growth of several cancer cells.	Li et al. (2012)
Mangifera indica	Mangiferin	It is a nutritional supplement, and prolongs the quality of life through increasing free-radical scavenging mechanism.	Ganogpichayagrai et al. (2017)
Nervilia fordii	Labdane diterpenoid	Its extracts have prominent anticancer effects when administered to S-180 mice and H-22 mice models and increase the survival rate of cancer-bearing mice.	Tang et al. (2014)
Oroxylum indicum	Oroxindin	Studies have proved its anticancer potential and cytotoxicity against Hep2 cell lines at a concentration of 0.05%.	Sithisarn et al. (2017)
Picrorrhiza kurroa	Picroside	It is a commonly used hepatoprotective agent and inhibits liver cancer growth formed due to exposure of chemicals in animal studies.	Kumar et al. (2016)
Rubia cordifolia	Hexapeptides, quinones	It has shown prominent antitumor activity by binding to eukaryotic 80S ribosomes and resulting in inhibition of aminoacyl-tRNA binding and peptidyl-tRNA translocation, which is a necessary mechanism for protein synthesis.	Gong et al. (2017)
Salvia miltiorrhiza	Tanshinone-I	A potential anticancer effect of tanshinone I was observed on breast cancer cells.	Park et al. (2017)
Scutellaria	Apigenin, chrysin	The flavonoids exhibited potent antitumor activity with different mechanism of actions.	Kan et al. (2017)
Silybum marianum	Silymarini	Silymarin was studied against UV radiation-induced skin cancer in mice and found to reduce the chances of tumor incidence, tumor multiplicity, and tumor volume.	Yaldiz (2017)

(*Continued*)

TABLE 7.3 (*Continued*)
Selected Plants Providing Selective Antioxidant Activity (in Alphabetical Order)

Plant(s)	Active Moiety(ies)	Details	Reference
Tanacetumparthenium (Feverfew)	Pyrethrum	Causes an interruption in the DNA replication due to the presence of highly reactive epoxide, lactone ring, and methylene groups. Further, it inhibits thymidine in DNA and causes intracellular thiol depletion, mitochondrial dysfunction, endoplasmic reticulum stress, and oxidative stress.	Pareek et al. (2011)
Vernonia amygdalina	Garcinia bioflavonoids	Mice were injected subcutaneously with 106 MCF-7 breast cancerous cells in the flank area and monitored for tumor growth.	Howard et al. (2017)
Withania somnifera	Withaferin-A	It has shown a significant tumor reducing activity in carcinomas like carcinoma of nasopharynx and mammary adenocarcinoma.	Henley et al. (2017)
Zingiber officinale	Zingiberine	Pre-application of its ethanolic extract resulted in inhibition of epidermal edema and hyperplasia.	Danwilai et al. (2017)

products with improvised know-how of nanostructured drug-delivery technologies, have tremendous potential to result in improved patient response and compliance.

A variety of nanocarrier systems are available for incorporation of herbal actives, selection of which will be dictated by the nature of the herbal active (i.e., whether extract or isolated compound), its physicochemical properties, water solubility, dose, route of administration, sterilization methods, stability, and scalability to commercial levels. Presently, several isolated active phytoconstituents have been fabricated into nanocarrier systems to improve their functionality and efficacy. However, if we have to move with the concepts of Ayurveda and traditional medicine, the thrust must be to focus on use of whole extracts instead of the isolated components, as the isolated compounds may have less efficacy *vis-à-vis* the whole plant or plant part extract. Isolation of the active component may be a tedious and expensive process with no definite assurance of efficacy. Nonetheless, use of whole plant extracts will put forth challenges of their detailed characterization, pharmacokinetic and biodistribution studies, and selection of chemicals or biomarkers when extracts are being used. Further, modern nanophytochemical research would be incomplete without concerted outcomes from bioactives with high therapeutic values, as well as their nanocarriers, on elucidation of the mechanisms of action, titration of their accurate dose, and determination of their pharmacokinetic parameters.

Polyherbal formulations form the major drive of traditional medicines due to the cumulative and synergistic effect of various phytocomponents against varied diseased conditions. The ultimate challenge is to formulate nanophytomedicines comprising whole plant extracts and multicomponent or polyherbal nanostructured drug delivery-systems for more effective usage.

Another major concern to be addressed in the future will be toxicity and safety issues of these nanostructured systems. Due to their nanometric size, nanocarrier-based systems are able to transcend all of the biological barriers within the cells, and the toxicity associated with this phenomenon has not so far been investigated in requisite detail. Further, research studies involving toxicity and safety issues arising due to the use of nanophytomedicines are still in their early stages of development. More stringent tests are required to assess the safety and toxicity of these nanosystems before their commercial use, since the health and safety of humans is of primary concern. Both short- and long-term toxicity studies need to be undertaken, in this context, in order to ensure the safety of these newer delivery systems.

REFERENCES

Alfadda AA, Sallam RM. Reactive oxygen species in health and disease. *J Biomed Biotechnol.* 2012; 2012: 1–14.

Andreu GP, Delgado R, Velho JA, Curti C, Vercesi AE. Iron complexing activity of mangiferin, a naturally occurring glucosylxanthone, inhibits mitochondrial lipid peroxidation induced by Fe2+-citrate. *Eur J Pharmacol.* 2005; 513: 47–55.

Annapoorna M, Sudheesh Kumar PT, Lakshman LR, Vinoth-Kumar L, Nair SV, Jayakumar R. Biochemical properties of *Hemigraphis alternata* incorporated chitosan hydrogel scaffold. *Carbohydr Polym.* 2013; 92: 1561–1565.

Aqil F, Khan MSA, Owais M, Ahmad I. Effect of certain bioactive plant extracts on clinical isolates of beta-lactamase producing methicillin resistant *Staphylococcus aureus. J Basic Microbiol.* 2005; 45: 106–114.

Arokiyaraj S, Saravanan M, Udaya Prakash NK, Arasu VM, Vijayakumar B, Vincent S. Enhanced antibacterial activity of iron oxide magnetic nanoparticles treated with Argemone mexicana L. leaf extract: An *in vitro* study. *Mater Res Bull.* 2013; 48: 3323–3327.

Bennett SL, Melanson DA, Torchiana DF, Wiseman DM, Sawhney AS. Next-generation hydrogel films as tissue sealants and adhesion barriers. *J Card Surg.* 2003; 18(6): 494–499.

Benzie IFF. Evolution of antioxidant defence mechanism. *Eur J Nutr.* 2000; 39: 53–61.

Cao X, Fu M, Wang L, Liu H, Deng W, Qu R et al. Oral bioavailability of silymarin formulated as a novel 3-day delivery system based on porous silica nanoparticles. *Acta Biomater.* 2012; 8: 2104–2112.

Chattopadhyay I, Biswas K, Bandyopadhyay U, Banerjee RK. Turmeric and curcumin: Biological actions and medicinal applications. *Curr Sci.* 2004; 87: 44–53.

Chavla G, Bansal AK. Improved dissolution of a poorly water soluble drug in solid dispersions with polymeric and non-polymeric hydrophilic additives. *Acta Pharm.* 2008; 58: 257–274.

Dai L, Chang DW, Baek J-B, Lu W. Carbon nanomaterials for advanced energy conversion and storage. *Small.* 2012; 8:1130–1166.

Danwilai K, Konmun J, Sripanidkulchai B, Subongkot S. Antioxidant activity of ginger extract as a daily supplement in cancer patients receiving adjuvant chemotherapy: A pilot study. *Cancer Res.* 2017; 9: 11–18.

Daud NH, Aung CS, Hewavitharana AK, Wilkinson AS, Pierson JT, Roberts-Thomson SJ et al. Mango extracts and the mango component mangiferin promote endothelial cell migration. *J Agri Food Chem.* 2010; 58: 5181–5186.

De Marino S, Festa C, Zollo F, Nini A, Antenucci L, Raimo G et al. Antioxidant activity and chemical components as potential anticancer agents in the olive leaf (*Olea europaea* L. cv Leccino.) decoction. *Anticancer Agents Med Chem.* 2014; 14: 1376–1385.

Dolatabadi JEN, Jamali AA, Hasanzadeh M, Omidi Y. Quercetin delivery into cancer cells with single walled carbon nanotubes. *Int J of Biosc, Biochem and Bioinfo.* 2011; 1: 1–5.

Fibach E, Rachmilewitz EA. The role of antioxidants and iron chelators in the treatment of oxidative stress in thalassemia. *Ann N Y Acad Sci.* 2010; 1202: 10–16.

Frankova A, Janatova A, Tauchen J, Kokoska L. *In vitro* antimicrobial and antioxidant activity of extracts from six chemotypes of medicinal cannabis. *Planta.* 2014; 81: S1–S381.

Ganogpichayagrai A, Palanuvej C, Ruangrungsi N. Antidiabetic and anticancer activities of *Mangifera indica* cv. Okrong leaves. *J Adv Pharm Technol Res.* 2017; 8: 19–24.

Garcia-Rivera D, Delgado R, Bougarne N, Haegeman G, Berghe WV. Gallic acid indanone and mangiferin xanthone are strong determinants of immunosuppressive anti-tumour effects of *Mangifera indica* L. bark in MDA-MB231 breast cancer cells. *Cancer Lett.* 2011; 305: 21–31.

Garg NK, Tyagi RK, Sharma G, Jain A, Singh B, Jain S et al. Functionalized lipid–polymer hybrid nanoparticles mediated codelivery of methotrexate and aceclofenac: A synergistic effect in breast cancer with improved pharmacokinetics attributes. *Mol Pharm.* 2017; 14: 1883–1897.

Ghantous A, Tayyoun AA, Lteif GA, Saliba NA, GaliMuhtasib G, El-Sabban M et al. Purified salograviolide A isolated from *Centaurea ainetensis* causes growth inhibition and apoptosis in neoplastic epidermal cells. *Int J Oncol.* 2008; 32: 841–849.

Gong XP, Sun YY, Chen W, Guo X, Guan JK, Li DY et al. Anti-diarrheal and anti-inflammatory activities of aqueous extract of the aerial part of *Rubia cordifolia. BMC Complement Altern Med.* 2017; 17: 20.

Han L, Fu Y, Cole AJ , Liu J, Wang J. Co-encapsulation and sustained-release of four components in ginkgo terpenes from injectable PELGE nanoparticles. *Fitoterapia.* 2012; 83: 721–731.

Henley AB, Yang L, Chuang KL, Sahuri-Arisoylu M, Wu LH, Bligh SW et al. *Withania somnifera* root extract enhances chemotherapy through 'priming'. *PLoS One.* 2017; 12: e0170917.

Howard CB, McDowell R, Feleke K, Deer E, Stamps S, Thames E et al. Chemotherapeutic vulnerability of triple-negative breast cancer cell-derived tumors to pretreatment with *Vernonia amygdalina* aqueous extracts. *Anticancer Res.* 2017; 36: 3933–3943.

Huang GC, Lee CJ, Wang KT, Weng BC, Chien TY, Tseng SH et al. Immunomodulatory effects of *Hedysarum polybotrys* extract in mice macrophages, splenocytes and leucopenia. *Molecules.* 2013; 18: 14862–14875.

Iosio T, Voinovicha D, Perissutti B, Serdoz F, Hasaa D, Dall' Acquac GS et al. Oral bioavailability of silymarin phytocomplex formulated as self-emulsifying pellets. *Phytomedicine.* 2011; 18: 505–512.

Iranshahy M, Iranshahi M. Traditional uses, phytochemistry and pharmacology of asafoetida (Ferula assa-foetida oleo-gum-resin)—A review. *J Ethnopharmacol.* 2011; 134: 1–10.

Jain A, Kesharwani P, Garg NK, Jain A, Jain SA, Jain AK et al. Galactose engineered solid lipid nanoparticles for targeted delivery of doxorubicin. *Colloids Surf B Biointerfaces.* 2015; 134: 47–58.

Jain A, Sharma G, Kushwah V, Garg NK, Kesharwani P, Ghoshal G et al. Methotrexate and beta-carotene loaded-lipid polymer hybrid nanoparticles: A preclinical study for breast cancer. *Nanomedicine (Lond).* 2017a; 12: 1851–1872.

Jain A, Sharma G, Kushwah V, Thakur K, Ghoshal G, Singh B, Jain S, Shivhare US, Katare OP. Fabrication and functional attributes of lipidic nanoconstructs of lycopene: An innovative endeavour for enhanced cytotoxicity in MCF-7 breast cancer cells. *Colloids Surf B.* 2017b; 152: 482–491.

Jain A, Thakur D, Ghoshal G, Katare OP, Singh B, Shivhare US. Formation and functional attributes of electrostatic complexes involving casein and anionic polysaccharides: An approach to enhance oral absorption of lycopene in rats *in vivo. Int J Biol Macromol.* 2016; 93: 746–756.

Jain AK, Jain A, Garg NK, Jain A, Jain SA, Tyagi RK et al. Adapalene loaded solid lipid nanoparticles gel: An effective approach for acne treatment. *Colloids Surf B Biointerfaces*. 2014; 121: 222–229.

Jain AK, Thanki K, Jain S. Co-encapsulation of tamoxifen and quercetin in polymeric nanoparticles: Implications on oral bioavailability, antitumor efficacy, and drug-induced toxicity. *Mol Pharm*. 2013; 10: 3459–3474.

Jeevitha D, Amarnath K. Chitosan/PLA nanoparticles as a novel carrier for the delivery of anthraquinone: Synthesis, characterization and *in vitro* cytotoxicity evaluation. *Colloids Surf B Biointerfaces*. 2013; 101: 126–134.

Jin Y, Xin R, Ai P, Chen D. Self-assembled drug delivery systems. *Int J Pharm*. 2008; 350: 330–337.

Kan X, Zhang W, You R, Niu Y, Guo J, Xue J. *Scutellaria barbata* D. Don extract inhibits the tumor growth through down-regulating of Treg cells and manipulating Th1/Th17 immune response in hepatoma H22-bearing mice. *BMC Complement Altern Med*. 2017; 17: 41.

Kavitha M, Manivasagam T, Essa MM, Tamilselvam K, Selvakumar GP, Karthikeyan S et al. Mangiferin antagonizes rotenone: Induced apoptosis through attenuating mitochondrial dysfunction and oxidative stress in SK-N-SH neuroblastoma cells. *Neurochem Res*. 2014; 39: 668–676.

Ke-pan L, Wang L, Yang L, Yang B, Chao D, Wang Y. Preparation, pharmacokinetics, and tissue distribution properties of icariin-loaded stealth solid lipid nanoparticles in mice. *Chinese Herbal Medicines*. 2012; 4: 170–174.

Kesharwani P, Tekade RK, Gajbhiye V, Jain K, Jain NK. Cancer targeting potential of some ligand-anchored poly(propylene imine) dendrimers: A comparison. *Nanomedicine*. 2011; 7: 295–304.

Khan N, Akhtar MS, Khan BA, Braga VA, Reich A. Antiobesity, hypolipidemic, antioxidant and hepatoprotective effects of *Achyranthes aspera* seed saponins in high cholesterol fed albino rats. *Arch Med Sci*. 2015; 11: 1261–1271.

Kim GT, Tran NK, Choi EH, Song YJ, Song JH, Shim SM et al. Immunomodulatory efficacy of standardized *Annona muricata* (Graviola) leaf extract via activation of mitogen-activated protein kinase pathways in RAW 264.7 macrophages. *Evid Based Complement Alternat Med*. 2016; 2016: 2905127.

Kuhad A, Pilkhwal S, Sharma S, Tirkey N, Chopra K. Effect of curcumin on inflammation and oxidative stress in cisplatin-induced experimental nephrotoxicity. *J Agric Food Chem*. 2007; 55: 10150–10155.

Kumar AB, Prasanna H, Lakshman T, Prabha H, Kumar RS. Phytosomes as novel drug delivery system for herbal medicine –A review. *Syst Rev Pharm*. 2017; 8: 5–7.

Kumar AS, Peter YA, Nadeau JL. Facile biosynthesis, separation and conjugation of gold nanoparticles to doxorubicin. *Nanotechnology*. 2008; 19: 495101–495110.

Kumar R, Gupta YK, Singh S, Raj A. Anti-inflammatory effect of *Picrorhiza kurroa* in experimental models of inflammation. *Planta Med*. 2016; 82: 1403–1409.

Landete JM. Ellagitannins, ellagic acid and their derived metabolites: A review about source, metabolism, functions and health. *Food Res Int*. 2011; 44: 1150–1160.

Lee KY, Mooney DJ. Hydrogels for tissue engineering. *Chemical Reviews*. 2001; 101:1869–1880.

Lee YJ, Kang SJ, Kim BM, Kim YJ, Woo HD, Chung HW. Cytotoxicity of honeybee (*Apis mellifera*) venom in normal human lymphocytes and HL-60 cells. *Chem Biol Interact*. 2007; 169: 189.

Lertsutthiwong P, Noomun K, Jongaroonngamsang N, Rojsitthisak P, Nimmannit U. Preparation of alginate nanocapsules containing turmeric oil. *Carbohydr Polym*. 2008; 74: 209–214.

Li CJ, Tsang SF, Tsai CH, Tsai HY, Chyuan JH, Hsu HY. *Momordica charantia* extract induces apoptosis in human cancer cells through caspase- and mitochondria-dependent pathways. *Evid Based Complement Alternat Med*. 2012; 2012: 1–11.

Li H, Huang J, Yang B, Xiang T, Yin X, Peng W et al. Mangiferin exerts antitumor activity in breast cancer cells by regulating matrix metalloproteinases, epithelial to mesenchymal transition, and beta-catenin signaling pathway. *Toxicol Appl Pharmacol*. 2013; 272: 180–190.

Liang J, Yan H, Puligundla P, Gao X, Zhou Y, Wan X. Applications of chitosan nanoparticles to enhance absorption and bioavailability of tea polyphenols: A review. *Food Hydrocoll*. 2017; 69: 286–292.

Lin YS, Tsai CP, Huang HY, Kuo CT, Hung Y, Huang DM et al. Well-ordered mesoporous silica nanoparticles as cell markers. *Chem Mater*. 2005; 17: 4570–4573.

Liong M, Lu J, Kovochich M, Xia T, Ruehm SG, Nel AE et al. Multifunctional inorganic nanoparticles for imaging, targeting, and drug delivery. *ACS Nano*. 2008; 2: 889–896.

Liu X, Chiying A, Peng J, Xuesong L, Longhu W. Protective effects of cationic bovine serum albumin-conjugated PEGylated tanshinone IIA nanoparticles on cerebral ischemia. *Biomaterials*. 2013; 34: 817–830.

Loaye AS, Shegokar R, Muller RH. Production and characterization of antioxidant apigenin nanocrystals as a novel UV skin protective formulation. *Int J Pharm*. 2011; 420: 133–140.

Lu X, Ross CF, Powers JR, Aston DE, Rasco BA. Determination of total phenolic content and antioxidant activity of garlic (*Allium sativum*) and elephant garlic (*Allium ampeloprasum*) by attenuated total reflectance-Fourier transformed infrared spectroscopy. *J Agric Food Chem.* 2011; 59: 5215–5221.

Lv J, Wang Z, Zhang L, Wang HL, Liu Y, Li C et al. Mangiferin induces apoptosis and cell cycle arrest in MCF-7 Cells Both *in vitro* and *in vivo*. *J Anim Vet Adv.* 2013; 12: 352–359.

Namdari M, Eatemadi A, Soleimaninejad M, Hammed AT. A brief review on the application of nanoparticle enclosed herbal medicine for the treatment of infective endocarditis. *Biomed Pharmacother.* 2017; 87: 321–331.

Orhan N, Icoz UG, Altun L, Aslan M. Anti-hyperglycaemic and antioxidant effects of *Bidens tripartita* and quantitative analysis on its active principles. *Iran J Basic Med Sci.* 2016; 19: 1114–1124.

Padma VV, Kalaiselvi P, Yuvaraj R, Rabeeth M. Mangiferin induces cell death against rhabdomyosarcoma through sustained oxidative stress. *Integr Med Res.* 2015; 4: 66–75.

Pan LL, Wang AY, Huang YQ, Luo Y, Ling M. Mangiferin induces apoptosis by regulating Bcl-2 and Bax expression in the CNE2 nasopharyngeal carcinoma cell line. *Asian Pac J Cancer Prev.* 2014; 15: 7065–7068.

Pareek A, Suthar M, Rathore GS, Bansal V. Feverfew (*Tanacetum parthenium* L.): A systematic review. *Pharmacogn Rev.* 2011; 5: 103–110.

Park CH, Shin SH, Lee EK, Kim DH, Kim MJ, Roh HH et al. Magnesium lithospermate B from *Salvia miltiorrhiza* bunge ameliorates aging-induced renal inflammation and senescence via NADPH oxidase-mediated reactive oxygen generation. *Phytother Res.* 2017; 31: 721–728.

Pathak AK, Bhutani M, Guleria R, Bal S, Mohan A, Mohanti BK, Sharma A, Pathak R, Bhardwaj NK, Prasad KN, Kochupillai V. Chemotherapy alone vs. chemotherapy plus high dose multiple antioxidants in patients with advanced non small cell lung cancer. *J Am Coll Nutr.* 2005; 24: 16–21.

Ponarulselvam S, Panneerselvam C, Murugan K, Aarthi N, Kalimuthu K, Thangamani S. Synthesis of silver nanoparticles using leaves of *Catharanthus roseus* Linn. G. Don and their antiplasmodial activities. *Asian Pac J Trop Biomed.* 2012; 574–580.

Prabhu N, Raj DT, Gowri Y. Synthesis of silver phyto nanoparticles and their antibacterial efficacy. *Dig J Nanomater Biostructures.* 2010; 5: 185–189.

Prabuseenivasan S, Jayakumar M, Ignacimuthu S. *In vitro* antibacterial activity of some plant essential oils. *BMC Complement Altern Med.* 2006; 6: 39.

Qianjun H, Zhang J, Chen F, Guo L, Zhu Z, Shi J. An anti-ROS/hepatic fibrosis drug delivery system based on salvianolic acid B loaded mesoporous silica nanoparticles. *Biomaterials.* 2010; 31: 7785–7796.

Qiao H, Chen L, Rui T, Wang J, Chen T, Fu T et al. Fabrication and *in vitro/in vivo* evaluation of amorphous andrographolide nanosuspensions stabilized by d-alpha-tocopheryl polyethylene glycol 1000 succinate/ sodium lauryl sulfate. *Int J Nanomedicine.* 2017; 12: 1033–1046.

Rajakumar G, Abdul Rahuman A. Larvicidal activity of synthesized silver nanoparticles using *Eclipta prostrata* leaf extract against filariasis and malaria vectors. *Acta Tropica.* 2011; 118: 196–203.

Rajendran P, Rengarajan T, Nandakumar N, Divya H, Nishigaki I. Mangiferin in cancer chemoprevention and treatment: Pharmacokinetics and molecular targets. *J Recept Signal Transduct Res.* 2015; 35: 76–84.

Rajkumar GG, Jain A, Goyal M. Rapid green synthesis of silver nanoparticles (AgNPs) using (*Prunus persica*) plants extract: Exploring its antimicrobial and catalytic activities. *J Nanomed Nanotechnol.* 2017; 8: 452.

Razo-Rodríguez AC, Chirino YI, Sánchez-González DJ, Martínez-Martínez CM, Cruz C, Pedraza-Chaverri J. Garlic powder ameliorates cisplatin-induced nephrotoxicity and oxidative stress. *J Med Food.* 2008; 11: 582–586.

Romero-Benavides JC, Ruano AL, Silva-Rivas R, Castillo-Veintimilla P, Vivanco-Jaramillo S, Bailon-Moscoso N. Medicinal plants used as anthelmintics: Ethnomedical, pharmacological, and phytochemical studies. *Eur J Med Chem.* 2017; 129: 209–217.

Sandhu PS, Kumar R, Beg S, Jain S, Kushwah V, Katare OP et al. Natural lipids enriched self-nano-emulsifying systems for effective co-delivery of tamoxifen and naringenin: Systematic approach for improved breast cancer therapeutics. *Nanomedicine.* 2017; 13: 1703–1713.

Seung RS, Gu K, Jeong W, Jin-Chul K. *In vivo* hair growth-promoting efficacies of herbal extracts and their cubosomal suspensions. *J Industr and Eng Chem.* 2013; 19: 1331–1339.

Sharma PR, Mondhe DM, Muthiah S, Pal HC, Shahi AK, Saxena AK et al. Anticancer activity of an essential oil from *Cymbopogon flexuosus. Chem Biol Interact.* 2009; 179: 160–168.

Shimeda Y, Hirotani Y, Akimoto Y, Shindou K, Ijiri Y, Nishihori T, Tanaka K. Protective effects of capsaicin against cisplatin-induced nephrotoxicity in rats. *Biol Pharm Bull.* 2005; 28: 1635–1638.

Shrivastava SR, Shrivastava PS, Ramasamy J. Mainstreaming of Ayurveda, yoga, naturopathy, Unani, Siddha, and homeopathy with the health care delivery system in India. *J Tradit Complement Med.* 2015; 5: 116–118.

Singh A, Ahmad I, Akhter S, Jain GK, Iqbal Z, Talegaonkar S et al. Nanocarrier based formulation of thymoquinone improves oral delivery: Stability assessment, *in vitro* and *in vivo* studies. *Colloids Surf B Biointerfaces.* 2013; 102: 822–832.

Sithisarn P, Nantateerapong P, Rojsanga P. Screening for antibacterial and antioxidant activities and phytochemical analysis of *Oroxylum indicum* fruit extracts. *Molecules.* 2017; 21: 446.

Tang MQ, Zhai YJ, Li XY, Li LX, Bai LH. Effect of nutrient solution in soilless cultivation on growth, antioxidase activities and amino acid contents of *Nervilia fordii. Zhong Yao Cai.* 2014; 37: 2127–2130.

Tatlidede E, Sehirli O, Velioğlu-Oğünc A, Cetinel S, Yeğen BC, Yarat A, Süleymanoğlu S, Sener G. Resveratrol treatment protects against doxorubicin-induced cardiotoxicity by alleviating oxidative damage. *Free Radic Res.* 2009; 43: 195–205.

Thadanki M, Paruchuri SK, Suria PK. Overview of cubosomes: A nano particle. *IJRPC.* 2011; 1: 535–541.

Trewyn BG, Giri S, Slowing II, Lin VSY. Mesoporous silica nanoparticle based controlled release, drug delivery, and biosensor systems. *Chem Commun.* 2007; 31: 3236–3245.

Tripathi S, Kushwah V, Thanki K, Jain S. Triple antioxidant SNEDDS formulation with enhanced oral bioavailability: Implication of chemoprevention of breast cancer. *Nanomedicine.* 2016; 12: 1431–1443.

Tsai YM, Chao FC, Lie CL, Tung HT. Curcumin and its nano-formulation: The kinetics of tissue distribution and blood–brain barrier penetration. *Int J Pharm.* 2011; 416: 331–338.

Van der Linden HJ, Herber S, Olthuis W, Bergveld P. Stimulus-sensitive hydrogels and their applications in chemical (micro) analysis. *Analyst.* 2003; 128: 325–331.

Ven HV, Vermeersch M, Matheeussen A, Vandervoort J, Weyenberg W, Apers S et al. PLGA nanoparticles loaded with the antileishmanial saponin ß-aescin: Factor influence study and *in vitro* efficacy evaluation. *Int J Pharm.* 2011; 420: 122–132.

Venkatachalam P, Priyanka N, Manikandan K, Ganeshbabu I, Indiraarulselvi P, Geetha N et al. Enhanced plant growth promoting role of phycomolecules coated zinc oxide nanoparticles with P supplementation in cotton (*Gossypium hirsutum* L.). *Plant Physiol Biochem.* 2017; 110: 118–127.

Verma G, Bhattacharjee J, Hassan PA. Self assembled systems: Design and drug delivery perspectives. *BARC Newsletter.* 2012; 328: 69–73.

Vijaykumar R, Devi V, Adavallan K, Saranya D. Green synthesis and characterization of gold nanoparticles using extract of anti-tumor potent *Crocus sativus. Physica E.* 2011; 44: 665–671.

Wang K, Burban J, Cussler E. *Hydrogels as Separation Agents.* Responsive gels: Volume transitions II; 1993; pp. 67–79.

Wang T, Song H, Xiaonan X, Wang J, Hao A, Li T. Preparation of an anhydrous reverse micelle delivery system to enhance oral bioavailability and anti-diabetic efficacy of berberine. *Eur J Pharm Biopharm.* 2011; 44: 127–135.

Weakley SM, Wang X, Mu H, Leu J, Lin PH, Yao Q et al. Ginkgolide A-Gold nanoparticles inhibit vascular smooth muscle proliferation and migration *in vitro* and reduce neointimal hyperplasia in a mouse model. *J Surg Res.* 2011; 171: 31–39.

Wu T, Geng J, Guo W, Gao J, Zhu X. Asiatic acid inhibits lung cancer cell growth *in vitro* and *in vivo* by destroying mitochondria. *Acta Pharm Sin B.* 2017; 7: 65–72.

Yadav A, Lomash V, Samim M, Flora SJ. Curcumin encapsulated in chitosan nanoparticles: A novel strategy for the treatment of arsenic toxicity. *Chem Biol Interact.* 2012; 199: 49–61.

Yaldiz G. Effects of potassium sulfate [K2SO4] on the element contents, polyphenol content, antioxidant and antimicrobial activities of milk thistle [*Silybum Marianum*]. *Pharmacogn Mag.* 2017; 13: 102–107.

Yang JB, Khan M, He YY, Yao M, Li YM, Gao HW et al. Tubeimoside-1 induces oxidative stress-mediated apoptosis and G0/G1 phase arrest in human prostate carcinoma cells *in vitro. Acta Pharmacol Sin.* 2013; 37: 950–962.

Yang P, Lian Y, Yan-qiu H, Li-na M, Yue-sheng C, Lei Z et al. Study on targeting and *in vitro* Anti-oxidation of baicalin solid lipid nanoparticles. *Chinese Herbal Medicines.* 2012; 4(4): 335–339.

Zahir AA, Rahuman AA. Evaluation of different extracts and synthesised silver nanoparticles from leaves of Euphorbia prostrata against *Haemaphysalis bispinosa* and *Hippobosca maculate. Vet Parasitol.* 2012; 187: 511–520.

Zhang J, Chen P, Sun C, Hu X. Sonochemical synthesis of colloidal silver catalysts for reduction of complexing silver in DTR system. *Appl Catal A.* 2004; 266: 49–54.

Zhang J, Lva H, Jiang K, Gao Y. Enhanced bioavailability after oral and pulmonary administration of baicalein nanocrystal. *Int J Pharm.* 2011; 420: 180–188.

Zhao X, Wang P, Li M, Wang Y, Jiang X, Cui L et al. Functional characterization of a new tea (*Camellia sinensis*) flavonoid glycosyltransferase. *J Agric Food Chem*. 2017; 65: 2074–2083.

Zhao Y, Wang C, Chow AHL, Ren K, Gong T, Zhang Z et al. Self-nanoemulsifying drug delivery system (SNEDDS) for oral delivery of zedoary essential oil: Formulation and bioavailability studies. *Int J Pharm*. 2010; 383: 170–177.

Zhu Y, Shi J, Shen W, Dong X, Feng J, Ruan M et al. Stimuli-responsive controlled drug release from a hollow mesoporous silica sphere/polyelectrolyte multilayers core-shell structure. *Angew Chem Int Ed*. 2005; 44: 5083–5087.

8 Phytoconstituents from Neem with Multiple Activities

Nanocarriers for Enhancing its Therapeutic Potential

Suman Chaudhary, Rupinder Kaur Kanwar, Teenu Sharma, Bhupinder Singh, and Jagat Rakesh Kanwar**

CONTENTS

8.1 INTRODUCTION

Natural and herbal products, of late, have gained enormous importance as precise sources of treatment in traditional medicine in order to treat a variety of diseases, ranging widely from minor infections to malignant diseases (Dhiman and Chawla, 2005; Firenzuoli and Gori, 2007; Dias et al., 2012; Fernanda et al., 2016). There are various research reports documenting the several pharmacological activities exhibited by a number of herbal plants (Dey et al., 2012; Mishra et al., 2016; Chang et al., 2017).

The prevalence of evidence confirms that when used in conjugation with conventional medicine, herbs augment the desired medicinal benefits while mitigating the harmful ones (www.motherearthnews). Herbal medicines, being natural, are devoid of significant side effects and are less likely to cause dependency. Herbs seldom exhibits ill effects which mainly includes transient reactions such as hot flashes, dizziness, headache, indigestion, and rashes that can be controlled with dose reduction or discontinued use (Tavakoli et al., 2012). However, herb-drug interactions along with herb safety remain controversial with their usage. Thus, with the escalating usage of herbs, their potential abuse and toxicity effects need to be considered quite legitimately (Tachjian et al., 2010).

* Corresponding author.

Most of the evidence pertaining to the same remain anecdotal, regardless of the fact that herbal medicines have been extensively used in cancer care (Khurana et al., 2018). Therefore, there is significant need for assessment of their efficacy and safety employing clinical trials (Yu et al., 2017). Further, interactions of synthetic drugs with herbs form an essential factor for consideration while administering various herbal medicines (Benzie and Wachtel-Galor, 2011).

Various herbs containing a variety of active phytochemicals have been identified with significant pharmacological properties. Some of the most commonly used remedial herbs are tabulated in Table 8.1. These have been screened for wide applications in the field of medicine for their antibacterial, antifungal, antioxidant, and anticancer activities. The spread of drug-resistant pathogens has made it essential to adopt alternative methods, among which essential oils derived from these have gained increasing interest. These are obtained from different plant materials (seeds, flowers, buds, leaves, fruits, etc.) and have proven to be biologically active compounds (Prabuseenivasan et al., 2006). Along with this, different herbs contain innumerous flavonoids, sulfides, terpenoids, saponins, and plant sterols, imparting oncolytic properties in these natural herbs (Craig, 1999).

Medicinal herbs along with their derivative phytocompounds are increasingly gaining recognition as potential complementary treatment for cancer. As per a population-based survey in the UK, about one quarter of patients had consulted an herbal medicine practitioner in the past (Thomas and Coleman, 2004), although a Canadian study shows that 20% of breast cancer patients used at least one herbal medicine treatment in the past (Ekor, 2013). Nevertheless, American studies report rates well above 65%, which are considerably higher than those reported in the general population or among other cancer diagnostic groups (Tavakoli et al., 2012). Herbal medicines are proposed to exhibit anticancer activity by improving the immune system, inducing cell differentiation, inhibiting telomerases, and inducing apoptosis of cancer cells (Agbarya et al., 2014).

In recent years, *Azadirachta indica (A. indica)*, common name; neem, has attracted worldwide prominence due to its wide usage in pathological conditions. Indian lilac or *Azadirachta indica (A. indica)* belongs to the Maliaceae (mahogany) family and is a deciduous or evergreen, fast-growing plant that can reach up to 25 m in height. It thrives in tropical climates with an annual rainfall of 400-800 m and can grow in areas with temperatures as high as 50°C. Its origin lies in southern and south-eastern Asia along with the tropical and subtropical areas of Australia, America and Africa (Schmutterer, 1990). It is a fast-growing tree that can reach up to 5-11 m in height, during its first 5 years. By the tenth year, each tree is able to produce up to 50 kg fruit per year and, thus, becomes fully productive. Neem trees are robust and resilient to extreme conditions such as severe droughts and hot climates. It can even exist in stony, shallow, poor, and sandy soils, which otherwise don't support agricultural crops (Ascher, 1993). It is relatively free from pest attacks, mainly due to the presence of a multitude of pesticidal ingredients called triterpenes (including limonoids) such as azadirachtin, nimbin, salannin, meliantriol, and nimbolide. These active components are not only restricted to agricultural applications but also have significant medicinal applications. All neem tree constituents, such as its leaves, seeds, oil, flowers and bark oil, exhibit extensive agro-medicinal properties.

8.2 NEEM AS AN ANTICANCER AGENT

8.2.1 Neem Leaf

Neem leaf extract has widely been studied for its several medicinal and pharmacological activities because of the ease in availability and extraction (Figure 8.1). It is considered as the storehouse of bioactives and contains 0.13% essential oils, that are responsible for its distinctive smell. Unlike neem oil, neem leaf extract is majorly used in the field of medicine due to its inherent antimicrobial, anti-inflammatory, anticancer, and antifertility attributes (Verma and Mehata, 2016; Gupta et al., 2017). Various studies have demonstrated significant antimicrobial efficacy of Neem leaves in a wide variety of viruses, fungi, and bacteria. The virucidal activity of neem leaves, obtained through methanolic extraction, have long-been reported against group-B-Coxsackie viruses (Badam et al.,

TABLE 8.1
Plant-Derived Active Ingredients with Therapeutic Properties

Herb	Active Ingredients	Medicinal: Mechanism of Action	References
Eucalyptus	Eugenol,1,8-cineole	Eucalyptus: extracts from leaves, stems, and flowers exhibit antibacterial, antifungal, and anticancer activity. Anticancer activity: activates macrophages, producing cytokine products such as interleukins (ILs), and tumor necrosis factor (TNF), which causes killing of tumor cells.	Prabuseenivasan et al. (2006), Ashour (2008), Sakarkar and Deshmukh (2011), Islam et al. (2012)
Lemon grass (*Cymbopogon*)	Limonene, borneol, geraniol, geranyl acetate	Lemon grass: antifungal, anti-inflammatory, antibacterial, antimutagenic (its oil has promising anticancer activity and activates the apoptotic process by inhibiting tumor cell viability).	Sharma et al. (2009), Charles (2013)
Feverfew (*Tanacetum parthenium*)	Pyrethrum	Main active ingredients: lactone parthenolide and sesquiterpene are used for a migraine prophylaxis remedy. It inhibits platelet aggregation, as well as release of serotonin from leucocytes and blood platelets. Anticancer: causes an interruption in the DNA replication due to the presence of highly reactive epoxide, lactone ring, and methylene groups. Further, it inhibits thymidine in DNA and causes intracellular thiol depletion, mitochondrial dysfunction, endoplasmic reticulum stress, and oxidative stress.	Ernst and Pittler (2000), Pareek et al. (2011)
Mentha species	Menthol	*Mentha*: contains phenolic antioxidants that prevent the recurrence of cancer.	Mimica-Dukič et al. (2003), Dorman et al. (2003)
Turmeric (*Curcuma longa*)	ar-turmerone and turmerone	Anti-inflammatory: curcumin offers an anti-inflammatory effect by supressing NFkB activation through downregulation of inducible nitric oxide synthetase and cyclooxygenase-2. Anticancer: prevents cell-cycle progression and induces apoptosis. The apoptotic effect of curcumin is caused through downregulation of protein tyrosine kinase, protein kinase C, and bcl-2 mRNA expression. Further, it is antibacterial and anticoagulant as it inhibits collagen and adrenaline-induced platelet aggregation.	Chattopadhyay et al. (2004)
Hing (*Ferula asafoetida*)	Asafoetida (oleo-gum-resin)	Antiviral, antioxidant, antifungal, chemopreventive, and antimutagentic activity. Its bioactive compounds exhibit radical scavenging activity, which causes a decrease in oxidative parameters. Anticancer: presence of phenols induce phase II enzymes such as glutathione S-transferase and quinone reductase and inhibit polyamine biosynthesis and DNA synthesis.	Saleem et al. (2001), Iranshahy and Iranshahi (2011)
Henna (*Lawsonia inermis*)	Quinones	Immunomodulatory: promotion of T-lymphocyte proliferative response. Antidiabetic: exhibit hypoglycemic and hypolipidemic activity. Antibacterial, antifungal	Chaudhary et al. (2010)
Mock strawberry (*Duchesnea indica*)	Methoxy dehydrochlesterol	Anti-inflammatory: It blocks the activation of NF-κB and supresses the release of pro-inflammatory cytokines.	Zhao et al. (2008), Miller et al. (2012)
Onion (*Allium sativum*)	Allicin	Antifungal, antibacterial, and antiviral. Antitumor and antimutagenic: induces apoptosis through activation of caspase-3, generation of hydrogen peroxide, degradation of poly(ADP-ribose) polymerase and DNA fragmentation.	Reuter (1995)
Bitter Melon (*Momordica charantia*)	Momordin	Antidiabetic: hypoglycemic Antibacterial, antiviral: crude as well as purified fraction exhibits anticancer activity Anticancer: induces apoptosis by increasing caspase 3 activity. Extract of *M. charantia* augments Bax/Bcl-2, Bak/Bcl-2, or Bad/Bcl-2 and thus inhibits the growth of several cancer cells.	Grover and Yadav (2004), Li et al. (2012)

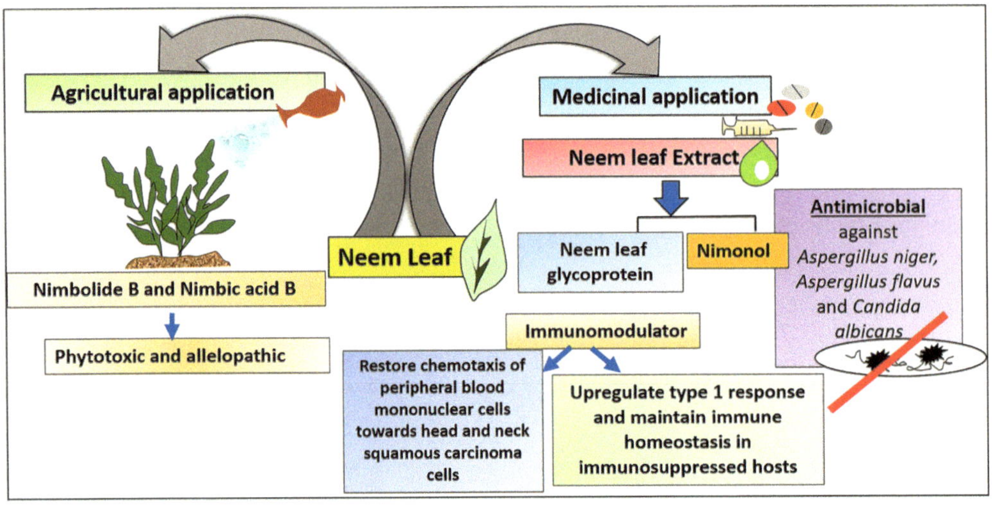

FIGURE 8.1 Pictorial depiction of medicinal and agricultural application of neem leaf components.

1999). Along with this, antifungal and antibacterial activity has been reported in which neem leaf extracts containing volatile sulfides and limonoids exerted high antimycotic activity against *Microsporidium, Epidermophyton, Trichosporon*, and *Candida* (Subapriya and Nagini, 2005). Further, oil extracted from neem leaves has been known to possess antibacterial and antifungal activity against *Staphylococcous aureus, Klebsiella pneumonia, P. vulgaris, S. typhi, S. dysenterae*, and *M. pyrigenes* (Asif, 2012). Such antibacterial activity of neem leaves and bark extracts have also been demonstrated against both gram-positive and -negative bacteria, including *Bacillus subtilis* and *Vibrio cholera* (Raut et al., 2014). These studies have successfully demonstrated the treasure trove of phytochemicals in neem leaves, with innumerous health benefits. However, the underlying metabolism, tissue distribution and toxicity profiles of these neem leaf components are lacking in current neem leaf–based therapeutics.

Considering the variable attributes of neem leaves, it also exhibits strong antifertility, antipyretic, analgesic, and anti-inflammatory activity (Subapriya and Nagini, 2005). Reports have indicated neem's apoptosis-inducing property, making it a potential candidate for developing reversible herbal contraceptives. It regulates fertility in females, by inducing granulosa cell apoptosis mediated by reactive oxygen species (ROS), causing deprivation of oocytes from nutrients, cell cycle proteins, and survival factors, which are necessary for follicular oocytes to achieve meiotic competency prior to ovulation (Sironmani, 2016). Further, it causes an increment in hydrogen peroxide (H_2O_2) levels, inducing activation of p53 along with overexpression of Bax protein. This triggers the activation of cytochrome c and subsequent induction of caspase-3 and caspase-9 activities, leading to oocyte apoptosis (Chaube et al., 2004). Although this property of neem leaves facilitates the regulation of fertility, it may also cause off-target effects. Therefore, it is important to consider the dose of neem leaf–based drugs during their oral administration for other pharmacological applications.

Neem extracts possess potential anticancer activity by inhibiting the onset of cancerous phenotypes occurring as a result of various procarcinogens and mutagens. It acts as an antimutagenic agent itself and also increases the antitumor potential of various drugs. One study examined the short-term chemopreventive effect and antioxidant activity of neem leaves. It was observed that neem leaf extract at all dose levels substantially reduced the formation of azoxymethane-induced aberrant crypt foci (ACF) and reduced proliferating cell nuclear antigen (PCNA) in the colon epithelium of rats. Even a high dose of 250 mg/kg body weight did not exhibit any acute toxicity in mice. However, the study lacked long-term examination of neem leaf–induced chemopreventive action, which necessitates in-depth evaluation (Arakaki et al., 2006).

Treatment with neem leaf extract moreover, exhibits hepato-protective efficacy. A skin tumor induces significant hepatic damage; this is characterized by an increment in the hepatic tissue injury marker enzymes and reduction in the hepatosomatic index, including alanine aminotransferase, alkaline transferase, and aspartate aminotransferase. Treatment with neem leaf extract reverses these alterations. It recedes oxidative stress in the hepatic tissue of a skin tumor by enhancing glutathione contents and by reducing the levels of lipid peroxidation. This evaluation was conducted using a crude extract of neem leaves. Nonetheless, an understanding of the components of neem leaves involved in this hepato-protective efficiency can aid in providing a more targeted approach for preventing hepatic damage (Koul et al., 2006).

Further, pretreatment of leukocytes with neem leaf preparations have been shown to inhibit 5-fluorouracil and cisplatin-induced apoptosis. It enhances the cytotoxicity by inducing the increase in the production of T cells and natural killer (NK) cells. Thus, neem leaf protein (NLP) can be used as an inexpensive substitute for granulocyte colony stimulating factor (GCSF) in reducing the side effects that are associated with cisplatin and 5-fluorouracil treatments (Ghosh et al., 2009). Another study conducted by Kumar et al. demonstrated the dose- and time-dependent cytotoxic potential of nimbolide, which is a limonoid of neem leaves and flowers, against human choriocarcinoma cells. It induces apoptosis by downregulating Bcl-2/Bax ratio, along with upregulating the expression of caspase-3, Apoptotic protease activating factor (Apaf)-1 and cleavage of poly(ADP-ribose) polymerase. Thus, it is validated that the apoptosis induced by nimbolide compound is mediated through mitochondrial pathway (Kumar et al., 2009). Consistent with the apoptosis-inducing potential of neem leaf, significant cytotoxic potential of neem leaf is also observed in various other cell lines, including leukemia (HL60), breast (SK-BR-3), lung (A549) and stomach (AZ521) cancer cell lines, with an IC_{50} value between 0.1 and 9.9 μm (Takagi et al., 2014). Although these studies have validated the anticancer potential of neem leaf, the studies lack complete evaluation of the dose-dependent toxicity associated with neem leaves. *In vivo* studies have indicated neem leaf–mediated decrease in heart rate and increased pulse rate in guinea pigs, along with genotoxicity as well as lethal toxicity in mice (Biswas et al., 2002). Therefore, critical evaluation of each fraction of neem leaves with potential cytotoxic properties is imperative. Such an analysis will not only facilitate a complete understanding of the mechanistic pathway involved but will also aid in preventing toxicity issues associated with neem leaves.

8.2.2 Neem Oil

Neem oil is extracted predominantly from the seeds and contains maximum limonoids, including azadirachtin, salannin, nimbin, cardenolide, nimbolide, and gedunin (Atawodi and Atawodi, 2009). Meliacin, a component of neem, is responsible for the bitter principle of neem seed oil. The limonoids are known to cause cancer cell specific cell-cycle arrest and apoptosis, suggesting that it could be a possible anticancer agent. However, the underlying mechanism involved in neem limonoid-mediated cell death is still undefined (Biswas et al., 2002). One study demonstrated this potential of neem oil, wherein the exposure of neem limonoids directly targeted the mitochondria, causing the release of pro-apoptotic proteins (cytochrome and apoptosis inducing factor) that induced caspase-dependent, as well as caspase-independent, apoptotic cell death, respectively. Additionally, microtubule-associated protein light chain 3 (LC3-II) accumulation in cancer cells was observed, suggesting that neem-induced cancer cell death may involve autophagy (Srivastava et al., 2012). However, no clinical trials have yet to be conducted using neem oil for cancer treatment; therefore further research is required to investigate the doses and duration of neem oil required for anticancer activities.

The immunomodulatory role of neem oil has been long investigated. Intraperitoneal treatment of mice with neem oil has been known to significantly enhance the phagocytic activity, along with elevation in the expression of Major histocompatibility complex (MHC) class-II antigens. Further, the treatment of spleen cells with neem oil has shown significant lymphocyte proliferation after three days of treatment, thus indicating the potential of neem oil as a nonspecific immunostimulant with an ability to activate a cell-mediated immune response (Upadhyay et al., 1992). Another study exploited the anti-inflammatory and antiarthritic property of neem seed oil by revealing the significant role of nimbidin tetranortriterpene as the active agent. Results indicated that oral administration of nimbidin inhibited macrophage migration to peritoneal cavities as a result of inflammatory stimuli. Further, it prevented phagocytosis along with respiratory burst stimulated by phorbol-12-myristate-13-acetate (PMA) in the cells. It also inhibited prostaglandin E_2 and nitric oxide production when stimulated by lipopolysaccharide. Thus, suggesting a significant role of nimbidin in inhibiting macrophage and neutrophil functions (Kaur et al., 2004). Although the studies have evidenced the immunomodulatory role of neem oil, its inherent, unpleasant, sharp odor makes it unpalatable and limits its use for pharmacological applications. Also, neem oil intoxication can produce nausea, diarrhea, encephalopathy, and acidosis in humans (Elakovich, 1996).

The antifertility properties of neem oil have been studied for decades. An early study conducted by Sinha et al., documented this attribute of neem oil wherein intravaginal application of neem oil prevented pregnancy as it exhibits significant spermicidal action (Sinha et al., 1984). This antifertility efficiency of neem oil was also validated in a recent study on cyclic albino rats in which a single dose of neem oil at varying durations on the mucosal surface of the vagina was able to disturb the reproductive cycle (Singh et al., 2014).

8.2.3 Neem Flower

From January to February, neem tree flowers are whitish-pink in color, small (5 mm long), pentamerous, and have a jasmine-like fragrance. As pharmacological agents, they offer a cheap, effective, widely accepted and available alternative as a contraceptive, which is equally nontoxic, noninvasive, long acting, and nonhormonal in action. Its potential as a female contraceptive was demonstrated in a study in which alcoholic extracts of the neem flower caused disruption in the estrous cycle in Sprague-Dawley rats by partially blocking ovulation. However, some level of toxicity was also observed in the treated rats as they suffered from diarrhea (Gbotolorun et al., 2008). Reports have also suggested the chemopreventive potential of neem flowers. Dietary feeding of neem flowers significantly increased the glutathione *S*-transferase (GST) activity along with subsequent reduction in hepatic P450-dependent monooxygenases. Reduction in the multiplicity of tumors per rat was observed, which was 44.0% for mammary gland tumors and 87.9% and 88.9% for benign and malignant tumors, respectively (Tepsuwan et al., 2002). However, the study was performed using crude extracts of the neem flower. Isolation of each component and their analysis is necessary to determine whether the chemopreventive effect was a result of one or multiple phytochemicals.

8.2.4 Neem Bark

Along with the other components of the neem tree, neem bark also possesses significant medical applications. The bark of neem contains proteins, amino acids, sugars, and oil. It is also rich in phenolic limonoids, triterpene, unsaturated sterol, and saponone. It has mainly been studied for its pharmacological action, including its gut-protection effect (Biswas et al., 2002). This pharmacological effect of bark is due to the presence of phenolic glycosides. It prevents the adhering of mucus and the depletion of endogenous glutathione, thus exhibiting gastroprotection against stress ulcers. One study indicated that neem bark significantly blocks lipid peroxidation and scavenges endogenous hydroxyl radicals, thus preventing oxidative damage, which is the main factor for ulcers. Further, it indicated that bark is more efficient than the known antiulcer antioxidants, including vitamin E, melatonin and desferrioxamine, thus validating the therapeutic potential of neem bark for controlling gastric ulcers and hyperacidity (Bandyopadhyay et al., 2002). This study was further evaluated in human subjects, in which aqueous extracts of neem bark was used to orally treat patients suffering from gastroduodenal ulcers and acid-related problems. Significant reduction in the secretion of gastric acid was observed when patients were administered 30 mg of aqueous neem bark daily. Overall, a 30-60 mg dose twice a day was able to completely heal duodenal ulcers (Bandyopadhyay et al., 2004).

8.3 ACTIVE PHYTOCONSTITUENTS OF NEEM

8.3.1 Azadirachtin

The bioactivity of neem seed oil is mainly attributed to its rich content of azadirachtin. Azadirachtin, along with nimbolide, exerts significant cytotoxic effects on the viability of the human cervical cancer cell line (HeLa). It induces p53-dependent p21 accumulation, along with cell-cycle arrest at the G0/G1 phase, and causes decrease in cell-cycle regulatory proteins including cyclin b, PCNA, and cyclin D1. Further, it causes modifications in the nuclear morphology and induces apoptotic signals via a mitochondrial pathway (Priyadarsini et al., 2010). However, both azadirachtin and nimbolide are capable of causing toxicity since a high dose of azadirachtin ($LD_{50} > 5000$ mg/kg bw) and intraperitoneal or intravenous administration of nimbolide can cause acute toxicity in experimental animals (Glinsukon et al., 1986; Raizada et al., 2001). Another study, conducted on the bioinsecticide azadirachtin A, indicated that it is not genotoxic and has an antiproliferative effect on the Chinese hamster ovary cells (CHO) and on human lymphocytes. It modulates the first and second (M1 and M2) division metaphases, thus interfering with the cell-cycle progression (Mosesso et al., 2012). Nonetheless, extensive investigations are required on the precise molecular mechanisms, pharmacokinetics, and toxicity of these neem limonoids to explore the therapeutic and preventive potential of neem in humans.

8.3.2 Nimbin and Nimbidin

Other components of *A. indica* with significant biological activity are nimbin and nimbidin. Nimbidin is a complex mixture of tetranortriterpenes extracted as the active component from the neem seed oil. It includes nimbin, nimbolide, nimbidinin, nimbinin, and nimbidic acid and possesses potential anti-inflammatory, antiulcer, diuretic, and antiarthritic potential (Pillai et al., 1984; Naik et al., 2014; Vimala and Gricilda Shoba, 2014). Oral administration of nimbidin has shown to significantly inhibit the functions of macrophages and neutrophils, during both *in vitro* as well as *in vivo* exposure, wherein it inhibits the phagocytic capacity, chemotaxis, release of inflammatory mediators, and lytic cells from these cells (Kaur et al., 2004). Several studies have documented the antimicrobial activity of nimbidin and nimbolide. Antifungal activity of neem leaf extracts was conducted on some human pathogens, including *Aspergillus niger, Aspergillus flavus*, and *Candida albicans, in vitro*. Analysis through high performance liquid chromatography (HPLC) indicated nimonol as the main antifungal component. However, nimonol as a separate component was less effective compared with the mixed fractions, indicating a potential synergism of the constituents present in Neem leaf for antifungal activity (Mahmoud et al., 2011). Evaluation of neem oil for its antifungal activity has demonstrated that the antifungal potential of neem oil is not a function of one component alone, but is due to a mixture of tetranortriterpenoids. An early study has demonstrated the antifungal activity of Neem oil against 13 phytopathogenic fungi, including *Fusarium oxysporum, Drechslera oryzae*, and *Alternaria tenuis*. This activity was prominent in fractions containing a mixture of compounds such as nimbin, salannin, azadiradione, and epoxyazadiradione. However, pure active compounds did not show appreciable antifungal activity, thus validating the synergistic/additive effects of the components (Govindachari et al., 1998).

8.3.3 Neem Leaf Glycoprotein (NLGP)

NLGP is a nontoxic component of neem leaves, extracted and investigated for its potential role as an immunomodulator. Importantly, neem leaf protein contains an active ingredient, glycoprotein, which is known to exhibit immunotherapeutic, immunoprophylactic, and immunostimulatory activity (Chakraborty et al., 2008). This glycoprotein has the ability to replace the conventional therapies that, although reduce the tumor mass, also promote immunosuppressive milieu and cause inhibition of tumor effector cells. The potential of this functional ingredient in exerting antitumor activity was demonstrated by Chakraborty et al., in head and neck squamous cell carcinoma during which dysregulation between CXCL10 and CXCR3 inhibits the chemotaxis of cytotoxic cells at the tumor site. It was observed that treatment with NLGP enhanced the secretion of interferon γ (IFNγ), along with the upregulation of CXCR3A, the subsequent downregulation of CXCR3B, and an increase in the ratio of the chemokine (C-X-C motif) receptor (CXCR) 3A/CXCR3B, thus restoring the chemotaxis of peripheral blood mononuclear cells (PBMCs) towards head and neck squamous cells (Chakraborty et al., 2008). The immunomodulatory role of NLGP was also investigated in terms of observed upregulation of early activation markers, and a cluster of differentiation (CD69) and CD45RO. Also, there was an increased secretion of T-helper (Th) 1 cytokine interferon-γ observed. Thus, the study effectively indicated the potential role of NLGP in upregulating a type 1 response and maintaining normal immune homeostasis in immunosuppressed hosts (Bose et al., 2009). A recent study has shown that NLGP impedes B16-melanoma in mice, thus enhancing their survival. Subcutaneous treatment of NLGP weekly for four weeks was able to restrict melanoma growth and maintain 35% tumor-free mice on day 60 post-treatment. It suggested a direct association of the anticarcinogenic effect of NLGP with the dominance of type 1 chemokines/cytokines, enhancement in the activity of $CD8^+$ T cells and downregulation of suppressive cellular functions (Barik et al., 2015). In conclusion, NLGP could efficiently diminish tumor growth through the activation of subdued T-cell function in the tumor microenvironment along with maintaining optimum type-1/regulatory balance. Another study evaluated this immunomodulatory role of nontoxic NLGP by pretreating a solid tumor with NLGP in mice. Results demonstrated vascular normalization in carcinoma and melanoma, along with reduction of vascular endothelial growth factor (VEGF), vascular endothelial growth factor receptor (VEGFR2), and CD31. The treatment caused $CD8^+$ T cells to infiltrate within the tumor, which lead to regulation of VEGF-VEGFR2 signalling, thus preventing aberrant neovascularization (Banerjee et al., 2014).

Although the immunomodulatory role of NLGP has been intensively investigated in the above studies, its application is restricted due to its sensitivity to temperature and pH. It denatures at a temperature of 100°C and on exposure to proteolytic enzymes (trypsin) that can limit its use in oral delivery, thus, causing an urgent need for an alternative strategy (Kundu et al., 2014).

8.4 ANTICANCER NANOTHERAPEUTICS OF NEEM

Nanotechnology is an emerging field of nanoscience, flourishing over the years due to its enormous application in various fields, including drug delivery, cancer diagnostics, and therapeutics (De Jong and Borm, 2008; Rathod et al., 2011). Its capability to encapsulate neem active ingredients has previously been studied and documented extensively as illustrated in Table 8.2.

Several carrier systems, including polymeric and lipidic nanoparticles (Figure 8.2), have gained enormous significance owing to their solid matrices, that protects the functional compounds like neem from degradation and facilitates modulation in the release profile if these bioactives (Wang et al., 2011). Encapsulation of functional ingredients and/or extractives in nanocarriers ensures improvement in stability and sustainability, and provides the controlled release of anticancer active ingredients, thus making them an ideal drug-delivery tool. Additionally, these nanoparticles offer improved biocompatibility, ability to functionalize or modify the surface, encapsulation of active ingredient, sustained release and biodegradability. One of the major challenges associated with the use of Neem is the high photosensitivity of azadirachtin, which tends to break down or isomerize in the presence of sunlight. Thus, encapsulation of neem oil/extracts into lipidic nanocarriers or polymeric membranes improves its efficiency (as these carrier systems protect the active ingredients, thereby preventing the loss of volatile compounds and increasing their stability in the environment) and allows controlled release (Riyajan and Sakdapipanich, 2009).

Recently, nanocarriers for natural products have been developed so as to control the release rate as well as targeting of bioactives through nanoencapsulation. This technique elevates the stability,

TABLE 8.2
List of Nanocarriers Used to Encapsulate Neem Active Components and Their Potential Medicinal Applications

Neem Components	Active Ingredient/Extract	Carrier	Nanoparticle Size	Potential Application	Reference
Neem oil	–	Nanocapsules	around 400 nm	Cytotoxicity and genotoxicity results indicated increase in toxicity of the neem nanocapsule formulations	Pasquoto-Stigliani et al. (2017)
Neem leaves	Neem leaf extract	Silver nanoparticles	2–100 nm	The apoptotic activity and antiproliferative activity were established *in vitro* with gastric cancer cells	Sironmani (2016)
Neem leaves	Aqueous extract of neem leaves	Silver nanoparticles (AgNPs)	approximately 4.0 nm	The AgNPs formed presented enhanced antimicrobial activity and indicated zone of inhibition against isolated bacteria (*Escherichia coli*).	Verma and Mehata (2016)
Neem	Leaf extract	Silver nanoparticles	up to 200 nm	Results indicated that AgNPs treated seeds exhibited improved germination rate and oxidative stress enzyme activity.	Banerjee et al. (2014)
Neem	Azadirachtin	Chemically conjugated glycol chitosan and octahydrogenated retinoic acid	312–410 nm	Anticancer: controlled release formulations exhibited significant inhibition of cell proliferation compared to free azadirachtin	Lu et al. (2013)
Neem oil	Oleic acidpalmitic acid, salicylic acid and linoleic acid	Solid lipid nanoparticles	221.6 ± 2.0 nm	Antibacterial efficacy against acne microbe *Propionibacterium acne*	Vijayan et al. (2013)
Neem oil	Azadirachtin	Nanoemulsion	31.03 ± 1.73 nm	Larvicidal effect against *Culex quinquefasciatus* responsible for zoonotic diseases including West Nile fever, avian malaria and lymphatic filariasis	Anjali et al. (2012)
Neem leaf	Neem leaf extract	Ag nanoparticles	10–100 nm	Antibacterial efficacy against *E.coli*, *Bacillus cereus*, *Bacillus subtilis*, *Enterobacter spp.*, *Klebsiella pneumoniae*, *Staphylococcus aureus*, *Vibrio cholera*, *Salmonella typhi*, and anticancer activity	Renugadevi and Aswini (2012)
Neem leaf	Neem leaf broth	Ag nanoparticles	440–1.6 nm.	Larvicidal activity against *Anopheles stephensi*, *Aedes aegypti,* and *Culex quinquefasciatus*, responsible for malaria, dengue, and filariasis.	Arjunan et al. (2012)
Neem oil	Azadirachtin Aza-A	Sodium alginate (Na-Alg), coated with starch and polyethylene glycol	1.28–1.49 mm	Can be used effectively for both agricultural and medicinal applications	Jerobin et al. (2012)
Neem leaf	Neem leaf broth	Au/Ag nanoparticles	Silver-450 nm Gold-550 nm	To develop environmentally friendly nanoparticles without the use of toxic chemicals during their synthesis	Shankar et al. (2004)

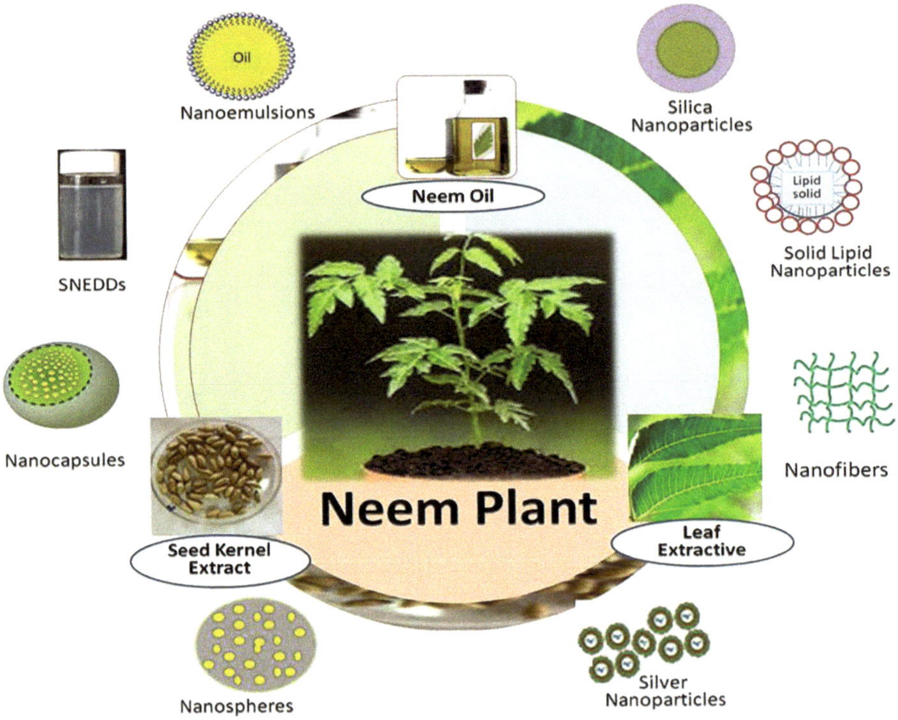

FIGURE 8.2 Different nanostructured delivery systems developed using various parts of neem plant.

sustainability and solubility of natural compounds and subsequently their efficacy. Therefore, in general, these novel nanocarrier formulations facilitates improved biological performance of herbal bioactives holistically for optimum delivery of the functional ingredients to its active site, along with improvement in the active moieties' photostability and thermostability.

Also, there are many reports suggesting the side effects of neem pesticides in terrestrial ecosystems rather than aquatic ecosystem. Therefore, various novel nanocarriers for neem extract/oil are developed so as to obtain systems in agriculture that are less toxic to the environment as well as human health, and that are also efficient biopesticides (de Oliveira et al., 2014).

8.5 CONCLUSIONS

Neem, lately, has been investigated for its myriad of pharmacological activities, as its phytoconstituents are capable of preventing and treating various chronic diseases. However, in this review, the anticancer potential of the various phytoconstituents in amalgamation to varied nanocarrier systems is being highlighted. The functional ingredients derived from leaves, oil, flowers, and bark of neem act through multiple apoptotic pathways by inhibiting the proliferation and migration of tumor. Additionally, one component of neem, NLGP, has immunomodulatory properties, which can be exploited to generate antitumor immunity. Nevertheless, the major challenges with the delivery of its functional ingredients and/or extractives include stability, controlled release, biocompatibility, and biodegradability. Thus, these challenges can be circumvented by the usage of several carrier systems, including polymeric and lipidic nanoparticles, which would not only protect the bioactive compounds like neem from degradation, but would also enable release profile modulations. Therefore, these components of neem either alone or in combination with advanced nanotherapy can provide a new generation of herbal therapeutics.

ACKNOWLEDGMENT

The authors would like to thank the National Health and Medical Research Council (AISRF), the Australia–India Strategic Research Fund (AISRF), and the Indian Council of Agriculture Research (ICAR) for their financial support. The authors declare no conflict of interest.

REFERENCES

Agbarya A, Ruimi N, Epelbaum R, Ben-Arye E, Mahajna J. Natural products as potential cancer therapy enhancers: A preclinical update. *SAGE Open Med.* 2014; 2. Doi: 2050312114546924.

Anjali C, Sharma Y, Mukherjee A, Chandrasekaran N. Neem oil (*Azadirachta indica*) nanoemulsion—A potent larvicidal agent against *Culex quinquefasciatus. Pest Manag Sci.* 2012; 68(2): 158–163.

Arakaki J, Suzui M, Morioka T, Kinjo T, Kaneshiro T, Inamine M et al. Antioxidative and modifying effects of a tropical plant *Azadirachta indica* (neem) on azoxymethane-induced preneoplastic lesions in the rat colon. *Asian Pac J Cancer Prev.* 2006; 7(3): 467.

Arjunan N K, Murugan K, Rejeeth C, Madhiyazhagan P, Barnard D R. Green synthesis of silver nanoparticles for the control of mosquito vectors of malaria, filariasis, and dengue. *Vector Borne Zoonotic Dis.* 2012; 12(3): 262–268.

Ascher K. Nonconventional insecticidal effects of pesticides available from the neem tree, *Azadirachta indica. Archives of Insect Biochemistry and Physiology.* 1993; 22(3–4): 433–449.

Ashour H M. Antibacterial, antifungal, and anticancer activities of volatile oils and extracts from stems, leaves, and flowers of *Eucalyptus sideroxylon* and *Eucalyptus torquata. Cancer Biol Ther.* 2008; 7(3): 399–403.

Asif M. Antimicrobial potential of *Azadirachta indica* against pathogenic bacteria and fungi. *J Pharmacogn Phytochem.* 2012; 1(4).

Atawodi S E, Atawodi J C. *Azadirachta indica* (neem): A plant of multiple biological and pharmacological activities. *Phytochem Rev.* 2009; 8(3): 601–620.

Badam L, Joshi S, Bedekar S. *In vitro* antiviral activity of neem (*Azadirachta indica.* A. Juss) leaf extract against group B coxsackie viruses. *The Journal of Communicable Diseases.* 1999; 31(2): 79–90.

Bandyopadhyay U, Biswas K, Chatterjee R, Bandyopadhyay D, Chattopadhyay I, Ganguly C K et al. Gastroprotective effect of neem (*Azadirachta indica*) bark extract: Possible involvement of H+-K+-ATPase inhibition and scavenging of hydroxyl radical. *Life Sci.* 2002; 71(24): 2845–2865.

Bandyopadhyay U, Biswas K, Sengupta A, Moitra P, Dutta P, Sarkar D et al. Clinical studies on the effect of neem (*Azadirachta indica*) bark extract on gastric secretion and gastroduodenal ulcer. *Life Sci.* 2004; 75(24): 2867–2878.

Banerjee S, Ghosh T, Barik S, Das A, Ghosh S, Bhuniya A et al. Neem leaf glycoprotein prophylaxis transduces immune dependent stop signal for tumor angiogenic switch within tumor microenvironment. *PloS One.* 2014; 9(11): e110040.

Barik S, Banerjee S, Sarkar M, Bhuniya A, Roy S, Bose A et al. Neem leaf glycoprotein optimizes effector and regulatory functions within tumor microenvironment to intervene therapeutically the growth of B16 melanoma in C57BL/6 mice. *Trials in Vaccinology.* 2015; 4: e80–e87.

Benzie I F F, Wachtel-Galor S. *Herbal Medicine: Biomolecular and Clinical Aspects.* 2nd edition. CRC Press: US; 2011.

Biswas K, Chattopadhyay I, Banerjee R K, Bandyopadhyay U. Biological activities and medicinal properties of neem (*Azadirachta indica*). *Curr. Sci.* 2002; 82(11): 1336–1345.

Bose A, Chakraborty K, Sarkar K, Goswami S, Haque E, Chakraborty T et al. Neem leaf glycoprotein directs T-bet–associated type 1 immune commitment. *Hum Immunol.* 2009; 70(1): 6–15.

Chakraborty K, Bose A, Pal S, Sarkar K, Goswami S, Ghosh D et al. Neem leaf glycoprotein restores the impaired chemotactic activity of peripheral blood mononuclear cells from head and neck squamous cell carcinoma patients by maintaining CXCR3/CXCL10 balance. *Int Immunopharmacol.* 2008; 8(2): 330–340.

Chang N, Luo Z, Li D, Song H. Indigenous uses and pharmacological activity of traditional medicinal plants in Mount Taibai, China. *Evid Based Complement Alternat Med.* 2017; 2017: 1–11.

Charles D J. Lemongrass. In: *Antioxidant Properties of Spices, Herbs and Other Sources.* Springer; 2013; pp. 377–383.

Chattopadhyay I, Biswas K, Bandyopadhyay U, Banerjee R K. Turmeric and curcumin: Biological actions and medicinal applications. *Curr Sci.* 2004; 87(1): 44–53.

Chaube S K, Shrivastav T G, Tiwari M, Prasad S, Tripathi A, Pandey A K. Neem (*Azadirachta indica* L.) leaf extract deteriorates oocyte quality by inducing ROS-mediated apoptosis in mammals. *Springer Plus.* 2004; 3(1): 464.

Chaudhary G, Goyal S, Poonia P. *Lawsonia inermis* Linnaeus: A phytopharmacological review. *Int J Pharm Sci Drug Res.* 2010; 2(2): 91–98.

Craig W J. Health-promoting properties of common herbs. *Am J Clin Nutr.* 1999; 70(3): 491s–499s.

De Jong W H, Borm P J. Drug delivery and nanoparticles: Applications and hazards. *Int J Nanomedicine.* 2008; 3(2): 133.

de Oliveira J L, Campos E V, Bakshi M, Abhilash P C, Fraceto L F. Application of nanotechnology for the encapsulation of botanical insecticides for sustainable agriculture: Prospects and promises. *Biotechnol Adv.* 2014; 32(8): 1550–1561.

Dey Y N, Ota S, Srikanth N, Jamal M, Wanjari M. A phytopharmacological review on an important medicinal plant —*Amorphophallus paeoniifolius. Ayu.* 2012; 33(1): 27–32.

Dhiman R, Chawla Y. Herbal medicines for liver diseases. *Dig Dis Sci.* 2005; 50(10): 1807–1812.

Dias D A, Urban S, Roessner U. A historical overview of natural products in drug discovery. *Metabolites.* 2012; 2(2): 303–336.

Dorman H D, Kosar M, Kahlos K, Holm Y, Hiltunen R. Antioxidant properties and composition of aqueous extracts from *mentha* species, hybrids, varieties, and cultivars. *J. Agric. Food Chem.* 2003; 51(16): 4563–4569.

Ekor M. The growing use of herbal medicines: Issues relating to adverse reactions and challenges in monitoring safety. *Front Pharmacol.* 2013; 4(1–10).

Elakovich S D. *The Neem Tree: Source of Unique Natural Products for Integrated Pest Management, Medicinal, Industrial, & Other Purposes* Edited by Heinrich Schmutterer (Giessen U., FGR). VCH: New York. 1995. xxi+ 680 pp. $125.00. ISBN 3-527-30054-6." *J Am Chem Soc.* 1996; 118(16): 3999–4000.

Ernst E, Pittler M. The efficacy and safety of feverfew (*Tanacetum parthenium* L.): An update of a systematic review. *Public Health Nutr.* 2000; 3(4a): 509–514.

Fernanda R S, Pinheiro N, Mernak M. Evidences of herbal medicine-derived natural products effects in inflammatory lung diseases. *Mediators Inflamm.* 2016; 2016: 1–14.

Firenzuoli F, Gori L. Herbal medicine today: Clinical and research issues. *Evid Based Complement Alternat Med.* 2007; 4(Suppl 1): 37–40.

Gbotolorun S, Osinubi A, Noronha C, Okanlawon A. Antifertility potential of neem flower extract on adult female Sprague-Dawley rats. *Afr. Health Sci.* 2008; 8(3): 168–173.

Ghosh D, Bose A, Haque E, Baral R. Neem (*Azadirachta indica*) leaf preparation prevents leukocyte apoptosis mediated by cisplatin plus 5-fluorouracil treatment in Swiss mice. *Chemotherapy.* 2009; 55(3): 137–144.

Glinsukon T, Somjaree R, Piyachaturawat P, Thebtaranonth Y. Acute toxicity of nimbolide and nimbic acid in mice, rats and hamsters. *Toxicol Lett.* 1986; 30(2): 159–166.

Govindachari T, Suresh G, Gopalakrishnan G, Banumathy B, Masilamani S. Identification of antifungal compounds from the seed oil of *Azadirachta indica. Phytoparasitica.* 1998; 26(2): 109–116.

Grover J, Yadav S. Pharmacological actions and potential uses of *Momordica charantia*: A review. *J Ethnopharmacol.* 2004; 93(1): 123–132.

Gupta S C, Prasad S, Tyagi A K, Kunnumakkara A B, Aggarwal B B. Neem (Azadirachta indica): An indian traditional panacea with modern molecular basis. *Phytomedicine.* 2017; 34: 14–20.

Iranshahy M, Iranshahi M. Traditional uses, phytochemistry and pharmacology of asafoetida (*Ferula assa-foetida* oleo-gum-resin)—A review. *J Ethnopharmacol.* 2011; 134(1): 1–10.

Islam F, Khatun H, Ghosh S, Ali M, Khanam J. Bioassay of Eucalyptus extracts for anticancer activity against Ehrlich ascites carcinoma (EAC) cells in Swiss albino mice. *Asian Pac J Trop Biomed.* 2012; 2(5): 394–398.

Jerobin J, Sureshkumar R, Anjali C, Mukherjee A, Chandrasekaran N. Biodegradable polymer based encapsulation of neem oil nanoemulsion for controlled release of Aza-A. *Carbohydr Polym.* 2012; 90(4): 1750–1756.

Kaur G, Sarwar Alam M, Athar M. Nimbidin suppresses functions of macrophages and neutrophils: Relevance to its antiinflammatory mechanisms. *Phytother Res.* 2004; 18(5): 419–424.

Khurana R K, Jain A, Jain A, Sharma T, Singh B, Kesharwani P. Administration of antioxidants in cancer: Debate of the decade. *Drug Discov Today.* 2018; 23(4): 763–770.

Koul A, Ghara A R, Gangar S C. Chemomodulatory effects of *Azadirachta indica* on the hepatic status of skin tumor bearing mice. *Phytother Res.* 2006; 20(3): 169–177.

Kumar G H, Mohan K C, Rao A J, Nagini S. Nimbolide a limonoid from *Azadirachta indica* inhibits proliferation and induces apoptosis of human choriocarcinoma (BeWo) cells. *Invest New Drugs.* 2009; 27(3): 246–252.

Kundu P, Barik S, Sarkar K, Bose A, Baral R, Laskar S. Chemical investigation of neem leaf glycoprotein used as immunoprophylactic agent for tumor growth restriction. *Int J Pharm Pharm Sci.* 2014; 7(2).

Li C-J, Tsang S-F, Tsai C-H, Tsai H-Y, Chyuan J-H, Hsu H-Y. *Momordica charantia* extract induces apoptosis in human cancer cells through caspase-and mitochondria-dependent pathways. *Evid Based Complement Alternat Med.* 2012; 2012: 261971.

Lu W, Lu M L, Zhang Q P, Tian Y Q, Zhang Z X, Xu H H. Octahydrogenated retinoic acid-conjugated glycol chitosan nanoparticles as a novel carrier of azadirachtin: Synthesis, characterization, and *in vitro* evaluation. *J Polym Sci A Polym Chem.* 2013; 51(18): 3932–3940.

Mahmoud D, Hassanein N, Youssef K, Zeid A. Antifungal activity of different neem leaf extracts and the nimonol against some important human pathogens. *Braz J Microbiol.* 2011; 42(3): 1007–1016.

Miller K I, Qing C, Sze D M Y, Neilan B A. Investigation of the biosynthetic potential of endophytes in traditional Chinese anticancer herbs. *PloS one.* 2012; 7(5): e35953.

Mimica-Dukić N, Božin B, Soković M, Mihajlović B, Matavulj M. Antimicrobial and antioxidant activities of three *mentha* species essential oils. *Planta medica.* 2003; 69(05): 413–419.

Mishra A, Seth A, Maurya S. Therapeutic significance and pharmacological activities of antidiarrheal medicinal plants mention in Ayurveda: A review. *J Intercult Ethnopharmacol.* 2016; 5(3): 290–307.

Mosesso P, Bohm L, Pepe G, Fiore M, Carpinelli A, Gäde G et al. Cytogenetic analyses of azadirachtin reveal absence of genotoxicity but marked antiproliferative effects in human lymphocytes and CHO cells *in vitro. Toxicol Lett.* 2012; 213(3): 361–366.

Naik M, Bhattacharya A, Behera R, Agrawal D, Dehury S, Kumar S. Study of anti-inflammatory effect of neem seed oil (*Azadirachta indica*) on infected albino rats. *J Health Res.* 2014; 1(3): 66.

Pareek A, Suthar M, Rathore G S, Bansal V. Feverfew (*Tanacetum parthenium* L.): A systematic review. *Phcog J.* 2011; 5(9): 103.

Pasquoto-Stigliani T, Campos E V, Oliveira J L, Silva C M, Bilesky-José N, Guilger M et al. Nanocapsules containing neem (Azadirachta Indica) oil: Development, characterization, and toxicity evaluation. *Sci Rep.* 2017; 7: 5929.

Pillai N, Santhakumari G, Laping J. Some pharmacological actions of 'nimbidin'-A bitter principle of *Azadirachta Indica*-A Juss (neem). *Ancient Sci Life.* 1984; 4(2): 88.

Prabuseenivasan S, Jayakumar M, Ignacimuthu S. *In vitro* antibacterial activity of some plant essential oils. *BMC Complem Altern M.* 2006; 6(1): 39.

Priyadarsini R V, Murugan R S, Sripriya P, Karunagaran D, Nagini S. The neem limonoids azadirachtin and nimbolide induce cell cycle arrest and mitochondria-mediated apoptosis in human cervical cancer (HeLa) cells. *Free Radic Res.* 2010; 44(6): 624–634.

Raizada R, Srivastava M, Kaushal R, Singh R. Azadirachtin, a neem biopesticide: Subchronic toxicity assessment in rats. *Food Chem Toxicol.* 2001; 39(5): 477–483.

Rathod K B, Patel M B, Parmar P K, Kharadi S R, Patel P V, Patel K S. Glimpses of current advances of nanotechnology in therapeutics. *Int J Pharm Pharma Sci.* 2011; 3: 8–12.

Raut R R, Sawant A R, Jamge B B. Antimicrobial activity of *Azadirachta indica* (neem) against pathogenic microorganisms. *J Acad Ind Res (JAIR).* 2014; 3(7): 327.

Renugadevi K, Aswini R V. Microwave irradiation assisted synthesis of silver nanoparticle using *Azadirachta indica* leaf extract as a reducing agent and *in vitro* evaluation of its antibacterial and anticancer activity. *Int J Nanomat Bio.* 2012; 2: 5–10.

Reuter H. Allium sativum and *Allium ursinum*: Part 2 pharmacology and medicinal application. *Phytomedicine.* 1995; 2(1): 73–91.

Riyajan S, Sakdapipanich J T. Development of a controlled release neem capsule with a sodium alginate matrix, crosslinked by glutaraldehyde and coated with natural rubber. *Polym Bull.* 2009; 63: 609–622.

Sakarkar D, Deshmukh V. Ethnopharmacological review of traditional medicinal plants for anticancer activity. *Int J PharmTech Res.* 2011; 3(1): 398–308.

Saleem M, Alam A, Sultana S. Asafoetida inhibits early events of carcinogenesis: A chemopreventive study. *Life Sci.* 2001; 68(16): 1913–1921.

Schmutterer H. Properties and potential of natural pesticides from the neem tree, *Azadirachta indica. Annu Rev Entomol.* 1990; 35(1): 271–297.

Shankar S S, Rai A, Ahmad A, Sastry M. Rapid synthesis of Au, Ag, and bimetallic Au core–Ag shell nanoparticles using neem (*Azadirachta indica*) leaf broth. *J Colloid Interface Sci.* 2004; 275(2): 496–502.

Sharma P R, Mondhe D M, Muthiah S, Pal H C, Shahi A K, Saxena A K et al. Anticancer activity of an essential oil from *Cymbopogon flexuosus. Chem Biol Interact.* 2009; 179(2): 160–168.

Sironmani T A. Therapeutic Potential of Neem Synthesized Silver Nanoparticles on Human Gastric Cancer Cells *in Vitro. WJNSE.* 2016; 6: 90–110.

Singh H R, Singh R R, Dubey A K, Jha S K. Effect of ivnt on ovarian weight, follicle number, estrus frequency and reversibility of reproductive cycle in albino rat. *Int J Pharm Sci Res.* 2014; 5(8): 3293.

Sinha K, Riar S, Tiwary R, Dhawan A, Bardham J, Thomas P et al. Neem oil as a vaginal contraceptive. *Indian J Med Res.* 1984; 79: 131–136.

Srivastava P, Yadav N, Lella R, Schneider A, Jones A, Marlowe T et al. Neem oil limonoids induces p53-independent apoptosis and autophagy. *Carcinogenesis*: 2012; bgs269.

Subapriya R, Nagini S. Medicinal properties of neem leaves: A review. *Curr Med Chem Anticancer Agents.* 2005; 5(2): 149–156.

Tachjian A, Maria V, Jahangir A. Use of herbal products and potential interactions in patients with cardiovascular diseases. *J Am Coll Cardiol.* 2010; 55(6): 515–525.

Takagi M, Tachi Y, Zhang J, Shinozaki T, Ishii K, Kikuchi T et al. Cytotoxic and melanogenesis-inhibitory activities of limonoids from the leaves of *Azadirachta indica* (neem). *Chem. Biodivers.* 2014; 11(3): 451–468.

Tavakoli J, Miar S, Majid Zadehzare M, Akbari H. Evaluation of effectiveness of herbal medication in cancer care: A review study. *Iran.* 2012; 5(3): 144–156.

Tepsuwan A, Kupradinun P, Kusamran W R. Chemopreventive potential of neem flowers on carcinogen-induced rat mammary and liver carcinogenesis. *Asian Pac J Cancer Prev.* 2002; 3(3): 231–238.

Thomas K, Coleman P. Use of complementary or alternative medicine in a general population in Great Britain. Results from the National Omnibus survey. *J of Pub Health.* 2004; 26: 152–157.

Upadhyay S N, Dhawan S, Garg S, Talwar G. Immunomodulatory effects of neem (*Azadirachta indica*) oil. *Int J Immunopharmacol.* 1992; 14(7): 1187–1193.

Verma A, Mehata M. Controllable synthesis of silver nanoparticles using Neem leaves and their antimicrobial activity. *J Radiat Res Appl Sci.* 2016; 9(1): 109–115.

Vijayan V, Aafreen S, Sakthivel S, Reddy K R. Formulation and characterization of solid lipid nanoparticles loaded Neem oil for topical treatment of acne. *J Acute Dis.* 2013; 2(4): 282–286.

Vimala G, Gricilda Shoba F. A review on antiulcer activity of few Indian medicinal plants. *Int J Food Microbiol.* 2014; 2014: 1–14.

Wang J J, Zeng Z W, Xiao R Z, Xie T, Zhou G L, Zhan X R et al. Recent advances of chitosan nanoparticles as drug carriers. *Int.* 2011; 6: 765–774.

Yu J Y, Gupta B, Park H. Preclinical and clinical studies demonstrate that the proprietary herbal extract DA-5512 effectively stimulates hair growth and promotes hair health. *Evid Based Complement Alternat Med.* 2017; 2017: 1–11.

Zhao L, Zhang S-L, Tao J-Y, Jin F, Pang R, Guo Y-J et al. Anti-inflammatory mechanism of a folk herbal medicine, *Duchesnea indica* (Andr) Focke at RAW264. *7 Cell Line Immunol Invest.* 2008; 37(4): 339–357.

9 Developing Taxanes for Oral Intake Employing Apt Nanocarriers

Rajneet Kaur Khurana, Teenu Sharma, Harpreet Kaur, Anchal, Kaisar Raza, and Bhupinder Singh**

CONTENTS

9.1 INTRODUCTION

The National Cancer Institute (NCI) of the US started a course to treat cancer through plant extracts in the year 1960. The researchers at the US Department of Agriculture gathered various plant parts including twigs, bark, and leaves of the Pacific yew tree (http://www.cancer.gov). Later, Wani *et al.* (1971) isolated paclitaxel from *Taxus brevifolia* belonging to the family Taxaceae. The sojourn of paclitaxel and other taxanes has been schematically presented in Figure 9.1. Bristol-Myers Squibb, NY, US, composed and marketed a paclitaxel ethanolic formulation in 1992, popularly called Taxol, which consisted of an emulsifier, namely, Cremophor EL (Gao and Morozowich, 2006). Since the early 1990s, it has been widely accepted clinically, exhibiting distinct prowess in retarding tumor growth. Being extracted from a completely natural source, and that too with low yield, there were chances of extinction of the yew tree plant (Butler, 2008) . This has led to the synthesis of a blockbuster semisynthetic compound in the chemo-dominated world. Docetaxel was originally derived from 10-deacetyl baccatin III, an inactive taxoid precursor, extracted from *Taxus baccata*, which is marketed and licensed as Taxotere by Sanofi-Aventis (Fauzee *et al.*, 2011).

These alkaloidal esters differ in physicochemical characteristics due to paclitaxel having a 3′-benzamide phenyl group and 10-acetyl group, whereas docetaxel has a 3′-$OC(CH_3)_3$ group and 10-OH group (Figure 9.2). Interestingly, owing to a small structural difference, docetaxel is reported to be quite a bit more soluble than paclitaxel. Also, the bulky and fused rings, along with the various functional moieties, are responsible for the high hydrophobicity and lattice energy of the taxanes. The pertinent information on various physicochemical properties of taxanes has been depicted in Table 9.1.

These diterpenoids promote microtubule polymerization and restrain microtubule disassembly, thus hampering cell division and reproduction (Pazdur *et al.*, 1993). Higher binding affinity of

* Corresponding author.

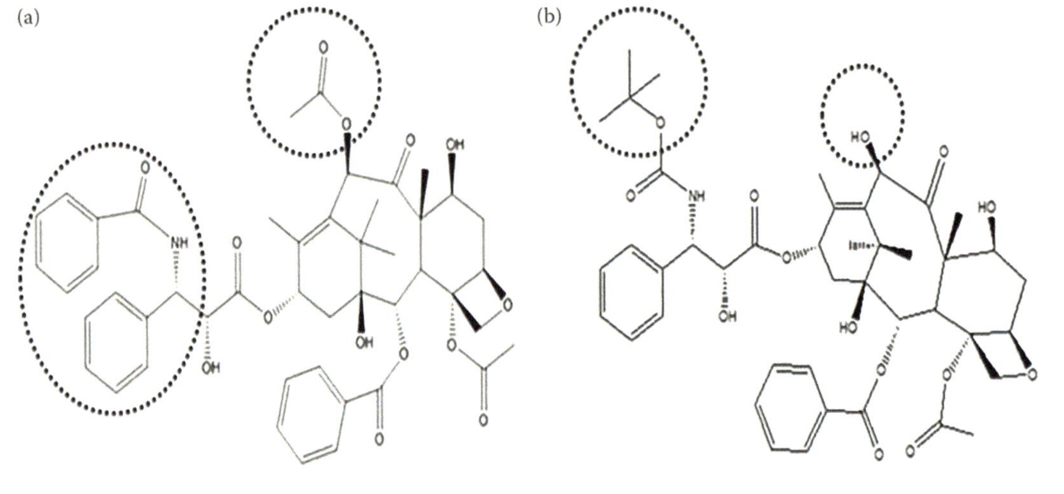

FIGURE 9.1 Sojourn of taxanes since the discovery of Taxol.

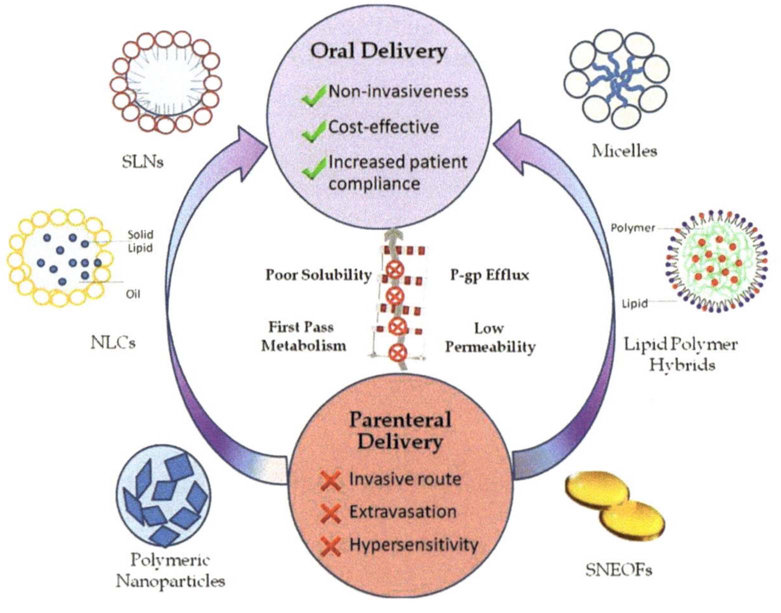

FIGURE 9.2 Structural representation of taxanes: (a) Paclitaxel and (b) Docetaxel.

TABLE 9.1
Structural and Physicochemical Differences amongst Paclitaxel and Docetaxel

	Physicochemical Differences of Taxanes	
Features	**Paclitaxel**	**Docetaxel**
Functional group	3′-benzamide phenyl group and 10-acetyl group	3′-$OC(CH_3)_3$ and 10-OH
Solubility in water	0.3–0.75 μg/mL	3–25 μg/mL
Binding affinity to microtubules	1 fold	1.9 fold

docetaxel to microtubules results in significantly higher cellular accumulation (Galletti *et al.*, 2007). Cellular uptake studies show three-fold higher uptake and decreased efflux of docetaxel (Verweij *et al.*, 1994). A notable observation is the difference in the pharmacokinetic profile of docetaxel and paclitaxel. The former shows linear pharmacokinetic behavior up to 115 mg/m^2; however, the latter is known to exhibit a nonlinear pharmacokinetic profile (Eisenhauer and Vermorken, 1998; Winefield *et al.*, 2008).

Other remarkable properties of these microtubule binding agents are their antiangiogenic and vascular-disrupting properties at relatively lower levels (Zou *et al.*, 2012), with a direct or indirect effect on the endothelial cells, resulting in the proposed vascular collapse, probably due to cascade of inhibition of their proliferation, the tube formation, and migration (Benelli *et al.*, 2009). Microtubule interaction with transient self-assembling protein complexes is essentially required for stability, motility, and cell-cell interactions; thus, its inhibition leads to a reduction in tumor growth (Schwartz, 2009).

9.2 SHORTCOMINGS OF EXISTING MARKETED PARENTERAL FORMULATIONS

Parenteral formulations of currently available taxanes tend to encounter a number of adverse effects:

- Emulsifiers employed to solubilize taxanes like Cremophor EL and Polysorbate 80 are known to cause severe side effects, such as fluid retention, histamine release, hypersensitivity, anaphylaxis, gastrointestinal toxicity, hyperlipidemia, fluctuating lipoprotein levels, erythrocyte sedimentation, and myelosuppression (Attili-Qadri *et al.*, 2013).
- In extreme cases, there could be several others side effects related to motor, sensory, and nervous systems, including irreversible peripheral neuropathies, axonal degeneration, demyelination, and inhibition of endothelial transcytosis (Hennenfent and Govindan, 2006).
- Also, it is speculated that solubilizers might decrease the usual access of the taxane drugs into the carcinogenic site, thus reducing the therapeutic efficacy.
- Another concern seems to be the entrapment of drug within polar micelles, resulting eventually in undesirable nonlinear pharmacokinetic profiles with decreased drug clearance and distribution volume.
- The packaging material could leach on interaction with solubilizers, thus mostly releasing di-(2-ethylhexyl) phthalate (DEHP). Premedication with corticosteroids and/or antihistamines could bypass these adverse effects.
- Use of Tween 80 is associated with hemolytic and cholestatic liver. Limiting the content of this agent may ultimately result in decreased drug loading potential of these formulations, thus requiring longer infusion time.

Therefore, a suitable route of administration for taxanes is highly probable to offer stellar advantages over parenteral administration (Surapaneni *et al.*, 2012).

9.3 CONFRONTS ASSOCIATED WITH ORAL CHEMOTHERAPY

Oral chemotherapy, being a highly sought-after domain these days, has to keep a close observation for maximum and minimum tolerable concentrations, especially for a drug with a narrow therapeutic index. Successful monitoring of the upper and lower concentrations of taxanes could certainly lead to reduced frequency of administration and stern adverse effects. In order to achieve the prolonged action, avoid fluctuations in drug exposure and maintain sustained taxane levels for cancerous cells; a noninvasive route and patient convenience is highly desirable. Further, oral therapy certainly decreases the total cost of hospitalization and other vital support by healthcare professionals (Thanki *et al.*, 2013).

Challenges related with oral intake of taxanes, that is, Class II drugs as per the Biopharmaceutical Classification System (BCS), have been highlighted in Box 9.1.

BOX 9.1 CHALLENGES ASSOCIATED WITH ORAL TAXANES AS BCS CLASS IV DRUGS

- Poor aqueous solubility
- Moderate intestinal permeability
- Poor bioavailability
- High hepatic first-pass metabolism by CYP3A4 and CYP2C8 enzymes
- Substrates for P-glycoprotein (P-gp) efflux transporters

Further, side effects such as odynophagia, dysphagia, vomiting, and nausea generally result in noncompliance of the approved treatment and these are notable concerns for the peroral usage of such agents (Bhattacharyya, 2010). Many times, these drugs are also reported to disrupt the gastrointestinal lining, which poses a concern about a multiday oral regimen, especially for chronic usage. In order to overcome the afore-stated hiccups, a lot of research efforts have been undertaken towards either synthesis of newer analogues of taxanes or fabrication of novel nanocarrier systems, which may surmount the other physicochemical challenges, as well as various absorption- and permeability-related hindrances (Fauzee *et al.*, 2011).

9.4 SECOND GENERATION ANALOGUES OF TAXANES

In tune with the discussion in the previous sections and to achieve higher cytotoxicity with a broader antitumor spectrum than existing taxanes, the idea of developing other newer generation taxane analogues has been put forward (Figure 9.3). Hence, chemists and researchers across the world have been putting tremendous efforts into synthesizing novel second-generation taxanes, such that not merely the physicochemical hiccups and drug-resistance mechanisms are surmounted, but various other toxicity issues related to formulations aspects could also be circumvented (Lavelle, 2002). Table 9.2 enlists the recently synthesized second-generation taxanes with superior efficacy. Some of the taxane analogues investigated under preclinical trials do not necessarily demonstrate

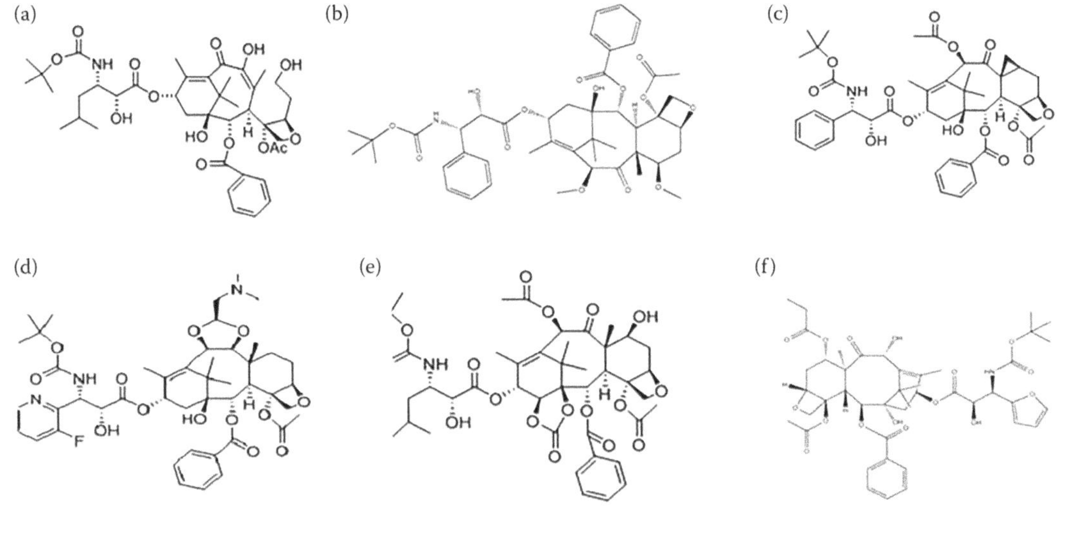

FIGURE 9.3 Structural representation of taxane analogues: (a) Secotaxane, (b) Cabazitaxel, (c) Larotaxel, (d) Tesetaxel , (e) Ortataxel, and (f) Milataxel.

TABLE 9.2
Second-Generation Taxanes

Second-Generation Taxane	Identification	Features	Company	Reference(s)
Secotaxane	IDN 5390	Its oral treatment and potent antimotility activity on endothelial cells has been reported as the most appropriate to exploit the antiangiogenic effects of cytotoxic drugs.	–	Pratesi *et al.* (2002)
Cabazitaxel	XRP-6258	It is a novel, orally bioavailable second-line semisynthetic derivative of taxanes, employed for metastatic castration-resistant prostate cancer and works by disrupting the microtubular network essential for mitotic and interphase cellular functions.	Approved by the US FDA in 2010 and is currently marketed under the trade name of Jevtana by Sanofi-Aventis	Channing and Emmanuel (2011), Bouchet and Galmarini (2010), Paller and Antonarakis (2011); Kümmel *et al.* (2017)
Larotaxel	XRP9881	Larotaxel dehydrate is a semi-synthetic derivative of taxane 10-deacetylbaccatin III with potential antineoplastic activities. It acts as a poor P-gp substrate and therefore is quite effective in multidrug resistant tumors of breast and prostate.	With Sanofi-Aventis and is in phase III trials	(http://www.cancer.gov; Kingston 2009)
Tesetaxel	DJ-927	Tesetaxel is a novel, orally absorbed, semisynthetic taxane. With oral administration, it assures to maintain high antitumor activity of the taxanes, particularly for treating colorectal and gastric cancers. Subsequently, it was found to be not that effective and was eventually withdrawn in 2006.	Daiichi Sankyo Co. Ltd.	Saif *et al.* (2011)
Ortataxel	IDN-5109, BAY 59-8862	Ortataxel is a novel analogue of paclitaxel which is effective orally. Due to association of severe neutropenia with administration of ortataxel, it was withdrawn from clinical trials by Bayer. It is indicated for non small cell lung cancer (NSCLC) and head and neck squamous cellcarcinomaG(HNSCC).	With Spectrum Pharmaceuticals, California, and is in phase II clinical trials	Tsao *et al.* (2011)
Milataxel	MAC-321	It is a novel analogue of docetaxel, which has shown its effectiveness in advanced solid malignancies and colorectal cancer.	With Taxolog and is in phase II trials and with Wyeth and is in phase I and II clinical trials	Ramanathan *et al.* (2008), Flores and Saif (2013)

(Continued)

TABLE 9.2 (*Continued*)
Second-Generation Taxanes

Second-Generation Taxane	Identification	Features	Company	Reference(s)
	BMS-275183	Subsequent phase I dose-escalation studies investigated dosing with several different dosing schedules with the rationale that different schedules would have different toxicity profiles.	Clinical phase I trials	Flores and Saif (2013)
4-carbonate derivative of paclitaxel	BMS-188797	It showed objective responses in ovarian and cervical cancer patients.	Phase I trial	(http://www.fiercebiotech.com)
Nab™-docetaxel	–	This is a nanoparticle albumin-bound docetaxel formulation related to Abraxane and is in early clinical studies for the treatment of hormone refractory prostate cancer and other solid tumors.	Abraxis BioScience	(http://www.fiercebiotech.com)
Taxoprexin	–	It is a combination of docosahexaenoic acid-paclitaxel, for treatment of NSCLC.	Luitpold Pharmaceuticals and is currently in phase III trials	Kingston (2009)

greater therapeutic indices, unless these agents are proven to be inherently safer to normal tissues. Therefore, effectiveness of such entities and most of the test drugs cannot be asserted solely due to the dose-response curve shifting, but the same needs to subsequently demonstrate superior antitumor activity in the preclinical studies and ultimately in clinics *vis-à-vis* the established agents (Rowinsky, 2002).

Currently, these taxanes, combined with monoclonal bodies, have proved to be more effective, owing to attainment of site-specific targeting (Osborne *et al.*, 2017; Spraggs *et al.*, 2017).

9.5 PHARMACEUTICALLY DEVELOPED NOVEL NANO-BASED ORAL FORMULATIONS OF TAXANES

Taxanes are being commonly prescribed for the management of diversity of cancers including non-small cell lung, ovarian, head, gastric, neck, and prostate cancers. Taxanes hold the second biggest global market of approximately 700 million USD, next only to cardiovascular agents (Jibodh *et al.*, 2013; Kumar *et al.*, 2016). Taxane resistance has also been observed in cancer patients, thus calling for the need for development of oral nanostructured drug-delivery systems (Hennenfent and Govindan, 2006). Table 9.3 highlights the various nanostructured delivery systems of taxanes, primarily of paclitaxel and docetaxel, reported in the literature.

9.6 PRECLINICAL STUDIES: TO ACCESS REPORTS OF USAGE IN DIFFERENT KINDS OF CANCER

Novel oral taxanes have been successfully explored in various preclinical phases alone and as part of combination regimens for several cancer models. These second-generation oral taxanes tend to circumnavigate the limitations associated with intravenous administration of paclitaxel and docetaxel. Oral chemotherapy offers stellar advantages, including easier administration in smaller and more frequent doses, while optimizing toxicity. Since most of the available data are from dose-ranging studies, comparisons of toxicity profiles of the novel oral taxanes with intravenous paclitaxel and docetaxel are difficult to extrapolate at this time. So far, they have substantially been used only in populations that have been pretreated, in patients with advanced cancers with poor prognoses, or during the course of preclinical trials at subtherapeutic doses. However, there were patients in these trials who demonstrated a definitive response to drugs under the study, despite previously failing taxane-based regimens. Phase I studies have established successful oral administration of some of these newly developed taxanes, along with much safer toxicity profiles (Rose *et al.*, 2001).

Currently, these agents are under study in various preclinical phases (Table 9.4), and need to be further unfolded for future meaningful outcomes. These novel taxanes may eventually prove to be effective alternatives to administer paclitaxel and docetaxel, or these may be established as potential taxane therapies for myriad cancer conditions.

9.7 EPILOGUE

In clinical practice, the mode of administration of the taxanes is invariably parenteral. According to a recent study, most of the patients receiving chemotherapy preferred the oral route over the invasive route, primarily due to improved convenience with no fear of needles, amalgamated with a hope for economic and affordable chemotherapy (Liu *et al.*, 1997).

Despite the potential promises, the challenges associated with the oral taxane delivery vouch significant attention. The oral absorption of these agents is hampered by the co-administered of drugs, food items, nutritional supplements, and herbal medications, which may unpredictably influence bioavailability (Khurana *et al.*, 2016). On the other hand, side effects—such as dysphagia, nausea,

TABLE 9.3
Literature Instances of Developed Nanocarrier Systems for Taxanes

S. No.	Nanocarrier System	Active Pharmaceutical Ingredient	Excipients Employed	Benefits of the Study	Reference(s)
1.	Micelles	Paclitaxel	Solutol HS 15, d-a-tocopheryl polyethylene glycol succinate, and bromotetrandrine	Super-antiresistant paclitaxel micelles showed a great potential for oral delivery against resistant breast cancer.	Zhang *et al.* (2016)
2.	Protein-grafted solid lipid nanoparicles	Paclitaxel	Wheat germ agglutinin, glyceryl monostearate, soy lecithin, poloxomer 188, and Tween 80	Biodistribution studies in rats revealed that these systems significantly improved the oral bioavailability and lung targetability, which could be due to the cumulative bioadhesive property of the nanocarrier system and the targeting ligand.	Pooja *et al.* (2016)
3.	Polymeric micelles	Docetaxel	Stearic acid-modified *Bletilla striata* polysaccharides copolymers	These showed good hemocompatibility and as a carrier for delivering hydrophobic drugs.	Guan *et al.* (2016)
4.	Biotin-modified micelles into alginate microparticles	Docetaxel	Biotin and alginate	The bioavailability of drug-loaded microparticles was 27.4-fold higher than that of free drug after oral administration and achieved 84.6% tumor inhibition.	Shen *et al.* (2016)
5.	Self-emulsifying drug-delivery systems (SEDDS)	Docetaxel	Capryol 90, Vitamin E TPGS, Gelucire 44/14, and Transcutol HP	SEDDS showed desired stability, enhanced oral bioavailability, and *in vitro* antitumor efficacy.	Valicherla *et al.* (2016)
6.	Self-nanoemulsifying drug-delivery systems (SNEDDS)	Paclitaxel	Sesame oil, labrasol, and sodium deoxycholate	5.8-fold and 6.3-fold enhancement in the oral bioavailability (Cmax and AUC) of the plain drug from SNEDDS.	Sandhu *et al.* (2015)
7.	Solid lipid nanoparticles	Docetaxel	D-alpha-tocopheryl poly(ethylene glycol 1000) succinate, tween 80, and tristearin	Enhanced intestinal absorption and lymphatic uptake with increased oral bioavailability.	Cho *et al.* (2014)
8.	Mixed micelles	Paclitaxel	Pluronic F127, d-α-tocopheryl polyethylene glycol 1000 succinate, soybean phosphatidylcholine, and sodium cholate	Improved intestinal permeability with enhanced oral bioavailability.	Zhao *et al.* (2014a,b)

(*Continued*)

TABLE 9.3 (*Continued*)
Literature Instances of Developed Nanocarrier Systems for Taxanes

S. No.	Nanocarrier System	Active Pharmaceutical Ingredient	Excipients Employed	Benefits of the Study	Reference(s)
9.	Nanocapsules	Docetaxel	Chylomicrons	Enhanced oral absorption and bioavailability.	Attili-Qadri *et al.* (2013)
10.	Polymeric nanoparticles	Docetaxel	Thiolated chitosan and polymethyl methacrylate	Enhanced oral bioavailability, longer half-life with higher cellular permeability.	Saremi *et al.* (2013)
11.	Polymeric nanoparticles	Docetaxel	Chitosan and L-glutathione	Sustained drug release, higher cytotoxicity, and cellular uptake.	Saremi *et al.* (2011)
12.	Polymeric micelles	Paclitaxel	Pluronic F127 and polyethylenimine	Superior permeability across intestinal barrier with improved oral bioavailability.	Li *et al.* (2013)
13.	Liposome	Paclitaxel	Polyacrylic acid, polyallylamine hydrochloride, phospholipid, and cholesterol	Enhanced oral bioavailability with reduced tumor growth.	Jain *et al.* (2012)
14.	Polymeric nanoparticles	Paclitaxel	Poly(ethylene glycol), methyl vinyl ether, and maleic anhydride	Increased oral permeability, prolonged release, and enhanced oral bioavailability.	Zabaleta *et al.* (2012)
15.	Polymeric micelles	Paclitaxel	Pluronic F127 and pluronic P188	Enhanced AUC and Cmax, superior oral bioavailability, and cytotoxicity with increased *in situ* intestinal absorption.	Dahmani *et al.* (2012)
16.	Niosomes	Paclitaxel	Tween 20, 60, Span 20, 40, 60, Brij 76, 78, 72, and cholesterol	Gastrointestinal stability was observed.	Bayindir and Yuksel (2010)
17.	Drug conjugates	Docetaxel	2′-hemisuccinate derivative, succinic anhydride, NHS, and EDC	Increased half-life and better antitumor efficacy.	Lee *et al.* (2009)
18.	Microemulsion	Docetaxel	Capryol 90, cremophor EL ,and transcutol	Increased solubility, permeability, and oral bioavailability.	Yin *et al.* (2009)
19.	Polymeric nanoparticles	Docetaxel	Vitamin E TPGS, montmorillonite, PLGA, and lactate	Longer half-life due to sustain release with enhanced oral bioavailability.	Feng *et al.* (2009)
20.	Microemulsion	Paclitaxel	N-benzylbenzamine, 14C polyethylene glycol 4000, capric acid, L-aphosphatidylcholine, polysobate 80, lecithin, and Caproyl 90	Enhanced *in situ* permeability and higher AUC.	Nornoo *et al.* (2009)

(*Continued*)

TABLE 9.3 (*Continued*)
Literature Instances of Developed Nanocarrier Systems for Taxanes

S. No.	Nanocarrier System	Active Pharmaceutical Ingredient	Excipients Employed	Benefits of the Study	Reference(s)
21.	Polymeric micelles	Paclitaxel	Methoxy poly(ethylene glycol), 2-hydroxy nicotinic acid, 4-vinylbenzyl chloride, and N,N,N′,N″,N″-pentamethyldiethylenetriamine	Sustained drug release for better therapeutic action.	Kim *et al.* (2008)
22.	Self-microemulsifying drug-delivery system	Paclitaxel	Vitamin E , deoxycholic acid sodium salt, D-tocopheryl polyethylene glycol succinate 1000, and Cremophor RH 40	Increased solubility, improved oral absorption, and higher bioavailability.	Yang *et al.* (2004)
23.	Polymeric nanoparticles	Docetaxel	PEG 2000, Gantrez AN, Glycine, and disodium edetate	Offered prolonged and sustained plasma levels even after oral administration.	Ruiz-Gatón *et al.* (2018)
24.	Nanocrystals	Docetaxel	Pluronic F127, and human serum albumin	Promising formulation for the delivery of poorly water-soluble drug to multidrug-resistant tumors.	Gad *et al.* (2018)

Abbreviations: NHS, N-Hydroxysuccinimide; EDC, 1-Ethyl-3-(3-dimethylaminopropyl)carbodiimide).

TABLE 9.4
Preclinical Trials Phases of Second-Generation Taxanes

Model/Subject	Drug(s)	Inference	Manufacturer	Preclinical Phase	Reference
Mdr1a P-glycoprotein knock-out mice	DHP107 (Paclitaxel)	Good bioavailability	DAE HWA Pharmaceutical Co., Ltd (Seoul, Republic of Korea)	Phase I/II	Malingré *et al.* (2001)
Murine M109 lung, C3H mammary 16/C, human A2780 ovarian (grown in mice and rats), and HCT/pk colon, CWR-22	BMS-275183	Greater cumulative dose tolerance and improved therapeutic outcome	Bristol-Myers Squibb (NY, USA)	Phase I	Rose *et al.* (2001)
Breast carcinoma model, MX-1, LoVo, IGROV/DDP, and U87	IDN 5109 (Ortataxel)	Improved preclinical profile in terms of efficacy and tolerability	Spectrum Pharmaceuticals (CA, USA)	Phase II	Polizzi *et al.* (2000)
Wild-type mice	Docetaxel and paclitaxel in combination with cyclosporin A	Enhanced oral bioavailability	The Netherlands Cancer Institute, Amsterdam	Phase II	Malingré *et al.* (2001)
Mouse xenografts with the cell lines KB-8-5, MX-1W, and HCT-15	MAC-321 (Milataxel)	Disease stabilization	Taxolog, Inc. (NJ, USA)	Phase II	Flores and Saif (2013)
Mice xenografts with DLD-1 (colon cancer) and DU4475 (breast cancer)	DJ-927 (Tesetaxel)	Effective against cell lines overexpressing P-glycoprotein	Genta Inc. (NJ, USA)	Phase IIa studies for breast and colon cancers and a Phase I study in combination	Flores and Saif (2013)
Patients with advanced solid tumors	ModraDoc001 (Docetaxel)	Safe, well-tolerated, and shows encouraging antitumor activity	Slotervaart Hospital (Amsterdam, The Netherlands)	Phase I	Moes *et al.* (2011)
Patients with inoperable, advanced, or metastasized pancreatic carcinoma	EndoTAG® -1 (Paclitaxel encapsulated in a cationic lipid complex delivery system)	Targeted anticancer effect while reducing chemotherapy side effects	Jules Bordet Institute; MediGene AG; SynCore Biotechnology	Phase II/III	Jabir *et al.* (2018)
Patients having histological or cytological evidence of locally advanced, metastatic, or recurrent NSCLC	Genexol-PM (Paclitaxel)	Overcoming taxane resistance by delivery of higher doses of paclitaxel and enhanced targeting to tumor tissues	Samyang Biopharmaceutic Corporation, Gyeonggi-do	Phase III	Jabir *et al.* (2018)

odynophagia, gastrectomy, vomiting, and diarrhea—generally associated with oral administration are the other major concern. Therefore, the regimen needs to be supplemented with antiulcer and antiemetic drugs prior to oral chemotherapy.

As with other therapies, oral chemotherapy also requires counselling of the patient and proper training of the caretaker. The stakeholders must be made to understand the purpose of the medication, its schedule, and the identification, managemen,t and reporting of the associated adverse effects, if any. The concern of missed dose needs to be further discussed. On the clinical front, the co-administration of P-gp/CYP450 inhibitors is generally promoted; however, the side effects and dose titration should also be considered (Molassiotis, 2004). Constant endeavor, accordingly, must be made by the pharmaceutical chemists and clinicians to reduce the toxicity profiles of existing taxanes, or to develop newer and safer analogues. Developing effective oral taxane formulations requires ensuring gastrointestinal stability of the drug and its high intracellular uptake by intestinal cells (Thanki *et al.*, 2013), in which case the evolving drug-delivery approaches can be a boon.

ACKNOWLEDGMENTS

Financial grants received in the form of a meritorious research fellowship by the University Grant Commission (UGC), New Delhi, India, to Rajneet Kaur Khurana, and Inspire fellowship by the Department of Science and Technology (DST), New Delhi, Government of India, to Teenu Sharma, for research assistance are gratefully acknowledged.

REFERENCES

Attili-Qadri S, Karra N, Nemirovski A, Schwob O, Talmon Y, Nassar T *et al.* Oral delivery system prolongs blood circulation of docetaxel nanocapsules via lymphatic absorption. *Proc Natl Acad Sci USA.* 2013; 110: 17498–503.

Bayindir ZS, Yuksel N. Characterization of niosomes prepared with various nonionic surfactants for paclitaxel oral delivery. *J Pharm Sci.* 2010; 99: 2049–60.

Benelli R, Monteghirfo S, Balbi C, Barboro P, Ferrari N. Novel antivascular efficacy of metronomic docetaxel therapy in prostate cancer: HnRNP K as a player. *Int J Cancer.* 2009; 124: 2989–96.

Bhattacharyya GS. Oral systemic therapy: Not all "win-win". *Indian J Med Paediatr Oncol.* 2010; 31: 1–3.

Biotech and Fierce: Genta's tesetaxel granted fast track designation by FDA for advanced gastric cancer. Available online at: http://www.fiercebiotech.com/press-releases/gentas-tesetaxel-granted-fast-track-designation-fda-advanced-gastric-cancer. (Accessed on 13 January, 2017).

Bouchet BP, Galmarini CM. Cabazitaxel, a new taxane with favorable properties. *Drugs Today (Barc).* 2010; 46: 735–42.

Butler MS. Natural products to drugs: Natural product-derived compounds in clinical trials. *Nat Prod Rep.* 2008; 25: 475–516.

Channing JP, Emmanuel SA. Cabazitaxel: A novel second-line treatment for metastatic castration-resistant prostate cancer. *Drug Des Devel Ther.* 2011; 5: 117–24.

Cho HJ, Park JW, Yoon IS, Kim DD. Surface-modified solid lipid nanoparticles for oral delivery of docetaxel: Enhanced intestinal absorption and lymphatic uptake. *Int J Nanomedicine.* 2014; 9: 495–504.

Dahmani FZ, Yang H, Zhou J, Yao J, Zhang T, Zhang Q. Enhanced oral bioavailability of paclitaxel in pluronic/LHR mixed polymeric micelles: Preparation, *in vitro* and *in vivo* evaluation. *Eur J Pharm Sci.* 2012; 47: 179–89.

Eisenhauer EA, Vermorken JB. The taxoids: Comparative clinical pharmacology and therapeutic potential. *Drugs.* 1998; 55: 5–30.

Fauzee NJ, Dong SZ, Wang Y. Taxanes: Promising anti-cancer drugs. *Asian Pacific J Cancer Prev.* 2011; 12: 837–51.

Feng SS, Mei L, Anitha P, Gan CW, Zhou W. Poly (lactide)–vitamin E derivative/montmorillonite nanoparticle formulations for the oral delivery of docetaxel. *Biomaterials.* 2009; 30: 3297–306.

Flores JP, Saif MW. Novel oral taxane therapies: Recent phase I results. *Clin Investig (Lond).* 2013; 3: 333–41.

Gad SF, Park J, Park JE, Fetih GN, Tous SS, Lee W, Yeo Y. Enhancing docetaxel delivery to multidrug-resistant cancer cells with albumin-coated nanocrystals. *Mol Pharm.* 2018. DOI: 10.1021/acs.molpharmaceut.7b00783

Galletti E, Magnani M, Renzulli ML, Botta M. Paclitaxel and docetaxel resistance: Molecular mechanisms and development of new generation taxanes. *Med Chem.* 2007; 2: 920–42.

Gao P, Morozowich W. Chemotherapeutic microemulsion compositions of paclitaxel with improved oral bioavailability. US 7115565 B2; 2006.

Guan Q, Sun D, Zhang G, Sun C, Wang M, Ji D *et al.* Docetaxel-loaded self-assembly stearic acid-modified *Bletilla striata* polysaccharide micelles and their anticancer effect: Preparation, characterization, cellular uptake and *in vitro* evaluation. *Molecules.* 2016; 21(12): E1641.

Hennenfent KL, Govindan R. Novel formulations of taxanes: A review. Old wine in a new bottle? *Ann Oncol.* 2006; 17: 735–49.

Jabir NR, Anwar K, Firoz CK, Oves M, Kamal MA, Tabrez S. An overview on the current status of cancer nanomedicines. *Curr Med Res Opin.* 2018; 34: 911–921.

Jain S, Kumar D, Swarnakar NK, Thanki K. Polyelectrolyte stabilized multilayered liposomes for oral delivery of paclitaxel. *Biomaterials.* 2012; 33: 6758–68.

Jibodh RA, Lagas JS, Nuijen B, Beijnen JH, Schellens JH. Taxanes: Old drugs, new oral formulations. *Eur J Pharmacol.* 2013; 717: 40–6.

Khurana RK, Mahajan M, Teenu, Kapoor S, Jain S, Singh B. The sojourn from parenteral to oral taxanes using nanocarrier systems: A patent review. *Recent Pat Drug Deliv Formul.* 2016; 10: 44–58.

Kim S, Kim JY, Huh KM, Acharya G, Park K. Hydrotropic polymer micelles containing acrylic acid moieties for oral delivery of paclitaxel. *J Control Release.* 2008; 132: 222–9.

Kingston DG. Tubulin-interactive natural products as anticancer agents. *J Nat Prod.* 2009; 72: 507–15.

Kumar P, Raza K, Kaushik L, Malik R, Arora S, Katare OP. Role of colloidal drug delivery carriers in taxane-mediated chemotherapy: A Review. *Curr Pharm Des.* 2016; 22(33): 5127–43.

Kümmel S, Paepke S, Huober J, Schem C, Untch M, Blohmer JU *et al.* Randomised, open-label, phase II study comparing the efficacy and the safety of cabazitaxel versus weekly paclitaxel given as neoadjuvant treatment in patients with operable triple-negative or luminal B/HER2-negative breast cancer (GENEVIEVE). *Eur J Cancer.* 2017; 30: 84: 1–8.

Lavelle F. New taxanes and epotholone derivatives in clinical trials. *Bull Cancer (Paris).* 2002; 89: 343–50.

Lee E, Kim H, Lee IH, Jon S. *In vivo* antitumor effects of chitosan-conjugated docetaxel after oral administration. *J Control Release.* 2009; 140: 79–85.

Li Y, Bi Y, Xi Y, Li L. Enhancement on oral absorption of paclitaxel by multifunctional pluronic micelles. *J Drug Target.* 2013; 21: 188–99.

Liu G, Franssen E, Fitch MI, Warner E. Patient preferences for oral versus intravenous palliative chemotherapy. *J Clin Oncol.* 1997; 15: 110–5.

Malingré MW, Beijnen JH, Schellens JHM. Oral delivery of taxanes. *Invest New Drug.* 2001; 19: 155–62.

Moes JJ, Koolen SL, Huitema AD, Schellens JH, Beijnen JH, Nuijen B. Pharmaceutical development and preliminary clinical testing of an oral solid dispersion formulation of docetaxel. *Int J Pharm.* 2011; 420: 244–50.

Molassiotis A. Challenges and opportunities when using oral chemo-therapy in the home setting: The example of capecitabine. *Eur J Oncol Nurs.* 2004; 8: S1–S3.

Nornoo AO, Zheng H, Lopes LB, Johnson-Restrepo B, Kannan K, Reed R. Oral microemulsions of paclitaxel: *In situ* and pharmacokinetic studies. *Eur J Pharm Biopharm.* 2009; 71: 310–7.

Osborne C, Challagalla JD, Eisenbeis CF, Holmes FA, Neubauer MA, Koutrelakos NW *et al.* Ixabepilone and carboplatin for hormone receptor positive/HER2-neu negative and triple negative metastatic breast cancer. *Clin Breast Cancer.* 2017. pii: S1526-8209(16)30521-3.

Paller CJ, Antonarakis ES. Cabazitaxel: A novel second-line treatment for metastatic castration-resistant prostate cancer. *Drug Devel Ther.* 2011; 5: 117–24.

Pazdur R, Kudelka AP, Kavanagh JJ, Cohen PR, Raber MN. The taxoids: Paclitaxel (taxol) and docetaxel (taxotere). *Cancer Treat Rev.* 1993; 19: 351–86.

Polizzi D, Pratesi G, Monestiroli S, Tortoreto M, Zunino F, Bombardelli E *et al.* Oral efficacy and bioavailability of a novel taxane. *Clin Cancer Res.* 2000; 6: 2070–4.

Pooja D, Kulhari H, Kuncha M, Rachamalla SS, Adams DJ, Bansal V *et al.* Improving efficacy, oral bioavailability, and delivery of paclitaxel using protein-grafted solid lipid nanoparticles. *Mol Pharm.* 2016; 13: 3903–12.

Pratesi GD, Laccabue C, Lanzi G, Cassinelli R, Supino M, Zucchetti R *et al.* An oral taxane candidate for protracted treatment schedules. *Br J Cancer.* 2002; 88: 965–72.

Ramanathan RK, Picus J, Raftopoulos H, Bernard S, Lockhart AC, Frenette G *et al.* A phase II study of milataxel: A novel taxane analogue in previously treated patients with advanced colorectal cancer. *Cancer Chemother Pharmacol.* 2008; 61: 453–8.

Rose WC, Long BH, Fairchild CR, Lee FY, Kadow JF. Preclinical pharmacology of BMS-275183, an orally active taxane. *Clin Cancer Res.* 2001; 7: 2016–21.

Rowinsky EK. Taxane analogues. *Clin Cancer Res.* 2002; 8: 2759–63.

Ruiz-Gatón L, Espuelas S, Larrañeta E, Reviakine I, Yate LA, Irache JM. Pegylated poly(anhydride) nanoparticles for oral delivery of docetaxel. *Eur J Pharm Sci.* 2018; 118: 165–175.

Saif MW, Sarantopoulos J, Patnaik A, Tolcher AW, Takimoto C, Beeram M. Tesetaxel, a new oral taxane, in combination with capecitabine: A phase I, dose-escalation study in patients with advanced solid tumors. *Cancer Chemother Pharmacol.* 2011; 68: 1565–73.

Sandhu PS, Beg S, Katare OP, Singh B. QbD-driven development and validation of a HPLC method for estimation of tamoxifen citrate with improved performance. *J Chromatogr Sci.* 2015; 54 (8): 1373–84.

Spraggs CF, Parham LR, Briley LP, Warren L, Williams LS, Fraser DJ *et al.* Characterisation of the HLA-DRB1*07:01 biomarker for lapatinib-induced liver toxicity during treatment of early-stage breast cancer patients with lapatinib in combination with trastuzumab and/or taxanes. *Pharmacogenomics J.* 2017. doi: 10.1038/tpj.2017.39

Saremi S, Atyabi F, Akhlaghi SP, Ostad SN, Dinarvand R. Thiolated chitosan nanoparticles for enhancing oral absorption of docetaxel: Preparation, *in vitro* and *ex vivo* evaluation. *Int J Nanomedicine.* 2011; 6: 119–28.

Saremi S, Dinarvand R, Kebriaeezadeh A, Ostad SN, Atyabi F. Enhanced oral delivery of docetaxel using thiolated chitosan nanoparticles: Preparation, *in vitro* and *in vivo* studies. *Bio Med Res Int.* 2013. Article ID 150478.

Schwartz EL. Anti-vascular actions of microtubule-binding drugs. *Clinical Cancer Res.* 2009; 15: 2594–601.

Shen Y, Hu M, Qiu L. Sequentially dual-targeting vector with nano-in-micro structure for improved docetaxel oral delivery *in vivo. Nanomedicine (Lond).* 2016; 11: 3071–86.

Surapaneni MS, Das SK, Das NG. Designing paclitaxel drug delivery systems aimed at improved patient outcomes: Current status and challenges. *ISRN Pharmacol.* 2012; 1–15.

Thanki K, Gangwal RP, Sangamwar AT, Jain S. Oral delivery of anticancer drugs: Challenges and opportunities. *J Control Release.* 2013; 170: 15–40.

Tsao AS, Liu S, Fujimoto J, Wistuba II, Lee JJ, Marom EM *et al.* Phase II trials of imatinib mesylate and docetaxel in patients with metastatic non-small cell lung cancer and head and neck squamous cell carcinoma. *J Thorac Oncol.* 2011; 6: 2104–11.

Valicherla GR, Dave KM, Syed AA, Riyazuddin M, Gupta AP, Singh A *et al.* Formulation optimization of docetaxel loaded self-emulsifying drug delivery system to enhance bioavailability and anti-tumor activity. *Sci Rep.* 2016; 6: 26895.

Verweij J, Clavel M, Chevalier B. Paclitaxel (taxol) and docetaxel (taxotere): Not simply two of a kind. *Ann Oncol.* 1994; 5: 495–505.

Wani MC, Taylor HL, Wall ME, Coggon P, McPhail AT. Plant antitumor agents. VI. The isolation and structure of taxol, a novel antileukemic and antitumor agent from *Taxus brevifolia. J Am Chem Soc.* 1971; 93: 2325–7.

Winefield RD, Entwistle RA, Foland TB, Lushington GH, Himes RH. Differences in paclitaxel and docetaxel interactions with tubulin detected by mutagenesis of yeast tubulin. *Med Chem.* 2008; 3: 1844–7.

Yang S, Gursoy RN, Lambert G, Benita S. Enhanced oral absorption of paclitaxel in a novel self-microemulsifying drug delivery system with or without concomitant use of P-glycoprotein inhibitors. *Pharm Res.* 2004; 21: 261–70.

Yin YM, Cui FD, Mu CF, Choi MK, Kim JS, Chung SJ *et al.* Docetaxel microemulsion for enhanced oral bioavailability: Preparation and *in vitro* and *in vivo* evaluation. *J Control Release.* 2009; 140: 86–94.

Zabaleta V, Ponchel G, Salman H, Agüeros M, Vauthier C, Irache JM. Oral administration of paclitaxel with pegylated poly (anhydride) nanoparticles: Permeability and pharmacokinetic study. *Eur J Pharm Biopharm.* 2012; 81: 514–23.

Zhang T, Luo J, Fu Y, Li H, Ding R, Gong T *et al.* Novel oral administrated paclitaxel micelles with enhanced bioavailability and antitumor efficacy for resistant breast cancer. *Colloids Surf B Biointerfaces.* 2016; 150: 89–97.

Zhao Y, Cui Y, Li Y, Li L. Stable phosphatidylcholine-bile salt mixed micelles enhance oral absorption of paclitaxel: Preparation and mechanism in rats. *J Drug Target.* 2014a; 22: 901–12.

Zhao Y, Li Y, Ge J, Li N, Li LB. Pluronic-poly (acrylic acid)-cysteine/Pluronic L121 mixed micelles improve the oral bioavailability of paclitaxel. *Drug Dev Ind Pharm.* 2014b; 40: 1483–93.

Zou Z, Xie L, Wei J, Yu L, Qian X, Chen J *et al.* Synergistic anti-proliferative effects of gambogic acid with docetaxel in gastrointestinal cancer cell lines. *BMC Complement Altern Med.* 2012; 12: 1–9.

10 Nanotechnological Interventions for Neurodegenerative Disorders Using Phytoactives

*Sumant Saini, Charan Singh, Shikha Lohan, Atul Jain, Eliana B. Souto, and Bhupinder Singh**

CONTENTS

10.1 INTRODUCTION

Present-day society is afflicted with a number of diseases, marked with different etiologies. Out of these umpteen health conditions, neurological disorders, especially neurodegenerative disorders, have sprung up as a major healthcare concern for various medical professionals worldwide (Sandhir *et al.*, 2015). Neurodegeneration is verily an umbrella term used for the conditions in which the vitality and functionality of the nerve or the nervous systems is affected in a partial or complete manner (McMurray, 2000). Afflicting damage to the cells of the nervous system, it comprises a number of disorders, including Alzheimer's disease (AD), Parkinson's disease (PD), Huntington's disease (HD), multiple sclerosis (MS),

* Corresponding author.

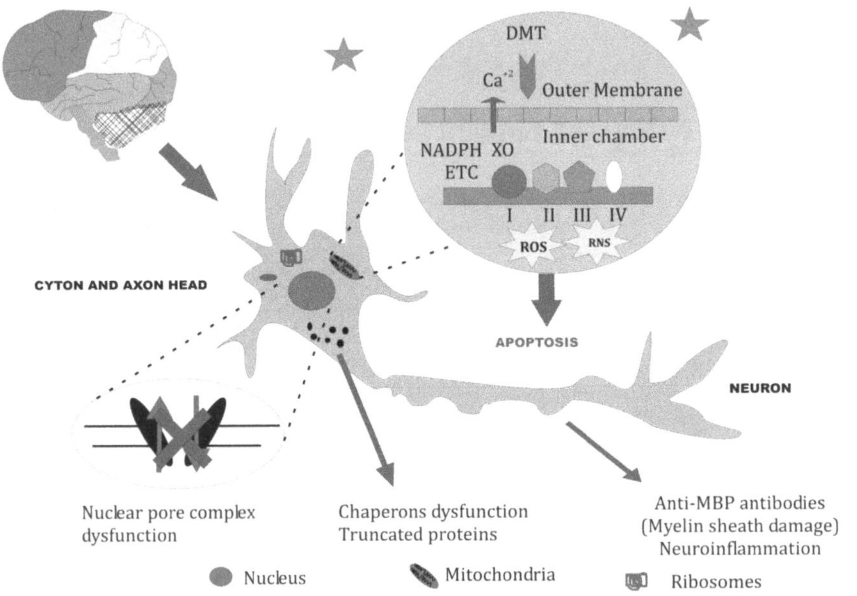

FIGURE 10.1 Molecular mechanism involved in the advent and progress of neurodegeneration.

and amyotrophic lateral sclerosis (ALS). The neurons depart from their normal physiology, anatomy, and, in some cases, both. Neurodegeneration, as understood from the various scientific endeavors around the world, is a cocktail of multiple etiologies. In most cases, the exact cause of the symptoms is unknown, and the disease progresses in an unpredictable fashion. This significantly changes the biochemical homeostasis of the body, whereas the levels of biochemicals are either increased or decreased, causing dysregulation of the functionality of the neurons, leading consequently to their apoptosis or loss of their functionality and vitality. Such deleterious damage can span over a group of neurons, to a neuronal tissue, or to the entire nervous systems, depending on the severity of damage. There is an array of biochemical parameters that are affected during the course of neurodegeneration (McMurray, 2000). The list of affected biochemicals include the biochemical mediators, as shown in Figure 10.1. Often, with the etiology of neurodegeneration being unknown and its progression being unpredictable, there is a progressive fall in the vitality and efficacy of the nervous system (Dong *et al.*, 2009).

10.2 PATHWAYS/FACTORS INVOLVED IN NEURODEGENERATION

10.2.1 Glutamate Pathways

Glutamate is an excitatory neurotransmitter in the central nervous system (CNS). Overexcitation of the system activates N-methyl-D-aspartate (NMDA) and α-amino-3-hydroxy-5-methyl-4-isoxazole propionic acid (AMPA) (Lee *et al.*, 2011). Overactivation of non-NMDA glutamate receptors tends to trigger autophagic lysosomal pathways. Moreover, autophagy and catabolic cascades involve proapoptotic proteins (such as p53) which, in turn, play a major role in neurodegeneration (Dong *et al.*, 2009).

10.2.2 Proteinopathy

Protein aggregation has been considered an important cause of neurodegeneration. Accumulation of the misfolded proteins—such as α-synuclein (in PD) and β-amyloid (in AD)—is reported to be elevated in the clinical conditions of neurodegeneration. Recent studies have emphasized that there are common molecular and cellular mechanisms that govern neurodegenerative cascades. There has been increasing evidence suggesting that there is an accumulation of the truncated proteins in patients with neurodegenerative disorders. Such insights can pave the way towards novel therapeutic strategies for effective management of neurodegenerative disorders (Jucker and Walker, 2013).

After the translational phase, the synthesized protein is subject to the post-translational modifications and folding of proteins. This particular step is vital as it helps the protein to orient its structure in such a manner that the active sites on their surface get exposed. This is carried out via a specialized class of proteins termed *chaperones*, which belong to the family of heat shock proteins (HSPs) (Muchowski and Wacker, 2005). Translation is a quite precise process, but in certain instances, an error might creep into the formation and/or folding of the proteins. Usually, the combined action of chaperones, the ubiquitin-proteasome system, and lysosome-mediated autophagy are sufficient in correcting any anomalies and removing these truncated proteins. However, sometimes a load of such misfolded proteins is beyond the capacity of the above-mentioned correction assembly of the cell. This leads to accumulation of the misfolded proteins inside the cell, which is a major cause of neurodegeneration. Patients with AD and PD disorders show a characteristic rise in the levels of such misfolded proteins in the neurons (Ciechanover and Kwon, 2017).

10.2.3 Free Radicals in Neurodegeneration

Among the various causes, reported or otherwise hypothesized, the role of free radicals remains the most crucial factor in the etiology of the neurodegeneration. There is an array of free radicals that are considered as the culprits for neurodegenerative cascades. Out of these, reactive oxygen species (ROS) and reactive nitrogen species (RNS) hold the lion's share. These are radicals with a lone pair of electrons which can trigger free-radical cascade (Sesti *et al.*, 2010; Khurana *et al.*, 2018).

10.2.3.1 ROS in Neurodegeneration

Oxygen is a unique molecule found abundantly in the cells. Its chemical structure reveals the presence of the lone pair of electrons ($1s^2 2s^2 2p^4$), the presence of which makes it susceptible to transform into an electron-deficient species, referred to as ROS. The cell metabolism produces ROS in the form of superoxide (O_2^-) and a hydroxyl radical (OH), as well as peroxide (O_2^{-2}) existing in the form of hydrogen peroxide (H_2O_2). This lone pair of electrons act as potent oxidants. ROS gives rise to a free-radical propagation mechanism that targets and adversely affects various biomolecules, causing damage to their structure and biochemistry, which may lead to an altered functionality (Henze *et al.*, 2005).

10.2.3.1.1 Sources of ROS in Neurodegeneration

Oxidative phosphorylation is a vital phenomenon occurring in biological systems. It is necessary for the bioenergetics of the cell. However, in the course of electron flow, leakage of a small fraction of electrons may occur from the electron transport chain (ETC) cascade; these electrons can enter the cytosol and initiate the free-radical chain-reaction mechanisms of ROS. The electrons react with molecular oxygen, leading to the synthesis of free radicals. Free radicals damage cells by oxidizing various biomolecules. Superoxides give rise to hydroxyl-free radicals that can hamper the normal neuron physiology. The hydroxyl radicals are produced via Fenton reaction by catalytic conversion from the ferrous to ferric iron radical (Korshunov *et al.*, 1997).

In this context, the mitochondrion serves as the pivotal organelle, with the ETC operating in the inner chamber. In mitochondrial physiology, fusion and fission processes occur simultaneously, and a disruption of this dynamic balance leads to erroneously working mitochondria, resulting in the symptoms of neurodegenerative disorders. Nicotinamide-Adenine Dinucleotide Phosphate (NADPH) oxidase is also responsible for increasing the ROS levels in the body. At the advent of neurodegeneration, there is an elevated level of NADPH oxidase. Excessive levels of ROS, as the result of an upregulated NADPH oxidase, causes the oxidation of various cell biomolecules, including DNA, via a redox-signal pathway. This leads to neuroinflammation, which has now been realized as the cardinal sign for the various neurodegenerative disorders (Hernandes and Britto, 2012). Another similar enzyme, xanthine oxidase (XO), is also considered as a key player in ROS-mediated mitochondrial neurodegeneration, as an enzyme associated with purine and pyrimidine metabolism. The enzyme catalyzes the conversion of hypoxanthine to uric acid, superoxide (O_2^-), and peroxide (O_2^{-2}) (Atlante *et al.*, 1997; Obuobi *et al.*, 2016).

10.2.3.2 RNS in Neurodegeneration

Similar to ROS, free radicals of nitrogen species (RNS)–such as nitrates (NO_3^-) and nitrites (NO_2^-)–contribute to the load of reactive species and subsequent neurodegeneration. ROS/RNS are involved in various capacities in the physiological setup, ranging from their subtle role in aging and carcinogenesis, to oxidative stress, and even to cell death via various pathways. Thus, RNS can trigger neurodegeneration by damaging the neuronal physiology and disturbing its biochemistry (Patel *et al.*, 1999).

10.2.4 Nuclear Pore Anomalies

The influx and efflux of proteins across the nuclear barrier play a key role in the normal functioning of the neurons. This process is controlled by a number of signals; these include nuclear localization signal class 1 (NLS 1) in protein import, nuclear export signal (NES) for protein export, importins and karyopherins for transport receptors, nuclear protein complexes (NPCs) such as nucleoporins (Nups), and Ran GTPase-activating protein (RanGAP). A balance between the influx and efflux of cargo exists at the nuclear pore. Impairment of this machinery by any means causes damage to the nuclear physiology, which subsequently induces neurodegeneration. Increase in the leakiness of the nuclear barrier affects the nucleocytoplasmic transport, changes protein localization across the nuclear coat, and is likely to be involved in Ca^{2+}-mediated apoptosis (Sheffield *et al.*, 2006; Bano *et al.*, 2010).

10.2.5 Inflammation in Neurodegeneration

Neuroinflammation has been recently recognized as another dimension in the etiology of various neurodegenerative disorders that have been under investigation for a long time. Scientists have found a distinctive role of inflammatory mediators and calcium in the progress of neurodegeneration (Currais *et al.*, 2014).

10.2.5.1 Involvement of Calcium

Over years of neurological research, calcium (Ca), in particular, has been identified as a key ion involved in neurodegeneration. It is an important mediator involved in neuronal physiology and integrity. Numerous scientific reports reveal the crucial role of Ca in synaptic activity, autophagy and cell degradation. Disturbance in Ca homeostasis can cause abnormal calcium transport and barrier functions, which, in turn, lead to neuroinflammation, apoptosis, and neurodegeneration (Patricia and Sabine, 2015; Feng and Yang, 2016).

10.2.5.2 Involvement of Inflammatory Mediators

The immune system has a profound effect on the nervous system physiology. Even minor dysregulation can trigger subtle neurological responses, which can lead to myriad pathologies. Restricting expression of secreted inflammatory molecules and receptor/signalling cascades in certain cell types is a key physiological regulatory strategy. Resident immune cells in the CNS include microglia and astrocytes, but microglia are the only CNS cells of hematopoietic origin (Lehnardt, 2010). Early in embryonic development, myeloid progenitors differentiate into primitive macrophage progenitors in the yolk sac. Later, these cells engraft into the CNS, where they differentiate into mature microglia, which are capable of local expansion and maintenance of a population of phagocytic monocytes for the life of the cell. Under physiologic conditions, microglia largely exhibit a ramified morphology with highly mobile processes optimized for sampling the local environment. Interleukin-1, interleukin-6, and tumor necrosis factor- α (TNF-α) are the major cytokines contributing to the neuroinflammation via Th-1 and Th-2 pathways. Various interleukin antagonists have also been shown to reverse the action of neuroinflammation, thereby mitigating the progression of the neurodegenerative processes (McManus and Heneka, 2017).

10.2.6 Effect of Metals

Metals play a crucial role, as they are part of the enzymes. These are also referred to as trace elements as they are present in very minute quantities. Increased levels of these in the CNS, beyond the permissible limits, can trigger the process of their accumulation in the neuronal milieu. This phenomenon leads to a multitude of neurodegenerative disorders, such as AD, PD, HD, and MS (Chen *et al.*, 2016; Tosi *et al.*, 2008). These transition metals also act as catalysts to various oxidative reactions, which oxidize and destroy the biomolecules, ultimately damaging the physiology and architecture of the whole nervous system (Goldberg and Lippard, 2017).

10.3 THERAPEUTIC MANAGEMENT APPROACHES FOR NEURODEGENERATIVE DISORDERS

In the pursuit to tackle the scourge of neurodegeneration, various therapeutic strategies have been proposed. The approaches range from the conventional surgical procedures to the latest pain-free, noninvasive techniques. Over the years, a quantum of research has been conducted, which not only improved the efficacy of the treatment methodologies, but also provided the medical experts with the tools that can furnish an early and timely diagnosis of medical conditions (Marianecci *et al.*, 2017).

In this context, to deliver the desired cargo across the BBB, novel strategies have been devised and put to the practical use (Mandal *et al.*, 2017). These include

1. Disruption of the BBB
2. Receptor-mediated brain targeting
3. Bypassing the BBB via the nasal route, through the olfactory epithelium
4. Usage of nanoscale delivery systems

There are certain unique characteristics that ought to be present as the prerequisites for a drug-delivery system to qualify for neurotargeting agents (Figure 10.2).

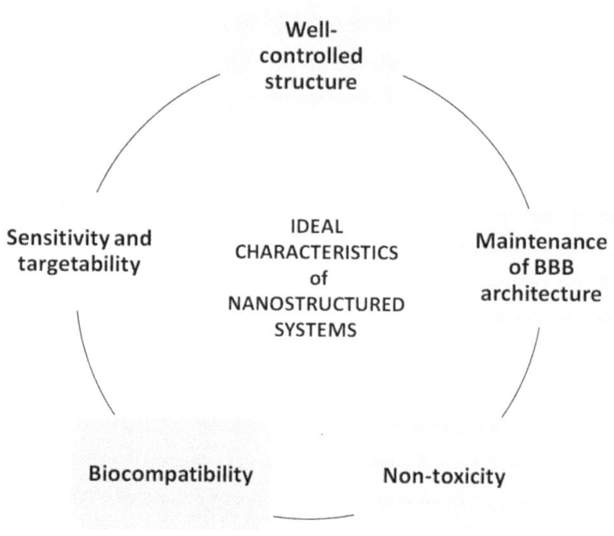

FIGURE 10.2 Ideal characteristics of a nanocarrier for neurodegeneration.

10.3.1 Use of Plant Bioactives

Nature has a huge share of contribution to the therapeutic products for the various ailments. Out of a plethora of molecules explored for neurodegenerative disorders, there are umpteen reports on bioactives belonging to diverse pharmacognostic categories explored for the said purpose, as enumerated in Table 10.1. Out of these, the flavonoids constitute the most vital class of secondary

TABLE 10.1
Select Literature Instances of Sources of Bioactives with Neuroprotective Activity

Disease	Bioactive	Dose	Biological Source	Reference(s)
AD	Paeoniflorin	100 μM	*Paeonia sinjiang*	Ma *et al.* (2018)
	Fisetin	25 mg/kg	Acacia, apple, grapes	Currais *et al.* (2014)
	Apigenin	50 μM	Parsley, celery, celeriac	Zhao *et al.* (2013)
	Kaempferol	50 μM	Apples, grapes, tomatoes	Park *et al.* (2011)
	Morin	0.1–1, 1, 10 μM	Guava, osage orange	Gong *et al.* (2011)
	Quercetin	0.1–1 μM	Citrus fruits, apples, red wine. Olive oil, grapes	Jiménez-Aliaga *et al.* (2011)
	Rutin	0.1–1 μM	Buckwheat, orange, lemon	Jiménez-Aliaga *et al.* (2011)
	Anthocyanin	100 μM	All plants	Liu *et al.* (2010)
	Glycitein	100 μg/mL	Soya bean	Gutierrez-Zepeda *et al.* (2005)
	Genistein	40 μM	Chickpeas (garbanzo beans), tofu, soymilk	Zeng *et al.* (2004)
	Myricetin	0.1–1 μM	Black tea, guava, papaya	Ono *et al.* (2003)
	Maidenhair tree	10–100 μg/mL	*Ginkgo biloba*	Bastianetto *et al.* (2000)
PD	Sulforaphane	5 μM	Brussels sprout, cabbage, cauliflower, bok choy	Morroni *et al.* (2018)
	Acacetin	10 mg/kg per day	Black locust, damiana, silver birch	Kim *et al.* (2012)
	Quercetin	50, 100, and 200 mg/kg body wt.	Citrus fruits, apples, red wine, olive oil, grapes	Lv *et al.* (2012)
	Rutin	25 mg/kg body wt	Citrus fruits	Khan *et al.* (2012)
	Genistein	10 mg/kg	Soya bean	Baluchnejadmojarad *et al.* (2009)
	ECGC	2 and 10 mg/kg	Carob flour, nuts, fruits, green tea	Nie *et al.* (2002)
	Catechin	0.5 and 1 mg/kg	Green tea	Levites *et al.* (2001)
	Tangeretin	10 and 20 mg/kg	Citrus peels	Datla *et al.* (2001)
HD	Naringin	80 mg/kg body wt	Grapefruit and other citrus fruits	Gopinath and Sudhandiran (2012)
	Quercetin	25 mg/kg body wt	Citrus fruits, apples, red wine. olive oil, grapes	Sandhir and Mehrotra (2013)
	EGCG	10, 20 and 40 mg/kg	Carob flour, nuts, fruits, green tea	Levites *et al.* (2001)
MS	EGCG	10 μg/mL	Green leaf tea	Herges *et al.* (2011)
	Fisetin	20–80 μM	Strawberries, mangoes, cucumber	Hendriks *et al.* (2003)
	Luteolin	20–80 μM	Broccoli, green pepper, parsley, thyme	Hendriks *et al.* (2003)
	Quercetin	20–80 μM	Citrus fruits, apples, red wine, olive oil, grapes	Hendriks *et al.* (2003)

Abbreviations: AD, Alzheimer's disease; PD, Parkinson's disease; HD, Huntington's disease; MS, Multiple sclerosis; ECGC, Epigallocatechin gallate.

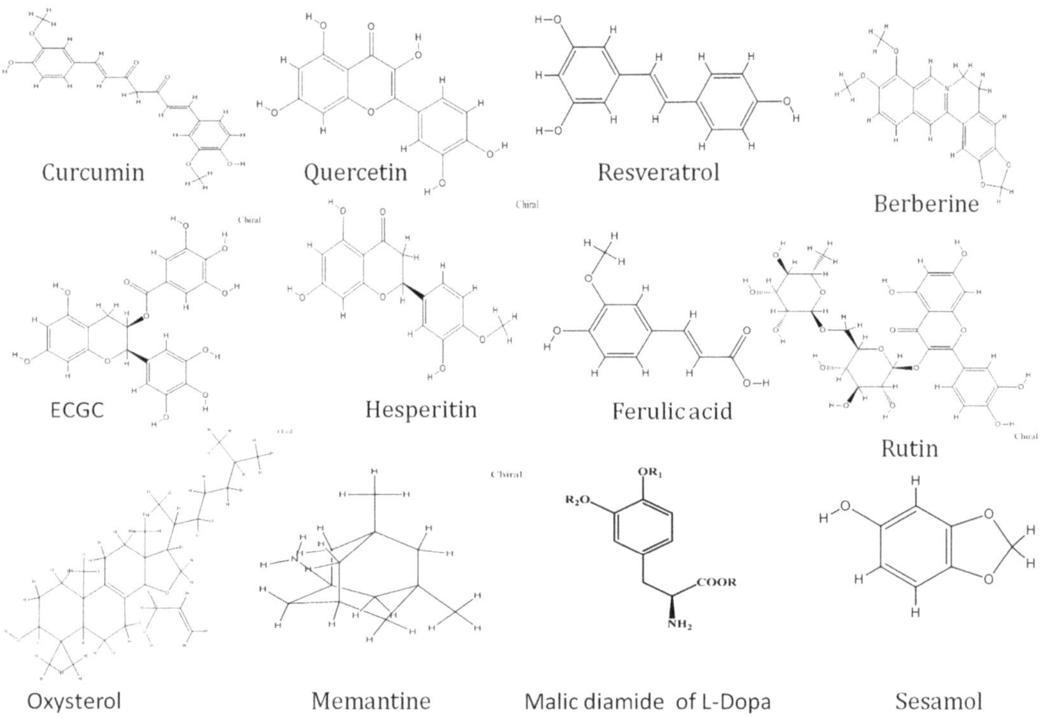

FIGURE 10.3 Some plant bioactives with reported neuroprotective properties.

metabolites that have come up in a significant way (Solanki *et al.*, 2015). Chemical structures of various plant bioactives, which have demonstrated promising potential in the management of neurodegenerative disorders, are illustrated in Figure 10.3.

10.3.2 Nanotechnological Interventions

The nanotechnology-driven approach has proved its mettle in biomedical research in attaining safe and effective therapeutic outcomes. The ultramicroscopic size of the obtained systems grants them novel and improved attributes in comparison with the pure bioactives or their microscaled counterparts. The domain of nanobiomedicine has been provided with the major successful application of nanotechnology in the development of biomedical delivery systems and devices (Kumar *et al.*, 2016).

Considering the multimodal applicability of such engineered nanoscaled systems, these have also been explored in the management of various neurodegenerative disorders. The merit associated with these nanostructured systems is that they can easily move across the hitherto impermeable BBB and can effectively act at the target site in a precise manner, resulting eventually in improved access and modulatory potential of the bioactives.

The present management therapy of the neurodegenerative disorders includes the use of various drugs, bioactives, biologics, and other compounds, which when used in the naïve form *in vivo*, result in a subtherapeutic response. This can be primarily attributed to the presence of the BBB. Therefore, use of "nanowagons" for the same has been shown to overcome this problem and improve the therapeutic efficacy of bioactives. The flexibility to engineer the surface of these nanoscaled systems provides an additional opportunity for use for diverse applications. Figure 10.4 depicts a panoramic view of various nanocarriers employed to ameliorate the obstacles in the therapeutic management of neurodegeneration using plant bioactives (Fernandes *et al.*, 2010). A sizable amount of such variegated nanosystems are discussed in the following sections.

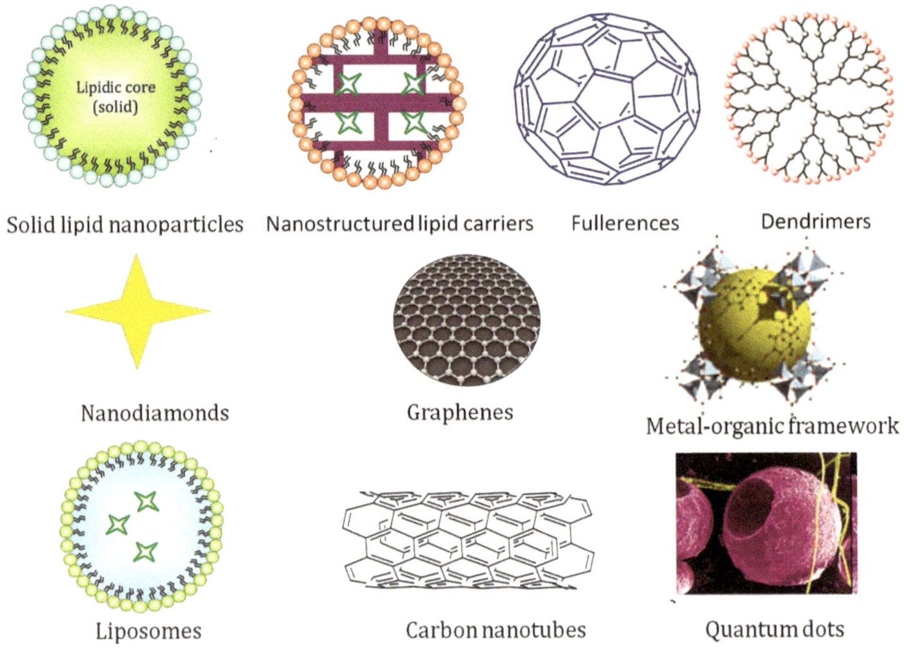

FIGURE 10.4 Various nanostructured systems employed for enhanced site-specific delivery of the plant bioactives in managing neurodegenerative disorders.

10.3.2.1 Lipid-Based Systems

10.3.2.1.1 Solid Lipid Nanoparticles

Introduced in 1991, solid lipid nanoparticles (SLNs)—comprising a solid lipid(s), surfactant(s), cosurfactant (optional)—successfully emerged as an alternative drug-delivery system over emulsions, liposomes, suspensions, and others. SLNs are composed of lipids that are solid at room temperature. These systems can be tailored as per the requirements by surface decoration with myriad types of ligands to exhibit targeting characteristics (Cunha *et al.*, 2017). By replacing a certain amount of solid lipid with a liquid lipid (up to 30% of the nanoparticle [NP] matrix), a new type of lipid nanoparticles can be prepared called nanostructured lipid carriers (NLCs) (Doktorovova *et al.*, 2016). The liquid lipid enhances the solubility of lipophilic drugs, improving the loading capacity and encapsulation efficiency of lipid NPs. This ultimately results in enhanced bioavailability and reduction in the adverse effects associated thereafter (Li *et al.*, 2017). The mechanism by which these nanolipids are internalized by the brain varies with the change in system composition. Besides providing a hydrophobic composition, in some cases, the employed carrier may additionally limit binding to P-gp and avoid efflux, eventually leading to higher brain-drug concentration (Koziara *et al.*, 2004). Preferential BBB uptake can also be hypothesized owing to the surface modifications with positive charge inducers. Numerous studies have proved that the positively charged lipid NPs are preferentially taken up by brain endothelial cells (Reddy *et al.*, 2004; Blasi *et al.*, 2007).

10.3.2.1.2 Liposomes

Liposomes are vesicular drug-delivery systems based on a phospholipid bilayer, with the capacity to load a variety of chemically different molecules (Texeira *et al.*, 2017). Since their discovery in 1961 by Alec D. Bangham (Pattni *et al.*, 2015), liposomes have been exploited for a set of applications, including site-specific delivery. Surface decoration with various agents has been carried out with the aim to improve their targetability (e.g., surface tethering of arginine-glycine-aspartate [RGD], lysine-aspartic acid-glutamic acid-leucine [KDEL] peptides) and/or circulation time (e.g., surface

PEGylation of the liposomes). Moreover, their small size below 100 nm makes them an eligible to smuggle the cargo across the BBB (Ko *et al.*, 2009).

10.3.2.2 Carbon Nanomaterials

Carbon-based nanoengineering has been a global trend in drug-delivery research. Because of its unique allotropic attributes, carbon exists itself in a variety of physical forms, each with special qualities and potential that can be harnessed to come up with a novel nanowagon that safe as well as effective. The superfamily of carbon comprises a diverse genre of carbon nanosystems. These include carbon nanotubes (CNTs), fullerenes, graphenes, and nanodiamonds, among others. Considering the purpose of brain targeting for neurodegenerative disorders, these carbon nanosystems have shown exceptional capability in terms of targetability, suitability, and carrier potential for an assortment of cargos (Kumar *et al.*, 2016). Table 10.2 provides a succinct account on various nanosystems employed to detect the levels of different neurotransmitters as biomarkers for neurodegenerative diseases.

10.3.2.2.1 Carbon Nanotubes (CNTs)

CNTs are the tubular nanoscaled systems that have been used to target at the cellular level. Their hollow structure enhance the loading capability of CNTs, whereas the large surface area and optical properties widen the applications of CNTs in biomedical research. After certain modifications in the chemistry of CNTs using various chemical processes, functionalized CNTs can be obtained, as pristine CNTs have an issue of toxicity and aggregation in the biological milieu (Amiri *et al.*, 2013). Literature also documents the theranostic potential of CNTs. Owing to their unique size, area, and structure, they can easily traverse the BBB and have proved experimentally to be instrumental in neurodegenerative disorders (Zhang *et al.*, 2011; Lohan *et al.*, 2017).

10.3.2.2.2 Graphenes

Graphenes, as evident from the name, is akin to graphite. These sp^2 hybridized honeycomb lattices have found wide application in several domains, for example, biomedicine theranostics, biosensing, energy, and advanced functional materials. These are unique in terms of internal geometry, as they possess very high surface areas (order of 2600 m^2/g) and exceptionally high electrical and thermal conductivities. Moreover, the surface chemistry of graphenes has a unique electronic structure,

TABLE 10.2
Select Literature Instances of Nanosystems in Biosensing of Various Neurodegenerative Disorders

Nanosystem	Primary Objective	Attributes	Reference
MWCNTs	Dopamine sensing	High detection sensitivity of NADH (17.73 μM), epinephrine (22.28 μM), and dopamine (2.210 μM)	da Silva *et al.* (2017)
CNTs	Ascorbic acid sensing in rats with AD	Improved and fast detection and measurement of ascorbic acid levels in the different parts of the brain in μM concentration	Zhang *et al.* (2017)
Quadruplex-integrated DNA	Dopamine sensing	High dopamine sensitivity with a lower detection limit of 110 μM	Morales *et al.* (2015)
MOFs	Dopamine sensing	A sensitive method for detecting dopamine within 30 μM to 4 mM concentration range	Zhao *et al.* (2015)

Abbreviations: AD, Alzheimer's disease; CNTs, Carbon nanotubes; DNA, Deoxyribonucleic acid; MOFs, Metallic organic frameworks; MWCNTs, Multiwalled carbon nanotubes; NADH, Nicotinamide adenine dinucleotide.

which makes them quite convenient to modify the surface topology and functionalize as per the requirements (Zhu *et al.*, 2010).

Graphenes also have been used in drug delivery, as they have demonstrated to be excellent carriers for drugs, proteins, DNA, and many other types of biomolecules. Recent strides in the biomedical domain have led to the use of graphene nanosystems as the platform technology for the development of nanoscale and precise nanosensors. These have used been in biological processes such as enzyme immobilization (John *et al.*, 2015).

10.3.2.2.3 Fullerenes

Buckminsterfullerene (C_{60}) is a unique category of carbon allotrope which has proved to have antioxidant activity. Moreover, the carboxylated form of fullerenes, carboxy-fullerenes have demonstrated neuroprotection against the neurotoxic, oxidative, and excitotoxic insult to the neurons. Fullerenes have also been explored as efficient carriers with a variety of theranostic cargos loaded and shuttled across the BBB (Dugan *et al.*, 2001).

10.3.2.2.4 Nanodiamonds

Carbon allotropes, such as nanodiamonds, have been explored for their theranostic potential, a high degree of fluorescence, and ability to resolve to a very minute level. These properties make them perfect candidates for neuronal theranostic applications (Alawdi *et al.*, 2017). Moreover, absence of toxicity of these carbon nanosystems and their resistance to *in vivo* degradation for an extended period of time make them perfect biomedical imaging tools using magnetic resonance imaging (MRI) (Haziza *et al.*, 2017).

10.3.2.3 Polymeric Systems

Nanosystems of various architectural forms and compositions have been investigated for the treatment of neurodegeneration. There is a variety of biodegradable polymers (for example, from natural, synthetic, and/or semisynthetic sources) that can be used for the development of polymeric nanosystems to target drugs to cross the BBB. Scientists have explored various types of polylactic-co-glycolic acid (PLGA) polymers, polysaccharides, and polypeptides for brain targeting. Memantine was loaded in biodegradable polylactic-co-glycolic (PLGA) NPs, produced by a double emulsion method and surface-coated with polyethylene glycol (PEGylated) (Jain *et al.*, 2015; Sánchez-López *et al.*, 2018). Behavioral tests carried out on transgenic APPswe/PS1dE9 mice demonstrated a more efficient memory-impairment reduction when using the developed nanoparticles *vis-à-vis* free memantine. Histological studies confirmed that MEM-PEG-PLGA NPs reduced β-amyloid plaques and the associated inflammation characteristics of Alzheimer's disease. Experimental strides have unearthed the potential of such polymeric nanosystems to be explored for myriad theranostic applications in the milieu of neurodegeneration. Polymeric systems have shown to form a polymer-drug conjugate rather than getting encased in an outer core (Costantino and Boraschi, 2012). These polymeric systems possess good CNS biocompatibility, which has been exploited for crossing the BBB (Emerich *et al.*, 1999). Moreover, such polymeric structures can easily be appended with a number of ligands. Based on the type of ligand used, a variety of transcellular transport processes across the BBB can be distinguished. The most common one, diffusion, mainly involves the transfer of small hydrophobic molecules into the brain. Carrier-mediated transport occurs mainly for glucose, amino acids, nucleosides, and many therapeutics, whereas regulatory molecules such as insulin, leptin, interleukins, and various nutrients are taken up using receptor-mediated transcytosis (Georgieva *et al.*, 2014).

10.3.2.3.1 Dendrimers

Dendrimers are hyperbranched, tree-like nanosystems with multiple layers, recognized as promising drug-delivery systems attributed to their high targeting potential, low toxicity, and high order of specificity. Functionalization of dendrimers grant them the various theranostic attributes (Benseny-Cases *et al.*, 2012). Dendrimers have shown tremendous potential not only for diagnosis but also for the treatment of many CNS diseases such as AD. Ultralow concentrations of protein biomarkers (e.g. ADDL-amyloid-ß-derived diffusible ligands), which have been implicated in the pathogenesis of AD, can be detected by means of dendrimers. Some of the dendrimers were demonstrated to cross the BBB, whereas studies have revealed that dendrimers possess the intrinsic ability to localize in cells associated with neuroinflammation (activated microglia and astrocytes) and thus can be used in neuroinflammation therapy (Mignani *et al.*, 2017).

10.3.2.3.2 Metal-Organic Frameworks (MOFs)

MOFs are porous, hybrid nanomaterials comprising three-dimensional arrays of the various metals tethered to an organic linker. MOFs have been functionalized with various types of diagnostic and imaging agents that make them promising candidates to be explored for biomedical imaging (positron-emission tomography [PET] imaging and MRI) contrast agent. Apart from these, they have also been used as nanosensors for various chemicals. In the drug-delivery milieu, MOFs have shown an exceptional capability with high drug-loading efficacy and ability to carry across the barriers (Horcajada *et al.*, 2010). There are a few scientific reports in which these nanoporous systems have been explored in the domain of neurological theranostics (Hayne *et al.*, 2014). Some scientific reports have highlighted the specificity of these MOFs to the truncated proteins of the brain in various neurodegenerative conditions, thus making them promising candidates for nanoneuromedicine (Beg *et al.*, 2017).

10.3.2.3.3 Semiconducting Nanoparticles and Quantum Dots

These fluorescent nanocrystals constitute a unique genre of nanoscaled systems. Quantum dots (QDs) have sizes of the order of Bohr's radius (0.529 Å). Owing to such small size, some stellar properties are exhibited by these nanosystems in terms of their fluorescence (Qu *et al.*, 2017). QDs have been extensively explored for their imaging properties. They have been modified and put to use in the biological milieu as biosensors and molecular probes. Their composition makes them resistant to any damage caused by the high energy radiations during the fluorescence, unlike the fluorescent dyes that exhibit the bleaching effect. Bioconjugated QDs have been exploited as biovectors, attributed to their fluorescent activity across the BBB (Huang *et al.*, 2017). Table 10.3 enlists different imaging agents employed in neurodegenerative disorders.

TABLE 10.3
Select Literature Instances of Nanosystems in Imaging of Various Neurodegenerative Disorders

Nanosystem	Ligand	Attributes/Key Findings	Reference
Magnetic NPs	Curcumin	Revealed that Cur-MNPs were co-localized with amyloid plaques, thus establishing the potential of Cur-MNPs for noninvasive diagnosis of AD using MRI	Cheng *et al.* (2015)
Liposomes	Methoxy-XO4	Proved binding of drug and liposomes to the amyloid plaques, thereby causing degradation of plaques for improved imaging	Tanifum *et al.* (2012)
Liposomes	Apomorphine and QDs	Revealed that QD-loaded liposomes enter into bEND3 cells via clathrin- and caveolae-mediated endocytosis	Wen *et al.* (2012)

Abbreviations: AD, Alzheimer's disease; bEND, Brain endothelial; Cur, Curcumin; MNPs, Magnetic nanoparticles; MRI, Magnetic resonance imaging; NPs, Nanoparticles; QDs, Quantum dots.

TABLE 10.4
Select Literature Instances of Nanosystems in Therapeutics of Various Neurodegenerative Disorders

Nanosystem(s)	Bioactive	Attributes/Key Findings	Reference
NLCs	Ferulic acid	Improved therapeutic potential in cerebral stroke	Hassanzadeh *et al.* (2018)
Nanoemulsion	Quercetin	Improved brain targeting potential and efficacy	Ahmad *et al.* (2018)
NPs	Quercetin	Enhanced bioavailability and decreased Aβ-induced oxidative stress	Han *et al.* (2018)
Polymeric NPs	Memantine	Reduced β-amyloid plaques and the associated inflammation in AD	Sánchez-López *et al.* (2018)
Chitosan-coated lipid NPs	Resveratrol	Improved bioavailability and specific nose to brain delivery	Trotta *et al.* (2018)
SLN	Resveratrol	Reduced ROS levels, lipid peroxidation and protein carbonyls	Yadav *et al.* (2018)
Nanocrystals	Hesperitin	Improved cognition and reduced oxidative stress	Moghaddam and Zare (2018)
CNTs	Berberine	Improved therapeutic potential of polysorbate/phospholipid-coated MWCNTs	Lohan *et al.* (2017)
Liposomes	H102 Peptide	2.92-folds augmentation in the AUC *vis-à-vis* pure solution	Zheng *et al.* (2015)
SPIONs	Curcumin	Improved Aβ plaques detection in brain via better contrast in MRI	Cheng *et al.* (2015)
Congo Red/Rutin-MNPs	Rutin	Targeted delivery of rutin, along with precise detection of Aβ plaques and reduced oxidative stress	Hu *et al.* (2015)
Magnetic NPs	Quercetin	Improved antioxidant activity against copper-induced neurodegeneration	Nday *et al.* (2015)
SLNs	Sesame	Decreased oxidative stress and improved cognition	Sachdeva *et al.* (2015)
NPs	Curcumin	Increased cellular uptake of NPs-Cur 50:50 *vis-à-vis* curcumin coupled with inhibition of Akt and Tau proteins in SK-N-SH cells	Djiokeng Paka *et al.* (2015)
Fullerenes	—	Enhanced neuroprotection via activation of dopamine receptors using C_{60}HyFn	Vorobyov *et al.* (2015)
Silica NPs	Quercetin	Reversal of copper-induced neurodegeneration	Nday *et al.* (2015)
Liposomes	Curcumin	Enhanced affinity towards amyloid deposits and impeded amyloid aggregation.	Mourtas *et al.* (2014)
Nanoemulsions	Resveratrol	Improved levels of antioxidants and decrease in oxidative stress	Pangeni *et al.* (2014)
Liposomes	ApoE-derived peptide	Improved permeability and therapeutic efficacy	Bana *et al.* (2014)

(Continued)

TABLE 10.4 (*Continued*)
Select Literature Instances of Nanosystems in Therapeutics of Various Neurodegenerative Disorders

Nanosystem(s)	Bioactive	Attributes/Key Findings	Reference
Polymeric NPs	Oxysterols	Improved permeability across BBB eventually leading to improved anti-AD potential	Testa *et al.* (2014)
Nanocapsules	Quercetin	Decreased oxidative stress in ischemic reperfusion injury with less neuronal damage	Ghosh *et al.* (2013)
Lipid core Nanocapsules	Resveratrol	Decreased in Aβ triggered neuroinflammation	Frozza *et al.* (2013)
Polymeric NPs	Curcumin	Enhanced neurogenesis by internalization into the hippocampal neural stem cells	Tiwari *et al.* (2013)
Lipid core nanocapsules	Resveratrol	Improved cognition with activation of astrocytes and microglial cells	Hoppe *et al.* (2013)
Lipid core nanocapsules	Resveratrol	Decrease in neuroinflammatory markers concentration	Frozza *et al.* (2013)
Nanocomposites	Alginate–curcumin	Improved cognition	Siddique *et al.* (2013)
Polymeric NPs	Curcumin	Improved cognition and BBB permeability	Cheng *et al.* (2013)
Polymeric NPs	Curcumin	Improved cognition, targeting potential and BBB permeability	Mathew *et al.* (2012)
Fullerenes	Curcumin	Enhanced antioxidant and anti-aggregative properties	Makarova *et al.* (2012)
SLNs	Quercetin	Improved cognition	Dhawan *et al.* (2011)
NLCs	ECGC	Improved cognition	Smith *et al.* (2010)
Polymeric NPs	Resveratrol	Decreased oxidative stress, improved cognition	Lu *et al.* (2009)
Nanocapsules	Quercetin	Improvement in age-related cerebral oxidative injury	Das *et al.* (2008)
Liposomes	Maleic- and fumaric-diamides of (O, O-diacetyl)-L-Dopa-methyl ester	Increased bioavailability	Di Stefano *et al.* (2006)
CNTs	Agarose	Enhanced neuroregeneration	Mattson *et al.* (2000)

Abbreiviations: AD, Alzheimer's disease; Akt, Protein kinase B; AUC, Area under curve; BBB, Blood–brain barrier; bEND, Brain endothelial; CNTs, Carbon nanotubes; EGCG, Epigallocatechin gallate; MRI, Magnetic resonance imaging; NLCs, Nanostructured lipid carriers; NPs, Nanoparticles; ROS, Reactive oxygen species; SLNs, Solid lipid nanoparticles; SPIONs, Superparamagnetic iron oxide NPs.

10.3.2.4 Phospholipid Complexes

Use of a drug-phospholipid complex has been another approach for molecules that have difficulty in traversing the BBB. Conjugation of drugs with the phospholipid layers aims to increase the solubility of the former, as well as their permeability through biological barriers. Such complexes are commonly referred to as phytosomes. Various studies have suggested marked improvement in the therapeutic and biopharmaceutical attributes of molecules for their brain targeting, when formulated in these complexes. Literature also documents the efficacy of such complexes in the treatment and management of neurodegenerative diseases (Kassem *et al.*, 2017). Table 10.4 lists the different nanoscaled formulations of bioactives with improved neuroprotective potential.

10.4 CONCLUSIONS AND FUTURE PERSPECTIVES

With the advancement of neurological sciences, newer challenges crop up, which need to be addressed and handled aptly. In such endeavors, nanobiomedicine plays a vital role, especially in the management of various neurodegenerative disorders. The present scenario has unleashed the successful outcomes only as a tip of the iceberg. Efforts are still needed to unearth the hidden potential of nanomedicine as the promising tool that can revolutionize nanoneuromedicine. Taken together, the hiccups associated with the bioactives for their use in the various neurodegenerative disorders can be rationally circumvented using platform technology of nanocarriers.

The field of nano-antioxidants is still in its nascent stages, whereas comprehensive research could prompt building of potential nano-antioxidants for the treatment and handling of neurodegenerative disorders. The eventual fate of nano-antioxidants will rely upon the noetic outline of the selected nanosystems in view of expanding and deepening information on biological pathways. Therefore, nano-antioxidants hold a unique place in our potential armamentarium against a wide range of neurodegenerative diseases that involve free radical–induced damage. Together with the appropriate site-specific nanosystem, a synergistic effect may be achieved when combined with the antioxidant molecules too.

REFERENCES

Ahmad N, Ahmad R, Naqvi AA, Alam MA, Ashafaq M, Abdur Rub R *et al.* Intranasal delivery of quercetin-loaded mucoadhesive nanoemulsion for treatment of cerebral ischaemia. *Artif. Cells Nanomed. Biotechnol.* 2018; 46: 717–29.

Alawdi SH, El-Denshary ES, Safar MM, Eidi H, David MO, Abdel-Wahhab MA. Neuroprotective effect of nanodiamond in Alzheimer's disease rat model: A pivotal role for modulating NF-κB and STAT3 signaling. *Mol. Neurobiol.* 2017; 54: 1906–18.

Amiri H, Saeidi K, Borhani P, Manafirad A, Ghavami M, Zerbi V. Alzheimer's disease: Pathophysiology and applications of magnetic nanoparticles as MRI theranostic agents. *ACS Chem. Neusci.* 2013; 4: 1417–29.

Atlante A, Gagliardi S, Minervini GM, Ciotti MT, Marra E, Calissano P. Glutamate neurotoxicity in rat cerebellar granule cells: A major role for xanthine oxidase in oxygen radical formation. *J. Neurochem.* 1997; 68: 2038–45.

Baluchnejadmojarad T, Roghani M, Nadoushan MR, Bagheri M. Neuroprotective effect of genistein in 6-hydroxydopamine Hemi-parkinsonian rat model. *Phytother. Res.* 2009; 23: 132–5.

Bana L, Minniti S, Salvati E, Sesana S, Zambelli V, Cagnotto A *et al.* Liposomes bi-functionalized with phosphatidic acid and an ApoE-derived peptide affect Aβ aggregation features and cross the blood-brain barrier: Implications for therapy of Alzheimer disease. *Nanomed. Nanotech. Biol. Med.* 2014; 10: 1583–90.

Bano D, Dinsdale D, Cabrera-Socorro A, Maida S, Lambacher N, McColl B, Ferrando-May E, Hengartner MO, Nicotera P. Alteration of the nuclear pore complex in Ca^{2+}-mediated cell death. *Cell Death Differ.* 2010; 17: 119–33.

Bastianetto S, Ramassamy C, Doré S, Christen Y, Poirier J, Quirion R. The Ginkgo biloba extract (EGb 761) protects hippocampal neurons against cell death induced by β-amyloid. *Eur. J. Neurosci.* 2000; 12: 1882–90.

Beg S, Rahman M, Jain A, Saini S, Midoux P, Pichon C *et al.* Nanoporous metal organic frameworks as hybrid polymer–metal composites for drug delivery and biomedical applications. *Drug Discov. Today.* 2017; 22: 625–37.

Benseny-Cases N, Klementieva O, Cladera J. Dendrimers antiamyloidogenic potential in neurodegenerative diseases. *New J. Chem.* 2012; 36: 211–6.

Blasi P, Giovagnoli S, Schoubben A, Ricci M, Rossi C. Solid lipid nanoparticles for targeted brain drug delivery. *Adv. Drug Deliv. Rev.* 2007; 59: 454–77.

Chen P, Miah MR, Aschner M. Metals and neurodegeneration. *F1000Res.* 2016; 5: 1366.

Cheng KK, Chan PS, Fan S, Kwan SM, Yeung KL, Wáng YX *et al.* Curcumin-conjugated magnetic nanoparticles for detecting amyloid plaques in Alzheimer's disease mice using magnetic resonance imaging (MRI). *Biomaterials.* 2015; 44: 155–72.

Ciechanover A, Kwon YT. Protein quality control by molecular chaperones in neurodegeneration. *Front. Neurosci.* 2017; 11: 185.

Costantino L, Boraschi D. Is there a clinical future for polymeric nanoparticles as brain-targeting drug delivery agents? *Drug Discov. Today.* 2012; 17: 367–78.

Cunha S, Amaral MH, Lobo JS, Silva AC. Lipid nanoparticles for nasal/intranasal drug delivery. *Crit. Rev. Ther. Drug Carrier Syst.* 2017; 34: 257–82.

Currais A, Prior M, Dargusch R, Armando A, Ehren J, Schubert D *et al.* Modulation of p25 and inflammatory pathways by fisetin maintains cognitive function in Alzheimer's disease transgenic mice. *Aging Cell.* 2014; 13: 379–90.

da Silva LV, Lopes CB, da Silva WC, de Paiva YG, dos Santos Silva FD, Lima PR *et al.* Electropolymerization of ferulic acid on multi-walled carbon nanotubes modified glassy carbon electrode as a versatile platform for NADH, dopamine and epinephrine separate detection. *Microchem. J.* 2017; 133: 460–7.

Das S, Mandal AK, Ghosh A, Panda S, Das N, Sarkar S. Nanoparticulated quercetin in combating age related cerebral oxidative injury. *Curr. Aging Sci.* 2008; 1: 169–74.

Datla KP, Christidou M, Widmer WW, Rooprai HK, Dexter DT. Tissue distribution and neuroprotective effects of citrus flavonoid tangeretinin a rat model of Parkinson's disease. *Neuroreport.* 2001; 12: 3871–5.

Dhawan S, Kapil R, Singh B. Formulation development and systematic optimization of solid lipid nanoparticles of quercetin for improved brain delivery. *J. Pharm. Pharmacol.* 2011; 63: 342–51.

Di Stefano A, Sozio P, Iannitelli A, Marianecci C, Santucci E, Carafa M. Maleic-and fumaric-diamides of (O, O-diacetyl)-L-Dopa-methylester as anti-Parkinson prodrugs in liposomal formulation. *J. Drug Target.* 2006; 14: 652–61.

Djiokeng Paka G, Doggui S, Zaghmi A, Safar R, Dao L, Reisch A *et al.* Neuronal uptake and neuroprotective properties of curcumin-loaded nanoparticles on SK-N-SH cell line: Role of poly (lactide-co-glycolide) polymeric matrix composition. *Mol. Pharm.* 2015; 13: 391–403.

Dong XX, Wang Y, Qin ZH. Molecular mechanisms of excitotoxicity and their relevance to pathogenesis of neurodegenerative diseases. *Acta Pharmacol. Sin.* 2009; 30: 379.

Doktorovova S, Kovačević AB, Garcia ML, Souto EB. Pre-clinical safety of solid lipid nanoparticles and nanostructured lipid carriers: Current evidence from *in vitro* and *in vivo* evaluation. *Eur. J. Pharm. Biopharm.* 2016; 108: 235–52.

Dugan LL, Lovett EG, Quick KL, Lotharius J, Lin TT, O'Malley KL. Fullerene-based antioxidants and neurodegenerative disorders. *Parkinsonism Relat. Disord.* 2001; 7: 243–6.

Emerich DF, Tracy MA, Ward KL, Figueiredo M, Qian R, Henschel C *et al.* Biocompatibility of poly (DL-lactide-co-glycolide) microspheres implanted into the brain. *Cell Transplant.* 1999; 8: 47–58.

Feng X, Yang J. Lysosomal calcium in neurodegeneration. *Messenger (Los Angel).* 2016; 5: 56–66.

Fernandes C, Soni U, Patravale V. Nano-interventions for neurodegenerative disorders. *Pharmacol. Res.* 2010; 62: 166–78.

Frozza RL, Bernardi A, Hoppe JB, Meneghetti AB, Battastini AM, Pohlmann AR *et al.* Lipid-core nanocapsules improve the effects of resveratrol against Aβ-induced neuroinflammation. *J. Biomed. Nanotechnol.* 2013; 9: 2086–104.

Georgieva JV, Hoekstra D, Zuhorn IS. Smuggling drugs into the brain: An overview of ligands targeting transcytosis for drug delivery across the blood-brain barrier. *Pharmaceutics.* 2014; 6: 557–83.

Ghosh A, Sarkar S, Mandal AK, Das N. Neuroprotective role of nanoencapsulated quercetin in combating ischemia-reperfusion induced neuronal damage in young and aged rats. *PLoS One.* 2013; 8: e57735.

Goldberg JM, Lippard SJ. Challenges and opportunities in brain bioinorganic chemistry. *Acc. Chem. Res.* 2017; 50: 577–9.

Gong EJ, Park HR, Kim ME, Piao S, Lee E, Jo DG, Chung *et al.* Morin attenuates tau hyperphosphorylation by inhibiting GSK3β. *Neurobiol. Dis.* 2011; 44: 223–30.

Gopinath K, Sudhandiran G. Naringin modulates oxidative stress and inflammation in 3-nitropropionic acid-induced neurodegeneration through the activation of nuclear factor-erythroid 2-related factor-2 signalling pathway. *Neuroscience.* 2012; 227: 134–43.

Gutierrez-Zepeda A, Santell R, Wu Z, Brown M, Wu Y, Khan I, Link CD *et al.* Soy isoflavone glycitein protects against beta amyloid-induced toxicity and oxidative stress in transgenic Caenorhabditis elegans. *BMC Neurosci.* 2005; 6: 54.

Han Q, Wang X, Cai S, Liu X, Zhang Y, Yang L *et al.* Quercetin nanoparticles with enhanced bioavailability as multifunctional agents toward amyloid induced neurotoxicity. *J. Mater. Chem. B* 2018; 6(9): 1387–93.

Hassanzadeh P, Arbabi E, Atyabi F, Dinarvand R. Ferulicacid-loaded nanostructured lipid carriers: A promising nanoformulation against the ischemic neural injuries. *Life Sci.* 2018; 193: 64–76.

Hayne DJ, Lim S, Donnelly PS. Metal complexes designed to bind to amyloid-β for the diagnosis and treatment of Alzheimer's disease. *Chem. Soc. Rev.* 2014; 43: 6701–15.

Haziza S, Mohan N, Loe-Mie Y, Lepagnol-Bestel AM, Massou S, Adam MP *et al.* Fluorescent nanodiamond tracking reveals intraneuronal transport abnormalities induced by brain-disease-related genetic risk factors. *Nat. Nano.* 2017; 12: 322–8.

Hendriks JJ, de Vries HE, van der Pol SM, van den Berg TK, van Tol EA, Dijkstra CD. Flavonoids inhibit myelin phagocytosis by macrophages; a structure–activity relationship study. *Biochem. Pharmacol.* 2003; 65: 877–85.

Henze C, Earl C, Sautter J, Schmidt N, Themann C, Hartmann A *et al.* Reactive oxidative and nitrogen species in the nigrostriatal system following striatal 6-hydroxydopamine lesion in rats. *Brain Res.* 2005; 1052: 97–104.

Herges K, Millward JM, Hentschel N, Infante-Duarte C, Aktas O, Zipp F. Neuroprotective effect of combination therapy of glatiramer acetate and epigallocatechin-3-gallate in neuroinflammation. *PloS One.* 2011; 6: e25456.

Hernandes MS, Britto LR. NADPH oxidase and neurodegeneration. *Curr. Neuropharmacol.* 2012; 10: 321–7.

Hoppe JB, Coradini K, Frozza RL, Oliveira CM, Meneghetti AB, Bernardi A *et al.* Free and nanoencapsulated curcumin suppress β-amyloid-induced cognitive impairments in rats: Involvement of BDNF and Akt/GSK-3β signaling pathway. *Neurobiol. Learn. Mem.* 2013; 106: 134–44.

Horcajada P, Chalati T, Serre C, Gillet B, Sebrie C, Baati T *et al.* Porous metal-organic-framework nanoscale carriers as a potential platform for drug delivery and imaging. *Nat. Mater.* 2010; 9: 172.

Hu B, Dai F, Fan Z, Ma G, Tang Q, Zhang X. Nanotheranostics: Congo Red/Rutin-MNPs with enhanced magnetic resonance imaging and H_2O_2-responsive therapy of Alzheimer's disease in APPswe/PS1dE9 transgenic mice. *Adv. Mater.* 2015; 27: 5499–505.

Huang N, Cheng S, Zhang X, Tian Q, Pi J, Tang J *et al.* Efficacy of NGR peptide-modified PEGylated quantum dots for crossing the blood–brain barrier and targeted fluorescence imaging of glioma and tumor vasculature. *Nanomed. Nanotech. Biol. Med.* 2017; 13: 83–93.

Jain A, Jain A, Garg NK, Tyagi RK, Singh B, Katare OP *et al.* Surface engineered polymeric nanocarriers mediate the delivery of transferrin–methotrexate conjugates for an improved understanding of brain cancer. *Acta Biomater.* 2015; 24: 140–51.

Jiménez-Aliaga K, Bermejo-Bescós P, Benedí J, Martín-Aragón S. Quercetin and rutin exhibit antiamyloidogenic and fibril-disaggregating effects *in vitro* and potent antioxidant activity in APPswe cells. *Life Sci.* 2011; 89: 939–45.

John AA, Subramanian AP, Vellayappan MV, Balaji A, Mohandas H, Jaganathan SK. Carbon nanotubes and graphene as emerging candidates in neuroregeneration and neurodrug delivery. *Int. J. Nanomedicine.* 2015; 10: 4267.

Jucker M, Walker LC. Self-propagation of pathogenic protein aggregates in neurodegenerative diseases. *Nature.* 2013; 501: 45–51.

Kassem AA, El-Alim SH, Basha M, Salama A. Phospholipid complex enriched micelles: A novel drug delivery approach for promoting the antidiabetic effect of repaglinide. *Eur. J. Pharm. Sci.* 2017; 99: 75–84.

Khan MM, Raza SS, Javed H, Ahmad A, Khan A, Islam F *et al.* Rutin protects dopaminergic neurons from oxidative stress in an animal model of Parkinson's disease. *Neurotox. Res.* 2012; 22: 1–15.

Khurana RK, Jain A, Jain A, Sharma T, Singh B, Keshwani P. Administration of antioxidants in cancer: Debate of the decade. *Drug Discov. Today.* 2018; 23: 763–70.

Kim HG, Ju MS, Ha SK, Lee H, Lee H, Kim SY *et al.* Acacetin protects dopaminergic cells against 1-methyl-4-phenyl-1, 2, 3, 6-tetrahydropyridine-induced neuroinflammation *in vitro* and *in vivo. Biol. Pharm. Bull.* 2012; 35: 1287–94.

Ko YT, Bhattacharya R, Bickel U. Liposome encapsulated polyethylenimine/ODN polyplexes for brain targeting. *J. Control. Release.* 2009; 133: 230–7.

Korshunov SS, Skulachev VP, Starkov AA. High protonic potential actuates a mechanism of production of reactive oxygen species in mitochondria. *FEBS Lett.* 1997; 416: 15–21.

Koziara JM, Lockman PR, Allen DD, Mumper RJ. Paclitaxel nanoparticles for the potential treatment of brain tumors. *J. Control. Release.* 2004; 99: 259–69.

Kumar P, Modi G, Choonara YE. Nano (neuro) medicinal interventions for neurodegenerative disorders: A meta-analysis of concurrent challenges and strategic solutions. In *Nanotechnology and Drug Delivery.* Arias JL, (ed.). CRC Press. 2016; 1: 284.

Lee SG, Kim K, Kegelman TP, Dash R, Das SK, Choi JK *et al.* AEG-1 promotes glioma-induced neurodegeneration by increasing glutamate excitotoxicity. *Cancer Res.* 2011; 71: 6514–23.

Lehnardt S. Innate immunity and neuroinflammation in the CNS: The role of microglia in toll-like receptor-mediated neuronal injury. *Glia.* 2010; 58: 253–63.

Levites Y, Weinreb O, Maor G, Youdim MB, Mandel S. Green tea polyphenol (-)-epigallocatechin-3-gallate prevents N-methyl-4-phenyl-1, 2, 3, 6-tetrahydropyridine-induced dopaminergic neurodegeneration. *J. Neurochem.* 2001; 78: 1073–82.

Li Q, Cai T, Huang Y, Xia X, Cole SP, Cai Y. A Review of the structure, preparation, and application of NLCs, PNPs, and PLNs. *Nanomaterials.* 2017; 7: 122.

Liu LL, Sheng BY, Yan YF, Gong K, Ma T, Zhao NM *et al.* Protective effect of anthocyanin against the oxidative stress in neuroblastoma N2a cells. *Prog. Biochem. Biophys.* 2010; 37: 779–85.

Lohan S, Raza K, Mehta SK, Bhatti GK, Saini S, Singh B. Anti-Alzheimer's potential of berberine using surface decorated multi-walled carbon nanotubes: A preclinical evidence. *Int. J. Pharm.* 2017; 530: 263–78.

Lu X, Ji C, Xu H, Li X, Ding H, Ye M *et al.* Resveratrol-loaded polymeric micelles protect cells from Aβ-induced oxidative stress. *Int. J. Pharm.* 2009; 375: 89–96.

Lv C, Hong T, Yang Z, Zhang Y, Wang L, Dong M *et al.* Effect of quercetin in the 1-Methyl-4-phenyl-1, 2, 3, 6-tetrahydropyridine-induced mouse model of Parkinson's disease. *Evid. Based Complement. Alternat. Med.* 2012: 1–6.

Ma XH, Duan WJ, Mo YS, Chen JL, Li S, Zhao W *et al.* Neuroprotective effect of paeoniflorin on okadaic acid-induced tau hyperphosphorylation via calpain/Akt/GSK-3β pathway in SH-SY5Y cells. *Brain Res.* 2018; 1690: 1–11.

Makarova EG, Gordon RY, Podolski IY. Fullerene C60 prevents neurotoxicity induced by intrahippocampal microinjection of amyloid-β peptide. *J. Nanosci. Nanotechnol.* 2012; 12: 119–26.

Mandal A, Bisht B, Pal D, Mitra AK. Diagnosis and drug delivery to the brain: Novel strategies. In *Emerging Nanotechnologies for Diagnostics, Drug Delivery and Medical Devices.* Mitra AK, Cholkar K, Mandal A, (eds.). Netherland, Elsevier. 2017; 59–83.

Marianecci C, Rinaldi F, Hanieh PN, Di Marzio L, Paolino D, Carafa M. Drug delivery in overcoming the blood–brain barrier: Role of nasal mucosal grafting. *Drug Des. Devel. Ther.* 2017; 11: 325.

Mathew A, Fukuda T, Nagaoka Y, Hasumura T, Morimoto H, Yoshida Y *et al.* Curcumin loaded-PLGA nanoparticles conjugated with Tet-1 peptide for potential use in Alzheimer's disease. *PLoS One.* 2012; 7: e32616.

Mattson MP, Haddon RC, Rao AM. Molecular functionalization of carbon nanotubes and use as substrates for neuronal growth. *J Mol. Neurosci.* 2000; 14: 175–82.

McManus RM, Heneka MT. Role of neuroinflammation in neurodegeneration: New insights. *Alzheimers Res. Ther.* 2017; 9: 14.

McMurray CT. Neurodegeneration: Diseases of the cytoskeleton? *Cell Death Differ.* 2000; 7: 861–5.

Mignani S, Bryszewska M, Zablocka M, Klajnert-Maculewicz B, Cladera J, Shcharbin D *et al.* Can dendrimer based nanoparticles fight neurodegenerative diseases? Current situation versus other established approaches. *Prog. Polym. Sci.* 2017; 64: 23–51.

Moghaddam AH, Zare M. Neuroprotective effect of hesperetin and nano-hesperetin on recognition memory impairment and the elevated oxygen stress in rat model of Alzheimer's disease. *Biomed. Pharmacother.* 2018; 97: 1096–101.

Morales JM, Skipwith CG, Clark HA. Quadruplex integrated DNA(quID) nanosensors for monitoring dopamine. *Sensors.* 2015; 15(8): 19912–24.

Morroni F, Sita G, Djemil A, D'Amico M, Pruccoli L, Forti G *et al.* Comparison of adaptive neuroprotective mechanisms of sulforaphane and its interconversion product erucinin *in vitro* and *in vivo* models of Parkinson's Disease. *J. Agric. Food Chem.* 2018; 66(4): 856–65.

Mourtas S, Lazar AN, Markoutsa E, Duyckaerts C, Antimisiaris SG. Multifunctional nanoliposomes with curcumin–lipid derivative and brain targeting functionality with potential applications for Alzheimer disease. *Eur. J. Med. Chem.* 2014; 80: 175–83.

Muchowski PJ, Wacker JL. Modulation of neurodegeneration by molecular chaperones. *Nat. Rev. Neurosci.* 2005; 6: 11–22.

Nday CM, Halevas E, Jackson GE, Salifoglou A. Quercetin encapsulation in modified silica nanoparticles: Potential use against Cu (II)-induced oxidative stress in neurodegeneration. *J. Inorg. Biochem.* 2015; 145: 51–64.

Nie G, Cao Y, Zhao B. Protective effects of green tea polyphenols and their major component, (–)-epigallocatechin-3-gallate (EGCG), on 6-hydroxydopamine-induced apoptosis in PC12 cells. *Redox Rep.* 2002; 7: 171–7.

Obuobi S, Karatayev S, Chai CL, Ee PL, Mátyus P. The role of modulation of antioxidant enzyme systems in the treatment of neurodegenerative diseases. *J. Enzyme Inhib. Med. Chem.* 2016; 31: 194–204.

Ono K, Yoshiike Y, Takashima A, Hasegawa K, Naiki H, Yamada M. Potent anti-amyloidogenic and fibril-destabilizing effects of polyphenols *in vitro*: Implications for the prevention and therapeutics of Alzheimer's disease. *J. Neurochem.* 2003; 87: 172–81.

Pangeni R, Sharma S, Mustafa G, Ali J, Baboota S. Vitamin E loaded resveratrol nanoemulsion for brain targeting for the treatment of Parkinson's disease by reducing oxidative stress. *Nanotechnology.* 2014; 25: 485102.

Park SE, Sapkota K, Kim S, Kim H, Kim SJ. Kaempferol acts through mitogen-activated protein kinases and protein kinase B/AKT to elicit protection in a model of neuroinflammation in BV2 microglial cells. *Br. J. Pharmacol.* 2011; 164: 1008–25.

Patel RP, McAndrew J, Sellak H, White CR, Jo H, Freeman BA, Darley-Usmar VM. Biological aspects of reactive nitrogen species. *Biochim. Biophys. Acta.* 1999; 1411: 385–400.

Patricia G-S, Sabine H. Autophagic pathology and calcium deregulation in neurodegeneration. In *Current Topics in Neurotoxicity.* Fuentes JM, (ed.). Springer. 2015; 9: 247–66.

Pattni BS, Chupin VV, Torchilin VP. New developments in liposomal drug delivery. *Chem. Rev.* 2015; 115: 10938–66.

Qu W, Zuo W, Li N, Hou Y, Song Z, Gou G, Yang J. Design of multifunctional liposome-quantum dot hybrid nanocarriers and their biomedical application. *J. Drug Target.* 2017; 25: 661–72.

Reddy LH, Sharma RK, Chuttani K, Mishra AK, Murthy RR. Etoposide-incorporated tripalmitin nanoparticles with different surface charge: Formulation, characterization, radiolabeling, and biodistribution studies. *AAPS J.* 2004; 6: 55–64.

Sachdeva AK, Misra S, Kaur IP, Chopra K. Neuroprotective potential of sesamol and its loaded solid lipid nanoparticles in ICV-STZ-induced cognitive deficits: Behavioral and biochemical evidence. *Eur. J. Pharmacol.* 2015; 747: 132–40.

Sandhir R, Mehrotra A. Quercetin supplementation is effective in improving mitochondrial dysfunctions induced by 3-nitropropionic acid: Implications in Huntington's disease. *Biochim. Biophys. Acta.* 2013; 1832: 421–30.

Sandhir R, Yadav A, Sunkaria A, Singhal N. Nano-antioxidants: An emerging strategy for intervention against neurodegenerative conditions. *Neurochem. Int.* 2015; 89: 209–26.

Sánchez-López E, Ettcheto M, Egea MA, Espina M, Silva AM, Calpena AC, Camins A, Carmona N, Souto EB, García ML. Sustained-released Memantine PEGylated PLGA Nanoparticles: *In vitro* & *in vivo* targeting the blood-brain-barrier for the treatment of Alzheimer's disease. *J. Nanotech.* 2018; 16: 32.

Sesti F, Liu S, Cai SQ. Oxidation of potassium channels by ROS: A general mechanism of aging and neurodegeneration? *Trends Cell Biol.* 2010; 20: 45–51.

Sheffield LG, Miskiewicz HB, Tannenbaum LB, Mirra SS. Nuclear pore complex proteins in Alzheimer disease. *J. Neuropathol. Exp. Neurol.* 2006; 65: 45–54.

Siddique YH, Khan W, Singh BR, Naqvi AH. Synthesis of alginate-curcumin nanocomposite and its protective role in transgenic Drosophila model of Parkinson's disease. *ISRN Pharmacol.* 2013; 2013(Article ID 794582): 8 pages.

Smith A, Giunta B, Bickford PC, Fountain M, Tan J, Shytle RD. Nanolipidic particles improve the bioavailability and α-secretase inducing ability of epigallocatechin-3-gallate (EGCG) for the treatment of Alzheimer's disease. *Int. J. Pharm.* 2010; 389: 207–12.

Solanki I, Parihar P, Mansuri ML, Parihar MS. Flavonoid-based therapies in the early management of neurodegenerative diseases. *Adv. Nutr.* 2015; 6: 64–72.

Tanifum EA, Dasgupta I, Srivastava M, Bhavane RC, Sun L, Berridge J. Intravenous delivery of targeted liposomes to amyloid-β pathology in APP/PSEN1 transgenic mice. *PloS One.* 2012; 7: e48515.

Texeira MC, Carbone C, Souto EB. Beyond liposomes: Recent advances on lipid based nanostructures for poorly soluble/poorly permeable drug delivery. *Prog. Lipid Res.* 2017; 68: 1–11.

Testa G, Gamba P, Badilli U, Gargiulo S, Maina M, Guina T *et al.* Loading into nanoparticles improves quercetin's efficacy in preventing neuroinflammation induced by oxysterols. *PloS One* 2014; 9: e96795.

Tiwari SK, Agarwal S, Seth B, Yadav A, Nair S, Bhatnagar P *et al.* Curcumin-loaded nanoparticles potently induce adult neurogenesis and reverse cognitive deficits in Alzheimer's disease model via canonical Wnt/β-catenin pathway. *ACS Nano.* 2013; 8: 76–103.

Tosi G, Costantino L, Ruozi B, Forni F, Vandelli MA. Polymeric nanoparticles for the drug delivery to the central nervous system. *Expert Opin. Drug Deliv.* 2008; 5: 155–74.

Trotta V, Pavan B, Ferraro L, Beggiato S, Traini D, Des Reis LG *et al.* Brain targeting of resveratrol by nasal administration of chitosan-coated lipid microparticles. *Eur. J. Pharm. Biopharm.* 2018; 127: 250–9.

Vorobyov V, Kaptsov V, Gordon R, Makarova E, Podolski I, Sengpiel F. Neuroprotective effects of hydrated fullerene C60: Cortical and hippocampal EEG interplay in an amyloid-infused rat model of Alzheimer's disease. *J. Alzheimers Dis.* 2015; 45: 217–33.

Wen CJ, Zhang LW, Al-Suwayeh SA, Yen TC, Fang JY. Theranostic liposomes loaded with quantum dots and apomorphine for brain targeting and bioimaging. *Int. J. Nanomedicine.* 2012; 7: 1599–611.

Yadav A, Sunkaria A, Singhal N, Sandhir R. Resveratrol loaded solid lipid nanoparticles attenuate mitochondrial oxidative stress in vascular dementia by activating Nrf2/HO-1 pathway. *Neurochem. Int.* 2018; 112: 239–54.

Zeng H, Chen Q, Zhao B. Genistein ameliorates β-amyloid peptide (25–35)-induced hippocampal neuronal apoptosis. *Free Radic. Biol. Med.* 2004; 36: 180–8.

Zhang L, Liu F, Sun X, Wei GF, Tian Y, Liu ZP *et al.* Engineering carbon nanotube fiber for real-time quantification of ascorbic acid levels in a live rat model of Alzheimer's disease. *Anal. Chem.* 2017; 89: 1831–7.

Zhang W, Zhang Z, Zhang Y. The application of carbon nanotubes in target drug delivery systems for cancer therapies. *Nanoscale Res Lett.* 2011; 6(1): 555.

Zhao C, Liu Y, Li Y. Colorimetric and fluorometric assays for dopamine with a wide concentration range based on Fe-MIL-88NH2 metal-organic framework. *Anal. Chem.* 2015; 31: 1035–9.

Zhao L, Wang JL, Liu R, Li XX, Li JF, Zhang L. Neuroprotective, anti-amyloidogenic and neurotrophic effects of apigenin in an Alzheimer's disease mouse model. *Molecules.* 2013; 18: 9949–65.

Zheng X, Shao X, Zhang C, Tan Y, Liu Q, Wan X *et al.* Intranasal H102 peptide-loaded liposomes for brain delivery to treat Alzheimer's disease. *Pharm. Res.* 2015; 32: 3837–49.

Zhu Y, Murali S, Cai W, Li X, Suk JW, Potts JR *et al.* Graphene and graphene oxide: Synthesis, properties, and applications. *Adv. Mater.* 2010; 15(22): 3906–24.

11 Cellulose-Based NanoBioMaterials

Michael Ioelovich, Sumant Saini, Teenu Sharma, and Bhupinder Singh*

CONTENTS

11.1 INTRODUCTION

Cellulose is known to be the most widespread and renewed biopolymer in nature (Chen, 2014). Furthermore, cellulose is a nanostructured polymer, as it contains nanofibrils, which are made up of nanocrystallites and noncrystalline (amorphous) nanodomains. For thousands of years, cellulose absorbents and bandages are used in the domains of personal care and medicine. Since the twentieth century, cellulosic powders and modified celluloses began to be produced, whereas in the twenty-first century, various kinds of smart and multifunctional nanomaterials were joined to these products to find wider applications in biology, medicine, hygiene, and many other areas. Moreover, cellulosic carriers containing various therapeutically active substances have been developed which open additional application areas of this natural polymer. Biodegradability of cellulose makes this biopolymer unique for production of environmentally friendly biomedical materials and products.

This chapter describes and discusses the structure, properties, and main application areas of natural, artificial, and modified celluloses, as well as micro- and nanoparticles, nanofibrils, and nanofilaments.

11.2 CELLULOSE AS A NATURAL NANOSTRUCTURED BIOPOLYMER

Cellulose is a linear, semicrystalline polysaccharide (Ioelovich, 2004). Macromolecules of natural cellulose consist of thousands of repeated anhydroglucose units (AGUs), connected by means of 1,4-β-glycosidic bonds. Each of such units is a six-membered ring in a "chair" shape and contains three hydroxyl groups, imparting cellulose with increased hydrophilicity, sorption ability, and reactivity. For instance, this polysaccharide can form polyethers and polyesters, as well as polymeric carbonyl, carboxyl, and other derivatives. The average number of AGUs in polymer chains or the average degree

* Corresponding author.

of polymerization (DP) depends on type of natural source and method of isolation of cellulose (Siqueira et al., 2010). For instance, DP of native cellulose in wood can achieve approximately 10,000 AGUs. DP of native cotton cellulose can be 15,000–20,000, while DP of cellulose in *Valonia* algae is in the range of 25,000–30,000. On the other hand, DP of bacterial cellulose is, relatively, quite less, in the range of 4000–8000. However, when exposed to high temperatures, oxidants, acids, cellulolytic enzymes, and some other reagents, this polysaccharide is easily depolymerized by cleavage of glycosidic bonds in accessible noncrystalline domains of cellulose, which leads to reduction in DP value. Therefore, the typical DP value of cellulose specimens isolated from plant materials is within only 800–2000.

As is known, the total amount of cellulose in natural biomass is estimated at 10^{12} tons (Klemm et al., 2005) and tends to increase by about 10% annually (Field et al., 1998). Natural cellulose is found in terrestrial plants, algae, and some marine animals; furthermore, cellulose microfibrils are biosynthesized by certain microorganisms (Ioelovich, 2016a; Dos Santos et al., 2016). The main commercial source of cellulose is wood, containing 44%–50% cellulose and, to a lesser extent, cotton, containing more than 90% cellulose (Young and Rowell, 1986; Hon, 1996; Ioelovich, 2016a).

Three industrial pulping methods are used currently to isolate cellulose from wood or other plant materials: kraft, soda, and sulfite cooking (Sjöström, 1993). In the pulping process, noncellulosic components are removed, whereas cellulose is released in the form of fiber pulp. Plant cellulose is biosynthesized by means of the terminal enzyme complex (TEC) having the shape of a nanorosette (Saxena and Brown, 2005; Lerouxel et al., 2006; Yan et al., 2008). Each rosette contains six subunits of cellulose-synthase. According to modern concepts, the process of cellulose biosynthesis includes three stages: polycondensation, crystallization, and extrusion (Brown and Saxena, 2000). The activated monomer, uracil-diphosphate glucose (UDP glucose), is located in the cytoplasm of the cell, whence this monomer enters in the rosette. Synthesis of cellulose macromolecules is carried out by polycondensation of glucose under catalytic effect of TEC, followed by crystallization of chain segments, formation of nanofibrils, and drawing (extrusion) of these nanofibrils through the orifice of the rosette onto the surface of cell membrane.

Cellulosic macromolecules are biosynthesized by the bacterium *G. xylinus* by means of TEC inside the bacterium utilizing nutrient sugar solutions (Brown et al., 1976; Ross et al., 1991; Czaja et al., 2006a). Then, the macromolecule bunches are passed through nanopores of the membrane, crystallized, and nanofibrils formed, which are then pushed into the liquid medium. The nanofibrils of bacterial cellulose intertwine and form gel-like pellicles filled with immobilized water. Biosynthesis of macromolecules and formation of nanofibrils in tunicates and algae also occur with the assistance of TEC. However, in contrast to rosette TEC of plants, terminal enzyme complexes of bacteria, tunicate, and algae contain linearly arranged subunits of synthase (Moon et al., 2011).

The first studies suggested that elementary fibrils in cellulose have a constant lateral size of 3.5 nm (Manley, 1964; Mühlenthaler, 1969). However, further studies have shown that the lateral size of elementary fibrils in celluloses of various origins is variable, from 3–4 nm for cellulose of plant biomass to 15–20 nm for *Valonia* cellulose (Moon et al., 2011; Ioelovich, 2014a,b,c,d, 2018a,b,c) (Table 11.1). The elementary nanofibrils, having a large surface area, are prone to aggregation with the formation of fibrillar bundles, called microfibrils.

Various models of supramolecular structure of cellulose were developed to explain properties of this biopolymer. In the early twentieth century, a model of extended crystalline micelles was proposed (Zugenmaier, 2009). But later, this model was rejected due to lack of chains between micelles, which was contrary to the mechanical properties of natural cellulose fibers. Various models of elementary fibrils with folded chains (Tönnesen and Ellefsen, 1960; Dolmetsch, 1961; Manley, 1964) contradicted the mechanical properties of cellulosic fibers. Furthermore, the results of structural studies revealed that elementary fibrils of cellulose contain straightened chains. Further, various models of straightened fibrils were proposed, for example, a model of "fringed" fibrils (Hearle, 1963) containing the elongated single cellulose crystals with a thick amorphous fringe on their surface.

Since the model of "fringed" fibrils was contrary to structural characteristics of natural celluloses, Verlhac and coauthors (1990) have proposed a revised model of elementary fibrils having a thin

TABLE 11.1
Average Lateral Size of Elementary Fibrils (D_{ef}) and Degree of Crystallinity (X) for Celluloses of Various Sources

Source	D_{ef}, nm	X
Herbaceous plants	3–4	0.50–0.55
Wood	3–4	0.60–0.62
Flax	4–5	0.66–0.67
Cotton	5–6	0.67–0.69
Bacterial cellulose	7–8	0.75–0.80
Tunicate	10–15	0.80–0.85
Valonia algae	15–20	0.80–0.85

monomolecular amorphous layer on the surface of crystalline fibrils. The problem was that models with surface amorphous layers could not explain the mechanism of cellulose hydrolysis to a level-off value and splitting of fibrils with formation of small nanocrystallites. Subsequently, O'Sullivan (1997) also noted that crystallites do not contain amorphous surface layers. Despite these shortcomings, Verlhac's model continues to be used, for example, for estimation of crystallinity of cellulose (Moon et al., 2011).

Recently, a more real model of elementary fibrils has been proposed (Ioelovich, 2014a,b,c,d, 2016a). This crystallinity-mesomorphous model with the paracrystalline monomolecular surface layer (CMPL model) contains longitudinally alternating crystallites (cr), noncrystalline noncrystalline domains (nc), and thin surface paracrystalline layers (pc), as depicted in Figure 11.1.

The CMPL model adequately describes most of the cellulose properties. Moreover, such a model validates the possibility of isolation of micro- and nanocrystallites after splitting of noncrystalline domains (NCDs) of cellulose by hydrolysis. Presence of paracrystalline layers on the surface of crystallites explains distortion of the crystalline lattice, possibility deuteration of the crystalline surface, and recombination of radicals on this surface, and some other properties (Ioelovich et al., 2010, 2016a). As it follows from the LCMP model, lateral size (D) of crystallites and NCDs in natural cellulose should be the same as lateral size of the elementary nanofibril (D_{ef}), though these sizes depend on the origin of the cellulose (Tables 11.1 and 11.2). Also based on the LCMP model, average length (L_{nc}) of NCDs can be estimated, as follows (Equation 11.1):

$$L_{nc} = k\, L_c\, (X^{-1} - 1) \tag{11.1}$$

where, k≈1.125 is coefficient; X is degree of crystallinity, and L_c is average length of crystallites.

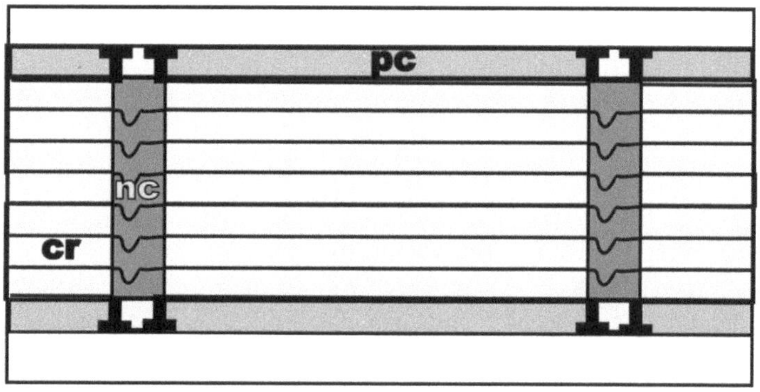

FIGURE 11.1 Improved model of elementary fibril.

TABLE 11.2
Average Lateral Size (D) and Length (L) of Crystallites and Noncrystalline Domains (NCDs) of Natural Celluloses

Source	X	D, nm	L_c, nm	L_{nc}, nm
Herbaceous plants	0.50–0.55	3–4	50–60	55–60
Wood	0.60–0.62	3–4	60–80	50–55
Flax	0.66–0.67	4–5	80–90	45–50
Cotton	0.67–0.69	5–6	90–100	40–45
Bacterial cellulose	0.75–0.80	7–8	120–140	35–40

TABLE 11.3
Structural Characteristics of Cellulose Isolated from Spruce Wood by Various Pulping Methods

Pulping Method	X	L_c, nm	D_c, nm
Non-delignified	0.62	72	3.6
Organosolv	0.62	72	4.4
Sulphite	0.63	74	6.3
Kraft	0.64	76	7.0

Abbreviations: X, degree of crystallinity; L_c, length; D_c, lateral size.

TABLE 11.4
Structural Characteristics of Isolated Celluloses

Isolated Cellulose	X	L_c, nm	D_c, nm
Kraft spruce cellulose	0.64	76	7.0
Kraft birch cellulose	0.63	68	6.2
Soda flax cellulose	0.69	98	6.5
Soda cotton cellulose	0.70	100	7.8

Abbreviations: X, degree of crystallinity; L_c, length; D_c, lateral size.

After isolation of cellulose from plant materials, crystallinity degree, length of nanocrystallites, and NCDs hardly change, but lateral size of nanocrystallites can increase due to involvement of the process of lateral co-crystallization (Tables 11.3 and 11.4).

As is well-known, free surface energy (G^o_s) of crystallites is a function of specific surface (S_{sp}), as in Equation 11.2:

$$G^o_s = \sigma V S_{sp} \tag{11.2}$$

where, σ is surface energy and V is molar volume of crystallite.

Since small nanocrystallites are characterized with increased specific surface, their free surface energy will be high and the cause of aggregative instability of nanocrystallites. To achieve a more stable state, the free surface energy must decrease (Ioelovich, 2016b) as per Equation 11.3:

$$\Delta G^o_s = \sigma V \Delta S_{sp} < 0 \tag{11.3}$$

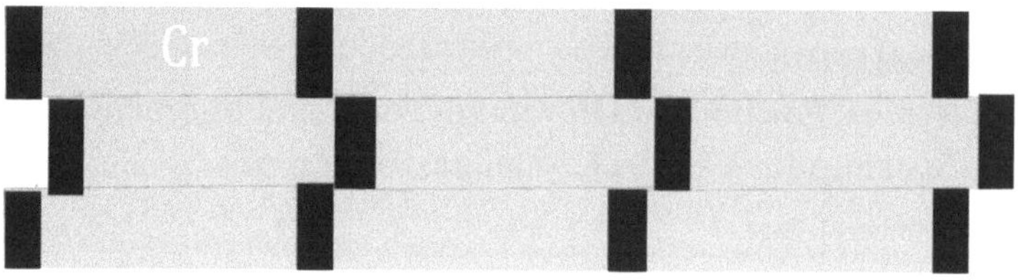

FIGURE 11.2 Fibrillar aggregate of isolated cellulose.

TABLE 11.5
Structural Characteristics of Modified Cotton Cellulose

Treatment	CA	X	D_c, nm	L_c, nm
Untreated	CI	0.70	7.8	100
Ball milling for 5 h	CI	0.45	4.1	43
Hydrolyzed with 2.5 M HCl	CI	0.75	10.0	110
Mercerization	CII	0.55	5.5	52

Abbreviations: CA, crystalline allomorph; X, degree of crystallinity; L_c, length; D_c, lateral size.

It implies that small nanocrystallites can aggregate via lateral co-crystallization to form large crystallites with decreased area of specific surface (Ioelovich, 2014a,b,c,d). Since a high temperature and plasticizing alkali medium facilitate the co-crystallization process, the kraft delignification process yields larger crystalline aggregates. The lateral co-crystallization process causes aggregation of elementary nanofibrils with the formation of bundles called microfibrils (Figure 11.2). Mechanical (e.g., ball milling, etc.) or physicochemical treatments (e.g., hydrolysis, mercerization, etc.) lead to changes in the structure of isolated cellulose (Table 11.5).

The structural investigations showed that cellulose crystallites have various crystalline allomorphs, Iα, Iβ, II, III, and IV (Sugiyama et al., 1991; Krässig, 1993; O'Sullivan, 1997; Zugenmaier, 2008). Triclinic allomorph Iα (space group P1) occurs in crystallites of algae and bacterial celluloses, while more stable monoclinic allomorph Iβ (space group $P2_1$) is dominant in cellulose crystallites of higher plants and tunicates. It was found that the specific gravity of Iβ crystallites (1.62–1.63 g/cm^3) is slightly higher than that of Iα crystallites (1.59–1.60 g/cm^3). Various allomorphs of cellulose crystallites have different space groups, sizes, and volumes of crystalline unit cells, as well as different interplanar spacings (d) and positions of diffraction peaks ($2\theta^o_{hkl}$) (Langan et al., 2001; Zugenmaier, 2008; Ioelovich, 2014a,b,c,d) (Table 11.6).

Due to the presence hydroxyl groups in AGUs, cellulose can form various derivatives. The most abundant natural derivative is chitin, which consists of acetyl-2-aminodeoxyglucose units (Rinaudo, 2006; Aranaz et al., 2009). The total amount of this biopolymer in nature is estimated at 10^{11} tons (Roberts, 1998). Chitin is found in marine animals, insects, and some fungi. The main commercial sources of chitin are marine animals such as crabs, shrimps, and lobsters. To isolate chitin, shells of these animals are treated by acid to remove calcium carbonate, followed by alkaline extraction of proteins and bleaching.

Chitin is a linear semicrystalline nitrogenated polysaccharide (Blackwell, 1973; Watthanaphanit and Rujiravanit, 2010). Primary supramolecular structure of this biopolymer consists of nanoscale fibrils with a diameter of 2–5 nm (Revol and Marchessault, 1993; Chen et al., 2008). Similar to cellulose, each fibril of chitin contains crystallites, separated by noncrystalline domains. Crystallites of chitin can have orthorhombic α-allomorph, monoclinic β-allomorph, and rare γ-allomorph (Rudall and Kenchington, 1973; Gardner and Blackwell, 1975). The unstable crystalline β- and γ-forms are

TABLE 11.6
Typical Positions of Diffraction Peaks and Interplanar Spacings for Various Crystalline Allomorphs

Allomorph Type	Index	$2\theta^o_{hkl}$	d, nm
CIβ	$(1\bar{1}0)$	14.6–14.8	0.600
	(110)	16.2–16.6	0.540
	(200)	22.4–22.6	0.393
CII	$(1\bar{1}0)$	12.0–12.2	0.730
	(110)	19.8–20.2	0.443
	(020)	21.8–22.2	0.400
CIII	$(1\bar{1}0)$	11.8–12.0	0.740
	(110 & 020)	19.8–21.2	0.422
CIV	$(1\bar{1}0)$ & 110)	15.4–15.8	0.570
	(200) & (020)	22.0–22.4	0.400

Abbreviations: d, interplanar spacings; θ^o_{hkl}, positions of diffraction peaks.

accessible for molecules of polar liquids, moreover, these crystalline allomorphs can be converted into more stable α-allomorph after some treatments (Saito et al., 1997, 2000).

Chitin is used for commercial production of chitosan, another known nitrogenated derivative of cellulose. Natural sources of native chitosan are some fungi, for example, *Mucoraceae* (Roberts, 1998; Rinaudo, 2006). Chitosan is a linear low crystalline nitrogenated polysaccharide, built from repeated 1,4-β-2-glucosamine links. Crystallites of chitosan have orthorhombic unit cells, belonging to space group of $P2_12_12_1$. Using improved methods of structural studies, the actual crystallinity degree of nitrogenated polysaccharides was determined as 0.70–0.75 for chitin and 0.55–0.60 for chitosan samples (Ioelovich, 2014a,b). In addition to natural derivatives, various synthetic mono-, di- and tri-substituted cellulose derivatives may also be prepared (Figure 11.3). In some cases, the substitution degree of cellulose can be less than one.

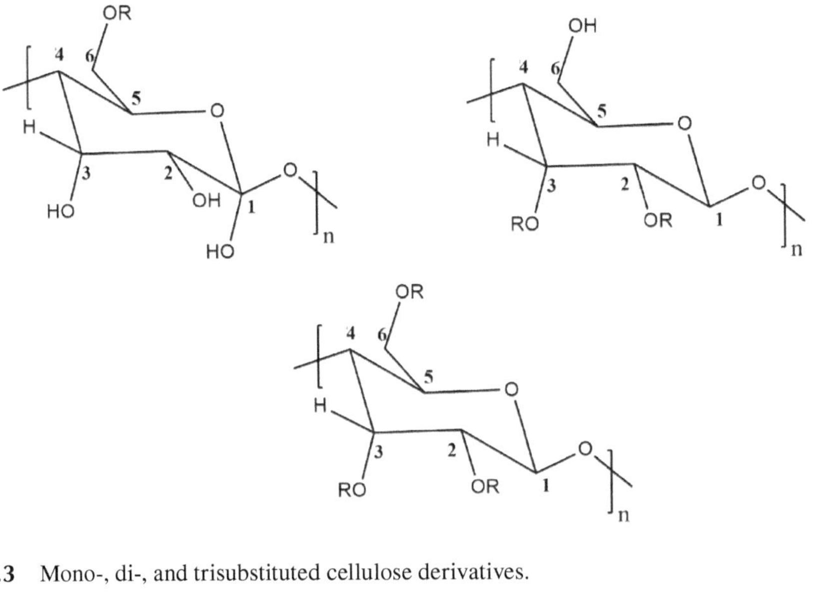

FIGURE 11.3 Mono-, di-, and trisubstituted cellulose derivatives.

11.3 PREPARATION OF CELLULOSE-BASED NANOBIOMATERIALS

11.3.1 Manufacturing of Biomedical Cellulosic Materials

The main plant sources for manufacturing of biomedical materials are cotton and wood. To isolate pure cellulose fibers from native cotton fibers, the initial feedstock is cleaned from impurities and then is subjected to soda pulping and bleaching. Soda pulping is performed in a boiler at a temperature between 130°C and 150°C under pressure of 0.3–0.5 MPa using 2%–4% solution of caustic soda (Rogovin, 1974). Then, the pure fibers are bleached by chlorine dioxide or hydrogen peroxide, washed, and dried. The dried cotton fibers are combed to obtain a canvas, which is subjected to carding to produce cotton wool. Purified and bleached cotton fibers also are used for manufacturing medical gauze and bandages.

To isolate wood pulp, the method of kraft pulping is typically used. Wood chips are cooked in a 6%–8% solution of sodium hydroxide containing an additive of sodium sulfide at a temperature of 170°C–175°C under a pressure of 0.8–1.0 MPa (Sjöström, 1993). To remove the residual lignin and hemicelluloses, technical kraft pulp undergoes multistage bleaching process, which includes chlorination, alkaline extraction, bleaching with hypochlorite, and additional bleaching with chlorine dioxide (Dence and Reeve, 1996). The bleached pulp is used for manufacturing of various personal care products. Further, this pulp can also be used as feedstock to produce regenerated cellulose, artificial fibers, and films.

The technology of viscose fibers includes several steps (Serkov, 1981): mercerization; oxidative depolymerization of alkali-cellulose; xanthation of alkali-cellulose; dissolution of cellulose xanthate in alkali to obtain "viscose"; ripening, filtration, and degassing of viscose; wet spinning using round spinnerets and drawing to produce the viscose rayon filaments; and finally, washing, drying, and optional cutting to obtain staple viscose fibers. Technology of cuprammonium rayon fibers uses cuprammonium hydroxide solvent (Kaneko, 1990), whereas technology of lyocell fibers uses monohydrate of N-methylmorpholine-N-oxide as the solvent (Borbely, 2008).

The main shortcoming of the conventional rayon fibers is a reduced water absorption, below 100%. Recently, new types of rayon fibers with increased absorption ability were developed (Roggenstein, 2011), including the following types: (i) hollow viscose fibers, which after contact with an aqueous medium fill completely with the liquid, resulting in an increased intrinsic absorption of 250%; (ii) hollow viscose fibers with introduced specific anionic functional groups, which impart to fibers a high absorption value of 400%; (iii) hollow thin-walled viscose fibers containing both anionic functional groups and gel-forming additives, which show superabsorbent properties.

Manufacturing technology of artificial cellulose films (i.e., membranes) is similar to technology of rayon fibers with the difference that slit-shape spinnerets are used at the wet-spinning step, and two-directional drawing is applied. Besides, the man-made films can contain glycerol and some other plasticizers. The cellulose rayon fibers and artificial films are characterized by reduced crystallinity, 0.35–0.40, and low DP, typically of 300–400. Nevertheless, these products are resistant to organic solvents, as well as to dilute aqueous solutions of acids and alkalis.

The conventional dressings are considered as less effective and more adhesive, such that these strongly adhering dressings cause damage to the wound. Therefore, a lot of studies have shown that an effective dressing should close and heal the wound, maintain its high humidity, and should be breathing, non-sticky, and atraumatic (Odland and Ross, 1968; Brown and Zitelli, 1995; Moon and Crabtree, 2003). Highly effective dressings described by Moon and Crabtree (2003) consisted of two or three functional layers.

11.3.2 Preparation of Cellulose Derivatives

A natural, nitrogenated cellulose derivative, chitin, has been isolated from shells of various sea animals (Zelencova et al., 2015). For this purpose, the chosen material is cleaned from dirt and ground. To remove minerals, the ground sample is treated with hot dilute hydrochloric acid and then washed. To remove proteins, the demineralized sample is extracted with dilute alkali at elevated temperatures, followed by washing. The residual lipids are extracted with a boiling alcohol/hydrocarbon mixture for 1 h. Finally, chitin is dried to constant weight.

The extracted chitin can serve as a feedstock for preparation of chitosan (Zelencova et al., 2015). For this purpose, the chitin sample is treated with 40%–50% NaOH at 130°C–150°C for 3–4 h, followed by washing and drying. Due to the basic amino groups in its structure, chitosan neutralizes weak acids. Also, this biopolymer exhibits chelating properties and can sorb multivalent cations (Dutta et al., 2004).

Amongst the natural and diverse synthetic cellulose derivatives, water-soluble ethers are of great importance (Obolenskaya et al., 1991). For instance, to prepare carboxymethyl cellulose, an initial cellulose material is cut into small pieces, which are mixed with a 10-fold amount of isopropanol. The 30% NaOH solution is slowly added while stirring to achieve the final alkali concentration of 7% in the reaction system. After 1 h of alkalization, monochloroacetic acid is added while stirring in a ratio 1.5:1 with respect to weight of the initial sample, and the reaction system is heated at 55°C for 4 h. The solid product is separated from the liquid phase, washed with 70% methanol and with 90% acetic acid, rinsed with absolute methanol, and then dried.

Methyl cellulose is synthesized using a reaction between alkali-cellulose and chloromethane. The initial cellulose is cut into small pieces, which are mixed with a 20-fold of amount of 40% NaOH for 1 h. The swollen cellulose is pressed to obtain alkali to cellulose ratio of 3, crushed and kept in air-atmosphere at 35°C for 1 h to reduce DP. The alkali-cellulose is placed into the autoclave, which is purged with nitrogen. After adding of a 3-fold amount of chloromethane to cellulose weight, the autoclave is closed and heated to 70°C for 5 h. The obtained product is isolated from liquid phase, washed, and dried.

Hydroxyethyl cellulose is synthesized using the reaction between alkali-cellulose and ethylene oxide. The initial cellulose is cut into small pieces, which are mixed with a 20-fold amount of 17.5% NaOH for 1 h. The swollen cellulose is pressed to obtain alkali to cellulose ratio of 3 and crushed. The alkali-cellulose is placed in an autoclave, which is purged with nitrogen. After adding liquid ethylene oxide in the ratio 1:1 to cellulose weight, the autoclave is closed, and the reaction system is treated at 45°C for 2 h. Finally, the obtained product is washed, purified, and dried. Reactions of selective oxidation of cellulose are used to prepare 2,3-dialdehyde, 6-monocarboxyl, 2,3-dicarboxyl and 2,3,6-tricarboxyl celluloses (Obolenskaya et al., 1991; Höglund, 2015). To obtain the dialdehyde cellulose (DALC), initial cellulose can be treated with 10% water solution of sodium metaperiodate ($NaIO_4$) at a temperature between 20°C and 50°C and pH between 3 and 4 for 10 h.

Monocarboxyl cellulose (MCAC) is produced by oxidizing initial cellulose feedstock with nitrogen dioxide or solutions of dinitrogen tetroxide in organic solvents, for example, in CCl_4 (Bichkovsky et al., 1993). To obtain the dicarboxyl cellulose (DCAC), sample of DALC is oxidized with 4% sodium chlorite ($NaClO_2$) for 24 h. Tri-carboxyl cellulose (TCAC) can be prepared by oxidation of DALC or DCAC with nitrogen dioxide (Sinha and Vasudevan, 1984).

Oxidized celluloses (DALC, MCAC, DCAC, and TCAC) are used as carriers for various therapeutic active substances (TAS). Due to the presence of aldehyde groups, DALC is able to bind TAS-containing amino groups, for example, proteolytic enzymes, amino acids, and other TAS (Filatov and Ryltsev, 1999; Filatov et al., 2002) (Equation 11.4):

$$\text{O=CH-CEL-CH=O} + \text{H}_2\text{N-R} \rightarrow \text{O=CH-CEL-CH(OH)NH-R} \quad (11.4)$$

The acidic groups of carboxylated celluloses can bind the basic functional groups of TAS, for example, lysine, according to Equation 11.5:

$$\text{CEL-COOH} + \text{H}_2\text{N-(CH}_2)_4\text{CHNH}_2\text{COOH} \rightarrow \text{CEL-COO}^{-}\,{}^{+}\text{H}_3\text{N-(CH}_2)_4\text{CHNH}_2\text{COOH} \quad (11.5)$$

Furthermore, these carboxylated celluloses are able to chemosorb various inorganic and organic cations (Equations 11.6 and 11.7):

$$\text{CEL-COO}^- + \text{CAT}^+ \rightarrow \text{CEL-COO}^{-+}\,\text{CAT} \tag{11.6}$$

$$2\ \text{CEL-COO}^- + \text{CAT}^{++} \rightarrow \text{CEL-COO}^{-}\ {}^{+}\text{CAT}^{+}\,{}^{-}\text{OOC-CEL} \tag{11.7}$$

11.3.3 Manufacturing of Micro- and Nanocellulose Materials

Micro- and nanocelluloses, as well as micro- and nanoderivatives, have small dimensions, enhanced specific surfaces, high thickening ability, increased binding ability, and other characteristics. Due to such unique features, these micro-and nanocelluloses can find applications in biomedicine, pharmaceutics, cosmetics, and personal care. In recent years, diverse types of micro- and nanomaterials—particles, fibrils, and filaments—have been produced and characterized. Besides, the possibility of commercial use of these materials was also studied.

11.3.3.1 Microparticles of Cellulose and Cellulose Derivatives

Microcrystalline cellulose (MCC) is produced by depolymerization of cotton or wood cellulose with dilute (1.5–2.5 M) solutions of mineral acids at elevated temperatures to achieve a low constant DP of 120–300 (Nada et al., 2009; Ioelovich, 2016a, 2018b). For example, pieces of pure cellulose are put in a reactor, which is filled with a 10-fold amount of 10% hydrochloric acid. The reactor is closed, and the reaction system is heated at 100°C for 1 h and then cooled. The hydrolyzed cellulose is discharged, separated from the acidic phase by filtration or centrifugation, washed, and concentrated to 30%–40% solid content. The wet product is bulk-dried and disaggregated to rod-like MCC particles with a length of 50–100 μm and a diameter of 10–30 μm (Figure 11.4a). Another technology is that the wet product is dispersed in water and spray-dried to obtain elliptical MCC particles with average diameter from 50 to 200 μm, as aggregates of rod-like MCC particles (Figure 11.4b).

It should be noted that the isolated cellulose has aggregated elementary fibrils via co-crystallization of adjacent crystallites. Therefore, after hydrolysis of chain segments in the noncrystalline (nc) domains and bulk-drying, micron-sized crystalline aggregates, that is, MCC particles are formed instead of nanocrystallites (Figure 11.5). MCC particles have low DP (120–300), increased crystallinity (0.72–0.75), and lateral sizes of crystallites (8–10 nm). Since MCC particles are uncharged and relatively large, dilute aqueous dispersions of MCC are unstable and stratified under storage into two phases, in which the bottom phase is MCC precipitate and the top phase is water. However, when the concentration of MCC dispersion approaches 10%, its viscosity increases and contributes towards phase stability.

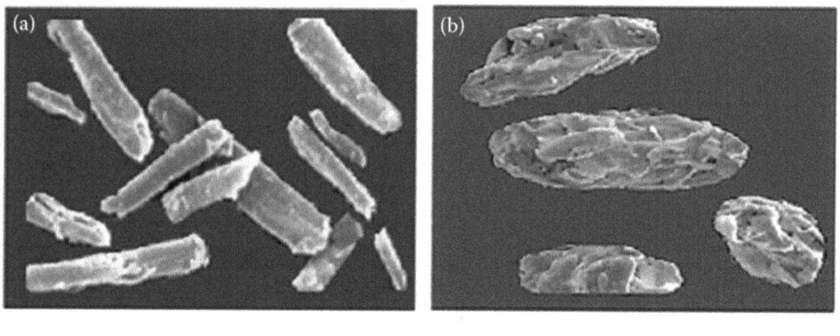

FIGURE 11.4 Optical image of (a) rod-like and (b) elliptical particles of microcrystalline cellulose.

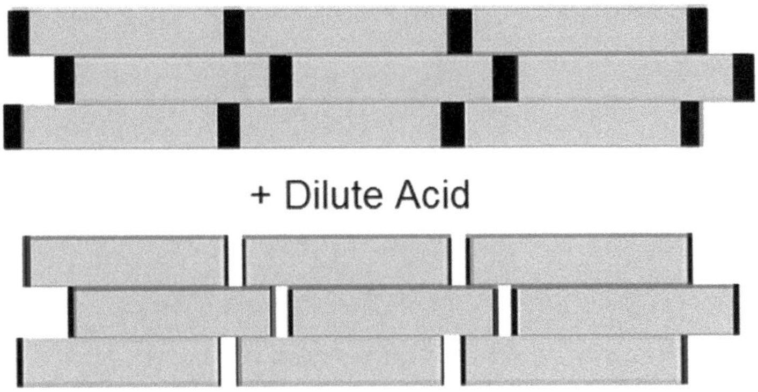

FIGURE 11.5 Scheme of MCC formation.

Some cellulose derivatives may also be prepared in the form of microparticles. For instance, chitosan powder undergoes thermocatalytic processing, thereby the microparticles with particle size of 1–15 μm are formed (Laka and Chernavskaya, 2006). Another method is based on sol-gel process, where a chitosan solution in acetic acid is treated with alkali solution to pH 7.5 with rapid stirring, which leads to precipitation of gel microparticles having a size of 1–10 μm (Struszcyk et al., 2004).

11.3.3.2 Nanosized Cellulose and Cellulose Derivatives

Unlike the MCC process, a method for isolation of crystalline nanoparticles (CNPs) requires four times more concentrated mineral acid, that is, 8–9 M (Said et al., 2004; Ioelovich and Leykin, 2006; Habibi et al., 2010; Qian et al., 2017; Kargarzadeh et al., 2018). Use of hydrochloric acid for cellulose hydrolysis produces uncharged nanoparticles, which precipitate from aqueous and nonaqueous dispersions. Therefore, to achieve the phase-stable dispersions, sulfuric acid is mainly used as it provides the negatively charged CNPs.

After hydrolysis of cellulose with dilute 2–3 M acid, the strong lateral intercrystallite contacts remain intact, despite hydrolysis of NCDs and achieving the levelling-off degree of polymerization (LODP), and as a consequence, microcrystalline particles are formed instead of nanocrystals (Ioelovich and Leykin, 2006) (Figure 11.5).

On the other hand, when hydrolysis of cellulose is carried out with sufficiently concentrated, 8–9 M sulfuric acid, then not only hydrolysis of NCDs, but also corrosion of intercrystallite contacts between neighboring nanofibrils and sulfonation of their surface occurs (Ioelovich, 2014a,b,c,d, 2016a, 2018c). Acid corrosion leads to replacement of strong crystalline contacts by weak amorphous contacts, which break down after mechanical or ultrasound comminution in aqueous medium with liberation of rod-like nanocrystalline particles (Figure 11.6).

To isolate nanocrystalline particles (NCPs), various concentrations of acid (50%–65%), treatment temperatures (20°C–70°C), and duration (30 min to 24 h) can be used. The optimal conditions to produce NCPs include concentration of sulfuric acid of 60%, temperature of 45°C, and duration of 60 min (Ioelovich, 2014d). To obtain free nanocrystallites, the aqueous dispersion of NCPs can be disintegrated using a high-pressure homogenizer or ultrasound disperser. Dry nanopowder is produced using freeze-drying of concentrated dispersion of NCPs, followed by comminuting and screening.

Size and shape of CNPs isolated from different cellulose sources were studied by methods of electron and atom force microscopy, laser light scattering, roentgen- and electronoscopy, and others. (Klemm et al., 2005; Hubbe et al., 2008; Habibi et al., 2010; Kalia et al., 2011; Li and Ragauskas, 2011; Peng et al., 2011; Ioelovich, 2014a,b,c,d; Foster et al., 2018; Kaur et al., 2018) The crystalline nanoparticles of various celluloses are rod-like and have a lateral size of 4–40 nm, length of 100–1000 nm, and an aspect ratio of 20–50 (Figure 11.7, Table 11.7). The aqueous dispersions of CNPs,

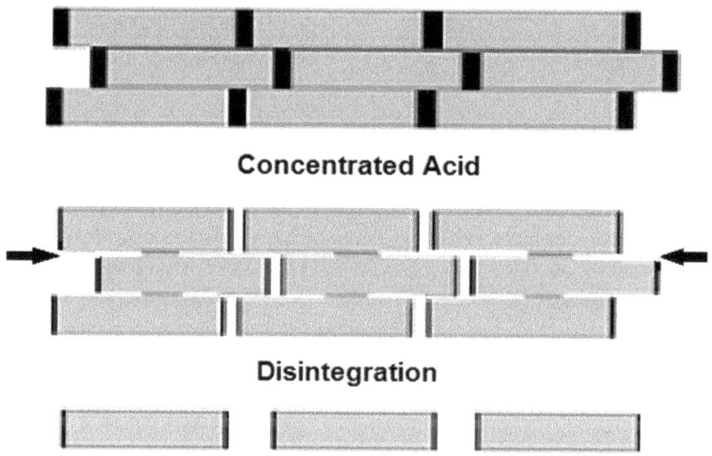

FIGURE 11.6 Scheme of isolation of rod-like CNPs (arrows denote the intercrystalline corrosion places).

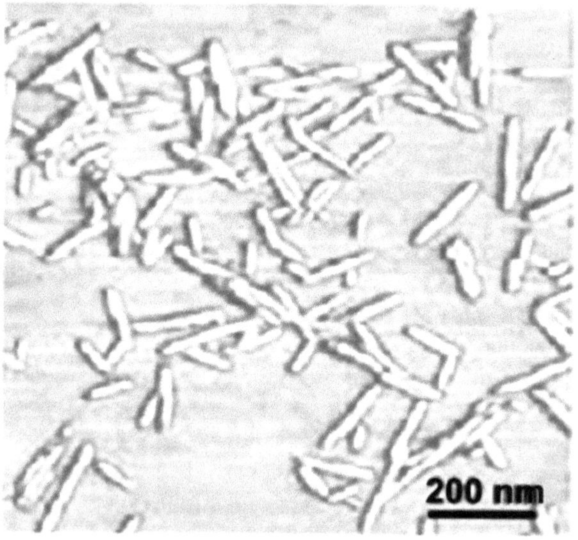

FIGURE 11.7 Atomic force microscopy image of rod-like CNP.

TABLE 11.7
Average Lateral Size (D_{CNP}) and Length (L_{CNP}) of Isolated Crystalline Nanoparticles of Various Origins

Cellulose Origin	D_{CNP}, nm	L_{CNP}, nm
Herbaceous plants	4–5	100–200
Wood cellulose pulp	5–10	100–300
Bast fibers	5–10	100–300
Cotton fibers	10–20	100–300
Bacterial cellulose	10–30	300–1000
Tunicate	20–40	500–1000
Algae	20–40	500–1000

having negatively charged sulfonic groups, are characterized by phase stability, specific rheological behavior, and birefringence (Araki et al., 1999; Habibi et al., 2010; Li and Ragauskas, 2011; Lin and Dufresne, 2014; Foster et al., 2018).

Dry nanocrystalline particles of cellulose are characterized by high crystallinity of 75%–80% and increased specific gravity of 1.57–1.58 g/cm^3. The tapped density of nanopowder is 0.25–0.30 g/cm^3 and the bulk density is 0.1–0.2 g/cm^3. Due to high crystallinity, dry cellulose nanoparticles have low sorption ability of water vapor. Furthermore, aggregation of small nanoparticles after conventional drying reduces the specific surface area to 1–2 m^2/g. Special drying methods are required—drying under supercritical conditions, low-temperature drying after solvents exchange, or freeze-drying methods—to prevent aggregation of nanoparticles and achieve increased area of specific surface. The cellulose nanocrystalline nanoparticles can be derivatized using reactions of etherification, esterification, oxidation, grafting, hydrophobization, cationization, etc. (Hubbe et al., 2008; Habibi et al., 2010; Ioelovich, 2014a,b,c,d; Rocha et al., 2018). Specific feature of CNPs is the dependence of reactivity on the size of nanoparticles (Ioelovich, 2014a,b,c,d).

Along with crystalline, amorphous nanoparticles (ANPs) of cellulose can also be prepared by a simple method, which includes dissolving of cellulose in concentrated (65%–67%) sulfuric acid, hydrolysis of dissolved cellulose at room temperature, regeneration of amorphous cellulose from the solution, and subsequent ultrasound or mechanical comminution in aqueous medium (Ioelovich, 2014d). Thus, the production process of ANPs includes stages of dissolving, depolymerization, sulfonation, regeneration of the amorphous cellulose flocs, and their comminution in aqueous medium. To obtain a dry nanopowder, the concentrated water dispersion of ANPs is freeze-dried, comminuted, and screened. The isolated round ANPs have an average diameter of 80–120 nm (Figure 11.8a). These particles are negatively charged and characterized by high amorphicity and low degree of polymerization (Table 11.8).

Dilute aqueous dispersions of ANPs are phase-stable due to the presence of increased content of negatively charged groups. Increased concentration of aqueous dispersions of ANPs leads to rise in viscosity, until a gel consistence is reached. ANPs of cellulose possess a high accessibility, sorption ability, and 100% enzymatic digestibility (Ioelovich, 2014d). The ANPs can be derivatized faster and deeper than CNPs. If introduced functional groups are acidic sulfonic groups, then the amorphous nanoparticles are capable of immobilizing the therapeutically active substances containing basic functional groups.

Nitrogenated cellulose derivatives, chitin and chitosan, can also be converted into nanoparticles. Rod-like crystalline nanoparticles of chitin were prepared by hydrolysis of isolated chitin with boiling dilute (10%–12%) of hydrochloric acid and subsequent ultrasonic or mechanical comminution in an aqueous medium (Morin and Dufresne, 2002; Mincea et al., 2012). Due to positively charged

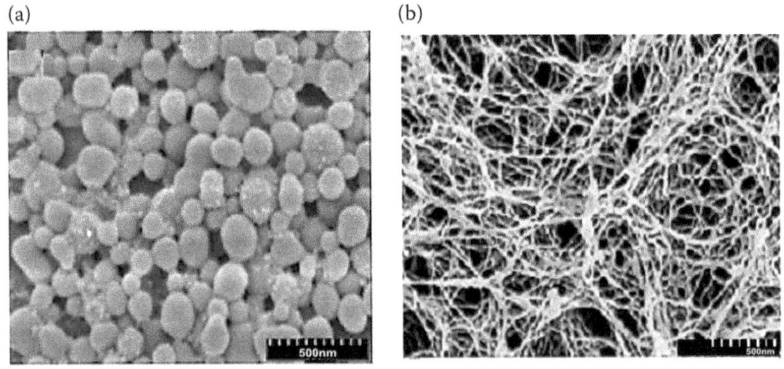

FIGURE 11.8 Scanning electron microscopy image of (a) round amorphous nanoparticles of cellulose and (b) tangled network of cellulose nanofibrils.

TABLE 11.8
Characteristics of the Cellulose Nanoparticles

ANP	CNP	Characteristics
CII	CI	Allomorph type of crystallites
20–25	75–77	Degree of crystallinity, %
80–120	200 × 10	Dimensions of particles, nm
60–70	130–150	Average DP
1.50–1.51	1.57–1.58	Specific gravity, g/cm^3
20–40	100–150	Area of specific surface, m^2/g
180–210	50–60	SO_3H-groups, meq/kg
70–90	45–50	Reducing groups, meq/kg

Abbreviations: DP, degree of polymerization; CNP, crystalline nanoparticles; ANP, amorphous nanoparticles.

particles containing protonated amino groups, dispersions of chitin nanoparticles are considered phase-stable (Marchessault et al., 1959).

To obtain chitosan nanoparticles (ChNPs), a sol-gel technique was used. This technique is based on the electrostatic interaction between negatively charged groups of polycation (chitosan) and negatively charged groups of polyanion (e.g., tripolyphosphate) in dilute solutions (Calvo et al., 1997; Tang et al., 2007). As a result, spherical ChNPs are formed with a diameter of 40–250 nm, depending upon the preparation condition. Decreasing the concentrations of chitosan solution and molecular weight as well as increasing the stirring speed facilitate the formation of smaller nanoparticles having a diameter of 40–60 nm.

Natural nanofibrils of bacterial cellulose (BC) are produced by bacteria such as *G. xylinus* (Klemm et al., 2005, 2006, 2009). Synthesis of nanofibrils by fermentation process was described by Gea and coauthors (2011). An aqueous medium, containing nutrients and glucose as a carbon source, was sterilized at 121°C for 2 h. A strain of *G. xylinus* was added to the stationary medium and fermented at 28°C and pH 4 for three weeks. The produced BC pellicles were washed with water, immersed overnight in dilute alkali solution, bleached with sodium hypochlorite and finally washed to neutral pH. To obtain nanofibrils of BC, other carbon sources also can be used, such as fructose, xylose, and galactose, as well as cellulose hydrolyzate, which contains glucose and some other reducing sugars.

Biosynthesis of nanofibrils is carried out by means of enzymatic complex of BC using an active monomer (e.g., uridine diphosphate glucose) attached to the end of the growing cellulose molecules (Klemm et al., 2006). The formed chains have an average DP of 4000–8000. Further, the cellulose chains crystallize and form the long and thin nanofibrils, which come out of the bacterial cell through pores into the liquid medium.

The following structural constituents were found in BC: elementary nanofibrils with lateral size of 8–10 nm, nanofibrillar bundles (i.e., microfibrils) with lateral size of 20–60 nm, and nanoribbons having a width of 100–150 nm (Klemm et al., 2009; Ioelovich, 2014a,b,c,d; Keshk, 2014). These constituents interlace and form a three-dimensional porous network that is capable of absorbing an extremely high amount of water. The resulting water-swollen fibrillar network obtained by stationary fermentation has the appearance of a hydrogel pellicle, which nevertheless, is relatively stronger and tougher (Scionti, 2010). After drying, the mechanical properties of the pellicle increase, and typical modulus reaches to 10–30 GPa and tensile strength to 200–250 MPa (Yamanaka et al., 1989). Dry BC is chemically pure, weakly hydrophilic, and a highly crystalline material (crystallinity of 75%–80%), with crystallites of BC consisting mainly of Iα allomorph (Klemm et al., 2009; Ioelovich, 2014c).

To produce artificial cellulose nanofibrils (NFCs), the initial cellulose material is intensively disintegrated in aqueous medium (Henriksson et al., 2007; Hubbe et al., 2008; Siro and Plackett, 2010; Ankerfors, 2012). The best feedstock for production of NFCs is delignified short-fiber cellulose, containing hemicelluloses too. However, the fibrillation process of cellulose fibers is too long and expensive. To reduce the energy consumption, chemical or biological pretreatments of the starting cellulose material are carried out. The known chemical pretreatments are tetramethylpiperidine-1-oxyl (TEMPO)-mediated oxidation and partial carboxymethylation (Saito et al., 2006; Lavoine et al., 2012), whereas the main biological pretreatment is a weak enzymatic hydrolysis of cellulose by endoglucanase (Henriksson et al., 2007; Henriksson, 2008; Janardhnan and Sain, 2011; Ankerfors, 2012). It should be noted that, unlike chemical pretreatment, the enzymatic pretreatment is environmentally friendly. As a consequence of disintegration of pretreated cellulose, a tangled network of thin and long nanofibrils, having a lateral size of 20–40 nm and an aspect ratio of 50–100, gets formed (Figure 11.8b). The high aspect ratio of NFCs is the cause for a low percolation threshold, whereby the nanofibrils show a good ability to form a continuous network in liquid medium or in polymer matrix.

Aqueous dispersions of NFCs have unique rheological properties such as thixotropy, shear-thinning characteristics, and pseudo-plasticity (Paakko et al., 2007; Lavoine et al., 2012; Ankerfors et al., 2013). The aqueous dispersions of uncharged or low-charged nanofibrils can form a stable gel when the concentration of NFCs is above 2%. The carboxylated NFCs turns into a stable gel, also in dilute dispersions ($C < 2\%$), when the pH value shifts to the acidic area (Fujisawa et al., 2011). The dry films of NFCs are characterized by a specific surface area of 0.5–1 m^2/g, modulus of 7–10 GPa, tensile strength of 170–230 MPa, and break elongation of 5%–7% (Henriksson, 2008; Fukuzumi, 2011). Moreover, these films are optically translucent (Dufresne, 2013). Unlike films, aerogels of NFC are rather weak, porous, and opaque materials (Paakko et al., 2008).

Along with nanoparticles and nanofibrils, nanofilaments with a diameter of 100–1000 nm may also be prepared using an electrospinning technique (Shin et al., 2001; Quan et al., 2010; Vargas, 2010; Zhao et al., 2011). Various cellulose solvents have been used for preparation of nanofilaments, such as lithium chloride (LiCl)/dimethylacetamide (DMAc), N-methylmorpholine-N-oxide (NMMO), and ionic liquids (Kulpinski, 2005; Kim et al., 2006; Frey, 2008; Quan et al., 2010). Structural studies show that cellulose nanofilaments, prepared from solutions in LiCl/DMAc, were amorphous, while the nanofilaments obtained from solutions in NMMO were semicrystalline. The electrospun nanofilaments can also be produced from chitosan, carboxymethyl cellulose, and other derivatives of cellulose. For instance, solutions of chitosan in dilute acetic or trifluoroacetic acid were used for electrospinning to obtain nanofilaments with a diameter ranging between 60 and 100 nm (Ohkawa et al., 2004; Schiffman and Schauer, 2007; Zhao et al., 2011).

11.3.4 Preparation of Biocarriers

As is known, diverse therapeutically active substances (TASs) are widely used in medicine. However, enzymes, hormones, vitamins, and some others TASs are unstable and decompose under the effect of elevated temperatures, unfavorable pH, and reactive ingredients. Furthermore, some ingredients, liquid drugs, and remedies are decomposed after administration of enzymes or other TASs. An additional problem is that common TASs do not provide a slow-release effect.

To improve stability and other properties, a TAS should be attached to a natural biocarrier, such as cellulose, by interaction with specific functional groups (Ioelovich, 2015). A special chemical modification is needed to transform the starting cellulose material into a biocarrier (Figure 11.9).

Due to specific reactive groups, the cellulosic biocarriers are able to attach certain types of TASs, thereby increasing their stability and providing them with a slow-release effect. Biocarriers having aldehyde groups fasten TAS-containing amine groups, for example, proteolytic enzymes (Equation 11.8):

$$\text{CEL-CH=O} + \text{H}_2\text{N-R} \rightarrow \text{CEL-CH(OH)NH-R} \quad (11.8)$$

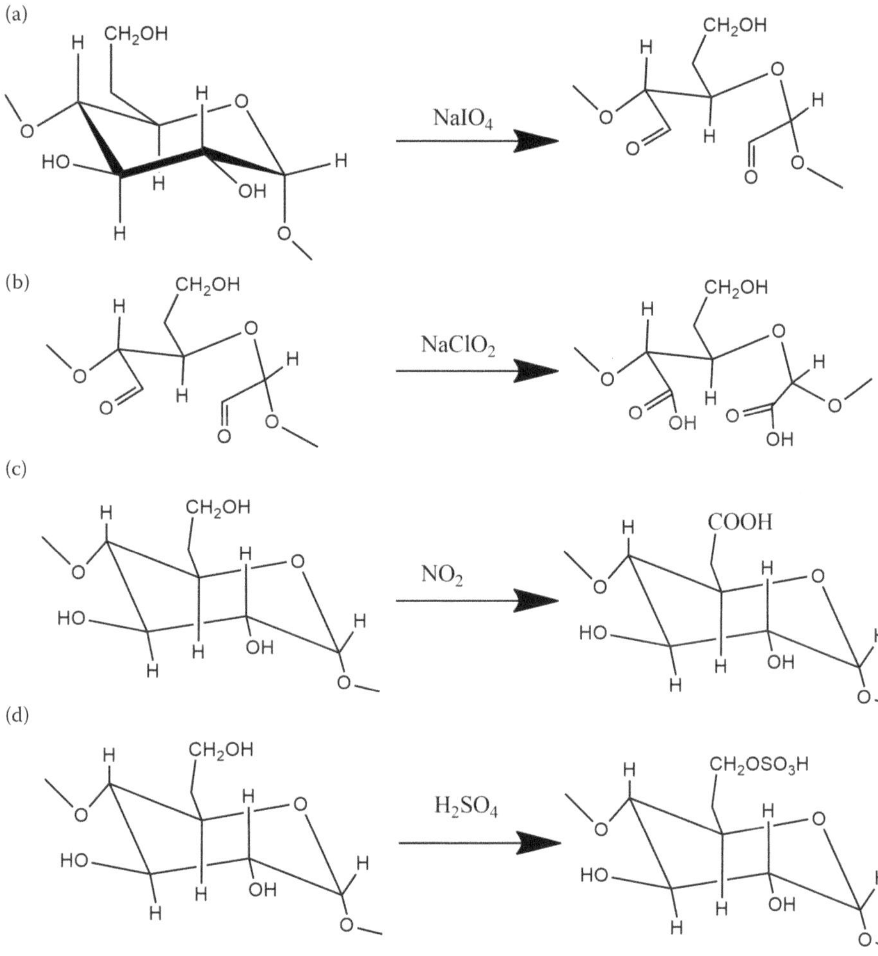

FIGURE 11.9 Ways of chemical transformation of cellulose into biocarrier; (a) introduction of aldehyde groups to form dialdehyde cellulose (DALC); (b) selective oxidation of DALC to form dicarboxyl cellulose (DCAC); (c) selective oxidation of primary hydroxyl groups of cellulose to form monocarboxyl cellulose (MCAC); and (d) introduction of sulfonic group to form cellulose sulfate (SAC).

Cellulose carriers having acidic groups, carboxylic and sulfonic, can bind TAS-containing basic functional group, for example, such amino acids as lysine or arginine (Equations 11.9 and 11.10):

$$\text{CEL-COOH} + \text{H}_2\text{N-(CH}_2)_4\text{CHNH}_2\text{COOH} \rightarrow \text{CEL-COO}^{-}\,{}^{+}\text{H}_3\text{N-(CH}_2)_4\text{CHNH}_2\text{COOH} \quad (11.9)$$

$$\text{CEL-SO}_3\text{H} + \text{H}_2\text{N-(CH}_2)_4\text{CHNH}_2\text{COOH} \rightarrow \text{CEL-SO}_3^{-}\,{}^{+}\text{H}_3\text{N-(CH}_2)_4\text{CHNH}_2\text{COOH} \quad (11.10)$$

Besides, acidic carriers are able to chemosorb various inorganic and organic cations (Equations 11.11 and 11.12), for example:

$$\text{CEL-COOH} + \text{NC}_5\text{H}_5 \rightarrow \text{CEL-COO}^{-}\,{}^{+}\text{HNC}_5\text{H}_5 \quad (11.11)$$

$$\text{CEL-SO}_3\text{H} + \text{NC}_5\text{H}_5 \rightarrow \text{CEL-SO}_3^{-}\,{}^{+}\text{HNC}_5\text{H}_5 \quad (11.12)$$

Biocarriers containing basic groups, such as chitin and chitosan, can bind, for example, aspartic amino acid (Equation 11.13):

$$Ch\text{-}NH_2 + HOOC\text{-}CH_2\text{–}NH_2COOH \rightarrow Ch\text{-}NH_3^{+}\,{}^{-}OOC\text{-}CH_2\text{–}NH_2COOH \quad (11.13)$$

Among various carriers, ANPs are mostly advantageous, because the formation of specific functional groups occurs directly during the preparation process of this nanocarrier. Moreover, ANPs acquire two specific functional groups, sulfonic and aldehyde, which provide the attachment of TASs of different types. For example, being an acidic carrier, ANPs may attach TAS-containing basic functional groups and cations with formation of ionic bonds (Equations 11.14 and 11.15):

$$ANP\text{-}SO_3H\ +\ NC_5H_5 \rightarrow ANP\text{-}SO_3^{-}\,{}^{+}HNC_5H_5 \quad (11.14)$$

$$2ANP\text{-}SO_3H\ +\ Cat^{++} \rightarrow ANP\text{-}SO_3^{-}\,{}^{+}Cat^{+-}\,O_3S\text{-}ANP \quad (11.15)$$

On the other hand, this nanocarrier contains end aldehyde groups, which allow binding TAS-containing amino groups (Equation 11.16):

$$ANP\text{-}CH{=}O + H_2N\text{-}R \rightarrow ANP\text{-}CH(OH)NH\text{-}R \quad (11.16)$$

In addition to cellulose, nitrogenated cellulose derivative, chitosan in a form of micro- and nanoparticles is also used as a biocarrier for binding acidic TASs (e.g., ketoprofen and diclofenac) (Equation 11.17):

$$CEL\text{-}NH_2 + HOOC\text{-}R \rightarrow CEL\text{-}NH_3^{+}\,{}^{-}OOC\text{-}R \quad (11.17)$$

11.4 APPLICATION AREAS OF CELLULOSE-BASED NANOBIOMATERIALS

For thousands of years, cellulose served for care and cure needs in the form of fibrous materials such as cotton, flax, ramie, hemp, and others. Since then and until today, cellulose-based materials are widely used in biomedicine and other areas. For example, the bleached pulp is used for manufacturing various personal care products, such as toilet paper, napkins, towels, and absorbent pads.

Rayon filaments are used for the production of strong bandages, while nonwoven materials manufactured from staple man-made fibers can be applied as absorbents in personal care, cosmetics, and medicine. The artificial cellulose membranes are also used for blood dialysis (Broek, 1993). Diverse dressing types are used to protect and cure wounds and burns and accelerate the healing process. The novel "smart" dressings are submicroporous, bactericidal, occlusive, and have barrier properties against microorganisms, high absorption abilities, permeability to oxygen and water vapor, and other desired properties. However, the conventional dressings, made up of natural and regenerated cellulose fibers, do not meet the modern requirements and should be improved further.

New types of rayon fibers with increased absorption ability were developed (Roggenstein, 2011), such as hollow viscose fibers. After contact with an aqueous medium, these fill completely with the liquid, resulting in an increased intrinsic absorption of 250%. Hollow viscose fibers with introduced specific anionic functional groups, for example, carboxymethyl, impart fibers with a high absorption value of 400%, while hollow thin-wall viscose fibers containing both anionic functional groups and gel-forming additives show superabsorbent properties with an absorption value close to 1000%. These innovative rayon fibers greatly extend the application scope of cellulosic absorbents in biomedicine and other areas. Promising materials for dressings can be made up of electrospun nanofilaments of chitosan or its composites due to combination of microporous structure, oxygen permeability, high absorption ability, nontoxicity, and antibacterial, analgesic and hemostatic properties (Aranaz et al., 2009; Naseri et al., 2014).

Studies have shown that effective dressings should be functional, bactericidal, occlusive, breathable, nonsticky, and atraumatic (Odland and Ross, 1968; Belozky and Breitman, 2000). Highly effective dressings, described by Moon and Crabtree (2003), consisted of two functional layers, inner cellulose-based atraumatic layer that can contain a therapeutic agent, absorbent or hydrogel, and an external occlusive but vapor-permeable layer. Highly effective dressings having three functional layers were also proposed, where the inner porous layer was non-adherent, the middle cellulose layer was absorbing and therapeutic, and whereas the external layer was occlusive but vapor-permeable. The inner layer of two-layered dressings and the middle layer of two-layered dressings may be made of modified cellulose gauze containing aldehyde and carboxylic and other functional groups, causing a hemostatic effect.

Furthermore, various proteolytic enzymes, antibiotics, and other types of TASs can be attached to functional groups of the modified gauze (Filatov and Ryltsev, 1999; Filatov et al., 2002; Liu and Min, 2009). The improved dressings with an attached TAS showed high efficiency in cleaning wounds and burns from dead cells, and cessation of bleeding and acceleration of the healing process (Belozky and Breitman, 2000).

Nanofibrils of BC have great potential for applications in biomedicine and cosmetics. A dimensionally stable hydrogel, comprising chemically pure nanofibrils of BC and immobilized water, finds wide medical application (Fontana et al., 1990; Czaja et al., 2006a,b; Klemm et al., 2009; Torres et al., 2012). A hydrogel of a BC-containing additive of various TASs (antibiotics, antiseptics, analgesics, etc.) was proposed to accelerate the healing process of wounds (Trovatti et al., 2011; Wei et al., 2011; Berndt et al., 2013). Currently, commercial wound dressings based on BC have been produced, such as XCELL, DERMAFILL, BIOPROCESS, and BIOFILL. Clinical examinations proved that treatment of non-healing ulcers by means of the wound dressing DERMAFILL accelerates the wound closure four times earlier than standard treatment duration (Portal et al., 2009). Trials of BC dressings also showed a clear acceleration of the epithelialization process of burns (Lina et al., 2011). Due to its unique features, the hydrogel of BC is successfully used in cosmetics to prepare moisturizing and cleaning masks (Klemm et al., 2006).

Artificial cellulose nanofibrils can be used as modifying agents in cosmetic and hygiene remedies to improve their phase stability and rheological properties (Paakko et al., 2008; Xhanari et al., 2011; Ankerfors, 2012). Dry films of NFCs find application in diagnostics (Orelma et al., 2012).

MCC has already been known for over 50 years. It is white crystalline powder, produced by the hydrolysis of NCD of cellulose with dilute solutions of hydrochloric or sulfuric acid, to achieve the degree of polymerization of 120–300, followed by bulk or spray drying, comminution, and screening (Fleming et al., 2001; Nada et al., 2009). Currently, MCC is common pharmaceutical excipient that is used as a filler, binder, and diluent for tablets, and also as a suspension aid. For these applications, food machinery corporation (FMC) produces MCC under the Avicel brand. For binding of tablets, Avicel-PH is used with an average particle size from 20 to 200 μm, depending upon the tableting technology. To prepare suspensions, a colloidal type of Avicel RC/CL is proposed, containing an additive of carboxymethyl cellulose (CMC) or another water-soluble polymeric thickener.

Prospective application area of CNPs is multifunctional filler for dentifrices due to their gentle sensation, faster cleaning, negligible enamel damage, higher compatibility, and nice texture of pastes at low CNP-loading. Besides, CNPs can be used in cosmetics as gentle cleaning and peeling agents too.

Amorphous nanoparticles of cellulose (ANPs) impart to aqueous cosmetic and medical remedies the increased viscosity and phase stability. Moreover, acidic ANPs display the pronounced hemostatic properties. Due to the presence of sulfonic groups, ANPs can serve as promising nanocarriers to bind bactericides (e.g., ZnO), therapeutically active basic amino substances (e.g., H_2N-R), and some other TASs, as follows (Equations 11.18 and 11.19):

$$\text{ANP-SO}_3\text{H} + \text{ZnO} \rightarrow \text{ANP-SO}_3^{-}{}^{+}\text{ZnOH} \quad (11.18)$$

$$\text{ANP-SO}_3\text{H} + \text{H}_2\text{N-R} \rightarrow \text{ANP-SO}_3^{-}{}^{+}\text{H}_3\text{N-R} \quad (11.19)$$

For instance, after binding the basic amino acid, arginine, by ANPs, the obtained conjugate can be used as a slow release aid to treat herpes and other skin lesions (Witte and Barbul, 2003; Stechmiller, 2005). Trypsin and other proteolytic enzymes may also be attached to the ANP-carrier. The remedies based on ANPs and their nano-conjugates are able to penetrate into small wounds and injuries accessible for nanoparticles only.

Nitrogenated cellulose derivative, chitosan, is a nontoxic, biocompatible, and biodegradable biopolymer, which exhibits a complex of diverse therapeutic properties, such as hemostatic, antimicrobial, and antioxidative (Koide, 1998; Kumar, 2000; Kumar et al., 2004). Due to complex and unique properties, chitosan has a promising potential for approbation in tissue engineering and manufacture of effective wound dressings. It is known that hemostatic bandages made up of chitosan are used in armies of several counties. Chitosan-based micro-and nanocarriers with attached TAS show its prolonged therapeutic efficiency (Bodek, 2000, 2004; Aranaz et al., 2009; Masarudin et al., 2015).

Other cellulose derivatives, ethers, esters, and oxycelluloses find diverse applications in biomedicine, pharmaceutics, cosmetics, and personal care. Cellulose acetate phthalate (CAP) is one of the most effective aids used as a protective coating for enteric capsules or tablets to impart them resistance to the acidic environment of the stomach (Edgar, 2007; Rowe et al., 2009). Carboxymethyl cellulose (CMC) is used as a binder for tablets, coating agent, thickener, and suspending aid (Kamel et al., 2008). Methylcellulose (MC) serves mainly as thickener and emulsifying and suspending agent in various flowable remedies (Glass and Haywood, 2006; Rowe et al., 2009). Some other soluble ethers — hydroxyethyl cellulose, hydroxypropyl cellulose, and hydroxypropylmethyl cellulose—also find application in cosmetics, hygiene, and medicine (Kamel et al., 2008).

Mono-, di-, and tricarboxyl celluloses are used as biocarriers for basic TASs, e.g. for the antibiotic gentamicin (Liu and Min, 2009; Lochman et al., 2010). Furthermore, these oxycelluloses exhibit hemostatic properties—for example, the hemostatic agent "Surgicel" made of monocarboxyl cellulose, or "Supercel" made up of 2,3,6-tricarboxyl cellulose. Additional applications of oxycelluloses are formulations of antiacne, fungicide, and other remedies (Banker and Kumar, 1995).

11.5 CONCLUSIONS

Fibers, micro- and nanoparticles of cellulose and cellulose derivatives, are biomaterials that are popularly used in medicine, cosmetics, hygiene, and some other areas. Vital instances of smart multifunctional (absorbing, cleansing, healing, etc.) materials are biocarriers made up of cellulose or its derivatives with attached TASs, such as antibiotics, proteolytic enzymes, amino acids, and many other TASs. A nitrogenated cellulose derivative, chitosan, is characterized by biocompatibility, biodegradability, and nontoxicity, and it possesses multipurpose therapeutic properties such as analgesic, hemostatic, bactericidal, and antioxidative.

Novel smart dressings with multiple functional layers have been described. These dressings can contain bacterial nanocellulose, artificial nanofibrils, chitosan, carboxymethyl cellulose, or a composite of various biopolymers. The therapeutic layers made of chitosan or oxycellulose with attached TAS (e.g., proteolytic enzymes, antibiotics, etc.) are successfully used to clean the wounds from necrotic cells, absorb exudate, reduce bleeding, and kill bacteria. Cellulose fibers and nanofibers are used for production of absorbing nonwoven bandages for the treatment of wounds, injuries, and burns.

Nanoparticles of cellulose and its derivatives are promising nanocarriers for diverse kinds of TASs, which are used in formulations of liquid remedies for the treatment of thin and deep injures, wounds, and burns, accessible only by nanoparticles. A specific feature of cellulose and some of its derivatives is biodegradability that makes materials based on these bio-sources friendly to the environment. Despite immense potential, the widespread use of new cellulose bio- and nanomaterials is limited due to lack of their apt regulations. Further progress in development of these materials requires obtaining the requisite permissions and approvals for applications of the new bionanomaterials in biomedicine, pharmaceutics, cosmetics, or personal care.

REFERENCES

Ankerfors M. *Microfibrillated cellulose: Energy efficient preparation techniques and key properties*. Innventia: Stockholm. (ISBN 978-91-7501-464-7); 2012.

Ankerfors M, Lindstrom T, Soderberg D. *Development for boosting the implementation of microfibrillated cellulose in papermaking*. Innventia: Stockholm; 2013. SPCI Convention.

Araki J, Wada M, Kuga S, Okana T. Influence of surface charge on viscosity behavior of cellulose microcrystal suspension. *J. Wood Sci.* 1999; 45: 258–61.

Aranaz I, Mengibar M, Harris R, Panos I, Miralles B et al. Functional characterization of chitin and chitosan. *Current Chem. Biol.* 2009; 3: 203–30.

Banker GS, Kumar V. Microfibrillated oxycellulose. US5405953 A1; 1995.

Belozky S, Breitman R. *Wounds and dressings*. DDB Publ: Ramat-Gan; 2000.

Berndt S, Wesarg F, Wiegand C, Kralisch D, Müller F. Antimicrobial porous hybrids consisting of bacterial nanocellulose and silver nanoparticles. *Cellulose*. 2013; 20: 771–83.

Bichkovsky PM, Kaputzy FN, Yurkshtovich TP. Sorption of anticancer drugs by cellulose oxidized with nitrogen oxide (IV). *News Belarus Acad. Sci.* 1993; 3: 41–5.

Blackwell J. *Biopolymers*. Academic Press: New York; 1973.

Bodek KH. Study of Rheological properties of microcrystalline chitosan hydrogels used as drug carriers. *Polimery*. 2000; 45: 818–25.

Bodek KH. Study on the stability of microcrystalline chitosan systems with selected non-steroidal anti-inflammatory drugs. *Polimery*. 2004; 49: 29–35.

Borbely E. Lyocell, The new generation of regenerated cellulose. *Acta Polytech. Hungarica*. 2008; 5: 11–8.

Broek AP. *Characterization of hemodialysis membranes: Structure and function*. Alfa Publ: Hengelo; 1993.

Brown CD, Zitelli JA. Choice of wood dressings and ointments. *Otolaryngol. Clinic North Am.* 1995; 28: 1081–91.

Brown RM Jr., Saxena IM. Cellulose biosynthesis: A model for understanding the assembly of biopolymers. *Plant Physiol. Biochem.* 2000; 38: 57–67.

Brown RM Jr., Willison HM, Richardson CL. Cellulose biosynthesis in *Acetobacter xylinum*: Visualization of the site of synthesis and direct measurement of the *in vivo* process. *Proc. Nat. Acad Sci. USA*. 1976; 73: 4565–9.

Calvo P, Remunan-Lopez C, Vila-Jato JL, Alonso MJ. Novel hydrophilic chitosan-polyethylene oxide nanoparticles as protein carrier. *J. Appl. Poly Sci.* 1997; 63: 125–32.

Chen H. *Biotechnology of cellulose: Theory and practice*. Chem. Ind. Press: Beijing; 2014.

Chen PY, Lin AYM, McKittrick J, Meyers MA. Structure and mechanical properties of crab exoskeletons. *Acta Biomater.* 2008; 4: 587–96.

Czaja WK, Krystynowicza A, Bieleckia S, Brown RM Jr. Microbial cellulose - the natural power to heal wounds. *Biomater.* 2006a; 27: 145–51.

Czaja WK, Young DJ, Kawecki M, Brown RM. The future prospects of microbial cellulose in biomedical applications. *Biomacromolecul.* 2006b; 8: 1–12.

Dence CW, Reeve DW. *Pulp bleaching: Principles and practice*. Tappi Press: Atlanta; 1996.

Dolmetsch H. Fractionation of non-esterified cellulose. *Koll. Z.* 1961; 176: 138–9.

Dos Santos FA, Iulianelli GCV, Tavares MIB. The use of cellulose nanofillers in obtaining polymer nanocomposites: Properties, processing, and applications. *Mater. Sci. Appl.* 2016; 7: 257–94.

Dufresne A. Nanocellulose: A new ageless bionanomaterial. *Mater. Today* 2013; 16: 220–7.

Dutta PK, Dutta J, Tripathi VS. Chitin and chitosan: Chemistry, properties and applications. *J. Sci. Ind. Res.* 2004; 63: 20–31.

Edgar KJ. Cellulose esters in drug delivery. *Cellulose*. 2007; 14: 49–64.

Field CB, Behrenfeld MJ, Randerson JT, Falkowski P. Primary production of the biosphere: Integrating terrestrial and oceanic components. *Science*. 1998; 281: 237–40.

Filatov VN, Ryltsev VV. Method for producing dressings with healing properties. RU2131268 A61; 1999.

Filatov VN, Ryltsev VV, Simenhaus Z. Wound dressing. US6500799 B2; 2002.

Fleming K, Gray DG, Matthews S. Cellulose crystallites. *Chem. Eur. J.* 2001; 7: 1831–5.

Fontana JD, Souza AM, Fontana CK, Torriani IL, Moreschi JC et al. Acetobacter cellulose pellicle as a temporary skin substitute. *Appl. Biochem. Biotech.* 1990; 24–25: 253–64.

Foster EJ, Moon RJ, Agarwal UP, Bortner MJ, Bras J et al. Current characterization methods for cellulose nanomaterials. *Chem. Soc. Rev.* 2018; 47: 2609–79.

Frey MW. Electrospinning cellulose and cellulose derivatives. *Poly. Rev.*. 2008; 48: 378–91.

Fujisawa S, Okita Y, Fukuzumi H, Saito T, Isogai A. Preparation and characterization of TEMPO-oxidized cellulose nanofibril films with free carboxyl groups. *Carbohydr. Polym.* 2011; 84: 579–83.

Fukuzumi H. *Studies on structure and properties of tempo-oxidized cellulose nanofibril films.* Tokyo University: Tokyo; 2011.

Gardner KH, Blackwell J. Refinement of the structure of β-chitin. *Biopolymers.* 1975; 14: 1581–95.

Gea S, Reynolds CT, Roohpour N, Wirjosentono B, Soykeabkaew N et al. Investigation into the structural, morphological, mechanical and thermal behavior of bacterial cellulose after a two-step purification process. *Biores. Technol.* 2011; 102: 9105–10.

Glass BD, Haywood A. Stability considerations in liquid dosage forms extemporaneously prepared from commercially available products. *J. Pharm Pharm Sci.* 2006; 9: 398–426.

Habibi Y, Lucia LA, Rojas OJ. Cellulose nanocrystals chemistry, self-assembly and applications. *Chem. Rev.* 2010; 110: 3479–500.

Hearle JWS. The fine structure of fibers and crystalline polymers – fringed fibril structure. *J. Appl. Polym. Sci.* 1963; 7: 1175–92.

Henriksson M. *Cellulose nanofibril networks and composites – Preparation, structure and properties.* KTH: Stockholm; 2008.

Henriksson M, Henriksson LA, Berglund G, Lindström T. An environmentally friendly method for enzyme-assisted preparation of microfibrillated cellulose nanofibers. *European Polym. J.* 2007; 43: 3434–41.

Höglund E. *Production of dialdehyde cellulose and periodate regeneration: Towards feasible oxidation processes.* University Press: Karlstadt; 2015.

Hon DNS. *Chemical modification of lignocellulosic material.* Marcel Dekker: New York; 1996.

Hubbe MA, Rojas OJ, Lucia LA, Sain M. Cellulosic nanocomposites – A review. *Bioresources.* 2008; 3: 929–80.

Ioelovich M. *Cellulose nanostructured natural polymer.* LAP: Saarbrücken; 2004.

Ioelovich M. Structure and physicochemical properties of nitrogenated derivatives of cellulose. *Am. J. Biosci.* 2014a; 2: 6–12.

Ioelovich M. Crystallinity and hydrophility of chitin and chitosan. *Res. Rev. J. Chem.* 2014b; 3: 7–14.

Ioelovich M. Study of enzymatic hydrolysis of bacterial nanocellulose. *ChemXpress.* 2014c; 7: 63–8.

Ioelovich M. Peculiarities of cellulose nanoparticles. *Tappi. J.* 2014d; 13: 45–52.

Ioelovich M. Preparation and application of nanoscale cellulose biocarriers. *CibTech J. Biotechnol.* 2015; 4: 19–24.

Ioelovich M. Models of supramolecular structure and properties of cellulose. *J. Polym. Sci.* 2016a; 58: 925–43.

Ioelovich M. Study of phase transitions of cellulose nanocrystallites. *Int. Sci. J.* 2016b; 8: 87–94.

Ioelovich M. Determination of distortions and sizes of cellulose nanocrystallites. *Res. J. Nanosci. Eng.* 2018a; 2: 1–5.

Ioelovich M. Green technology of microcrystalline cellulose. *SITA* 2018b; 20: 3–7.

Ioelovich M. Green chemistry of micro-and nanoparticles of plant biomass. *South Asian Res. J. Natural Products.* 2018c; 1: 1–10.

Ioelovich M, Leykin A. Formation of nanostructure of microcrystalline cellulose. *Cellulose Chem Technol.* 2006; 40: 313–7.

Ioelovich M, Leykin A, Figovsky O. Study of cellulose paracrystallinity. *Bioresources.* 2010; 5: 1393–407.

Janardhnan S, Sain M. Targeted disruption of hydroxyl chemistry and crystallinity in natural fibers for the isolation of cellulose nanofibers via enzymatic treatment. *Bioresources.* 2011; 6: 1242–50.

Kalia S, Dufresne A, Cherian BM, Kaith BS, Averous L et al. Cellulose-based bio- and nanocomposites: A review. *Int. J. Polym. Sci.* 2011; 2011: 1–35.

Kamel S, Ali N, Jahangir K, Shah SM, El-Gendy AA. Pharmaceutical significance of cellulose: A review. *Express Polym. Letter.* 2008; 2: 758–78.

Kaneko Y. Cuprammonium rayon and cuprammonium spunbond nonwoven. *Lenziger Berichte.* 1990; 69: 78–82.

Kargarzadeh H, Mariano M. Gopakumar D, Ahmad I, Thomas S et al. Advances in cellulose nanomaterials. *Cellulose.* 2018. DOI: 10.1007/s10570-018-1723-5.

Kaur M, Kumari S, Sharma P. Chemically modified nanocellulose from rice husk: Synthesis and characterization. *Advances in Research.* 2018; 13: 1–11.

Keshk SM. Bacterial cellulose production and its industrial applications. *Bioproc. Biotech.* 2014; 4: 1–10.

Kim CW, Kim DS, Kang SY, Marquez M, Joo YL. Structural studies of electrospun cellulose nanofibers. *Polymer.* 2006; 47: 5097–107.

Klemm D, Heublein B, Fink H-P, Bohn A. Cellulose: Fascinating biopolymer and sustainable raw material. *Angew. Chem.* 2005; 44: 2–37.

Klemm D, Schumann D, Kramer F, Hessler N, Hornung M et al. Nanocelluloses as innovative polymers in research and application. *Adv. Polym. Sci.* 2006; 205: 49–96.

Klemm D, Schumann D, Kramer F, Hessler N, Koth D, Sultanova B. Nanocellulose materials – Different cellulose, different functionality. *Macromol. Symp.* 2009; 280: 60–71.

Koide SS. Chitin-chitosan: Properties, benefits and risks. *Nut. Res.* 1998; 18: 1091–101.

Krässig H. *Cellulose: Structure, accessibility and reactivity.* Gordon and Breach Publ: Amsterdam; 1993.

Kulpinski P. Cellulose nanofibers prepared by the N-methylmorpholine-N-oxide method. *Appl. Polym. Sci.* 2005; 98: 1855–9.

Kumar MNVR. A review of chitin and chitosan applications. *React. Funct. Polym.* 2000; 46: 1–27.

Kumar MNVR, Muzarelli RAA, Muzarelli C, Sashiwa H, Domb AJ. Chitosan chemistry and pharmaceutical perspectives. *Chem. Rev.* 2004; 104: 6017–84.

Laka M, Chernavskaya S. Preparation of chitosan powder and investigation of its properties. *Proc. Estonian Acad. Sci.* 2006; 55: 78–84.

Langan P, Nishiyama Y, Chanzy H. X-ray structure of mercerized cellulose II at 1 Å resolution. *Biomacromolec.* 2001; 2: 410–6.

Lavoine N, Desloges I, Dufresne A, Bras J. Microfibrillated cellulose – its barrier properties and applications in cellulosic materials: A review. *Carbohydrate Polym.* 2012; 90: 735–64.

Lerouxel O, Cavalier DM, Aaron H, Liepman AH, Keegstra K. Biosynthesis of plant cell wall polysaccharides - A complex process. *Plant Biology.* 2006; 9: 621–30.

Li Y, Ragauskas A. Cellulose nano-whiskers as reinforcing filler in polyurethanes. *Adv. Div. Ind. Appl. Nanocompos.* 2011; 3: 17–36.

Lin N, Dufresne A. Surface chemistry, morphological analysis and properties of cellulose nanocrystals with gradiented sulfation degrees. *Nanoscale.* 2014; 6: 5384–93.

Lina F, Yue Z, Jin Z, Guang Y. Bacterial cellulose for skin repair materials. In: *Biomedical engineering – Frontiers and challenges.* In Tech: Rijeka; 2011; pp. 249–74.

Liu H-B, Min X-T. Polymers containing covalently bound antibiotic agents. RU2465917 A61; 2009.

Lochman P, Plodr M, Páral J, Šmejkal K. Nanofiber micro-dispersed oxidized cellulose as a carrier for topical antimicrobials: First experience. *Surg. Infect.* 2010; 11: 29–32.

Manley R. Fine structure of native cellulose microfibrils. *Nature* 1964; 204: 1155–7.

Marchessault RH, Morehead FF, Walter NM. Liquid crystal systems from fibrillar polysaccharides. *Nature.* 1959; 184: 632–3.

Masarudin MJ, Cutts SM, Evison BJ, Phillips DR, Pigram PJ. Factors determining the stability, size distribution, and cellular accumulation of small, monodisperse chitosan nanoparticles as candidate vectors for anticancer drug delivery: Application to the passive encapsulation of 14-C-doxorubicin. *Nanotech. Sci. Appl.* 2015; 8: 67–80.

Mincea M, Negrulescu A, Ostafe V. Preparation, modification, and applications of chitin nanowhiskers: A review. *Rev. Adv. Mater. Sci.* 2012; 30: 225–42.

Moon CH, Crabtree TG. New wound dressing techniques to accelerate healing. *Curr. Treatm. Opt. Infect. Dis.* 2003; 5: 251–60.

Moon RJ, Martini A, Nairn J, Simonsen J, Youngblood J. Cellulose nanomaterials review: Structure, properties and Nanocomposites. *Chem. Soc. Rev.* 2011; 40: 3941–94.

Morin A, Dufresne A. Nanocomposites of chitin whiskers from *Riftia* tubes and poly(caprolactone). *Macromolec.* 2002; 35: 2190–9.

Mühlenthaler R. Fine structure of natural polysaccharide systems. *J. Polym. Sci.* 1969; 28: 305–16.

Nada A-A, El-Kady MY, Abd El-Sayed ES, Amine FM. Preparation and characterization of microcrystalline cellulose. *Bioresources.* 2009; 4: 1359–71.

Naseri N, Algan C, Jacobs V, John M, Oksman K, Mathew AP. Electrospun chitosan-based nanocomposite mats reinforced with chitin nanocrystals for wound dressing. *Carbohydr. Polym.* 2014; 109: 7–15.

Obolenskaya AV, Elnitskaya ZP, Leonovich AA. *Handbook of cellulose and wood Chemistry.* Ecology: Moscow; 1991.

Odland G, Ross R. Human wound repair. *J. Cell. Biol.* 1968; 39: 135–51.

Ohkawa K, Cha DI, Kim H, Nishida A, Yamamoto H. Electrospinning of chitosan. *Macromol. Rapid Commun.* 2004; 25: 1600–5.

Orelma H, Filpponen I, Johansson L-S, Osterberg M, Rojas OJ, Laine J. Surface functionalized nanofibrillar cellulose film as a platform for immunoassays and diagnostics. *Biointerphas.* 2012; 7: 61–72.

O'Sullivan AC. *Cellulose*: The structure slowly unravels. *Cellulose.* 1997; 4: 173–207.

Paakko M, Ankerfors M, Kosonen H, Nykanen A, Ahola S, Osterberg M. Enzymatic hydrolysis combined with mechanical shearing and high-pressure homogenization for nanoscale cellulose fibrils and strong gels. *Biomacromol.* 2007; 8: 1934–41.
Paakko M, Vapaavuori J, Silvennoinen R, Kosonen H, Ankerfors M, Lindstrom T. Long and entangled native cellulose I nanofibers allow flexible aerogels and hierarchically porous templates for functionalities. *Soft Matter.* 2008; 4: 2492–9.
Peng BL, Dhar N, Liu HL, Tam KC. Chemistry and applications of nanocrystalline cellulose and its derivatives: A nanotechnology perspective. *Canad. J. Chem. Eng.* 2011; 9999: 1–16.
Portal O, Clark WA, Levinson DJ. Microbial cellulose wound dressing in the treatment of non-healing lower extremity ulcers wounds. *Compend Clinic Res Practic.* 2009; 21: 1–3.
Qian S, Zhang H, Sheng K. Cellulose nanowhiskers from moso bamboo residues: Extraction and characterization. *Bioresources.* 2017; 12: 419–33.
Quan SL, Kang SG, Chin IJ. Characterization of cellulose fibers electrospun using ionic liquid. *Cellulose.* 2010; 17: 223–30.
Revol JF, Marchessault RH. *In vitro* chiral nematic ordering of chitin crystallites. *Int. J. Biol. Macromol.* 1993; 15: 329–35.
Rinaudo M. Chitin and chitosan: Properties and applications. *Progr. Polym. Sci.* 2006; 31: 603–32.
Roberts G. *Chitin chemistry.* MacMillan: London; 1998.
Rocha I, Ferraz N, Mihranyan A, Strømme M, Lindh J et al. Sulfonated nanocellulose beads as potential immunosorbents. *Cellulose.* 2018; 25: 1899–910.
Roggenstein W. Viscose fibres with new functional qualities. *Lenzinger Berichte.* 2011; 89: 72–7.
Rogovin ZA. *Fundamentals of chemistry and technology of chemical fibers.* Chemistry: Moscow; 1974.
Ross P, Mayer R, Benziman M. Cellulose biosynthesis and function in bacteria. *Microbiol. Rev.* 1991; 55: 35–58.
Rowe RC, Sheskey PJ, Quinn ME. *Handbook of pharmaceutical excipients.* Pharmac. Press: London; 2009.
Rudall KM, Kenchington W. The α-chitin system. *Biol. Rev.* 1973; 40: 597–639.
Said MA, Samir A, Alloin F, Paillet M, Dufresne A. Tangling effect in fibrillated cellulose reinforced nanocomposites. *Macromolecul.* 2004; 37: 4313–6.
Saito T, Nishiyama T, Putaux JL, Vignon M, Isogai A. Homogeneous suspensions of individualized microfibrils from TEMPO-catalyzed oxidation of native cellulose. *Biomacromol.* 2006; 7: 1687–91.
Saito Y, Putaux J-L, Okano T, Gaill F, Chanzy H. Structural aspects of the swelling of β-chitin in HCl and its conversion into α-chitin. *Macromolecul.* 1997; 30: 3867–73.
Saito Y, Okano T, Gaill F, Chanzy H, Putaux J-L. Structural data on the intra-crystalline swelling of β-chitin. *Int. J. Biol. Macromol.* 2000; 28: 81–8.
Saxena IM, Brown RM Jr. Cellulose biosynthesis: Current views and evolving concepts. *Ann Bot.* 2005; 96: 9–21.
Schiffman JD, Schauer CL. One-step electrospinning of crosslinked chitosan fibers. *Biomacromolecul.* 2007; 8: 2665–7.
Scionti G. *Mechanical properties of bacterial cellulose implants.* Chalmers Tech University: Göteborg; 2010.
Serkov AT. *Viscose rayon fibers.* Chemistry: Moscow; 1981.
Shin YM, Hohman MM, Brenner MP, Rutledge GC. Experimental characterization of electrospinning: The electrically forced jet and instabilities. *Polymer.* 2001; 42: 9955–67.
Sinha TJ, Vasudevan P. Blood-cellulosics interactions. *Biomater. Med. Devices Artif Organs.* 1984; 12: 273–87.
Siro I, Plackett D. Microfibrillated cellulose and new nanocomposite materials: A review. *Cellulose.* 2010; 17: 459–94.
Siqueira G, Bras J, Dufresne A. Cellulosic bionanocomposites: A review of preparation, properties and applications. *Polym.* 2010; 2: 728–65.
Sjöström E (1993). *Wood chemistry, fundamentals and applications.* Academic Press: London; 1993.
Stechmiller JK. Arginine supplementation and wound healing. *Nutr. Clin. Pract.* 2005; 20: 52–61.
Struszcyk H, Niekraszewicz A, Kucharska M, Urbanowski A, Wisniewska-Wrona M. Process for preparing chitosan particles. US6740752 B2; 2004.
Sugiyama J, Vuong R, Chanzy H. Electron diffraction study on the two crystalline phases occurring in native cellulose from an algal cell wall. *Macromolecul.* 1991; 24: 4168–75.
Tang Z-X, Qian J-Q, Shi L-E. Preparation of chitosan nanoparticles as carrier for immobilized enzyme. *Appl. Biochem. Biotech.* 2007; 136: 77–96.
Tönnesen B, Ellefsen O. Chain folding – a possibility to be considered in connection with the cellulose molecule. *Norsk Skogind.* 1960; 14: 266–9.

Torres FG, Commeaux S, Troncoso OP. Biocompatibility of bacterial cellulose based biomaterials. *J. Funct. Biomater.* 2012; 3: 864–78.
Trovatti E, Silva NH, Duarte IF, Rosado CF, Almeida IF et al. Biocellulose membranes as supports for dermal release of lidocaine. *Biomacromolecul.* 2011; 12: 4162–8.
Vargas NH. *Aligned cellulose nanofibers prepared by electrospinning.* University Press: Lulea; 2010.
Verlhac C, Dedier J, Chanzy H. Availability of surface hydroxyl groups in Valonia and bacterial cellulose. *J. Polym. Sci. A.* 1990; 28: 1171–7.
Watthanaphanit A, Rujiravanit R. *Handbook of carbohydrate polymers: Development, properties and applications.* Nova Publ: New York; 2010.
Wei B, Yang G, Hong F. Preparation and evaluation of a kind of bacterial cellulose dry films with antibacterial properties. *Carbohydr Polym.* 2011; 84: 533–8.
Witte MB, Barbul A. Arginine physiology and its implication for wound healing. *Wound Repair Regen.* 2003; 11: 419–23.
Xhanari K, Syverud K, Stenius P. Emulsions stabilized by microfibrillated cellulose: The effect of hydrophobization, concentration and o/w ratio. *Dispers. Sci. Technol.* 2011; 32: 447–52.
Yamanaka S, Watanabe K, Kitamura N, Iguchi M, Mitsuhashi S et al. The structure and mechanical properties of sheets prepared from bacterial cellulose. *J. Mater. Sci.* 1989; 24: 3141–5.
Yan SP, Wang QY, Yang CP. Research advances in the plant cellulose biosynthesis. *J. An. Agric. Sci.* 2008; 36: 9049–51.
Young RA, Rowell RM (1986). *Cellulose: Structure, modification and hydrolysis.* Wiley: New York.
Zelencova L, Erdogan S, Baran T, Kaya M. Chitin extraction and chitosan production from Chilopoda (*Scolopendra cingulata*) with identification of physicochemical properties. *J. Polym Sci. A.* 2015; 57: 437–44.
Zhao LM, Shi LE, Zhang ZL, Chen JM, Shi DD et al. Preparation and application of chitosan nanoparticles and nanofibers. *Brazil. J. Chem. Eng.* 2011; 28: 353–62.
Zugenmaier P. *Crystalline cellulose and cellulose derivatives: Characterization and structures.* Springer-Verlag: Berlin, Heidelberg; 2008.
Zugenmaier P. Contribution to the historical development of macromolecular chemistry – exemplified on cellulose. *Cellulose Chem Technol.* 2009; 43: 351–78.

12 Nanotechnology Could Revolutionize Therapy for Age-Related Macular Degeneration

*Afrah Jalil Abd, Shahdeep Kaur, Ripandeep Kaur, Rupinder Kaur Kanwar, Bhupinder Singh, and Jagat Rakesh Kanwar**

CONTENTS

12.1 INTRODUCTION

Blindness and visual impairment are the third most-feared complicated medical conditions, after cardiac diseases and cancer (Fritsche et al., 2014). Worldwide, age-related macular degeneration (AMD), which affects the macula (i.e., peripheral retina) responsible for central vision, is the leading cause of irreversible visual loss among elderly people (Toto et al., 2016). This disorder involves many stages according to its severity, though loss of vision results from photoreceptors' degeneration and damage to the retinal pigment epithelium (RPE), with these clinical manifestations resembling the advanced stage of disease (Holz et al., 2012). AMD is a multifactorial disease as it gets influenced by several risk factors (Klein et al., 2010). Notably, AMD initiates as a few accumulations of drusen, the biological by-products of the

* Corresponding author.

retina cycle, between the retinal pigment epithelium and Bruch's membrane, and progresses gradually to either geographic atrophy or choroidal neovascularization in one or both eyes.

There is no successful cure, though the available therapeutic strategies slow the progression of this disorder, which, unfortunately, is followed by relapse (Kent, 2014). Nanotechnology is an emerging field involving minute particles in the nanometer scale (Birch and Liang, 2007). Nanoparticles could be synthesized using various combinations and encapsulated by specific adapted formulations. Thus, they could form the alternative strategy for posterior eye injection therapy, if they are designed in such a way that could reach the distal point of the eye and exert therapeutic efficacy. In this chapter, we shall review the mechanism and risk factors regarding the AMD disorder and discuss the nanotherapeutic strategies and the models used to study the underlying complicated mechanisms of this disorder.

12.2 RECENT PROTEOMIC CONCEPT REGARDING AMD PATHOGENESIS

Despite the intensive clinical and basic studies conducted on this multifactorial disorder, the pathogenesis of AMD still remains to be fully understood (McConnell and Silvestri, 2005; Nowak, 2005; Birch and Liang, 2007; Ambati et al., 2013). Oxidative stress is considered as the prime factor that drives the progression of AMD (Ambati et al., 2013). The aging process is accompanied by defective mitochondria and other organelles, in addition to the insufficient removal of the macromolecules by the lysosomal apparatus (Arjamaa et al., 2009; Wang et al., 2009). The main concept regarding AMD is the imbalance between mitochondrial damage and DNA repair. This leads to accumulation of biological by-products such as lipofuscin and elevation of autophagy and exocytotic activity, subsequently releasing the biological garbage out of the cells in the presence of endosomal sorting complexes required for the transport (ESCRT), such as tumor susceptibility gene (TSG), in the form of exosomes (Kang et al., 2014). Exosomes mostly have digested and partially digested proteins, in addition to the genetic materials (Atienzar et al., 2016), which contribute to the formation of drusen. However, the issue becomes more complicated if those exosomes are coated with complement C3, followed by formation of membrane attack complex (MAC), which plays a significant role in retinal pigment epithelium (RPE) degeneration and development of geographic atrophy (i.e., dry form).

The other accepted hypothesis regarding the progression of the late stage of AMD, that is, choroidal neovascularization (CNV), has been confirmed as well (Blaauwgeers et al., 1999; Atienzar et al., 2016), oxidative stress creates a hypoxic environment within RPE. Hypoxic RPE secretes vascular endothelial growth factor (VEGF), which binds to its receptors, VEGFR1 and VEGFR2, on the choroid and activates tube formation. These blood vessels, which originate from the choroid, are incomplete and constantly leaking beneath RPE, thus making the latter incompetent, followed by retinal detachment. Although the dry form is more common than CNV, the wet form is responsible for 90% of blindness among AMD patients.

Exosomes appear to have a critical role in the initiation and progression of CNV. Recent investigations confirm exosomal cargo from a stressed RPE holds genetic information regarding expression of VEGFR1 and VEGFR2 on the choroid to facilitate tube formation and subsequent angiogenesis (Amir et al., 2013). Considering the previous information, exosomal targeting could serve as molecular biomarkers to diagnose AMD and act as the basis for development of new cures to treat the advanced stages of AMD.

12.3 RISK FACTORS

AMD is supposed to be the leading cause of blindness in developing countries. An intensive understanding of AMD risk factors will assist in reducing the opportunity for its progression and consequently blindness. Most researchers have accepted smoking as a significant cause, although other factors, apart from heredity and age, remain controversial (Smith et al., 2001). AMD incidence, prevalence, and progression are interrelated by specific features. For instance, large drusen formation, indistinct drusen, and pigmentary abnormality increase with age, while its prevalence is related to

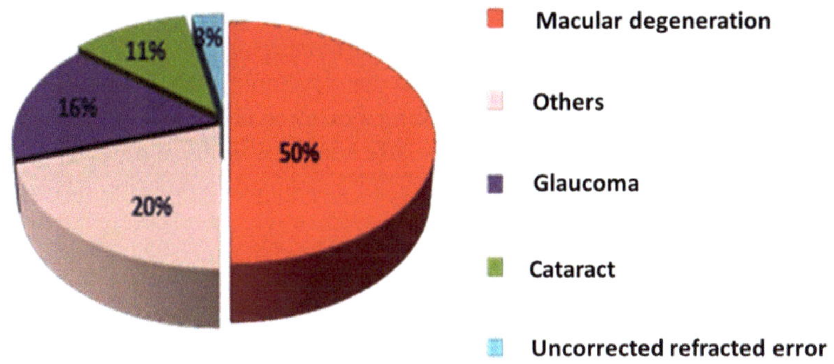

FIGURE 12.1 Cause of blindness among Australian population; AMD represents the highest rate as the leading cause of vision loss. (Based on Economics A. *Report for Vision 2020*, 2010: 1417–1436.)

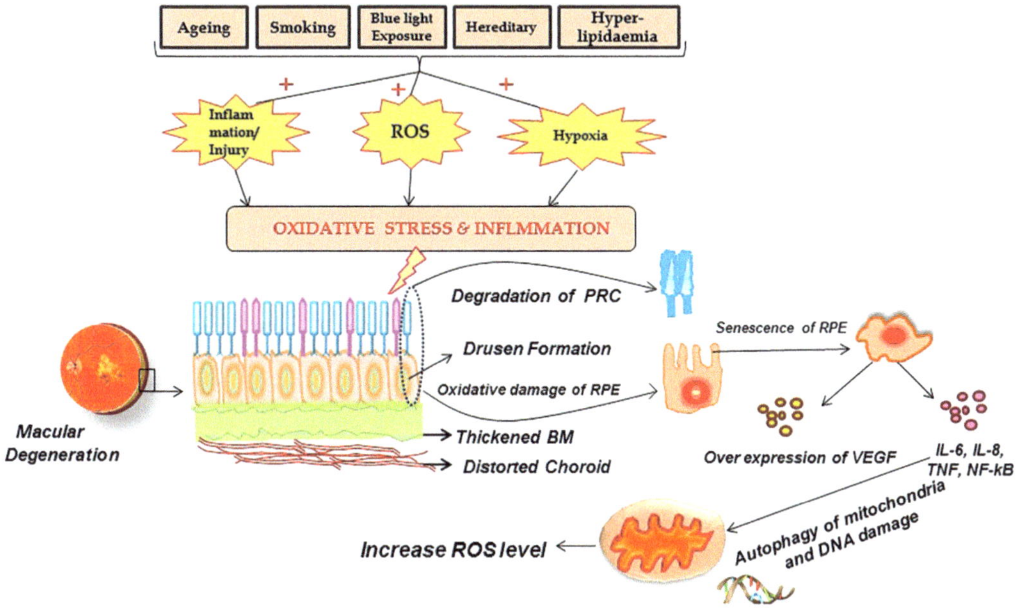

FIGURE 12.2 Illustration of different factors involved in AMD pathogenesis. (PRC: photoreceptor cells; BM: Bruch's membrane)

ethnicity (Cotter et al., 2006). Both the Macular Photocoagulation Study (Jampol and Tielsch, 1992) and the Barbados study (Schachat et al., 1995) assume that late AMD is more common among Caucasians and Asians than in African and Hispanic populations. Moreover, the visual problem encountered by the Australian population is mainly related to AMD (Figure 12.1). It was estimated in 2010 that over 100,000 Australians had wet form of AMD in one or both eyes, and this number is projected to double by 2030 (Economics, 2010). Additionally, several related risk-modifiable factors have also been investigated (Tomany et al., 2004; Hurley et al., 2008). These are schematically represented in Figure 12.2 and summarized in the following sections.

12.3.1 Smoking

Smoking is considered as the prime modifiable risk factor, which plays a significant role in AMD progression. This factor generally affects several processes speculated to be involved in AMD pathogenesis—for example, immune activation, antioxidant level depression, reduction in choroidal blood flow, retinal luteal pigments, and detoxification rate for RPE. However, data gathered reported a high incidence of pack years smoked to be the cause of early and late AMD induction. Therefore, smokers of more than 20 years are at greater risk for developing AMD than nonsmokers. Hence, once AMD is initiated with early stages of smoking, cessation is highly recommended (Khan et al., 2006). Hypertension and cardiovascular disease are supposed to have a distinct relation to AMD progression as well. Nevertheless, the results for an examined population have shown reflective outcomes (Olea and Tuñón, 2011).

12.3.2 Level of Antioxidants

Low systemic levels for antioxidants could participate in AMD induction. Initially, a high dietary level of vitamins C and E, β-carotene, and zinc had beneficial effects in reducing the progression of AMD development (Chew et al., 2012). Antioxidants are generally present in the macula, as these compounds have been proposed to protect the macula against AMD. Although the exact protection role remains obscure, it is assumed to be related to ameliorating the adverse effects for photochemical reaction (Guymer and Chong, 2006). Further, it has been shown that the continuous use of antioxidants inhibits the progression of AMD category 3 or 4 from 25% to 17%, but no benefit has been observed in lowering the risk of AMD progression for category 1 or 2 (Group, 2001). Several experiments have been conducted to investigate the possible effect of vitamin E and β-carotene utilization, but ophthalmological examinations revealed no beneficial effects within six years of supplementation (Taylor et al., 2002). Therefore, the protective mechanism remains only as a hypothesis and the adverse effects of high-dose supplementation for long periods remain unknown.

12.3.3 Diet

Research documents have identified the relationship between dietary fat intake and AMD incidence. Although this idea came from different sources, a link has been proposed between apolipoprotein E and AMD induction, as well as a relationship between cardiovascular disease and AMD, and a lipid-cholesterol pathway gene (Ruberti et al., 2003), such as hepatic triglyceride lipase (LIPC), cholesteryl ester transfer protein (CETP), and apolipoprotein E (APOE) heterogenicity (Pennington and DeAngelis 2016), Bruch's membrane is a semipermeable one, the function of which is to separate photoreceptors from their blood supply; the lipid content of this membrane normally builds up with age, the outer segments of photoreceptors contain a high percentage of docosahexaenoic acid (DHA) that is constantly shed and turned over during the visual cycle (Vinding et al., 1992). Therefore, the deficiency of DHA might impair the retinal function and promote AMD. Studies that attempt to find supportive evidence regarding the relationship between AMD and dietary fat intake are still very limited, and their outcomes are inconsistent and often conflicting (Guymer and Chong, 2006). Generally, it is clear that AMD disorder might be related to fat intake, yet specific recommendations need to be taken into account based on observational studies and their outcomes to buttress this fact.

12.3.4 Other Risk Factors

Other risk factors include aspirin utilization. Recent studies regarding aspirin intake refer to the presence of a possible link between AMD progression and patients using aspirin twice weekly for ten years (Klein et al., 2012). Others have shown a protective role of aspirin against AMD (Zhu et al., 2013). However, the available information at this stage, due to the inconclusive findings from this study, indicates that caution should be against taking this medication. The other risk factors that have been considered in various research projects include inflammation biomarkers such as the complement-reactive protein (Laine et al., 2007). Others involve an increase in waist-to-hip ratio of male, alcohol intake (Guymer and Chong, 2006), sunlight exposure (Khan et al., 2006), hormonal status (Feskanich et al., 2008), and status of vitamin B and D levels (Gopinath et al., 2013).

12.4 AMD MODELS

AMD is the common cause of visual impairment. Various animal models have been proposed to provide a better insight into the underlying pathological mechanism to fulfill the requirements for developing an effective medication, and also to serve as the backbone for testing new remedies. However, no model is perfect enough to represent what exactly happens in the individual eye, though the clinical features can be represented by several *in vitro* and *in vivo* models, which are shown in Tables 12.1 and 12.2. Most of the models have demands that mimic clinical features for AMD, especially for *in vitro* models. They offer the chance to recapitulate all the features of AMD and view the promise for considering the molecular mechanism and drug for AMD disease. Immortalized RPE-19 cells are the standard RPE cells used in several model experiments, as these cells are derived from stem cells and are considered a useful tool for AMD models. However, they behave as more functional RPE in models than pathologic model (Forest et al., 2015). Another model evaluates exudative AMD by using bovine choroidal endothelial cells (Hoffmann et al., 2006). This model provides the opportunity to examine drug toxicity and anti-angiogenic effects. Similarly, other kinds of isolated neovascular structure that mimics the rapid growth of blood vessels in exudative AMD can be represented by using 8–9 day old highly vascularized chorioallantoic membrane (Samkoe and Cramb, 2003). This model was used to investigate the therapeutic effect of photodynamic therapy on ocular vascularization, as both the other models were early attempts to create AMD models. However, AMD mainly affects macular cells, which could create distinct pathways.

TABLE 12.1
In Vitro Experimental Models Used to Induce Age-Related Macular Degeneration (AMD)

S. No	In Vitro AMD Models	
	Wet Form	**Dry Form**
1	Easily accessible monolayer-like neovascular net within a transparent matrix to evaluate the efficacy of drugs (Söderstjerna et al., 2014).	Growing RPE on coated flasks (Penfold et al., 2000).
2	Use of isolated neovascular structure 8- to 9-day-old chicken chorioallantoic membrane (Samkoe and Cramb, 2003).	ARPE-19 cells were grown on plastic and on collagen type I–coated membrane (Amin et al., 2004).
3	Bovine CECs involved in the process of CNV formation (Hoffmann et al., 2006).	Human adult RPE cells (ARPE-19 cells) are irradiated or treated with BSO at 100 M for 48 h (Jiang and Chiou, 2009).
4	Mice deficient ceruloplasmin and/or hephaestin due to retinal iron increment in RPE (Hahn et al., 2004).	RPE can be grown on 2-dimensional laminin-coated porous support. Complement is used to induce drusen like deposition (Johnson et al., 2011).
5	Using co-culture of ARPE-19 cell line with HUVECs on amniotic membrane (Hamilton et al., 2007).	–
6	Porous ultrathin membrane from silk fibroin that mimics the elasticity of Bruch's membrane in 3-dimensional tissue culture (Shadforth et al., 2012).	–
7	Tropoelastin can be incorporated into silk fibrin that mimics Bruch's membrane in 3-dimensional tissue culture (Feigl and Hutmacher, 2013).	–

Abbreviations: BSO, Buthionine; CEC, choroidal endothelial cells; CNV, choroidal neovascularization; HUVECs, human umbilical vein endothelial cells; RPE, retinal pigment epithelium; S, sulfoximine.

TABLE 12.2
In Vivo Experimental Models Used to Induce AMD

S No.	In Vivo AMD Models	
	Wet Form	**Dry Form**
1	Laser beam method (krypton red laser, green argon laser, or diode laser) (Jiang and Chiou, 2009).	ApoE3-Leiden by expressing the human apolipoprotein-E (Apo-E) 3-Leiden when using on a high fat/cholesterol diet or on a normal mouse chow for nine months (Kliffen et al., 2000).
2	Injection of adeno-associated viral vector encoding human VEGF-165 into the subretinal space of rat (Wang et al., 2003).	A transgenic mouse line (mcd/mcd) that express a mutated form of cathepsin D (Rakoczy et al., 2002).
3	Overexpression of VEGF with the retinal pigment epithelium using transgenic murine model. RPE (65) promoter coupled to murine VEGF (164) can be introduced into the genome of albino mice (Schwesinger et al., 2001).	Pigmented dystrophic Royal College of Surgeons rat during the first three months of life (Coffey et al., 2002).
4	Exposing postnatal 17-21 days mice into 750 % O_2 for five days then into room air to induce neovascularisation in the retina but not in the choroid (Smith et al., 1994).	Using a single injection of sterile 1% solution of NaIO3 in saline into the tail vein (Atmaca-Sonmez et al., 2006).
5	Prokineticin mice as a mitogen of fenestrated endothelium could induce CNV using the rhodopsin promoter (Tanaka et al., 2006).	Porcine eye used through debriding RPE to develop choriocapillaris atrophy by utilizing pars planavitrectomy and injecting mitomycin C into the subretinal space (Del Priore et al., 1996).
6	Nicotine administered orally in the drinking water (100 mg/mL) for four weeks (Bora et al., 2006).	Mice deficient in Cu, Zn-superoxide dismutase (SOD1) of different ages (Imamura et al., 2006).
7	Transvitral and trans-scleral injection of collagenase and hyaluronidase to induce choroidal neovascularization (Ryan, 1979).	Feeding mice with hydroquinone (HQ) or cigarette smoke component (Marin-Castano et al., 2006).
8	–	Mice deficient monocyte chemo-attractant protein-1 (Ccl-2) or chemokine receptor-2 (Ccr-2) (Ambati et al., 2003).
9	–	LDL receptor-deficient mice elicit AMD pathogenesis when feed on standard or high-fat, especially polyunsaturated, diet (Malek et al., 2005).
10	–	Blue-light exposure (14 mw/cm^2; bandwidth, 390–430 nm) used dark-adapted mice for two hours (Li et al., 1999).
11	–	

Abbreviations: AMD, age-related macular degeneration; CNV, choroidal neovascularization; HQ, hydroquinone; LDL, low-density lipoprotein; RPE, retinal pigment epithelium; VEGF, vascular endothelial growth factor.

The outer blood-retinal barrier has been efficiently utilized to illustrate drug capacity to modulate cellular permeability and stimulate the expression of adhesion for RPE, when grown on coated flasks (Penfold et al., 2000). Furthermore, it could assist in determining drug efficiency in reducing the incidence of wet AMD. This model could be used to obtain incredible results, especially with discovery studies, though it consumes long time and calls for specific requirements. A sub-RPE deposit as dry form model, on the other hand, is evaluated efficiently using electronic microscopy (Amin et al., 2004).

Hypoxia is assumed to be the primary factor that drives the pathogenesis regarding AMD (Lin et al., 2013). Several remedial attempts were made to ameliorate the negative effect of this disease (Martin et al., 2012; Reeves et al., 2012), followed by short-term relapse. It is proposed that these strategies were prehypoxic, since successful therapy requires considering both hypoxia and angiogenesis mechanisms. Several studies have been conducted to establish an *in vivo* study, since this is more critical because it represents what is happening inside the body. In vitro models could provide a general idea about the mechanism underlying AMD, while the *in vivo* ones are considered more significant and acceptable.

Induction of a stable wet *in vivo* model could be rapidly and easily developed using a laser beam method. A slit lamp delivery system is used to puncture the Bruch's membrane and stimulate CNV development in animals like rats, mice, or pigs (Jiang and Chiou, 2009). This model can induce CNV rapidly, though this is not practically feasible since the required slit lamp delivery system is cost-ineffective. Another method that could be utilized to induce CNV is by injecting adeno-associated viral vector encoding human VEGF-165 into the subretinal space of the rat (Wang et al., 2003). Precautions need to be exercised to avoid detachment of the retina due to the injection. Additional example is the transgenic murine model, which is established through overexpression of VEGF in RPE. This model increases the expression of VEGF protein in RPE and choroid, and promotes an inflammatory response (Schwesinger et al., 2001). Nevertheless, this model is time-consuming, because it requires at least four months' duration, in addition to the specific facilities Hypoxia-induced retinal ischemia was recorded to develop neovascularization in the eyes of mice. Briefly, 17-day postnatal mice were exposed to 750 % of O_2 for five days, followed by exposure to room air (Smith et al., 1994). The age and time of exposure can be modified according to different considerations. The results showed that this model might induce neovascularization in the retina, but not in the choroid.

Several studies have used drusen induction models to investigate the disease mechanisms and possible treatment strategies for dry AMD (Kliffen et al., 2000; Coffey et al., 2002; Jiang and Chiou, 2009). Drusen is a typical clinical manifestation in nonexudative AMD, located between the basement membrane of RPE and the Bruch's membrane, the prevalence and severity of drusen being proportionally related to age. Transgenic mice (APOE3-Leiden) can develop human apolipoprotein E (Apo-E) 3-leiden gene expression, when fed a high fat/cholesterol diet for nine months. However, this model can develop based on laminal deposit between the cell membrane of RPE and its basal membrane rather than drusen formation (Kliffen et al., 2000). Although this model mimics early stages of AMD, it seems rational to search for a less time-consuming model, particularly when the researchers have limited time available for the purpose. Similarly, low-density lipoprotein (LDL) receptor-deficient mice are capable of eliciting AMD pathogenesis, when fed a standard or high-fat, especially polyunsaturated fat, diet. Clinical manifestations include accumulation of lipid particles in Bruch's membrane and VEGF expression in the outer nuclear layer of the retina (Rudolf et al., 2005). Nevertheless, research for more strategies regarding AMD induction and progression is still required by finding other models that closely mimic the occurrences in the eyes of the aged, since this disease is more complicated. Other transgenic mice apolipoprotein B100 demonstrates a high frequency of moderate basal laminal deposit when exposed to blue-green light (Malek et al., 2005). Several other investigations were performed to create more suitable models, particularly for the purpose of therapeutic studies. A transgenic mice line (mcd/mcd) that overexpresses the mutated form of cathepsin D, can develop degeneration and shortening of photoreceptors, in addition to the accumulation of immune reactive photoreceptor breakdown product within RPE (Rakoczy et al., 2002). Other experiments for photoreceptor degeneration suggests using pigmented dystrophic Royal College of Surgeons rats in various treatment studies of RPE and AMD remedies. Due to photoreceptor genetic defects, rat photoreceptors typically die during the first three months of their life. Consequently, they can provide additional incredible models for studying the suggested AMD therapeutic systems (Coffey et al., 2002).

The mechanism and possible treatment strategies of AMD have been investigated using oxidative stress, as this is strongly associated with AMD and implicated in its pathogenesis. Sodium iodate ($NaIO_3$) is employed to induce progressive RPE degeneration, when injected as a single dose of sterile 1% solution into the mice tail vein (Atmaca-Sonmez et al., 2006). This model seems to be more suitable for assessment of retinal transplantation, since it creates permanent deficiency for RPE. More events related to oxidative damage can also be initiated by utilizing SOD1 knockout mice SOD1 −/− (superoxide dismutase-deficient mice) that might be beneficial to study the AMD mechanism, since these mice progressively accumulate lipofuscin during eight months, when the young ones are exposed to excessive light (Imamura et al., 2006). Blue-light exposure (14 mw/cm^2; bandwidth of 390–430 nm) could be used for dark-adapted mice for two hours on dilated pupils, diffuse white fluorescent light 15,000 lux (TLD-36 w/965 tubes, Philips), while Argon laser 488 nm blue-green light also possesses the same effect (Li et al., 1999). Feeding mice with hydroquinone (HQ) or a cigarette smoke component would promote oxidative damage and slow thickening of the Bruch's membrane, accumulating the deposit within it (Marin-Castano et al., 2006). Both of these models induce photoreceptor apoptosis. On the other hand, nicotine utilization in drinking water (100 mg/mL) can be used for four weeks to raise the severity of CNV in a murine model (Bora et al., 2006). It requires repetitive exposure for a minimum of four weeks. Another related study to oxidative stress (Li et al., 1999) illustrates the progressive accumulation of photoreactive pigment and formation of lipofuscin. Subsequent development of artificial aging can be triggered *in vitro* experimentally using bovine RPE and exposing them to ultraviolet light to develop artificial aging. This model was efficient in evaluating artificial RPE senescence at the cellular and biochemical level.

In order to study the retinopathy in mice, double knockout, ATP-binding cassette (ABC) transporter/retinol dehydrogenase 8 (Abca4-/-/ Rdh8-/-) mice exhibit retinopathy easily during their life clearance due to the absence of both retinal dehydrogenase (Rdh8) and ATP-binding cassette transporter 4 (Abca4) activities and subsequent delayed all-trans-retinal functions. These mice develop drusen accumulation after exposure to light that results in retinal degeneration (Puntel et al., 2015). Another model suggests that mice deficient in monocyte chemoattractant protein-1 (Ccl-2) or chemokine receptor-2 (Ccr-2) can develop typical clinical features of AMD, including lipofuscin accumulation beneath PRE plus CNV and photoreceptor atrophy (Ambati et al., 2003). Nevertheless, the mice need to be back-crossed ten times before using as an AMD model. Several models are presented that mimic the clinical features for AMD. However, more feasible models still require representing AMD signs, since others were either time-consuming or costly. Additional approaches are still required to yield a comprehensive view regarding the pathophysiological mechanism of AMD.

12.5 DIAGNOSIS OF AMD USING ANCILLARY TESTING

AMD could be easily diagnosed via clinical manifestation, however, accurate diagnosis requires using clinical tests since this disorder participates to a limited extant with other disorders; Table 12.3 illustrates the differential diagnosis of AMD. Ancillary testing could be used to provide an overview of the eye condition. This includes using optical coherence tomography, fundus autofluorescence imaging, fluorescein angiography, and indocyanine green angiography (Holz et al., 2012).

12.5.1 Optical Coherence Tomography

This technique involves noninvasive imaging to provide a high-resolution cross-sectional imaging of the fundus concerning RPE and the subretinal space. It could detect the stage of AMD, since it provides information about the actual amount of accumulated drusen and the degree of degeneration regarding photoreceptors and RPE, in addition to the presence of fluid within the macular region (Nicholson et al., 2015).

TABLE 12.3
Differential Diagnosis of AMD

Dry AMD	Wet AMD
Macular telangiectasia type II	Fundus flavimaculatus
Pattern dystrophy	Multifocal choroiditis
Acute posterior multifocal placoid pigment epitheliopathy	Ocular histoplasmosis syndrome
Adult vitelliform dystrophy	Polypoidalchoroidalvasculopathy
Cuticular drusen with or without a vitelliform lesion	Optic disc drusen
Central serous chorioretinopathy	Pathologic (progressive) myopia
Toxicity with Phenothiazines and Chloroquine	Idiopathic causes
Stargardt disease	Optic nerve head pits
Angioid streaks	Pattern dystrophies
Best's disease	Toxoplasmicretinochoroiditis, Sarcoidosis
Multifocal choroiditis	Choroidal osteoma
–	Serpiginous or geographic choroiditis
–	Angioid streaks
–	Traumatic choroidal rupture
–	Best's disease

Source: Kubicka-Trzaska A. *Klinika oczna.* 2004; 107: 147–155; Yanoff M, Jay SD. *Ophthalmology.* 4th edition. Elsevier. 2013; pp. 580–599, Jonisch J, Shah G. *Age-Related Macular Degeneration Diagnosis and Treatment.* Springer: New York. 2011; pp. 23–38; Holz F G et al. *Age-Related Macular Degeneration,* ISBN 978-3-642-22107-1. Springer-Verlag: New York. 2012; pp. 137–977.

12.5.2 Fundus Autofluorescence Imaging

Normal retinoid metabolism is accompanied by accumulation of natural fluorophores within RPE. It could provide noninvasive imaging to estimate the intrinsic macular fluorophores and predict dry AMD progression (Pilotto et al., 2013).

12.5.3 Fluorescein Angiography

This technique involves using a colored fluorophore (i.e., fluorescein sodium), injected into the antecubital vein, followed by dye appearance in the ocular vasculature. This technique is restricted to detecting the pathological changes within CNV, as the fluorescent dye demonstrates the anatomical alteration regarding the growth of new vessels and exhibits the leakage resulting from the breakdown of the blood-retinal barrier (Holz et al., 2012).

12.5.4 Indocyanine Green Angiography (IGA)

IGA involves using highly protein-bound fluorophore with a suitable wavelength to penetrate melanin of the choroid and RPE. This property is helpful to see retinal choroid anastomosis and to diagnose the presence of angiogenesis under RPE (Pilotto et al., 2013).

12.6 THERAPEUTIC STRATEGIES FOR AMD

The current recommended treatment for AMD patients does not provide a clear management for this deleterious disease. However, they just act to slow the progression of the disorder, since early vision loss can profoundly affect patient life and esteem. Here, the recent strategies have been summarized and evaluated in the following sections, depending on the recent clinical and research investigations,

12.6.1 Dry AMD Treatment

RPE-cell-replacement has great potential and offers considerable hope for the treatment of the dry form of AMD (Atmaca-Sonmez et al., 2006; Whiting et al., 2015). However, taking into account the obstacles faced and requirements for biomedical research and materials science integration, along with clinical applications and the long-term multidisciplinary approach to detect the safety profile, significant challenges might be encountered regarding this approach (Alexander et al., 2015). The Implantable Miniature Telescope was introduced along with a vision rehabilitation program to help the patient in performing daily activities with this device (Yonekawa and Wolfe, 2015).

12.6.2 Wet AMD

12.6.2.1 Photodynamic Therapy and Laser Photocoagulation

Both approaches principally act by introducing high energy to produce small burns, destroying abnormal blood vessels and preventing further leakage (Group, 1999; Luttrull et al., 2015). However, both techniques arrest repair and create a prehypoxic environment, thus fueling pro-angiogenesis, followed by relapse and further development of CNV (Kent, 2014).

12.6.2.2 Angiogenesis Inhibitors

CNV is responsible for 90% of blindness among AMD patients. The deleterious effects, accompanied in the wet form by leakage from fenestrated vessels extended from the choroid, cause separation of Bruch's membrane, the RPE, and the retina from each other, followed by generalized thickening of the retina, misalignment and degeneration of photoreceptors, and eventually fibrosis (Lim et al., 2012). The therapeutic strategy is administered by blocking VEGF activity (Heier et al., 2012), which is the key factor in angiogenesis, and varied therapeutic programs participate in the principle.

12.6.2.2.1 Classical Treatment of AMD

Several therapeutic agents are delivered via intravitreal injection (Figure 12.3a). The procedure starts with three sessions every month and continues according to the severity of disease; vision could be restored partially before photoreceptor degeneration, or it could prevent further eye damage. Bevacizumab (Avastin), Anecortave acetate (Retaane), and Vatalanib (Novartis) are delivered intravitreally using a special needle, though the procedure is risky and accompanied by several adverse effects (Augustin and Offermann, 2006; Scott et al., 2007; Martin et al., 2012; Soubrane Daguet et al., 2016).

12.6.2.2.2 Nanoparticles and Aptamer Injection

Treatment for AMD requires intraocular administration. Despite the achieved improvement for AMD therapy using the intravitreal approach, the issue for rapid clearance is still undesirable since it requires repeated injection and subsequent risk of surgery, as well as the probability of further requirement of doses to maintain drug release. In order to overcome this obstacle, sustained intraocular drug release has been developed using nanoparticles (Birch and Liang, 2007; Farjo and Ma, 2010; Elsaid et al., 2016). These microscopically sized spheres are made from artificial or natural polymers, as portrayed in Figure 12.3b. The aptamer Pegaptanib (Figure 12.3c) was approved by the US Food and Drug Administration (FDA) to treat AMD, as it has the ability to selectively bind with VEGF and inhibit vascular leakage. Although both the drugs showed some efficacy in slowing disease progression and improving vision, it could further reduce the frequency of intravitreal injection. Nanoparticles loaded with ganciclovir along with nanoparticle-mediated gene delivery were also administered into the vitreous chamber and have been reported to have a prolonged presence in the eye without toxic effects (Kanwar et al., 2014).

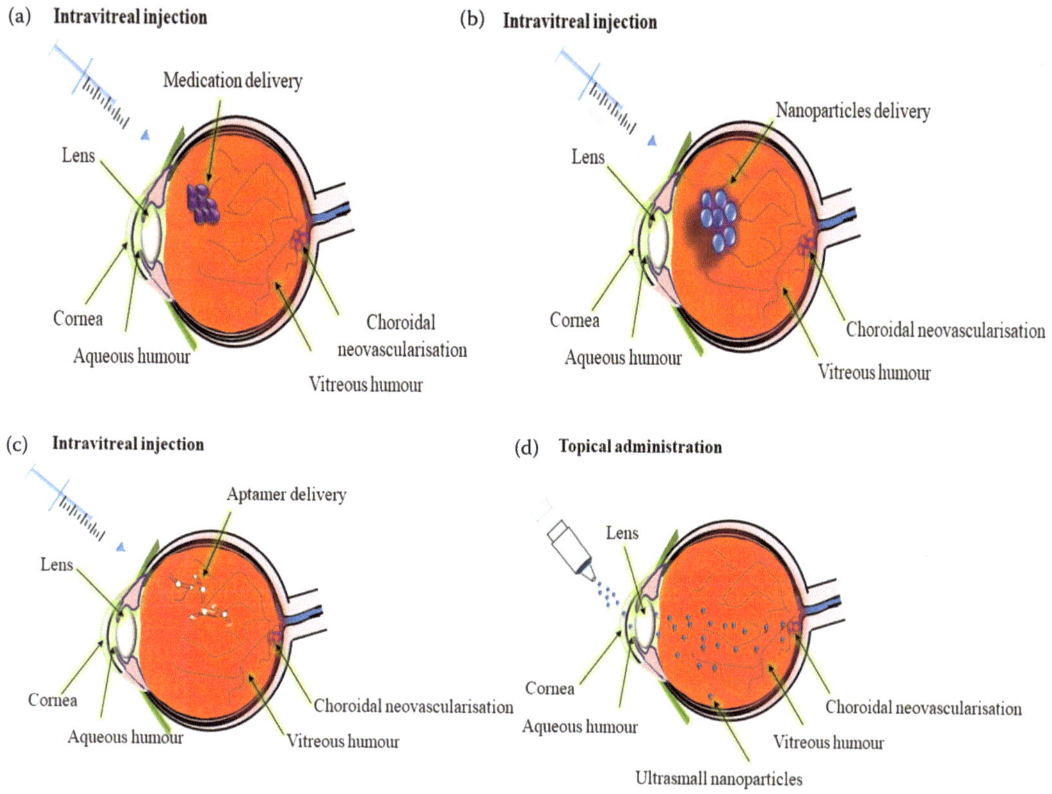

FIGURE 12.3 Therapeutic strategies used and suggested for AMD. (a) Usual intravitreal injection; (b) Nanoparticles delivery; (c) Aptamer used in treating AMD; (d) Suggested topical application by delivering ultrasmall nanoparticles.

12.6.2.2.3 Ultrasmall Chitosan Nanoparticles

A recent hypothesis by our laboratory introduces ultrasmall algal chitosan nanotherapy to treat eye disease for efficient delivery of bovine lactoferrin (Figure 12.3d). Newly developed nanoparticles were used to prevent carbendazim-induced toxicity, and the research outcomes confirm that the biodistribution of nanoparticles occurred through the whole ocular tissue (Sunkireddy et al., 2016). This could be considered a critical point in developing a new technique regarding retinal diseases and avoids the risk of accompanying intravitreal injection. Investigations are in progress for developing the process of preparing safe, effective, noninvasive therapy to treat AMD.

12.6.2.3 Bioactives

Identifying bioactives, which are helpful in the prevention and treatment of diseases, and to explore these as adjunct therapeutic agents is becoming a popular strategy in disease management. Many such bioactives have been studied for their preventive and therapeutic action in management of AMD. Three major carotenoid-based xanthophylls found in the eye are lutein, zeaxanthin, and meso-zeaxanthin. Of these, lutein and zeaxanthin are exclusively found in the retina and are concentrated in the macula. Apart from selective concentration in the macula, lutein protects photoreceptors from blue-light damage, as it selectively filters blue light, due to its peak absorption spectrum of 446 nm. Lutein causes CNV suppression by reduction in NF-κB-mediated inflammation (Nagai et al., 2007). Lutein and zeaxanthin also act as antioxidants and have the capability to scavenge singlet oxygen species and hence prevent oxidative damage, which is one of the main causes of pathogenesis of AMD. Due to these properties, many researchers (Tan et al., 2008; Ho et al., 2011; Ma et al., 2012) have studied the therapeutic potential of lutein and zeaxanthin to manage AMD. These studies reported an increase in visual activity with supplementation, and the effects were more pronounced in early AMD compared with late AMD. Chew et al. (2012) have also shown reduced progression of AMD in patients who were provided with xanthophyll supplements, as discussed in Section 12.3.2. Other bioactives which have been shown to prevent AMD are curcumin, anthocyanin, resveratrol, lycopene, and astaxanthin. Their mode of action can broadly be classified as antioxidant and/or anti-inflammatory, mediated through different cellular pathways. The dietary sources of these bioactives, along with their mechanism of action are summarized in Table 12.4. Apart from the above-mentioned bioactives, long-chain omega-3 fatty acids such as docosahexaenoic acid (DHA) and eicosapentaenoic acid (EPA) also help in vision improvement in AMD patients (Ho et al., 2011; Fu et al., 2017).

12.6.2.3.1 Nanoparticles of Bioactives

Most of the plant bioactives studied for AMD treatment, including lutein, curcumin, anthocyanin, resveratrol, lycopene, and astaxanthin (Figure 12.4), exhibit low bioavailability and poor aqueous solubility, owing to their hydrophobicity attributable to the presence of polyphenolic rings in their respective structures (Scheepens et al., 2010). To increase their bioavailability and to protect them against degradation, nanoencapsulation techniques have been employed by many researchers (Lacatusu et al., 2013; Zhao et al., 2014; Arunkumar et al., 2015; Lindner et al., 2015; Balata et al., 2016; Kamil et al., 2016; Neves et al., 2016; Surh et al., 2017; Tamjidi et al., 2018). Different type of nanocarriers—liposomes, nanostructured lipid nanocarriers (NLCs), solid lipid nanoparticles (SLNs), nanodispersions, nanoemulsions, polymeric nanoparticles using poly(lactic-co-glycolic acid) PLGA, and chitosan—have been studied for effective delivery of these bioactives, and the type of delivery systems explored for each of the bioactives is compiled as Table 12.3. Among these, the self-nanoemulsifying drug-delivery systems (SNEDDs) hold immense promise for delivery of these hydrophobic molecules due to their mono-disperse nature, reproducibility, and ease of synthesis. Due to the broad range of biological actions of these bioactives (antioxidant, anti-inflammatory, anticancer, antidiabetic, etc.), the prepared nanoformulations are mostly focused on increasing their *in vivo* bioavailability and preventing their *in vivo* degradation. Though the bioactives have been proved to help in AMD management, the activity of these bioactive-loaded nanoparticles in the treatment of AMD still remains to be explored and reported in the literature. A wide scope of these bioactive nanoparticles for AMD therapy makes it worth investigating and evaluating in the established *in vitro* and *in vivo* models of AMD, as elaborated in Section 12.4.

12.7 CONCLUSIONS

AMD is an important cause of vision impairment in the aged population. In terms of loss of the quality of life, economic impact, and loss of productivity, the disease may be considered as devastating. Besides the various causative factors, patients should be advised on the modifiable risk factors to reduce the incidence of this disease, rather than treating it later by changing lifestyle factors that might improve the patient prognosis and delay the disease progression.

Recently, several established therapeutic options, as well as promising surgical approaches, have been under evaluation to stabilize vision. Current treatments are seemingly quite effective, especially using plant bioactive, their nanoparticle, and aptamer regimes; for example, in pegaptanib it could reduce the cost considerably by up to 50 times and thus reduce the burden on patient and the healthcare system as well. Although studies, either *in vitro* or *in vivo*, are employed to obtain novel therapeutic options and prophylactic management for this disease, further investigations are required to provide definitive treatment options for both forms of AMD.

TABLE 12.4
Role of Various Bioactives in AMD: Sources, Mechanisms of Action and Nanocarriers

Bioactive	Sources	Mechanism of Action	Nanocarriers	References
Lutein and Zeaxanthin	Kale, spinach, corn, peanuts, broccoli, turnip greens, collard greens, and romaine	Lutein protects photoreceptors from blue-light damage, as it selectively filters blue light, due to its peak absorption spectrum of 446 nm. It also suppresses the degradation of IkB in the RPE choroid and thereby inhibits the NF-kB activation.	PLGA, polyvinylpyrrolidone, chitosan NPs, NEs, NDs, NCps, peptide NPs	Nagai et al. (2007), Kamil et al. (2016), Surh et al. (2017), do Prado Silva et al. (2017), Jiao et al. (2018)
Curcumin	Turmeric	Curcumin inhibits the activation of NF-κB by declining phosphorylation of IκBα. Besides, it ameliorates the cellular inflammatory genes, expression of many oxidative stress regulatory proteins such as platelet-derived growth factor, vascular endothelial growth factor, hemeoxygenase 1, superoxide dismutase 2, glutathione peroxidase 1, which, in turn, inhibit cellular inflammatory responses and protect RPE cells	SNEDDS, NLCs, SLNs, NPs, liposomes, dendrimers, NS, NCs, NDs, etc.	Maltzman et al. (1996), Mohanty et al. (2012), Chang et al. (2014)
Anthocyanin	Black currant, bilberry, blueberry, elderberry, concord grapes, cacao, pomegranate, schisandra, rose hip	Suppresses the activation of signal transducer and activator of transcription, which subsequently reduces the inflammation-related rhodopsin drop off. In addition, it also ameliorates the intracellular elevation of free-radical species, especially ROS and activated NF-κB.	PLGA-PEG NPs, serum albumin NPs, chitosan NPs, PEG-gold NPs	Miyake et al. (2012), Chen et al. (2014), He et al. (2016), Kim et al. (2017)
Resveratrol	Red grapes, itadori tea, blueberries, dark chocolate, peanuts	Attenuates the release of inflammatory cytokines such as transforming growth factor through activation of silent information regulator 2/ sirutin 1. Moreover, it also moderates the hypoxia induced through release of VEGF-A and VEGF-C in human retinal pigment epithelial cell.	SNEDDS, SLNs, albumin NPs, polysorbate 80-coated poly(lactide) NPs, PLA coated MSNs	Nagineni et al. (2014), Lindner et al. (2015), Balata et al. (2016), Neves et al. (2016), Shen et al. (2018)
Lycopene	Tomato, guava, watermelon, grapefruit, papaya, apricot	Neutralizes the free radicals, especially those derived from oxygen, thereby conferring protection against photoreceptor and RPE cell death due to oxidative stress	SNEDDS, SLNs, NLCs, NPs, liposomes	Sharma and Goswami (2011), Gul et al. (2015)
Astaxanthin	*Haematococcuspluvialis* (microalgae), wild Pacific sockeye salmon, krill, algae, red trout, shrimp, crab and lobster	Astaxanthin relaxes the ciliary muscles and suppresses TNF-α, thereby inhibits the NF-κB signaling pathways in the ciliary body. Besides, it also suppresses oxidative reactions, in the aqueous humor and reduces the VEGF levels in RPE Cells	SLNs, NLCs, liposomes, chitosan NPs	Hashimoto et al. (2016), Bhatt et al. (2016), Tamjidi et al. (2018) Wang et al. (2017)

Abbreviations: MSN, mesoporous silica nanoparticles; NCps, nanocapsules; NDs, nanodispersions; NE, nanoemulsion; NF-Kβ, nuclear factor-κβ; NLCs, nanostructured lipid carriers; NPs, nanoparticles; NS, nanosuspension; PEG-PLGA, polyethylene glycol–poly lactic acid-co-glycolic acid; RPE, retinal pigment epithelium; ROS, reactive oxygen species; SLNs, solid lipid nanocarriers; SNEDDS, self-nanoemulsifying drug-delivery systems; VEGF, vascular endothelial growth factor.

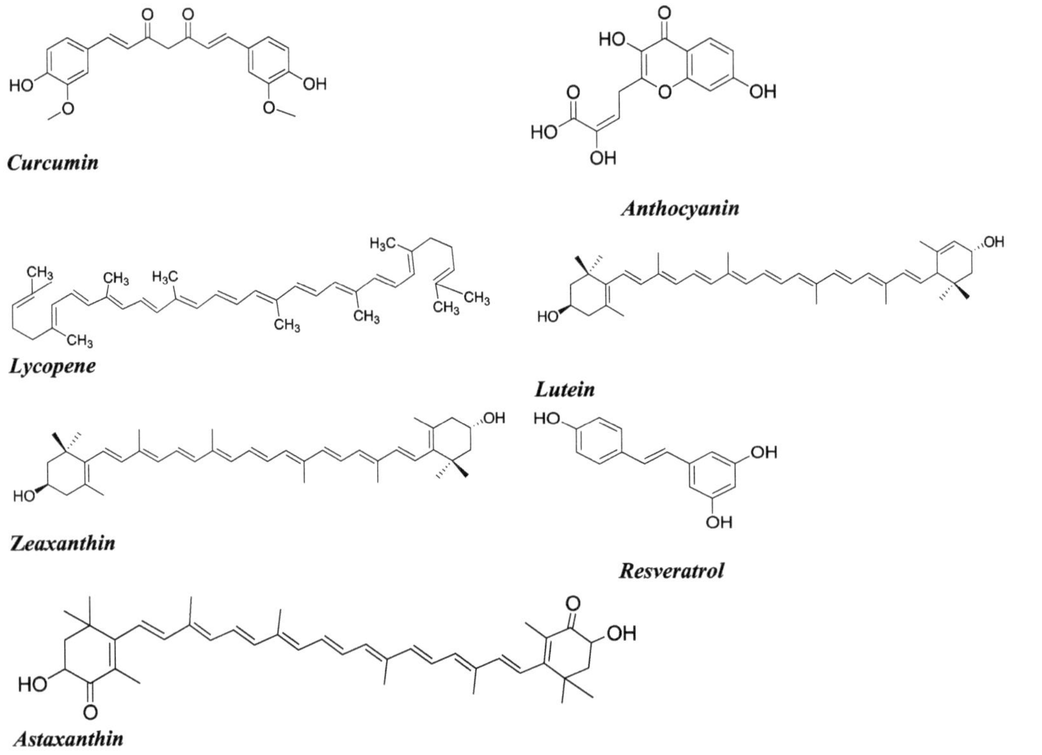

FIGURE 12.4 Chemical structures of different bioactives used in the treatment of AMD. (Abbasi A M, Shah M H, Khan M A. 2015. Phytochemicals and nutraceuticals. In: *Wild Edible Vegetables of Lesser Himalayas.* Springer, Cham.; Brouillard R. *Chemical Structure of Anthocyanins* (Vol. 1). Academic Press: New York. 1982; Gülçin İ. *Arch Toxicol.* 2011; 86(3):345–391; Petrou A L et al., *Antioxidants.* 2018; 7(3): 35.)

REFERENCES

Abbasi A M, Shah M H, Khan M A. 2015. Phytochemicals and nutraceuticals. In: *Wild Edible Vegetables of Lesser Himalayas.* Springer, Cham. https://link.springer.com/chapter/10.1007/978-3-319-09543-1_3

Alexander P, Thomson H A J, Luff A J, Lotery A J. Retinal pigment epithelium transplantation: Concepts, challenges, and future prospects. *Eye.* 2015; 29: 992–1002.

Ambati J, Anand A, Fernandez S, Sakurai E, Lynn B C, Kuziel W A. An animal model of age-related macular degeneration in senescent Ccl-2-or Ccr-2-deficient mice. *Nat Med.* 2003; 9: 1390–1397.

Ambati J, Atkinson J P, Gelfand B D. Immunology of age-related macular degeneration. *Nat Rev Immunol.* 2013; 13: 438–451.

Amin S, Chong, N V, Bailey T A, Zhang J, Knupp C, Cheetham M E. Modulation of sub-RPE deposits *in vitro*: A potential model for age-related macular degeneration. *Invest Ophth Vis Sci.* 2004; 45: 1281–1288.

Amir R H, Jiang G, Lu Q, Lu H, Kaplan H J, Zhan H G. Exosomes from retinal astrocytes contain antiangiogenic components that inhibit laser-induced choroidal neovascularization. *J Biol Chem.* 2013; 288: 28058–28067.

Arjamaa O, Nikinmaa M, Salminen A, Kaarniranta K. Regulatory role of HIF-1α in the pathogenesis of age-related macular degeneration (AMD). *Ageing Res Rev.* 2009; 8: 349–358.

Arunkumar R, Prashanth K V H, Manabe Y, Hirata T, Sugawara T, Dharmesh S M. Biodegradadable poly (lactic-co-glycolic acid)-polyethylene glycol nanocapsules: An efficient carrier for improved solubility, bioavailability, and anticancer property of lutein. *J Pharm Sci.* 2015; 104: 2085–2093.

Atienzar A S, Floresbellver M, Serranoheras G, Martinezgil N, Barcia J M, Aparicio S. Oxidative stress in retinal pigment epithelium cells increases exosome secretion and promotes angiogenesis in endothelial cells. *J Cell Mol Med.* 2016; 20: 1457–1466.

Atmaca-Sonmez P, Li Y, Yamauchi Y, Schanie C L., Idstad S T, Kaplan H J. Systemically transferred hematopoietic stem cells home to the subretinal space and express RPE-65 in a mouse model of retinal pigment epithelium damage. *Exp Eye Res.* 2006; 83: 1295–1302.

Augustin A J, Offermann I. Emerging drugs for age-related macular degeneration. *Expert Opin Emerg Dr.* 2006; 11: 725–740.

Balata G F, Essa E A, Shamard H A, Zaidan S H, Abourehab M A. Self-emulsifying drug delivery systems as a tool to improve solubility and bioavailability of resveratrol. *Drug Des Devel Ther.* 2016; 5:117–128.

Bhatt P C, Srivastava P, Pandey P, Khand W, Panda B P. Nose to brain delivery of astaxanthin-loaded solid lipid nanoparticles: Fabrication, radio labeling, optimization and biological studies. *RSC Adv.* 2016; 12: 1–11.

Birch D G, Liang F Q. Age-related macular degeneration: A target for nanotechnology derived medicines. *Int J Nanomed.* 2007; 2: 65–77.

Blaauwgeers H G, Holtkamp G M, Rutten H, Witmer A N, Koolwijk P, Partanen T A. Polarized vascular endothelial growth factor secretion by human retinal pigment epithelium and localization of vascular endothelial growth factor receptors on the inner choriocapillaris: Evidence for a trophic paracrine relation. *Am J Pathol.* 1999; 155: 421–428.

Bora P S, Kaliappan S, Xu Q, Kumar S, Wang Y, Kaplan H J. Alcohol linked to enhanced angiogenesis in rat model of choroidal neovascularization. *FEBS J.* 2006; 273: 1403–1414.

Brouillard R. Chemical Structure of Anthocyanins (Vol. 1). Academic Press: New York. 1982.

Chang Y C, Chang W C, Hung K H, Yang D M, Cheng Y H, Liao Y W. The generation of induced pluripotent stem cells for macular degeneration as a drug screening platform: Identification of curcumin as a protective agent for retinal pigment epithelial cells against oxidative stress. *Front Aging Neurosci.* 2014; 6: 1–12.

Chen J, Tao X, Zhang M, Sun A, Zhao L. Properties and stability of blueberry anthocyanin--bovine serum albumin nanoparticles. *J Sci Food Agric.* 2014; 9: 1781–1786.

Chew E Y, Clemons T, Sangiovanni J P, Danis R, Domalpall Y A, Mcbee W. The Age-Related Eye Disease Study 2 (AREDS2): Study design and baseline characteristics (AREDS2 report number 1). *Ophthalmology.* 2012; 119: 2282–2289.

Coffey P, Girman S, Wang S, Hetherington L, Keegan D, Adamson P. Long-term preservation of cortically dependent visual function in RCS rats by transplantation. *Nat Neurosci.* 2002; 5: 53–56.

Cotter S A, Varma R, Ying-Lai M, Azen S P, Klein R. Causes of low vision and blindness in adult Latinos: The Los Angeles Latino Eye Study. *Ophthalmology.* 2006; 113: 1574–1582.

Del Priore LV, Kaplan H J, Hornbeck R, Jones Z, Swinn M. Retinal pigment epithelial debridement as a model for the pathogenesis and treatment of macular degeneration. *Am J Ophthalmol.* 1996; 122: 629–643.

do Prado Silva J T, Geiss J M T, Oliveira S M, Brum E D S, Sagae S C, Becker D, Leimann F V et al. Nanoencapsulation of lutein and its effect on mice's declarative memory. *Mater Sci Eng C Mater Biol Appl.* 2017; 76:1005–1011.

Economics A. Clear focus: The economic impact of vision loss in Australia in 2009. *Report for Vision 2020,* 2010: 1417–1436.

Elsaid N, Jackson T L, Elsaid Z, Alqathama A, Somavarapu S. PLGA microparticles entrapping chitosan-based nanoparticles for the ocular delivery of ranibizumab. *Mol Pharm.* 2016; 13: 2923–2940.

Farjo K M, Ma J X. The potential of nanomedicine therapies to treat neovascular disease in the retina. *J Angiogenes Res.* 2010; 2:21–35.

Feigl B, Hutmacher D. Eyes on 3D-current 3D-biomimetic disease concept models and potential applications in age-related macular degeneration. *Adv Healthc Mater.* 2013; 2: 1056–1062.

Feskanich D, Cho E, Schaumberg D A, Colditz G A, Hankinson S E. Menopausal and reproductive factors and risk of age-related macular degeneration. *Arch Ophthalmol.* 2008; 126: 519–524.

Forest D L, Johnson L V, Clegg D O. Cellular models and therapies for age-related macular degeneration. *Dis Model Mech.* 2015; 8: 421–427.

Fritsche L G, Fariss R N, Stambolian D, Abecasis G R, Curcio C A, Swaroop A. Age-related macular degeneration: Genetics and biology coming together. *Annu Rev Genom Hum G.* 2014; 15: 151–171.

Fu Z, Liegl R, Wang Z, Gong Y, Liu C H, Sun Y. Adiponectin mediates dietary omega-3 long-chain polyunsaturated fatty acid protection against choroidal neovascularization in mice. *Invest Opthalmol Vis Sci.* 2017; 58: 3862–3870.

Gopinath B, Flood V M, Rochtchina E, Wang J J, Mitchell P. Homocysteine, folate, vitamin B-12, and 10-y incidence of age-related macular degeneration. *Am J Clin Nutr.* 2013; 98: 129–135.

Group AREDSR. A randomized, placebo-controlled, clinical trial of high-dose supplementation with vitamins C and E, beta carotene, and zinc for age-related macular degeneration and vision loss: AREDS report no. 8. *Arch Ophthalmol.* 2001; 119: 1417–1436.

Group ToA-RMDWPTS. Photodynamic therapy of subfoveal choroidal neovascularization in age-related macular degeneration with verteporfin: One-year results of 2 randomized clinical trials--TAP report 1. *Arch Ophthalmol.* 1999; 117: 1329.

Gülçin İ. Antioxidant activity of food constituents: An overview. *Arch Toxicol.* 2011; 86(3):345–391.

Gul K, Tak A, Singh A K, Singh P, Yousuf B, Wani A A. Chemistry, encapsulation, and health benefits of β-carotene - A review. *Cogent Food Agric.* 2015; 1: 1–12.

Guymer R H, Chong E. Modifiable risk factors for age-related macular degeneration. *Med J Australia.* 2006; 184: 455–458.

Hahn P, Qian Y, Dentchev T, Chen L, Beard J, Harris Z L. Disruption of ceruloplasmin and hephaestin in mice causes retinal iron overload and retinal degeneration with features of age-related macular degeneration. *P Natl Acad Sci USA.* 2004; 101: 13850–13855.

Hamilton R, Foss A, Leach L. Establishment of a human *in vitro* model of the outer blood–retinal barrier. *J Anat.* 2007; 211: 707–716.

Hashimoto H, Arai K, Hayashi S, Okamoto H, Takahashi J, Chikuda M. The effect of astaxanthin on vascular endothelial growth factor (VEGF) levels and peroxidation reactions in the aqueous humor. *J Clin Biochem Nutr.* 2016; 59: 10–15.

He B, Ge J, Yue P, Gao X. Loading of anthocyanins on chitosan nanoparticles influences anthocyanin degradation in gastrointestinal fluids and stability in a beverage. *Food Chem.* 2016; 221:1671–1677.

Heier J S, Brown D M, Chong V, Korobelnik J F, Kaiser P K, Nguyen Q D. Intravitreal aflibercept (VEGF trap-eye) in wet age-related macular degeneration. *Ophthalmology.* 2012; 119: 2537–2548.

Ho L, Van Leeuwen R, Witteman J C M. Reducing the genetic risk of age-related macular degeneration with dietary antioxidants, zinc, and ω-3 fatty acids: The Rotterdam study. *Arch Ophthalmol.* 2011; 129: 758–766.

Hoffmann S, Balthasar S, Friedrichs U, Ehren M, Ryan S J, Wiedemann P. Inhibitory effects of verapamil isomers on the proliferation of choroidal endothelial cells. *Graefe's Arch Clin Exp Ophthalmol.* 2006; 244: 376–381.

Holz F G, Pauleikhoff D, Spaide R F, Bird A C eds. *Age-Related Macular Degeneration*, ISBN 978-3-642-22107-1. Springer-Verlag: New York. 2012; pp. 137–977.

Hurley S F, Matthews J P, Guymer R H. Cost-effectiveness of ranibizumab for neovascular age-related macular degeneration. *Cost Eff Resour Alloc.* 2008; 6: 1–12.

Imamura Y, Noda S, Hashizume K, Shinoda K, Yamaguchi M, Uchiyama S. Drusen, choroidal neovascularization, and retinal pigment epithelium dysfunction in SOD1-deficient mice: A model of age-related macular degeneration. *P Natl Acad Sci.* 2006; 103: 11282–11287.

Jampol LM, Tielsch J. Race, macular degeneration, and the Macular Photocoagulation Study. *Arch Ophthalmol.* 1992; 110: 1699–1700.

Jiang W, Chiou GC. Development of age-related macular degeneration experimental models. *Int J Ophthalmol.* 2009; 2: 189–194.

Jiao Y, Zheng X, Chang Y, Li D, Sun X, Liu X. Zein-derived peptides as nanocarriers to increase the water solubility and stability of lutein. *Food Funct.* 2018; 9: 117–123.

Johnson L V, Forest D L, Banna C D, Radeke C M, Maloney M A, Hu J. Cell culture model that mimics drusen formation and triggers complement activation associated with age-related macular degeneration. *Proc Natl Acad Sci.* 2011; 108: 18277–18282.

Jonisch J, Shah G. Diagnosis of age-related macular degeneration. In: *Age-Related Macular Degeneration Diagnosis and Treatment.* Springer: New York; 2011; pp. 23–38.

Kamil A, Smith D E, Blumberg J B, Astete C, Sabliov C, Chen C O. Bioavailability and biodistribution of nanodelivered lutein. *Food Chem.* 2016; 192: 915–923.

Kang G Y, Bang J Y, Choi A J, Yoon J, Lee W C, Choi S. Exosomal proteins in the aqueous humor as novel biomarkers in patients with neovascular age-related macular degeneration. *J Proteome Res.* 2014; 13: 581–595.

Kanwar J R, Shankaranarayanan J S, Gurudevan S, Kanwar R K. Aptamer-based therapeutics of the past, present and future: From the perspective of eye-related diseases. *Drug Discov Today.* 2014; 19: 1309–1321.

Kent D L. Age-related macular degeneration: Beyond anti-angiogenesis. *Mol Vis.* 2014; 20: 46.

Khan J, Shahid H, Thurlby D, Bradley M, Clayton D, Moore A. Age related macular degeneration and sun exposure, iris colour, and skin sensitivity to sunlight. *Brit J Ophthalmol.* 2006; 90: 29–32.

Kim M J, Rehman S U, Amin F U, Kim M O. Enhanced neuroprotection of anthocyanin-loaded PEG-gold nanoparticles against Aβ1-42-inducedneuroinflammation and neurodegeneration via the NF-KB /JNK/GSK3β signaling pathway. *Nanomedicine.* 2017; 13: 2533–2544.

Klein B E, Howard K P, Gangnon R E, Dreyer J O, Lee K E, Klein R. Long-term use of aspirin and age-related macular degeneration. *JAMA*. 2012; 308: 2469–2478.

Klein R, Cruickshanks K J, Nash S D, Krantz E M, Nieto F J, Huang G H. The prevalence of age-related macular degeneration and associated risk factors. *Arch Ophthalmol*. 2010; 128: 750–758.

Kliffen M, Lutgens E, Daemen M J, De Muinck E D, Mooy C M, De Jong P T. The APO* E3-Leiden mouse as an animal model for basal laminar deposit. *Brit J Ophthalmol*. 2000; 84: 1415–1419.

Kubicka-Trzaska A. Differential diagnosis of exudative age-related macular degeneration with posterior pole choroidal tumours. *Klin Oczna*. 2004; 107: 147–155.

Lacatusu I, Mitrea E, Badea N, Stan R, Oprea O, Meghea A. Lipid nanoparticles based on omega-3 fatty acids as effective carriers for lutein delivery. Preparation and *in vitro* characterization studies. *J Funct Foods*. 2013; 5: 1260–1269.

Laine M, Jarva H, Seitsonen S, Haapasalo K, Lehtinen M J, Lindeman N. Y402H polymorphism of complement factor H affects binding affinity to C-reactive protein. *J Immunol*. 2007; 178: 3831–3836.

Li W, Yanoff M, Li Y, He Z. Artificial senescence of bovine retinal pigment epithelial cells induced by near-ultraviolet *in vitro*. *Mech Ageing Dev*. 1999; 110: 137–155.

Lim L S, Mitchell P, Seddon J M, Holz F G, Wong T Y. Age-related macular degeneration. *The Lancet*. 2012; 379: 1728–1738.

Lin C H, Li C H, Liao P, Tse L, Huang W, Cheng H, Cheng Y W et al. Silibinin inhibits VEGF secretion and age-related macular degeneration in a hypoxia dependent manner through the PI-3 kinase/Akt/mTOR pathway. *Brit J Pharmacol*. 2013; 168: 920–931.

Lindner G R, Santos D B, Colle D, Moreira E L G, Prediger R D, Farina M. Improved neuroprotective effects of resveratrol-loaded polysorbate 80-coated poly(lactide) nanoparticles in MPTP-induced Parkinsonism. *Nanomed*. 2015; 10: 1127–1138.

Luttrull J K, Chang D B, Margolis B W, Dorin G, Luttrull D K. Laser resensitization of medically unresponsive neovascular age-related macular degeneration: Efficacy and implications. *Retina*. 2015; 35: 1184–1194.

Ma L, Dou H L, Wu Y Q. Lutein and zeaxanthin intake and the risk of age-related macular degeneration: A systematic review and meta-analysis. *Brit J Nutr*. 2012; 107: 350–359.

Malek G, Johnson L, Mace B, Saloupis P, Schmechel D, Rickman D, Toth C et al. Apolipoprotein E allele-dependent pathogenesis: A model for age-related retinal degeneration. *P Natl Acad Sci USA*. 2005; 102: 11900–11905.

Maltzman J S, Carmen J A, Monroe J G. Transcriptional regulation of the Icam-1 gene in antigen receptor- and phorbol ester-stimulated B lymphocytes: Role for transcription factor EGR1. *J Exp Med*. 1996; 183:1747–1759.

Marin-Castano M E, Striker G E, Alcazar O, Catanuto P, Espinosa-Heidmann D G, Cousins S W. Repetitive nonlethal oxidant injury to retinal pigment epithelium decreased extracellular matrix turnover *in vitro* and induced sub-RPE deposits I. *Invest Ophth Vis Sci*. 2006; 47: 4098–4112.

Martin D F, Maguire M G, Fine S L, Ying G S, Jaffe G J, Grunwald J E. Ranibizumab and bevacizumab for treatment of neovascular age-related macular degeneration: Two-year results. *Ophthalmol*. 2012; 119: 1388–1398.

McConnell V, Silvestri G. Age-related macular degeneration. *Ulster Med J*. 2005; 74: 82–92.

Miyake S, Takahashi N, Sasaki M, Ozawa Y. Vision preservation during retinal inflammation by anthocyanin-rich bilberry extract: Cellular and molecular mechanism. *Lab Invest*. 2012; 92: 102–109.

Mohanty C, Das M, Sahoo S K. Emerging role of nanocarriers to increase the solubility and bioavailability of curcumin. *Expert Opin*. 2012; 9: 1347–1364.

Nagai K I, Nagai N, Ohgami K, Satofuka S, Ozawa Y, Tsubota K. Macular pigment lutein is antiinflammatory in preventing choroidal neovascularization. *Arterioscler Thromb Vasc Biol*. 2007; 2555–2562.

Nagineni C N, Raju R, Krishnasai K N, Kommineni V K, Cherukuri A, Kutty R K. Resveratrol suppresses expression of VEGF by human retinal pigment epithelial cells: Potential nutraceutical for age-related macular degeneration. *Aging Dis*. 2014; 5: 88–100.

Neves A R, Queiroz J F, Reis S. Brain-targeted delivery of resveratrol using solid lipid nanoparticles functionalized with apolipoprotein E. *J Nanobiotech*. 2016; 14: 1–11.

Nicholson B P, Nigam D, Toy B, Stetson P F, Agrón E, Jacobs-El N. Effect of ranibizumab on high-speed indocyanine green angiography and minimum intensity projection optical coherence tomography findings in neovascular age-related macular degeneration. *Retina (Philadelphia, Pa)*. 2015; 35: 58–68.

Nowak J. Role of lipofuscin in pathogenesis of age-related macular degeneration (AMD). *Mag Okul*. 2005; 2: 103–114.

Olea JL, Tuñón J. Patients with neovascular age-related macular degeneration in Spain display a high cardiovascular risk. *Eur J Ophthalmol.* 2011; 22: 404–411.
Penfold P, Wen L, Madigan M, Gillies M, King N, Provis J. Triamcinolone acetonide modulates permeability and intercellular adhesion molecule-1 (ICAM-1) expression of the ECV304 cell line: Implications for macular degeneration. *Clin Exp Immunol.* 2000; 121: 458–465.
Pennington KL, DeAngelis MM. Epidemiology of age-related macular degeneration (AMD): associations with cardiovascular disease phenotypes and lipid factors. *Eye and vision* 2016; 3: 34.
Petrou A L, Petrou P L, Ntanos T, Liapis A. A possible role for singlet oxygen in the degradation of various antioxidants. A meta-analysis and review of literature data. *Antioxidants.* 2018; 7(3): 35.
Pilotto E, Guidolin F, Convento E, Spedicato L, Vujosevic S, Cavarzeran F. Fundus autofluorescence and microperimetry in progressing geographic atrophy secondary to age-related macular degeneration. *Brit J Ophthalmol.* 2013; 97: 622–626.
Puntel A, Maeda A, Golczak M, Gao S Q, Yu G, Palczewski K. Prolonged prevention of retinal degeneration with retinylamine loaded nanoparticles. *Biomaterials.* 2015; 44: 103–110.
Rakoczy P E, Zhang D, Robertson T, Barnett N L, Papadimitriou J, Constable I J. Progressive age-related changes similar to age-related macular degeneration in a transgenic mouse model. *Am J Pathol.* 2002; 161: 1515–1524.
Reeves B, Harding S, Langham J, Grieve R, Tomlin K, Walker J. Verteporfin photodynamic therapy for neovascular age-related macular degeneration: Cohort study for the UK. *Health Technol Assess.* 2012; 16: 1–200.
Ruberti J W, Curcio C A, Millican C L, Menco B P, Huang J D, Johnson M. Quick-freeze/deep-etch visualization of age-related lipid accumulation in Bruch's membrane. *Invest Ophthalmol Vis Sci.* 2003; 44: 1753–1759.
Rudolf M, Winkler B, Aherrahou Z, Doehring L, Kaczmarek P, Schmidt-Erfurth U. Increased expression of vascular endothelial growth factor associated with accumulation of lipids in Bruch's membrane of LDL receptor knockout mice. *Brit J Ophthalmol.* 2005; 89: 1627–1630.
Ryan S J. The development of an experimental model of subretinal neovascularization in disciform macular degeneration. *Trans Am Ophthalmol Soc.* 1979; 77: 707–745.
Samkoe K S, Cramb D T. Application of an ex ovo chicken chorioallantoic membrane model for two-photon excitation photodynamic therapy of age-related macular degeneration. *J Biomed Opt.* 2003; 8: 410–417.
Schachat A P, Hyman L, Leske M C, Connell A, Wu S Y. Features of age-related macular degeneration in a black population. *Arch Ophthalmol.* 1995; 113: 728–735.
Scheepens A, Tan K, Paxton J W, 2010. Improving the oral bioavailability of beneficial polyphenols through designed synergies. *Genes & nutrition*, 5(1), p.75.
Schwesinger C, Yee C, Rohan R M, Joussen A M, Fernandez A, Meyer T N. Intrachoroidal neovascularization in transgenic mice overexpressing vascular endothelial growth factor in the retinal pigment epithelium. *Am J Pathol.* 2001; 158: 1161–1172.
Scott E N, Meinhardt G, Jacques C, Laurent D, Thomas A L. Vatalanib: The clinical development of a tyrosine kinase inhibitor of angiogenesis in solid tumours. *Expert Opin Inv Drug.* 2007; 16: 367–379.
Shadforth A M, George K A, Kwan A S, Chirila T V, Harkin D G. The cultivation of human retinal pigment epithelial cells on Bombyx mori silk fibroin. *Biomaterials.* 2012; 33: 4110–4117.
Sharma N, Goswami U C. Functioning of lycopene in mammalian system: A Review. *Proc Zool Soc.* 2011; 64: 1–7.
Shen Y, Cao B, Snyder NR, Woeppel K M, Eles J R, Cui X T. ROS responsive resveratrol delivery from LDLR peptide conjugated PLA-coated mesoporous silica nanoparticles across the blood-brain barrier. *J Nanobiotechnology.* 2018; 16: 13–27.
Smith L, Wesolowski E, Mclellan A, Kostyk S K, D'amato R, Sullivan R. Oxygen-induced retinopathy in the mouse. *Invest Ophthalmol Vis Sci.* 1994; 35: 101–111.
Smith W, Assink J, Klein R, Mitchell P, Klaver C C, Klein B E. Risk factors for age-related macular degeneration: Pooled findings from three continents. *Ophthalmol.* 2001; 108: 697–704.
Söderstjerna E, Bauer P, Cedervall T, Abdshill H, Johansson F, Johansson U E. Silver and gold nanoparticles exposure to *in vitro* cultured retina–studies on nanoparticle internalization, apoptosis, oxidative stress, glial-and microglial activity. *PloS One.* 2014; 9: e105359.
Soubrane Daguet G, Risard-Gasiorowski S, Massamba N. Evolving knowledge in pharmacologic treatments of age-related macular degeneration. *Dev Opthalmol.* Karger Publishers. 2016; 55: 1–6.
Sunkireddy P, Kanwar R K, Ram J, Kanwar J R. Ultra-small algal chitosan ocular nanoparticles with iron-binding milk protein prevents the toxic effects of carbendazim pesticide. *Nanomedicine.* 2016; 11: 495–511.

Surh J, Decker E A, McClements D J. Utilisation of spontaneous emulsification to fabricate lutein-loaded nanoemulsion-based delivery systems: Factors influencing particle size and colour. *Int J Food Sci Tech.* 2017; 52: 1408–1416.

Tamjidi F, Shahedi M, Varshosaz J, Nasirpour A. Stability of astaxanthin-loaded nanostructured lipid carriers in beverage systems. *J Sci Food Agric.* 2018; 98: 511–518.

Tan J S L, Wang J J, Flood V, Rochtchina E. Dietary antioxidants and the long-term incidence of age-related macular degeneration: The blue mountains eye study. *Ophthalmol.* 2008; 115: 334–341.

Tanaka N, Ikawa M, Mata N L, Verma I M. Choroidal neovascularization in transgenic mice expressing prokineticin 1: An animal model for age-related macular degeneration. *Mol Ther.* 2006; 13: 609–616.

Taylor H R, Tikellis G, Robman L D, Mccarty C A, Mcneil J J. Vitamin E supplementation and macular degeneration: Randomised controlled trial. *Brit Med J.* 2002; 325: 11.

Tomany S C, Wang J J, Van Leeuwen R, Klein R, Mitchell P, Vingerling J R. Risk factors for incident age-related macular degeneration: Pooled findings from 3 continents. *Ophthalmology.* 2004; 111: 1280–1287.

Toto L, Borrelli E, Di antonio L, Carpineto P, Mastropasqua R. Retinal vascular plexuses' changes in dry age-related macular degeneration, evaluated by means of optical coherence tomography angiography. *Retina (Philadelphia, Pa).* 2016; 36: 1566–1572.

Vinding T, Appleyard M, Nyboe J, Jensen G. Risk factor analysis for atrophic and exudative age-related macular degeneration. *Acta Ophthalmol.* 1992; 70: 66–72.

Wang A L, Lukas T J, Yuan M, Du N, Tso M O, Neufeld A H. Autophagy and exosomes in the aged retinal pigment epithelium: Possible relevance to drusen formation and age-related macular degeneration. *PloS One.* 2009; 4: e4160.

Wang F, Rendahl K G, Manning W C, Quiroz D, Coyne M, Miller S S. AAV-mediated expression of vascular endothelial growth factor induces choroidal neovascularization in rat. *Invest Ophthalmol Vis Sci.* 2003; 44: 781–790.

Wang Q, Zhao Y, Guan L, Zhang Y, Dang Q, Dong P. Preparation of astaxanthin-loaded DNA/chitosan nanoparticles for improved cellular uptake and antioxidation capability. *Food Chem.* 2017; 227: 9–15.

Whiting P, Kerby J, Coffey P, Da Cruz L, Mckernan R. Progressing a human embryonic stem-cell-based regenerative medicine therapy towards the clinic. *Phil Trans R Soc B.* 2015; 370: 20140375.

Yanoff M, Jay SD. *Ophthalmology*, 4th edition. Chapter 6.28. Elsevier; 2013. pp. 580–599.

Yonekawa Y, Wolfe J D. Monitoring age-related macular degeneration through an implantable miniature telescope using ultra–wide-field imaging. *JAMA Ophthalmol.* 2015; 133: e1588–e1588.

Zhao C, Cheng H, Jiang P, Yao Y, Han J. Preparation of lutein-loaded particles for improving solubility and stability by polyvinylpyrrolidone (PVP) as an emulsion-stabilizer. *Food Chem.* 2014; 156: 123–128.

Zhu W, Wu Y, Xu D, Li Y H, Jun B, Zhang X L. Aspirin use and risk of age-related macular degeneration: A meta-analysis. *PLoS One.* 2013; 8: 1–7.

13 Chitosan-Based Nanocarriers *A Promising Delivery System for Bioactives*

Sumit Sharma, V.R. Sinha, Amita Sarwal, and Rahul Shukla*

CONTENTS

13.1 INTRODUCTION

For a long time, polymers of natural origin have been utilized as pharmaceutical aids for formulation of several dosage forms and delivery systems. After cellulose, chitin is the next most abundantly found natural polysaccharide. Chitin was originally discovered from a mushroom in 1811 by Professor Henri Braconnot, a French scientist. Chitin has been attracting interest as a drug carrier due to its biodegradability, easy availability, biocompatibility, and mucoadhesive properties, though its use has been limited due to its poor aqueous solubility.

Chitin is a polysaccharide, which resembles cellulose in plants and collagen in the advanced animal kingdom. The cell walls of plants contain cellulose, while shells of insects and crustaceans contain chitin. Chemically, chitin is poly (1–4)-N-acetyl-d-glucosamine and can be prepared from different natural sources, though it is commercially obtained mainly form crab and shrimp shells. Depending upon its source, chitin is classified into three forms: the α-form is extracted from shrimp shells and crab; the β-form is extracted from mollusks; and the γ-form is extracted from cocoon fibers of the *Ptinus* beetle. Nitrogen content in chitin and chitosan amounts to around 6%–7% and 7%–9.5%, respectively. (Bernkop-Schnurch and Dunnhaupt, 2012).

Chitosan is obtained by deacetylation of chitin, where *N*-acetylglucosamine units are specifically deacetylated. Fungus and crustaceans are general sources of chitin. Possessing a positive charge, the latter biological polymer contains β-1,4 linkages. Chitosan get dissolved in aqueous acidic solutions, when the degree of deacetylation is about 50%. This happens because of the formation

* Corresponding author.

of polyelectrolyte due to protonation of the $-NH_2$ function on the C-2 position of thed-glucosamine repeat unit (Rinaudo, 2006).

Chitosan is a nontoxic and biocompatible natural polymer. It acquires gel-like consistency when it absorbs water and shows mucoadhesive characteristics on coming in contact with a mucosal membrane. This mucoadhesive property is due to the ionic interactions between the positive charge of amine units in chitosan and negative charge of mucin. As a consequence, chitosan intercalates within glycoprotein chains of mucous (Felt et al., 1998). Chitosan possesses good bioadhesive characteristics, which leads to increased drug bioavailability due to reduction in clearance of drug from the nasal cavity (Illum et al., 2001). Rocha et al. (2002) reported improved permeation of acidic molecules compared with basic molecules loaded into chitosan-based films.

Although various synthetic polymers (poly [lactic-co-glycolic acid] and polycaprolactone) and lipids are extensively studied for drug delivery as well as for other biomedical applications, such as scaffolds and prosthetics, but high cost of these polymers limits their use industrially to quite an extent (Lohan et al., 2013; Sinha et al., 2016; Kaur et al., 2017). Chitosan and its derivatives show wide applications in nanomedicine, artificial skin, drug delivery, pharmaceutical excipients, and tissue engineering (Sinha et al., 2004). Chitosan and its derivatives are widely used commercially in water-treatment plants, the paper industry, and the textile industry (Kumar, 2000). The further sections herein discuss chitosan as a potential nanocarrier for local delivery of challenging bioactives in detail. Also, the studies exhibiting the characteristic properties of chitosan have been described, thus inferring it as an apt polymer for being exploited as a nanocarrier in various domains of drug delivery.

13.2 CHITOSAN AS A NANOCARRIER FOR LOCAL DELIVERY

13.2.1 Site-Specific Drug Delivery to the Colon

Colonic delivery is intensively studied by many investigators for various plant bioactives—such as curcumin, proteins and peptides—for peroral treatment of local pathologies like colon cancer, pseudo-membranous colitis, and inflammatory bowel disorder. Various naturally occurring polysaccharides and synthetic polymers have been studied for colon delivery. Chitosan and its chemical modifications are intensively studied for colon targeting due to its good biocompatibility, yielding nontoxic products on biodegradation (Omid et al., 2018; Woraphatphadung et al., 2018).

The colon comprises a thick layer of mucosal membrane *vis-à-vis* other sites of gastrointestinal (GI) tract. Chitosan has a well-known mucoadhesive property, by virtue of which it adheres to the intestinal mucosa for longer periods. Moreover, chitosan has a property to open the tight junctions of epithelial membrane of the intestine temporarily and act as permeation enhancers for poorly permeable drugs. These are the reasons why chitosan is the first choice of polymer to be used for colon targeting. Chitosan nanoparticles with size range below 50 nm have been reported to get transported through the paracellular pathway and endocytosis. In addition, it has been reported in literature that chitosan also exhibits permeation enhancing property by opening the tight junctions of cell membrane reversibly. Also, studies report that the particulate carriers of chitosan with size range around 100 nm have higher deposition index in inflamed colonic tissues (Sharma and Sinha, 2018a). Hence, coupling nanotechnology with this promising polymer with unique properties for colon delivery adds additional advantages in view of specificity and therapeutic efficacy. Moreover, among various pharmaceutical techniques, chitosan nanoparticles can be delivered directly at distal parts of the intestine by incorporating particles into a hard gelatin capsule. The nanosystem-loaded hard gelatin capsules can further be dip-coated using appropriate polymeric solutions for site-specific delivery, particularly at the colonic site, as described in a study conducted by Sharma and Sinha (2018b).

Vancomycin, a glycopeptide antibiotic, is used to treat severe intestinal infections like pseudo-membranous colitis and enterocolitis caused by pathogenic bacteria like *Clostridium difficile* and *Staphylococcus aureus*. The problem associated with this drug is its acidic degradation in the stomach and hence lesser availability of active drug at the infection site, that is, ileo-caecal and colonic regions. Chitosan nanoparticles have demonstrated promising colonic delivery with improved antibacterial activity on *Staphylococcus aureus*. The smaller size of chitosan particles provides a larger surface area and better mucoadhesion due to improved electrostatic interactions of sulfonic acid and sialic acid of mucin and amino groups of chitosan. It has also been observed that nanoparticles obtained from spray-drying are of spherical shape and smooth architecture with greater encapsulation efficiency compared with particles obtained from the freeze-drying process. *In vitro* release studies revealed that chitosan nanoparticles prepared by ionic gelation with molar ratio of chitosan:tripolyphosphate as 8:1.5 and processed by spray-drying, had the ability to control the release of vancomycin. The percentage fractional amount released at pH 7.4 was found to be approximately 35% after 6 h, whereas it was approximately 4% at pH 2.0 (Cerchiara et al., 2015).

Further *in vivo* studies indicated difficulty to control burst release of drugs in the stomach and target colon because of high solubility of chitosan in gastric fluids, resulting in reduced colloidal stability at low pH. Therefore, to ameliorate the chitosan-based delivery system for colon targeting, investigators have attempted to cross-link chitosan with aldehydes such as glutaraldehyde in order to make chitosan acid insoluble. This approach further did not show effective colon targeting or a controlled-release profile. Makhlof et al. (2011) investigated pH-sensitive chitosan/hydroxypropyl methylcellulose phthalate (HPMCP) nanoparticles using insulin as the model drug. Chitosan nanoparticles formulated by ionic gelation method with different cross-linkers;that is, tripolyphosphate (TPP) and HPMCP were evaluated for colloidal stability (in 0.1M HCl) and drug release. HPMCP is an acid-insoluble polymer, which was used with the aim to prevent chitosan nanoparticles from dissociating at stomach pH. *In vitro* release studies have shown that chitosan/HPMCP nanoparticles had better colloidal stability in 0.1M HCl and release approximately 25% of insulin in 6h. Insulin release from chitosan/HPMCP nanoparticles was suggested to be governed by diffusion mechanism from pH-sensitive chitosan/HPMCP matrix, unlike in other polysaccharide conjugates, where release is due to dissolution or erosion of polymer. Chitosan/HPMCP nanoparticles showed approximately four-fold higher mucosal adhesion compared with chitosan/TPP nanoparticles, as evidenced by fluorescent labelling of chitosan and visualization under fluorescent microscopy. This shows the versatile property of chitosan, which enables fabricating compatible conjugates with various polymers.

13.2.2 Nasal Delivery

Nasal delivery is an alternate novel approach for circumventing first-pass metabolism, for dose reduction, potentiation of therapeutic efficacy and direct nose-to-brain delivery. As measured by computed tomography, the total surface area of the human nasal cavity is found to be approximately 150 cm^2 and volume is about 13 mL. The mucosal layer is replenished in about 10–20 minutes in normal humans, though the anterior nasal mucosa remains as such for long periods of time, resulting in viscous mucus gel in respiratory airways (Cole et al., 1999). Chitosan nanoparticulate systems have been successfully investigated to deliver various anti-migraine drugs, neurotransmitters, and antipsychotic drugs through the nasal route (Baltzley et al., 2014; Sharma et al., 2014; Barbi et al., 2015). Chitosan has always been a polymer of choice for nasal delivery due to its excellent mucoadhesive property, which ensures the residence time at the site. The nano-size of chitosan particulate below 100 nm as a carrier shows good transcellular absorption, whereas particles with less than 15 nm are reported to follow the paracellular pathway for absorption (Sharma et al., 2012). Furthermore, the positive charge on chitosan has an important role for mucoadhesion- and permeation-enhancing effects through the mucosal layer of the nasal membrane. In one study, it was found that the drug molecule with molecular weight of maximum of 50 KDa can easily be transported through the nasal paracellular pathway (Costantino et al., 2007). Also, by altering the tight junctions of nasal epithelial membrane with permeation enhancers, it was possible to translocate large molecules of up to 150 KDa. Chitosan has affinity to interfere with protein kinase C activity, which alters the functionality of tight junction of membrane proteins, zonal occludens complexes, and claudins, thus facilitating paracellular transport. A detailed mechanism of potential permeability of chitosan through the epithelial layer has been explained by Smith et al. (2005).

In a study by Wang and associates (2006), estradiol complexes with β-cyclodextrin were studied for direct nose-to-brain delivery by exploiting the potential of chitosan nanoparticles. *In vivo* studies evidently suggest that intranasal administration of chitosan nanoparticles (size range 265–274 nm with surface charge of about 25.0 mV) in rats have significantly enhanced availability of estradiol in cerebrospinal fluid *vis-à-vis* intravenous administration. This explains that chitosan nanoparticles result in better direct nose-to-brain delivery of estradiol, when administered intranasally compared with intravenously.

The pH of nasal mucosa in normal individuals is around 5.5–6.5, whereas it increases up to 7.2–8.3 in conditions such as rhinitis (England et al., 1999). Chitosan shows solubility at pH below 5.6, that is, selectively in acidic medium. For nasal drug delivery, therefore, a polymer should be soluble in nasal pH and simultaneously should have good mucoadhesive properties to provide good residence time by preventing the bioactive from predominant mucocilliary clearance. In this regard, chitosan is generally quaternized with methyl groups and synthesized as N,N,N-trimethyl chitosan, exhibiting superior solubility in basic and neutral pH.The nasal route has gained substantial interest in delivery of the hydrophilic drugs directly into the brain and provides a distinct advantage of bypassing the blood–brain barrier (BBB). Therefore, nasal route is an optimum route for certain peptides that are unable to cross the BBB or those degraded by enzymes in the GI tract. However, nasal secretions comprise a variety of peptidase enzymes, which are responsible for degradation of peptide molecules, and are delivered through the nasal route. Leucine-enkephalin (Leu-Enk), for instance, has nonaddictive opioid analgesic activity, but has limited absorption from the mucosal membrane and hence restricts the therapeutic application of this bioactive. Hence, considering all these factors, Leu-Enk (a neuropeptide)-loaded in N-trimethyl chitosan nanoparticles with size range of 443 ± 23 nm and zeta potential of $+15 \pm 2$ mV, were evaluated for brain delivery through the nasal route. N-trimethyl chitosan provided a distinctly better solubility and protection from enzyme degradation in nasal mucosa. Mucoadhesion of N-trimethyl chitosan ensures prolonged contact time and further nano-size ensured increased surface area with higher local drug concentration in a limited area of drug absorption in the nasal region. *Ex vivo* studies on porcine nasal mucosa supported the fact of chitosan nanoparticles as permeation enhancers by showing the gradual increase in cumulative amount of Leu-Enk permeated through the mucosal membrane compared with free Leu-Enk. Furthermore, microscopic examination of brain tissues has shown significant fluorescence with labelled drug (using 4-fluoro-7-nitrobenzofurazan) delivered through N-trimethyl chitosan nanoparticles, whereas no fluorescence was seen when drug was administered in buffer. One of the important reasons suggested by the study for better brain distribution of Leu-Enk is the hydrophilic nature of N-trimethyl chitosan, which prevents the phagocytic uptake of nanoparticles and ensures increased uptake into the brain (Kumar et al., 2013). However, no direct evidence-based detailed mechanism has been explained by the investigators.

13.2.3 Dentistry

Dental caries is the most common dental pathologic condition with which a large population is suffering, worldwide. Human oral microbiota consists of around 700 different bacteria and is not gender-specific. Certain bacteria, for example, *Streptococci*, have been identified that cause dental caries and can live in the acidic environment. Sugar digestion in the mouth yields lactic acid as a by-product, responsible to provide the favorable environment for the growth of pathogenic bacteria in the oral cavity. Many studies have evaluated and reported the antibacterial activity of chitosan along with its potential of bioadhesion and biocompatibility for oral application. The major mechanism postulated by researchers for the antibacterial activity of chitosan is the ability of chitosan nanoparticles to reduce the acid-tolerance of *S. mutans* and hence reduction in its growth. Moreover, chitosan nanoparticles cause leakage of bacterial cellular contents due to electrostatic interactions of positive charge of chitosan and negative charge of the bacterial cell membrane (Shahidi et al., 1999; Kim and Shin, 2013).

Aliasghari et al. (2016) have evaluated the potential of nano-chitosan as a bioactive for the antibacterial activity against four commonly known caries-causing pathogenic bacteria, that is,*Streptococcus species*: *S. salivarius, S. sanguis, S. sobrinus,* and *S. mutans*. It has been observed that chitosan nanoparticles have better performance with respect to inhibition-zone diameter and minimum-inhibitory and minimum-bactericidal concentrations. The potential reason explained for this significant antibactericidal activity compared with chitosan solution is ascribed to the nano-size, resulting in greater affinity towards cells. Although adherence between chitosan solution and chitosan nanoparticles had no significant difference on the listed bacteria, the authors suggested further detailed experimental studies to reveal the exact mechanism(s) and effectiveness of chitosan nanoparticles with respect to antigrowth activity.

13.2.4 Ocular Delivery

Conventional routes of drug delivery for the treatment and management of the diseased eye are hindered by poor intraocular penetration and/or rapid ocular elimination. In the conditions like diabetic retinopathy, retinoblastoma, retinitis pigmentosa, and choroidal neovascularization, the physiology of the eye is the main obstacle to deal with for better drug absorption. Therefore, in this context, nanotechnology has been the potential paradigm adopted for the treatment and management of eye disorders with special consideration to the factors such as increased contact time and drug permeation. Trimethylated derivative of chitosan (TMC) exhibits excellent mucoadhesive characteristics at neutral or basic pH. Moreover, tear fluid contains lysozyme, which is responsible for chitosan degradation and thus allows controlled release from the chitosan matrix. TMC degradation occurs at physiological pH of the eye, which allows designing of controlled release systems for ocular drug delivery. Chitosan has known properties to modulate the tight junctions between the cells and hence to promote drug absorption. Mucoadhesion of chitosan being greater in neutral or alkaline solutions is the prime reason that chitosan-based nanoparticles enhance biodistribution and enhance retention time onto eye. De Campos et al. (2001) have developed cyclosporine-loaded chitosan nanoparticles for ocular administration using an ionic gelation method, with the average size of chitosan nanoparticles as 273 nm with +37 mV surface charge. *In vivo* studies in rabbits showed higher drug concentration in the cornea and conjunctiva compared with drug suspension and drug suspension in chitosan solution up to 48 h. The proposed reason for enhanced retention was hypothesized as the interaction of negatively-charged, extra-ocular tissues with cationic polymer, that is, chitosan. Further, the nanoparticles of this cationic polymer ensure better distribution in the eye *vis-à-vis* chitosan solution.

In another study, freeze-dried chitosan nanoparticles coated with hyaluronic acid were developed for topical ocular delivery of dexamethasone with size range of around 400 nm. Hyaluronic acid was used to increase the mucoadhesive property of chitosan nanoparticles. The aim of using nanoparticles was to increase the contact time and surface area for attaining superior biodistribution. The large surface area of nanoparticles and inherent properties of chitosan as a permeation enhancer make it obligatory to use these as an ocular drug-delivery system. Further, pH, clarity, refractive index, surface tension, and viscosity are the critical parameters of the suspensions of dexamethasone–chitosan nanoparticles for ocular delivery, and were found promising for ocular use in terms of increasing precorneal residence and reducing clearance (Kalam, 2016).

Apart from increased residence time, mucoadhesion, and significantly higher biodistribution onto the eyes, the chitosan nanoparticles must have been evaluated for cytotoxicity, corneal irritation, appropriate particulate size range with low eye irritation and permeability coefficient. It was da Silva et al. (2015), who reported the in vitroevaluation of chitosan nanoparticlesagainst human corneal cell line and retina pigment epithelium for permeability coefficient and cytotoxicity. The chitosan nanoparticles developed were in the size range of 200–300 nm with zeta potential within the range of +20–+30 mV. The study suggested that hydrophilic drug (i.e., rosmarinic acid)-loaded chitosan nanoparticles as a promising delivery system for ocular application in oxidative eye conditions. It was reported that the nanoparticles ≤200 nm have the potential to permeate into the extra-ocular tissue and have an enhanced virtual half-life (Kompella et al., 2013). Chitosan derivatives are effective nanocarriers for ocular release of timolol. Drug nano encapsulation was performed via ionic cross-linking and gelation of the used carriers (Siafaka et al., 2015).

Transcorneal penetration is considered as the key challenge for efficacious ocular delivery of hydrophilic drugs. One of the most promising nanotechnological approaches with chitosan, therefore, studied is the development of nanostructured lipid carriers (NLCs), coated with chitosan oligosaccharides. Delayed clearance was observed using gamma scintigraphy, revealing bioadhesion of NLCs, coated with chitosan oligosachharide. Enhanced transcorneal penetration, that is, a 2.4-fold increase with respect to the reference, was achieved using the ocular drug-delivery system described earlier (Luo et al., 2011).

13.3 CHITOSAN AS A NANOCARRIER FOR VARIOUS NATURAL AND SYNTHETIC BIOACTIVES

13.3.1 Chitosan as a Nanocarrier for Plant Bioactives

Symbiosis between nanostructured delivery and plant bioactives is a stimulating technology. Chitosan has shown promising results as a nanocarrier for plant extracts and plant bioactive molecules. Table 13.1 encompasses some of the recent successful studies of chitosan-loaded plant bioactives in the field of nanomedicine. Plant bioactives are generally large, complex molecules, by virtue of which they exhibit low aqueous solubility and permeability. Mostly, the plant bioactives are known to have anticancer or antioxidant activity, for example, rosmarinic acid, ferulic acid, epigallo-catechin-3-gallate, etc. In this context, chitosan is one of the potential natural polymers for nanostructured delivery of various complex biomolecules resulting in improved bioavailability of such plant biomolecules. This is due to the presence of surface positive-charge amine groups on chitosan and transforming effects on cell membrane proteins such as protein kinase C, zonal occludens-1, and F-actin. Figure 13.1 depicts various mechanisms by which drug-loaded chitosan nanocarriers facilitate drug absorption. Owing to their membrane interaction and nano-size, the chitosan-based drug carriers promote cellular internalization of complex plant bioactives and act as absorption enhancers.

13.3.2 Chitosan as a Nanocarrier for Anticancer Molecules

Chemotherapy has always been a necessary and successful treatment for therapeutic management or eradication of cancer. Despite its immense potential, it is associated with certain drawbacks such as unbearable toxicity, limited access of anticancer bioactives to cancer cells, and nonspecific drug targeting to abnormal cells. Considering all these factors, chitosan nanoparticles are believed to be a potential candidate for effective treatment of cancer with localized delivery and enhanced retention time.

Chitosan and its derivatives have been documented to possess a bioadhesive property, which is due to the presence of positive charge on protonation of amino groups. The remarkable bioadhesion can be explained due to interaction between negative charge on the surface of cells and positive charge on chitosan, which generates a voltage gradient that enhances metastasis of cancerous cells. Moreover, chitosan has the ability to bind cancer cells and triggers the lymphatic system, which increases the phagocytic activity in cancerous cells and causes cell death. Studies have suggested the use of low molecular weight chitosan for effective antineoplastic activity because of better penetration and absorption. Further, the stimulation of definite metallo-proteins is hindered by inactivation of vascular endothelial growth factor.

TABLE 13.1
Various Instances of Advancement in Delivery of Plant Bioactives Employing Chitosan Nanocarriers

Plant Bioactives/ Extract	Use	Nanocarriers	Findings	References
Ferulic acid	Anticancer	Low molecular weight chitosan (85% deacetylated) nanoparticles fabricated with tripolyphosphate in the ratio of 5:1 by ionic gelation method	Enhanced cytotoxicity of ferulic acid-loaded nanoparticles against ME-180 cervical cell lines compared with native ferulic acid at the same concentration of 40μM.	Panwar et al. (2016)
Rosmarinic acid	Antioxidant	Low molecular weight chitosan nanoparticles fabricated with tripolyphosphate in the ratio of 7:1 by ionic gelation method	Protection from degradation by biological enzymes. Enhanced bioavailability compared with crude rosmarinic acid.	da Silva et al. (2015)
Turmeric oil	Anticancer	Chitosan-alginate nanocapsules in the ratio of 2:1 with an average size of 256 nm	Nanocapsules loaded with turmeric oil (0.4 mg/mL) showed 40% decline in A549 human lung adenocarcinoma epithelial cell lines viability, as evident by MTT assay, compared with free oil.	Natrajan et al. (2015)
Lemon grass oil	Anticancer	Chitosan-alginate nanocapsules in the ratio of 2:1 with an average size of 226 nm.	Nanocapsules loaded with turmeric oil (0.4 mg/mL) showed 20% decline in A549 human lung adenocarcinoma epithelial cell lines viability, as evident by MTT assay, compared withfree oil.	
Epigallo-catechin-3-gallate	Antioxidant	Epigallocatechin-3-gallate loaded chitosan–caseinophosphopeptide nanoparticles with average size of 143 nm.	Enhanced bioabsorption of epigallocatechin-3-gallate, when administered orally in chitosan–caseinophosphopeptide nanoparticles. Also, it showed significant enhanced antioxidant activity on HepG2 cells.	Hu et al. (2013)
Ethanol extract of *Ziziphusmauritiana* leaves	Immuno-restoratory	Ethanol extract of *Ziziphusmauritiana* leaves loaded chitosan nanoparticles.	Stable chitosan nanoparticles provide protection to bioactives against elevated pH of GI tract.	Bhatia et al. (2011)

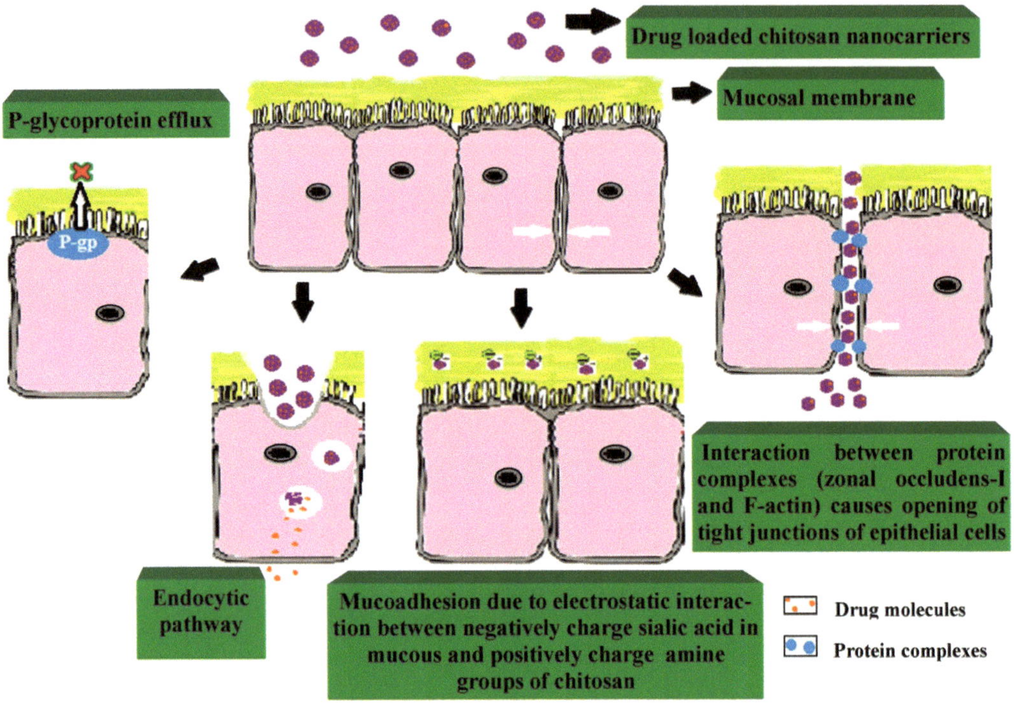

FIGURE 13.1 Various possible mechanisms for enhanced drug absorption usingchitosan nanocarriers.

Chitosan offers wide-ranging advantages in the formulation of nanostructured drug-delivery systems for cancer. Biocompatibility of chitosan makes it a potential biopolymer to form conjugated drugs having specific release characteristics. These are the reactive amino groups in the chemical structure of chitosan, which allow conjugation of certain bioactives with chitosan. Furthermore, nanoparticles have the ability to alter the biodistribution in cancerous tissues and pharmacokinetic release profile of drugs. For instance, nanoparticles of a chitosan-containing dextran–doxorubicin conjugate were administered in circulation. The developed chitosan nanoparticles have prolonged residence time in circulation due to the nano-size and enhanced hydrophilicity using dextran, which makes nanoparticles more stable in the biological milieu. Enhanced permeability and retention effect of these chitosan nanoparticles shows increased residence time and selective accumulation within the cancerous cells. Reduced side-effects, generally associated with doxorubicin, along with enhanced therapeutic efficacy, is evident from decreased tumor size and increased survival rates after 90 days, compared with groups containing doxorubicin and dextran–doxorubicin conjugate (Mitra et al., 2001).

Advancement in nanotechnology has explored delivery of nanoparticles at tumor sites, with the aim of increased retention time specifically for the antineoplastic drugs with a narrow therapeutic window and serious systemic side effects. For instance, gemcitabine, a nucleoside analogue, which acts by inhibiting DNA synthesis and causing apoptosis, is reported to exhibit antitumor activity in various clinical studies. Despite its antitumor efficacy, the drug is associated with systemic side effects, a short biological half-life of approximately 17 minutes, and poor cellular uptake. Parsian et al. (2016) has developed chitosan-coated iron oxide nanoparticles (CsMNPs), which gained attention with their biocompatibility, biodegradability, low toxicity, and target ability into tumor tissue under a magnetic field. Gemcitabine in chitosan magnetic nanoparticles increases the anticancer efficacy of the drug. Magnetic nanoparticles were synthesized by *in situ* precipitation of ferrous and ferric salts, and drug was loaded into the nanoparticles. Magnetic nanoparticles were coated with chitosan, which prevents their agglomeration. Moreover, it has been evidently shown that low molecular weight chitosan produces comparatively smaller nanoparticles with a porous and loose internal structure, allowing greater drug loading. The release of gemcitabine from CsMNPs was found to be higher at pH 4.2, which ensures maximum drug availability inside the tumor cells. This is attributed to pH-dependent decomposition of chitosan at acidic pH in tumor tissue.

13.3.3 Chitosan-Based Nanostructured Delivery for Gene Therapy, DNA, and si-RNA Carriers

Chitosan is cytocompatible with both animals and humans. It effectively binds with DNA present in saline or acetic acid solution, and thus protects DNA to some extent from nuclease degradation. This prevents the delivery of genes or other genetic materials to the desired area. One of such examples is delivery of insulin via the buccal route. Oral, nasal, and intramuscular routes are suggested as suitable for delivery of proteins, peptides, and genetic material by forming complexes with chitosan. *In vitro* and in vivogene transfer potential has been clearly observed in the case of chitosan-based DNA carriers. The ability of positive charge on chitosan helps in designing polyplexes with DNA-carrying negatively-charged phosphate groups.

Zhao et al. (2011) prepared chitosan nanoparticles by complex coacervation method, where chitosan solution in sodium acetate at pH 5.5 and plasmid solution in sodium sulfate with nitrogen:phosphorus (N:P) ratio 7:1 were heated separately at 55°C. After heating, both mixtures were vortexed in equal volumes and kept at room temperature for 1h. Advantages of this method for the preparation of plasmid-loaded chitosan nanoparticles are no need of organic solvents, no harsh conditions such as high frequency sonication, protection of genetic material from degradation, and ease of methodology. Clinical studies reported that DNA encapsulated with chitosan nanoparticles shows a new prospective for enhancing the therapeutic efficiency of DNA vaccines. Incorporation of chitosan-encapsulated DNA into the affected cells is directly linked to the N:P ratio, deacetylation and chain length of chitosan. Moreover, salt concentration used to form the complex, type of cell transfected, and its duration and time from transfection to harvest of the protein product are also important parameters to be considered. Study has suggested that N:P ratio has an important role in governing the stability of nanoparticles in terms of aggregation. Increase in N:P ratio results in more ionic charge on nanoparticles, which consequently results in higher attraction, causing instability. Hence, N:P ratio of 3 was optimized for stable nanoparticles. Immunological effectiveness of swine influenza DNA vaccine, encapsulated in chitosan nanoparticles, showed that cell-mediated humoral immunity was significantly enhanced by pDNA- chitosan nanoparticles, and a prolonged release profile was observed in BALB/c mice. Moreover, a stability study by agarose gel electrophoresis has revealed that chitosan nanoparticles were successfully able to protect the plasmid DNA from digestion by DNase 1. All these findings describe the feasibility of chitosan nanoparticles in effective delivery of genetic material and potentiate the immunological effectiveness of swine influenza DNA vaccine on immunization of mice. Further studies are recommended for ideal immunological effect with reduced inoculating frequency and vaccination dose.

In a recent study, chitosan is evaluated as a triblock copolymer micelle based on N-succinyl chitosan-poly-L-lysine-palmitic acid with an average size of 179 nm and zeta potential of +3.2 mV to co-deliver doxorubicin and si-RNA for effective treatment of cancer. Rapid degradation by RNase enzyme shortens the half-life and reduces renal clearance, and the poor cellular uptake (due to large molecular weight and negative charge) limits the therapeutic efficacy of si-RNA in the treatment of cancer. Herein, N-succinyl chitosan was used to provide hydrophobicity to the nanomicellesso as to prevent si-RNA degradation and enhanced cellular uptake. Moreover, the ability of triblock polymeric micelle to encapsulate both doxorubicin and si-RNA prevents multidrug resistance (MDR) by reducing doses of chemotherapeutic agents, increasing half-life with selective cellular localization, and bypassing the P-gp efflux pump. As discussed earlier, N:P ratio holds an important role in gene delivery through polymeric carrier, and here in this study, the ratio was determined as 20 for formation of stable nanomicelles. Confocal laser scanning microscopy (CLSM) results have revealed successful internalization of doxorubicin into the nucleus and si-RNA in the cytoplasm to interrupt the mRNA transcription. Authors have proved the effectiveness of N-succinyl chitosan-doxorubicin–si-RNA-nanomicelles for tumor-targeted delivery, higher antitumor activity and as a promising approach for cancer treatment that develops MDR (Zhang et al., 2016).

Novel chitosan-based nanoparticles loaded in liposomes have been pharmaceutically designed to deliver a plasmid into the living tissues. Size of the chitosan nanoparticles was optimized up to 65 nm, loaded with plasmid, and then incorporated into the lipid-based system in the form of liposomes, which results in the increased size up to 100 nm. Potential of transfection of plasmid through developed particulate delivery system was proved by *in vivo* studies on chorioallantoic membrane. Compared with chitosan nanoparticles, the authors confirmed better transfection of plasmid through enhanced expression of colored fluorescence protein, which was complexed with plasmid using CLSM. The theory suggested that increased transfection into the fibroblasts through liposomes loaded with chitosan nanoparticles was a consequence of the lipophilicity imparted by the lipid, as chitosan otherwise requires certain chemical modifications in order to increase the transfection efficiency under normal physiological conditions. Also, no damage on microvasculature of blood vessels and no sign of necrosis were observed, which could be due to decreasing the cationic charge on chitosan nanoparticles, rendering them more biocompatible (Baghdan et al., 2018).

13.4 SUMMARY

Chitosan is the most versatile natural polymer for controlled delivery used in oral, ocular, nasal, pulmonary, transdermal, and parenteral systems. In the present chapter, chitosan application with respect to formulation of delivery systems as nanocarrier of many bioactives has been discussed. Versatility of this polymer is clearly evident from its broad applications, ranging from its use as a supplement in weight gain to a successful drug carrier in biomedicine. Chitosan has been largely used for poorly soluble drugs, thereby increasing their dissolution rate and bioavailability. The wide-ranging applications of chitosan as nanocarriers for drug delivery are primarily attributable to its safe use as excipients in controlled release and targeting. With studies still being in progress within the twenty-first century, it is expected that chitosan along with its derivatives will hit the market predominating myriad medical applications.

In this review, many instances of the potential benefits of chitosan nanoparticles for efficient delivery and thus treatment of various ocular and nasal diseases have been described, both *in vitro* and *in vivo*, with enormouspotential of clinical translation. Solubility issues of chitosan were overcome by its quaternized hydrophilic derivative, that is, TMC, the inherent properties of which—such as its ability to degrade, cellular compatibility, adhesion, and wide pH range of solubility—make it a preferred polymer to be used in various pharmaceutical applications. Superior properties of TMC have staggered the researchers for formulating nanoparticles, and thus providing head over other materials.

For the last few decades, chitosan nanoparticles have been used in various research fields for enhancing absorption, tissue regeneration, and delivering drugs, genes, proteins, and peptides or vaccines. The instability and erratic absorption of chitosan nanoparticles warrants the modification of its surface. The certainty of chemical modifications in the structure of chitosan makes it potential candidate for development of chitosan-based nanoparticles. Modulating on the surface of nanoparticles with hydrophilic TMC results in improved aqueous stability and enhanced mucoadhesion and permeation characteristics of the nanoparticles. In addition, the feasibility of conjugating a hydrophobic ligand on the backbone of chitosan produces self-assembled nanoparticles with high encapsulation efficiency for poorly-soluble drugs. Grafting the targeting ligand on the chitosan nanoparticles permits active targeting again.

13.5 FUTURE PERSPECTIVES

Chitosan has proved its applications as a promising biomaterial in various pharmaceutical fields, including controlled release, protein delivery, and as an antibacterial agent. On the other hand, chitosan as a biopolymer poses a number of challenges that need to be resolved for its improved applications. One of the promising approaches exploited to potentiate and increase the utility of chitosan is enabled using nanotechnology. Chitosan as a nanobiomaterial has proved its application in effective site-specific drug targeting, as a drug carrier for various high molecular molecules, as permeation enhancer, and beyond. Despite numerous studies evidently establishing chitosan applicability as a biopolymer carrier in various pharmaceutical applications, there is still exhaustive research required to put it into a commercially viable form. More studies are required to establish the efficacy of nano-chitosan in the wound-healing processes, to supplement the data explaining the exact mechanisms for antibacterial effects of chitosan and the ability to deliver the genetic material. Additionally, the effective synthesizing methods for chemical modification and conjugation with improved desired physicochemical characteristics need to be elucidated. Further, *in vivo* studies have already established the utility of chitosan nanoparticles in diverse clinical applications in the future, particularly in the domains of cancer treatment, dentistry and topical delivery of various bioactives. This decade has witnessed extensive research on the development of chitosan-based nanoparticulate delivery systems, while the next decade is likely to translate the scientific investigations into clinical outcomes.

REFERENCES

Aliasghari A, Rabbani KM, Vaezifar FS, Rahimi H, Younesi, Khoroushi M et al. Evaluation of antibacterial efficiency of chitosan and chitosan nanoparticles on cariogenic streptococci: An *in vitro* study. *Iran J Microbiol.* 2016; 8: 93–100.

Baghdan E, Pinnapireddy SR, Strehlow B, Engelhardt KH, Schäfer J, Bakowsky U et al. Lipid coated chitosan-DNA nanoparticles for enhanced gene delivery. *Int J Pharm.* 2018; 535: 473–9.

Baltzley S, Mohammad A, Malkawi AH, Al-Ghananeem AM. Intranasal drug delivery of olanzapine-loaded chitosan nanoparticles. *AAPS PharmSciTech.* 2014; 15: 1598–602.

Barbi MS, Carvalho FC, Kiill CP, Barud HS, Santagneli SH, Ribeiro SJ et al. Preparation and characterization of chitosan nanoparticles for zidovudine nasal delivery. *J Nanosci Nanotechnol.* 2015; 15: 865–74.

Bernkop-Schnurch A, Dunnhaupt S. Chitosan-based drug delivery systems. *Eur J Pharm Biopharm.* 2012; 81: 463–9.

Bhatia A, Shard P, Chopra D, Mishra T. Chitosan nanoparticles as carrier of immunorestoratory plant extract: Synthesis, characterization and immunorestoratory efficacy. *Int J Drug Delv.* 2011; 3: 381–5.

Cerchiara T, Abruzzo AM, di Cagno, Bigucci F, Bauer-Brandl A, Parolin C et al. Chitosan based micro and nanoparticles for colon-targeted delivery of vancomycin prepared by alternative processing methods. *Eur J Pharm Biopharm.* 2015; 92: 112–9.

Cole AM, Dewan P, Ganz T. Innate antimicrobial activity of nasal secretions. *Infect Immun.* 1999; 67: 3267–75.

Costantino HR, Illum L, Brandt G, Johnson PH, Quay SC. Intranasal delivery: Physicochemical and therapeutic aspects. *Int J Pharm.* 2007; 337: 1–24.

da Silva SB, Ferreira D, Pintado M, Sarmento B. Chitosan-based nanoparticles for rosmarinic acid ocular delivery *In vitro* tests. *Int J BiolMacromolec.* 2015; 84: 112–20.

De Campos AM, Sanchez A, Alonso MJ. Chitosan nanoparticles: A new vehicle for the improvement of the delivery of drugs to the ocular surface. Application to cyclosporin A. *Int J Pharm.* 2001; 224: 159–68.

England RJ, Homer JJ, Knight LC, Ell SR. Nasal pH measurement: A reliable and repeatable parameter. *Clin Otolaryngol Allied Sci.* 1999; 24: 67–8.

Felt O, Buri P, Gurny R. Chitosan: A unique polysaccharide for drug delivery. *Drug Dev Ind Pharm.* 1998; 24: 979–93.

Hu B, Ting Y, Zeng X, Huang Q. Bioactive peptides/chitosan nanoparticles enhance cellular antioxidant activity of (-)-epigallocatechin-3-gallate. *J Agric Food Chem.* 2013; 61(4): 875–81.

Illum L, Jabbal-Gill I, Hinchcliffe M, Fisher AN, Davis SS. Chitosan as a novel nasal delivery system for vaccines. *Adv Drug Deliv Rev.* 2001; 51: 81–96.

Kalam MA. Development of chitosan nanoparticles coated with hyaluronic acid for topical ocular delivery of dexamethasone. *Int J Biol Macromolec.* 2016; 89: 127–36.

Kaur M, Sharma S, Sinha VR. Polymer based microspheres of aceclofenac as sustained release parenterals for prolonged anti-inflammatory effect. *Mater Sci Eng C Mater Biol Appl.* 2017; 72: 492–500.

Kim JS, Shin DH. Inhibitory effect on *Streptococcus mutans* and mechanical properties of the chitosan containing composite resin. *Restor Dent Endod.* 2013; 38(1): 36–42.

Kompella UB, Amrite AC, Pacha RR, Durazo SA. Nanomedicines for back of the eye drug delivery, gene delivery, and imaging. *Prog Retin Eye Res.* 2013; 36: 172–98.

Kumar M, Pandey RS, Patra KC, Jain SK, Soni ML, Dangi JS et al. Evaluation of neuropeptide loaded trimethyl chitosan nanoparticles for nose-to-brain delivery. *Int J Biol Macromolec.* 2013; 61: 189–95.

Kumar MNVR. A review of chitin and chitosan applications. *React Funct Polym.* 2000; 46: 1–27.

Lohan S, Sharma S, Murthy RSR. Formulation and evaluation of solid lipid nanoparticles of quetiapine fumarate and quetiapine hemifumarate for brain delivery in rat model. *Pharm Nanotechnol.* 2013; 1: 239–47.

Luo Q, Zhao J, Zhang X, Pan W. Nanostructured lipid carrier (NLC) coated with chitosan oligosaccharides and its potential use in ocular drug delivery system. *Int J Pharm.* 2011; 403: 185–91.

Makhlof A, Tozuka Y, Takeuchi H. Design and evaluation of novel pH-sensitive chitosan nanoparticles for oral insulin delivery. *Eur J Pharm Sci.* 2011; 42: 445–51.

Mitra S, Gaur U, Ghosh PC, Maitra AN. Tumour targeted delivery of encapsulated dextran-doxorubicin conjugate using chitosan nanoparticles as carrier. *J Control Release.* 2001; 74: 317–23.

Natrajan D, Srinivasan S, Sundar K, Ravindran A. Formulation of essential oil-loaded chitosan–alginate nanocapsules. *J Food Drug Anal.* 2015; 23(3): 560–8.

Omid NJ, Javan NB, Dehpour AR, Partoazar A, Tehrani MR, Dorkoosh F et al. *In-vitro* and *in-vivo* cytotoxicity and efficacy evaluation of novel glycyl-glycine and alanyl-alanine conjugates of chitosan and trimethyl chitosan nano-particles as carriers for oral insulin delivery. *Int J Pharm.* 2018; 535: 293–307.

Panwar R, Sharma AK, Kaloti M, Dutt D, Pruthi V. Characterization and anticancer potential of ferulic acid-loaded chitosan nanoparticles against ME-180 human cervical cancer cell lines. *Appl Nanosci.* 2016; 6(6): 803–13.

Parsian M, Unsoy G, Mutlu P, Yalcin S, Tezcaner A, Gunduz U et al. Loading of Gemcitabine on chitosan magnetic nanoparticles increases the anti-cancer efficacy of the drug. *Eur J Pharmacol.* 2016; 784: 121–8.

Rinaudo M. Chitin and chitosan: Properties and applications. *Prog Polym Sci.* 2006; 31: 603–32.

Rocha ANL, Dantas TNC, Fonseca JLC, Pereira MR. Permeation of drugs in chitosan membranes. *J Appl Polym Sci.* 2002; 84: 44–9.

Shahidi F, Arachchi JKV, Jeon YJ. Food applications of chitin and chitosans. *Trends Food Sci Technol.* 1999; 10(2): 37–51.

Sharma S, Lohan S, Murthy RSR. Potential of chitosan for nose-to-brain drug delivery. *Int J Pharm Sci Rev Res.* 2012; 16: 47–55.

Sharma S, Lohan S, Murthy RSR. Formulation and characterization of intranasal mucoadhesive nanoparticulates and thermo-reversible gel of levodopa for brain delivery. *Drug Dev Ind Pharm.* 2014; 40: 869–78.

Sharma S, Sinha VR. Current pharmaceutical strategies for efficient site-specific delivery in inflamed distal intestinal mucosa. *J Control Release.* 2018a; 272: 97–106.

Sharma S, Sinha VR. Liquid nanosize emulsion-filled enteric-coated capsules for colon delivery of immunosuppressant peptide. *AAPS PharmSciTech.* 2018b; 19: 881–5.

Siafaka PI, Titopoulou A, Koukaras EN, Kostoglou M, Koutris E, Karavas E et al. Chitosan derivatives as effective nanocarriers for ocular release of timolol drug. *Int J Pharm.* 2015; 495: 249–64.

Sinha VR, Sharma S, Silki, Kaur M, Sarwal A. *Nanoarchitectonics for Smart Delivery and Drug Targeting.* The Boulevard, Langford Lane, Kidlington, Oxford, OX5 1GB, Matthew Deans: United Kingdom; 2016; pp. 143–68.

Sinha VR, Singla AK, Wadhawan S, Kaushik R, Kumria R, Bansal K et al. Chitosan microspheres as a potential carrier for drugs. *Int J Pharm.* 2004; 274: 1–33.

Smith JM, Dornish M, Wood EJ. Involvement of protein kinase C in chitosan glutamate-mediated tight junction disruption. *Biomaterials.* 2005; 26: 3269–76.

Wang X, He H, Leng W, Tang X. Evaluation of brain-targeting for the nasal delivery of estradiol by the microdialysis method. *Int J Pharm.* 2006; 317: 40–6.

Woraphatphadung T, Sajomsang W, Rojanarata T, Ngawhirunpat T, Tonglairoum P, Opanasopit P et al. Development of chitosan-based pH-sensitive polymeric micelles containing curcumin for colon-targeted drug delivery. *AAPS PharmSciTech.* 2018; 19: 991–1000.

Zhang CG, Zhu WJ, Liu Y, Yuan ZQ, Yang SD, Chen WL et al. Novel polymer micelle mediated co-delivery of doxorubicin and P-glycoprotein siRNA for reversal of multidrug resistance and synergistic tumor therapy. *Sci Rep.* 2016; 6: 238–59.

Zhao K, Shi X, Zhao Y, Wei H, Sun Q, Huang T et al. Preparation and immunological effectiveness of a swine influenza DNA vaccine encapsulated in chitosan nanoparticles. *Vaccine.* 2011; 29(47): 8549–56.

14 Emergence of Nanovesicular Systems for Topical Delivery of Natural Bioactives

*Gajanand Sharma, Kanika Thakur, Akanksha Mahajan, Guneet Singh Randhawa, Bhupinder Singh, and Om Prakash Katare**

CONTENTS

14.1 INTRODUCTION

Natural bioactives consist of physiologically active substances obtained from natural sources that exert therapeutic effect upon the physiology of the organism. Bioactives comprise a varied group of substances including lipids, flavors, anti-infectives, antioxidants, tannins, carbohydrates, and plant extracts (Ubbink and Kruger, 2006; McClements *et al.*, 2007). Despite the availability of numerous delivery routes, delivery of bioactives still faces major hurdles, which is attributed to their unfavorable properties, such as high molecular mass, high rate of degradation, short plasma half-life, and tendency to form aggregates (Singh *et al.*, 2013, 2015). Topical delivery systems offer a promising solution to bioactive delivery across dermal strata with improved bioavailability and targeting efficacy.

Skin acts as a natural barrier in healthy and diseased states to the external stimuli, along with the presence of hair follicles and trans-barrier pathways. Stratum corneum (SC), the outermost, toughest skin layer, poses a formidable challenge to delivery of bioactives by the topical route (Desai *et al.*, 2010). The very efficient barrier function exerted by SC inhibits the passage of a vast majority of

* Corresponding author.

molecules across the skin layers, and hence, reduces their therapeutic efficacy. As a consequence, the development of topical delivery systems with improved permeation across skin barriers has been considered as one of the most daunting tasks by the pharmaceutical industry. The past few decades have witnessed a rapid surge in the application of nanotechnology in drug delivery that has made its mark in many fields, including dermal and transdermal delivery of cosmetic and pharmaceutical products (Puglia and Bonina, 2012). The overall objective of the current chapter is to provide an inclusive overview of the vesicular drug carriers for topical delivery of bioactives.

14.2 TOPICAL DELIVERY OF BIOACTIVES: POTENTIAL AND CHALLENGES

14.2.1 Physiology of Skin

Skin is the largest organ in the body with excellent biological barrier properties, contributing to around 4% to the body weight (Cevc and Vierl, 2010). Basically, skin consists of four layers, as depicted in Figure 14.1. SC, the outermost layer of the skin exposed to the external environment, which comprises keratin-rich cells, the keratinocytes, surrounded by a mixture of ceramides, fatty acids, and cholesterol, principally acts as the rate-controlling barrier (Gray *et al.*, 1982; Wertz *et al.*, 1989). The next layer is the dermis, with a thickness of about 0.1–0.5 cm, consisting of collagenous and elastin fibers. The covalent linkage of mucopolysaccharides with peptide chain results in the formation of proteoglycans, which, in turn, promotes the elasticity of skin. It contains an extensive vascular network required to provide nutrition and repair mechanism to the skin (Burgeson and Christiano, 1987). The underlying layer of the skin barrier is the hypodermis, consisting of fibrous connective tissue with fat and elastic fibers (Kanikkannan *et al.*, 2000). The fat cells are arranged in lobules within the hypodermis, which act as heat insulators. The hypodermis is connected to the dermis, further by collagen and elastin fibers.

14.2.2 Barrier Functions of Stratum Corneum (SC) and Penetration Pathways Into Skin

SC is the major obstructive barrier for topical delivery, which reduces dermal passage of varied active molecules across the skin layers (Wertz *et al.*, 1989; Raza *et al.*, 2014; Sharma *et al.*, 2017a). The literature reports a number of approaches to enhance the permeation of molecules across the skin barrier, including ultrasound waves, radiofrequency, iontophoresis, magnetophoresis, electroporation, and microneedles (Iqbal *et al.*, 2018). Despite their numerous advantages, these approaches have found limited application owing to their toxicity and lack of therapeutic feasibility. The potential role of vesicular carriers (e.g., liposomes, ethosomes, and transfersomes) for topical delivery has been reported in several studies (Katare *et al.*, 2010; Sharma *et al.*, 2016a,b, 2017b).

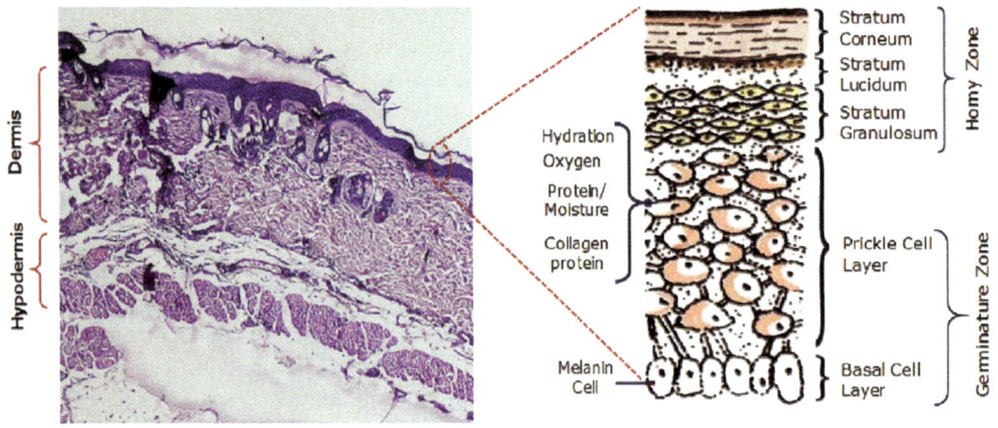

FIGURE 14.1 Physiology of normal skin delineating its vital layers.

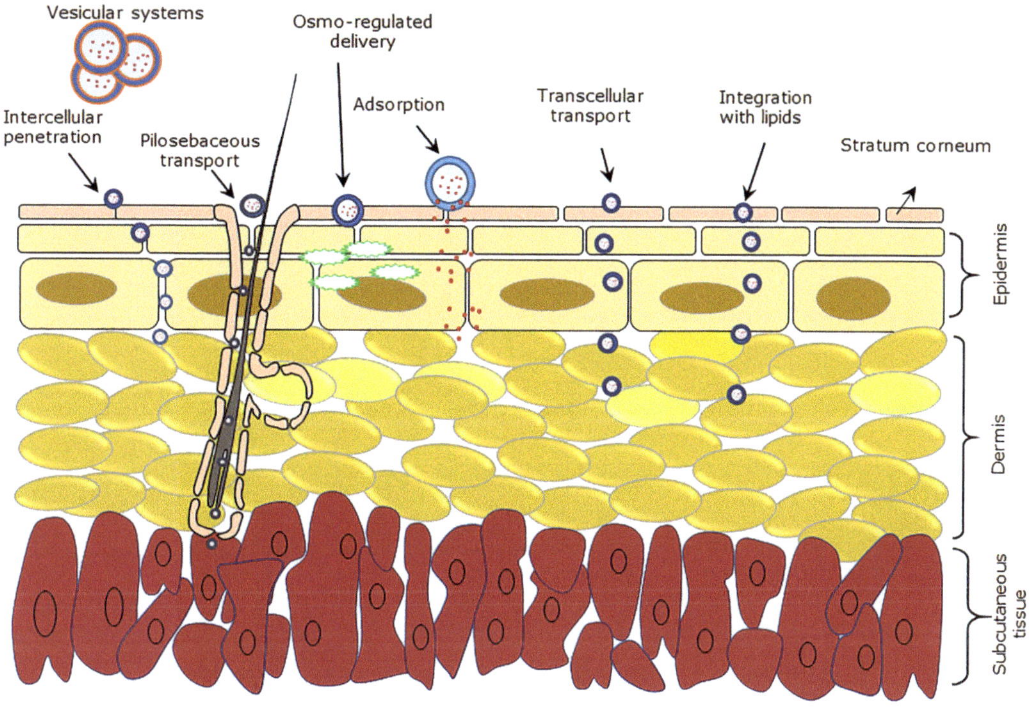

FIGURE 14.2 Various routes of penetration of vesicular systems across skin.

A number of drug molecules predominantly penetrate the skin by passing the horny layer, specifically using the pathways between the keratin cells. The physicochemical properties of drug molecules also dictate their penetration across the skin barrier. Changes in skin integrity in many disease conditions (e.g., atopic eczema and psoriasis) favor permeation of drug molecules through the SC. Penetration through the lipid layers of SC has been the primary pathway for a number of topically applied active molecules.

The molecules of varied size, including vesicles, can penetrate into the skin through different penetration pathways: the cellular pathway through the lipid layers, the transcellular pathway across the keratin cells, pilosebaceous-assisted pathway across hair follicles and sweat glands, osmo-regulated delivery through moisture cloud, adsorption through SC by diffusion, and integration with lipids (Desai *et al.*, 2010; Raza *et al.*, 2014). Figure 14.2 provides a panoramic view illustrating all such plausible pathways for penetration of bioactives across the skin tissue. Studies have also reported that the transappendageal route is one of the most efficient pathways for dermal permeation (Lademann *et al.*, 2001). Literature reports (Desai *et al.*, 2010) also suggest that hair follicles can also act as a pathway, thus enhancing the absorption of topically applied molecules. Besides, hair follicles can also serve as a center for the formation of a depot, thus adding to the storage capabilities of SC.

14.3 ADVENT OF VESICULAR SYSTEMS FOR THE TOPICAL DELIVERY OF BIOACTIVES

Vesicular systems have surfaced as prospective carriers for the controlled topical delivery of bioactives. The numerous advantages of vesicular drug-delivery systems include nontoxicity, biocompatibility, encapsulation of hydrophilic and hydrophobic moieties, and prolongation of the residence time of the drug (Jain *et al.*, 2005). Application of vesicles for topical delivery is attributed to their dual drug-carrier properties, that is, to deliver active moieties across the skin strata and to act as penetration enhancers concurrently (Pirvu *et al.*, 2010). The vital lipidic vesicular carriers include

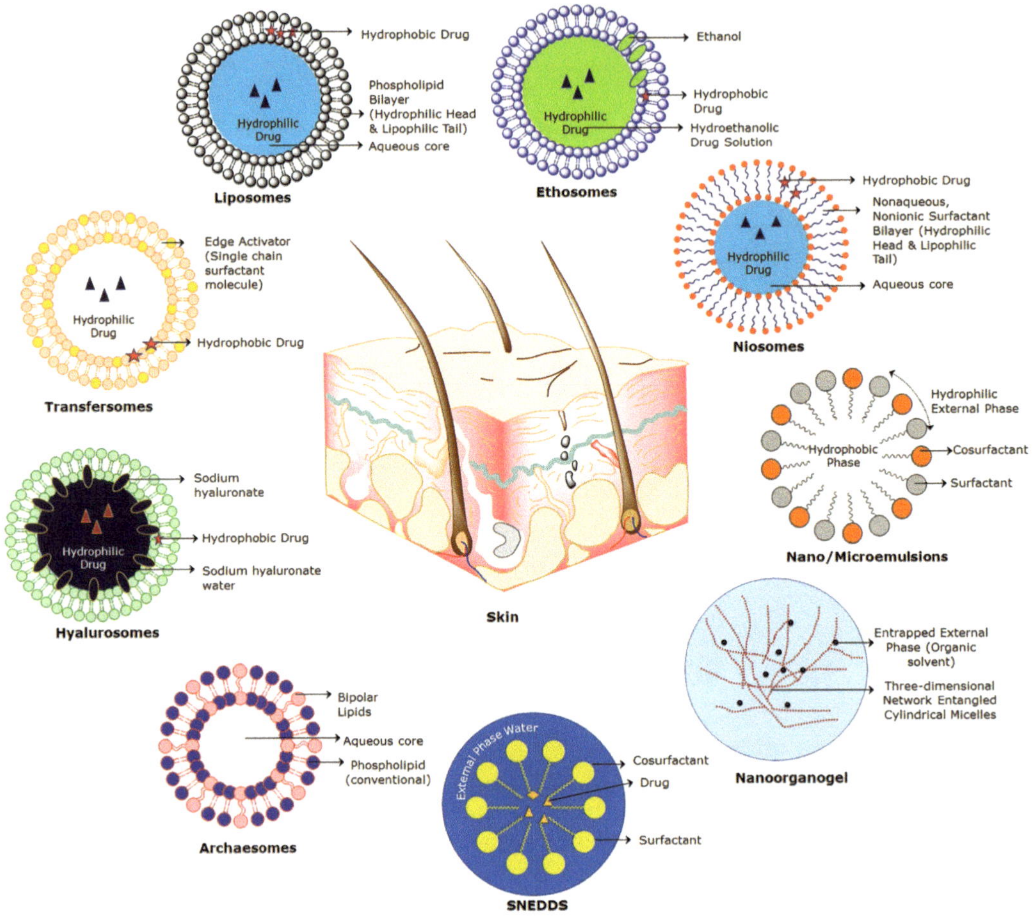

FIGURE 14.3 Principle types of vesicular systems employed for topical delivery.

liposomes, niosomes, transferosomes, ethosomes, and hyalurosomes, as portrayed in Figure 14.3. The varied lipid-based vesicular carriers, along with their merits and limitations, have been discussed in Table 14.1, while Table 14.2 provides a bird's eye account of various plant bioactives incorporated in different vesicular carriers, along with their composition.

14.3.1 Liposomes

Liposomes are the vesicles, composed of one or more lipid bilayers surrounding an aqueous layer containing hydrophilic and hydrophobic moieties (Singh *et al.*, 2015). The potential advantages of liposomes for topical delivery of bioactives include their biodegradability and nontoxic nature. Liposomes are versatile drug carriers with an ability to encapsulate hydrophilic and hydrophobic bioactives, ascribed to the amphiphilic properties of phospholipids. In particular, the liposomes have been studied extensively for topical delivery of lipophilic compounds, as they allow improved drug permeation and retention within the skin strata (Pierre and Dos Santos Miranda Costa, 2011; Sharma *et al.*, 2017b; Hatahet *et al.*, 2018). Liposomes also influence permeation across dermal strata by causing disruptions in the lipid arrangement of SC and promoting impaired barrier function, eventually leading to enhanced partitioning of drugs across skin layers (El Maghraby *et al.*, 2008; Prasanthi and Lakshmi, 2012; Franze *et al.*, 2018). As indicated earlier, a number of literature studies have reported the application of conventional various types of vesicular systems for topical/dermal delivery of various bioactives (Table 14.2).

TABLE 14.1
List of Vesicular Carriers Used for Topical Delivery of Bioactives

S. No.	Vesicular Carrier	Advantages	Disadvantages	References
1.	Liposomes	Biodegradability, nontoxicity, ability to encapsulate bioactives with different solubilities, permeation enhancing effect, avoidance of local and systemic effects	Leakage of entrapped drug molecule, fusion of vesicles leading to aggregation, stability issues of phospholipid due to oxidation and hydrolysis	Singh *et al.* (2015), Plangsombat *et al.* (2016), Sharma *et al.* (2017b), Kianvash *et al.* (2017), Doppalapudi *et al.* (2017)
2.	Transfersomes	Ability to encapsulate hydrophilic and lipophilic molecules, enhanced penetration across the skin barrier due to self-adaptability and high deformability, which would normally be a barrier for other particles of similar size	Expensive, toxicity issues of edge activators, stability problems due to oxidation and hydrolysis	Cosco *et al.* (2008), Raza *et al.* (2011), Sharma *et al.* (2016b)
3.	Ethosomes	Better penetration enhancing ability through the stratum corneum barrier, higher encapsulation efficiency for both hydrophobic and hydrophilic drugs, fluidity, malleability, and elasticity	Skin irritation due to excipients, stability problems due to oxidation and hydrolysis	Touitou *et al.* (2000), Sharma *et al.* (2016b)
4.	Niosomes	Biodegradability, biocompatibility, and nonimmunogenicity to the body, controlled sustained release of drugs due to depot formation	Low entrapment efficiency, physical instability	Manosroi *et al.* (2013), Singh *et al.* (2015), Yadav *et al.* (2016)
5.	Hyalurosomes	Avoidance of leakage, enhanced permeation capacity, mechanical resistance to the presence of sodium hyaluronate	Leakage of entrapped drug molecule, fusion of vesicles leading to aggregation, stability issues of phospholipid due to oxidation and hydrolysis	Manca *et al.* (2015), Castangia *et al.* (2016)

Raza *et al.* (2011) incorporated capsaicin (CAP) within the lipid bilayers of flexible membrane vesicles to improve its therapeutic efficacy in the treatment of pain-related muscle disorders. The CAP-loaded vesicles offered higher skin retention, thus proving the potential of liposomes for topical delivery. Similarly, CAP-loaded liposomal systems were also prepared by Gupta *et al.* (2016) and the skin retention studies revealed significantly higher retention across skin strata. Thus, both the studies concluded that liposomal delivery systems have immense potential to improve the therapeutic effectiveness through the topical delivery of CAP.

Chen *et al.* (2014) developed conventional liposomes of catechin for management of skin ailments and skin-aging conditions, and using porcine skin barrier, concluded that liposomes demonstrated highly significantly enhanced skin permeation and retention of catechin *vis-à-vis* catechin solution.

TABLE 14.2
List of Phytobioactive Molecules Incorporated in Different Vesicular Carriers, Along with their Lipid Core Composition

Vesicular Delivery System	Plant Bioactive	Composition	Remarks	References
Liposomes	Papain	Soy phosphatidylcholine, cholesterol	Papain-loaded liposomes reduced the hypertrophic scar and hence vesicular carrier proved as an excellent topical preparation for hypertrophic scar treatment.	Chen *et al.* (2017)
	Resveratrol with Psoralen	Cholesterol, sodium deoxycholate	RES with the aid of deformable liposomal carrier can be a potential therapeutic strategy for vitiligo treatment.	Doppalapudi *et al.* (2017)
	Curcumin and STAT3 siRNA	1, 2-dioleoyl-3-trimethylammonium propane, 1, 2-dioleoyl-sn-glycero-3-phospho-ethanolamine, C6 ceramide, and sodium cholate	Cationic liposomes can be utilized for topical iontophoretic co-delivery of curcumin and siRNA for effective treatment of skin diseases.	Jose *et al.* (2018)
	Quercetin and Resveratrol	Oleic acid, 12-O-tetradecanoylphorbol 13-acetate, Lipoid S75	Polyphenol vesicular formulation may be of vital value in the treatment of inflammatory and cancerous skin lesions.	Caddeo *et al.* (2016)
	Liquorice extract	Soy phosphatidylcholine, cholesterol	Liquorice extract loaded into vesicles protected 3T3 fibroblasts against oxidative stress compared with pure extract, which showed poor antioxidant activity.	Castangia *et al.* (2015)
	Resveratrol	Lipoid S100	Liposomal RES exhibited stronger anti-oxidative and anti-inflammatory activities than RES solution.	Jøraholmen *et al.* (2015)
	Curcumin	Soya lecithin, propylene glycol, cholesterol	Curcumin-loaded nanoliposomal formulation efficiently improved burn wound infections.	Kianvash *et al.* (2017)
	Asparagus racemosus root extracts	Phosphatidylcholine, cholesterol	AR liposomes exhibited substantial anti-inflammatory activity in terms of percentage inhibition of TNF-α which decreased at higher concentrations of AR.	Plangsombat *et al.* (2016)
	Catechin	Phosphatidylcholine, Sodium deoxycholate, Cholesterol	Deformable liposomes appear to be promising carriers for enhancing dermal delivery of catechin.	Chen *et al.* (2014)
	Curcumin	Soy phosphatidylcholine, cholesterol	Liposomal delivery systems for curcumin are promising formulations for the treatment of vaginal inflammation.	Basnet *et al.* (2012)
	Capsaicin	Soy phosphatidylcholine, Span 80, cholesterol	CAP-loaded liposomes may be a potential approach for effective topical management of psoriasis.	Raza *et al.* (2011), Gupta *et al.* (2016)

(Continued)

TABLE 14.2 (*Continued*)
List of Phytobioactive Molecules Incorporated in Different Vesicular Carriers, Along with their Lipid Core Composition

Vesicular Delivery System	Plant Bioactive	Composition	Remarks	References
Transfersomes	Capsaicin	Phosphatidylcholine, Tween 80	Improved tolerance of prepared vesicular formulation supported the suitability of transfersomal vesicles for topical administration of such highly irritant substance.	Sarwa *et al.* (2015)
	Ginsenoside from Red Ginseng	Soy phosphatidylcholine, Tween 80	Transfersomes provided significantly higher skin permeation of ginsenoside compared with ethosomes and conventional liposomes, proving their therapeutic potential.	Choi *et al.* (2015)
	Lycopene	Soy phosphatidylcholine, Tween 80	Lycopene-loaded deformable carriers demonstrated significant in vivo activity by inhibition of edema formation in anthralin-induced ear-swelling model.	Ascenso *et al.* (2013)
	Linoleic acid	Phospholipon® 100 G, Linoleic acid, Sodium cholate, Linoleic acid	The vesicular carriers could represent a potential system for the topical treatment of hyperpigmentation disorders.	Celia *et al.* (2012)
Ethosomes	Apigenin	Lipoid® S 75, Propylene glycol, Ethanol	Apigenin-loaded ethosomes showed higher skin deposition compared with liposomes both *in vitro* and in vivo. Ethosomes-mediated apigenin delivery produced the strongest effect on UVB-induced skin inflammation by suppressing COX-2 levels.	Shen *et al.* (2014)
	Psoralen	Lipoid S 75, ethanol	Enhanced permeation and skin deposition of psoralen delivered by ethosomes may help reduce toxicity and improve the efficacy of long-term psoralen treatment.	Zhang *et al.* (2014)
	Tetrandrine	Propylene glycol, ethanol, phosphatidylcholine	Topical administration of tetrandrine via ethosomes might offer favorable efficacy with reduced side effects, thus leading to improved patient compliance.	Fan *et al.* (2013)
	Triptolide	Dipalmitoylphosphatidylcholine, ethanol, cholesterol	Ethosomes might be promising carriers to enable the beneficial properties of triptolide to be safely delivered in a topical formulation	Chen *et al.* (2010)
	Ammonium glycyrrhizinate	Phosphatidylcholine, ethanol	Ethosomes proved as promising carriers for topical administration of ammonium glycyrrhizinate due to its enhanced delivery through the skin.	Paolino *et al.* (2005)

(Continued)

TABLE 14.2 (*Continued*)
List of Phytobioactive Molecules Incorporated in Different Vesicular Carriers, Along with their Lipid Core Composition

Vesicular Delivery System	Plant Bioactive	Composition	Remarks	References
Niosomes	Resveratrol	Span 80, cholesterol	RES-entrapped niosomal gel sufficiently reduced the edema and significantly prolonged therapeutic action *vis-a-vis* the marketed anti-inflammatory gel formulation.	Negi *et al.* (2017)
	Boswellic acid	Lecithin, cholesterol, surfactant	BA-loaded Proniosomal gel proved better in terms of absorption, bioavailability, and release kinetics.	Mehta *et al.* (2016)
	Resveratrol	Glycerol monooleate, phospholipid 90G, polyglyceryl-3 dioleate, cholesterol	Niosomes improved both dermal and transdermal delivery of RES in comparison with conventional liposomes.	Pando *et al.* (2013)
	Papain	Tween 61, cholesterol	The study demonstrated that elastic niosomes yielded superior topical delivery of papain as well as reduction of hypertrophic scars.	Manosroi *et al.* (2013)
	Capsaicin	Phospholipid, Tween 80, and Span 80	Niosome-encapsulated CAP demonstrated higher permeation and retention across skin strata.	Tavano *et al.* (2011)
Hyalurosomes	Phycocyanin	Soy phosphatidylcholine, Sodium hyaluronate	Phycocyanin-loaded hyalurosomes proved to be a suitable formulation for skin protection in cosmetic and pharmaceutical fields.	Castangia *et al.* (2016)
	Curcumin	Soy phosphatidylcholine, sodium hyaluronate	Hyaluronan immobilized the vesicles, thereby improving entrapment efficiency, vesicle stability, rheological properties, local drug availability and formulation, therapeutic activity, leading to faster wound healing.	Manca *et al.* (2015)
	Liquorice extract	Soy phosphatidylcholine, Phospholipid 90G	The proposed vesicular systems improved the local efficacy of the extract favoring the re-epithelialization process.	Castangia *et al.* (2015)
Archaeosomes	Ovalbumin	Total Polar Archaeo lipid, oybean phosphatidylcholine, odium cholate	Archaeosomes appeared as a suitable carrier for topical delivery of antioxidant phenolic compound and ovalbumin due to their characteristics of stability, entrapment efficiency, and antioxidant activity which were comparable to those obtained from classical phosphatidylcholine liposomes.	Carrer *et al.* (2014)
	Antioxidant phenolic extract obtained from olive mill waste	Archaeal lipids, soy phosphatidylcholine, cholesterol		Gonzalez-Paredes *et al.* (2011)

(*Continued*)

TABLE 14.2 (*Continued*)
List of Phytobioactive Molecules Incorporated in Different Vesicular Carriers, Along with their Lipid Core Composition

Vesicular Delivery System	Plant Bioactive	Composition	Remarks	References
Microemulsion	Curcumin and emu oil	Cremophor RH 40, Cremophor EL, Labrafil M2125CS, Labrafil M1944CS, Span 20, Span 80	The study demonstrated the potential of ME formulations containing curcumin and emu oil combination in the treatment of rheumatoid arthritis.	Jeengar *et al.* (2016)
	Kojic dipalmitate	Span 80, Tween 20, Liquid petrolatum	ME-based system may be a promising platform for the topical application of KDP in the treatment of skin disorders.	Goncalez *et al.* (2015)
	Boswellia carterii	Mygliol 812, Capmul MCM C8, Tween 80, Labrasol, Transcutol	The emulsion loaded formulation prolonged the anti-inflammatory activity that would serve as a better alternative for oral anti-inflammatory, avoiding the undesirable side effects on the gastrointestinal tract.	Mostafa *et al.* (2015)
	Glycyrrhiza glabra	Oleic acid, Ethanol, Tween 60	ME formulation could be effectively be used as a noninvasive transdermal formulation possessing highly potent antioxidant activity.	Mostafa *et al.* (2014)
	Hesperetin	Span 20, Span 80, Tween 80, Isopropyl myristate, Propylene glycol, Polyethylene glycol 400	Hesperetin-loaded ME showed significant topical whitening effect and diminished skin irritation when compared with the nontreatment group, indicating that it could be used as an effective whitening agent.	Tsai *et al.* (2010)
	Lycopene	BRIJ-propylene glycol, mono/diglycerides of capric and caprylic acids or triglycerides of the same fatty acids	The Lycopene loaded ME is an efficient and safe system to increase lycopene delivery to the skin and the antioxidant activity in the tissue.	Lopes *et al.* (2010)
Nanoemulsion	*Copaiba*	Sorbitan oleate, Polysorbate 80	Development of low-cost eco-friendly green natural-based nanoformulations with potential larvicidal activity, using a nanobiotechnology approach	Escarleth da *et al.* (2014)
	Phyllanthus urinaria	Tween 80, Span 80, Cetyl alcohol	The extract was successfully incorporated into a nanoemulsion delivery system. These formulations were able to neutralize reactive oxygen species and counteract oxidative injury induced by ultraviolet radiation and thereby ameliorate skin aging.	Mahdi *et al.* (2011)

Abbreviations: AR, Asparagus acemosus; BA, boswellic acid; COX, cyclooxygenase; KDP, kojic dipalmitate; ME, microemulsion; RES, resveratrol; NF-α, tumor necrosis factor alpha; UVB, ultraviolet B rays.

Jøraholmen *et al.* (2015) developed resveratrol (RES)-loaded liposomes, coated with chitosan, for dermal treatment of vaginal inflammation and associated infections. The biochemical investigations and cytokine studies supported the enhanced antioxidative and anti-inflammatory activity of coated vesicles as compared to the drug solution. Castangia *et al.* (2015) incorporated liquorice extract, obtained from ethanolic percolation of *Glycyrrhiza glabra* L. roots into liposomes, to facilitate topical delivery. The results of *in vitro* studies revealed that liquorice extract loaded into the lipid vesicles showed superior antioxidant activity and efficiency to pure glycyrrhizin solution.

Caddeo *et al.* (2016) explored the potential of liposomes for the combination delivery of naturally occurring polyphenols, namely, quercetin and RES. Co-administration of both the polyphenols incorporated into phospholipid vesicles resulted in improved *in vitro* cellular uptake, along with a greater ability to reduce reactive radicals in fibroblasts, compared with the administration of polyphenols alone. The study gave ample evidence in support of polyphenol vesicular delivery, which may prove to be of immense value in the treatment of inflammatory disorders. Papain is a proteolytic enzyme possessing antimicrobial and antioxidant properties with wide applications in dermatology for wound healing. However, the high molecular weight of papain poses a major problem for delivery across skin strata, thus making its topical application as difficult. Chen *et al.* (2017) investigated the pharmacodynamic efficacy of papain-loaded liposomes for the management of scars through topical application. Papain-loaded into liposomes resulted in higher cumulative amounts of the enzyme *vis-à-vis* papain solution, along with greater retention in skin layers. The findings demonstrated the potential of liposomes as an efficient topical scar treatment product.

Basnet *et al.* (2012) concluded that curcumin-loaded liposomes are prospective treatment strategies for vaginal inflammation. Further, curcumin was encapsulated into propylene glycol liposomes by Kianvash *et al.* (2017) for the management of deep burns in animals. The nanoliposomal formulation effectively improved injuries and burn wound infections. Positively charged liposomes have also been explored for the topical delivery of curcumin, along with anti-STAT3 (i.e., a signal transducer and activator of transcription 3) for the treatment of skin cancer. Curcumin-loaded liposomes penetrated up to a depth of 160 μm inside the skin strata after iontophoretic application. The study concluded high potential of cationic liposomes for effective topical delivery of curcumin-anti STAT3 in the treatment of skin diseases (Jose *et al.*, 2018).

14.3.2 Transfersomes

Transfersomes are the second generation liposomes developed to overcome the limitations of traditional liposomes. They are composed of lipids, surface active agent,s and an inner aqueous compartment within the lipid bilayer, which makes them capable of entrapping hydrophilic, as well as lipophilic, moieties. Transfersomes have been proposed to penetrate across the SC barrier, ascribed to their high elasticity and deformability (Choi and Maibach, 2005; Cosco *et al.*, 2008; Benson, 2017; Almeida *et al.*, 2018; Pathak *et al.*, 2018). Studies have also reported improved topical delivery of bioactives employing transfersomes as the carriers (Chen *et al.*, 2017). Lycopene possesses significant antioxidant properties including antiproliferative effects on cancerous cells. The extract containing lycopene obtained from tomatoes, was incorporated into transfersomes by Ascenso *et al.* (2013) and further characterized for cellular uptake. The vesicular formulation exhibited enhanced cellular uptake and topical delivery of lycopene, which reduced the level of anthralin-induced ear swelling by 97% *vis-à-vis* controls.

The potential of CAP as a topical antiarthritic agent was examined by encapsulating CAP within the transfersomal vesicular system. The *in vivo* antiarthritic activity study revealed that the developed formulation performed better than the marketed gel. Moreover, better skin compatibility of prepared vesicles further confirmed the aptness of the transfersomes for topical administration of such highly irritant substances (Sarwa *et al.*, 2015). Celia *et al.* (2012) investigated transfersomes for effective topical delivery of linoleic acid (LA) in the therapeutic management of melasma diseases. Permeation studies of LA-loaded transfersomes through human SC barrier exhibited significant accumulation in the skin membrane model. The study revealed that the vesicular carriers could represent a potential delivery vehicle for the topical management of hyperpigmentation disorders. Transfersomes have also been explored as a topical vehicle for red ginseng, which is frequently employed in the treatment of cancer, inflammatory disorders, diabetes mellitus, and acquired immunodeficiency syndrome (AIDS). The transfersomes demonstrated significantly higher skin permeation of ginsenoside compared with conventional liposomes, indicating that ginsenoside-loaded transfersomes can act as efficient carriers for topical therapeutics of inflammation, diabetes, and stress (Choi *et al.*, 2015).

14.3.3 Ethosomes

Ethosomes are elastic and vesicular carriers developed for enhanced delivery of bioactives due to their deformability and ability to penetrate the human skin. In recent times, ethosomes have emerged as suitable carriers for topical drug delivery with an inherent ability for effective intracellular delivery of both hydrophilic and lipophilic drugs (Touitou *et al.*, 2000; Paolino *et al.*, 2005; Sharma *et al.*, 2016b; Candido *et al.*, 2018; Zorec *et al.*, 2018). Paolino *et al.* (2005) studied the potential application of ethosomes for topical delivery of ammonium glycyrrhizinate. Ethosomes exhibited good dermal tolerance in humans with increased dermal permeation of glycyrrhizinate. Additionally, ethosomes also improved the anti-inflammatory properties of ammonium glycyrrhizinate in comparison with the ethanolic solution of the extract (Paolino *et al.*, 2005).

Apigenin, a flavonoid, possesses varied pharmacological properties, including antioxidant and anticancer effects (Patel *et al.*, 2007; Tong *et al.*, 2007). Attempts have also been made for topical delivery of apigenin by developing binary ethosomes, consisting of ethanol and propylene glycol (Shen *et al.*, 2014). The optimized ethosomes revealed superior skin targeting potential during their *in vitro* as well as *in vivo* conditions. In addition, ethosomes demonstrated the greatest effect on reduction of cyclooxygenase-2 levels in mouse skin inflammation. The study reported the potential advantages of apigenin-loaded ethosomes as a therapeutic approach for the management of skin inflammations (Shen *et al.*, 2014).

Fan *et al.* (2013) formulated tetrandrine-loaded ethosomes for the therapeutic management of edema, rheumatic disorders and inflammatory diseases. The ethosomal carriers generated significant improvement in therapeutic efficacy of tetrandrine on Freund's complete adjuvant-induced arthritis in comparison with liposomes (Fan *et al.*, 2013). The ethosomal formulation of triptolide reported the greatest *in vitro* retention (83.7%) in 24 h and significantly reduced erythema in pharmacodynamic model compared with other formulations. Ethosomes may prove to be a promising carrier that would enable the beneficial properties of triptolide to be safely delivered in a topical formulation (Chen *et al.*, 2010). Psoralen is a photosensitive molecule employed for the management of vitiligo and psoriasis. Application of psoralen-loaded ethosomes led to a 6.56-fold greater skin retention of psoralen compared with that of tincture (Zhang *et al.*, 2014). Ethosomal-encapsulated CAP revealed greater reduction of paw edema along with antinociceptive activity. The antiarthritic efficacy of ethosomes was greater than the marketed gel of capsaicin. The ethosomal-encapsulated CAP could prove to be an effective and safe topical delivery system for the management of arthritis and associated inflammatory disorders (Sarwa *et al.*, 2015).

14.3.4 Niosomes

Niosomes are nonionic surfactant-based vesicles, which are formed by the systemic association of nonionic amphiphiles in the aqueous media. These carriers are similar to liposomes and encapsulate hydrophilic moieties to serve as the drug carriers. Literature reports in the past have concluded significant permeation and photostability of various active moieties employing niosomal vesicles along with solubilizing matrix for controlled drug release (Schreier and Bouwstra, 1994; Teixeira *et al.*, 2017; Madni *et al.*, 2018; Priprem *et al.*, 2018a,b). Manosroi *et al.* (2013) encapsulated papain into niosomal gel for burn-related infections. The topical application of niosomal gel exhibited greater reduction of hypertrophic scars on rabbit ears compared with plain gel after 28 days of topical application. The histopathological studies revealed significant reduction in collagen fibers and scar fading treated with niosomal gel. The study demonstrated the significant potential of niosomes in enhanced permeation of papain and scar reduction. Pando *et al.* (2013) showed high cutaneous deposition of RES, when incorporated into glyceryl monostearate (Peceol) niosomes.

Tavano *et al.* (2011) prepared CAP-loaded niosomes composed of Tween 80 and Span 80 for topical delivery. The results concluded that niosomes could promote the topical delivery of CAP. Similarly, Negi *et al.* (2017) prepared RES-loaded niosomal hydrogel to improve its anti-inflammatory activity. The *ex vivo* studies revealed high permeation and retention of RES in skin strata, leading ultimately to improved anti-inflammatory activity of the niosomal gel in animals.

Boswellic acid (BA) is isolated from oleo gum of *Boswellia serrata* and employed as a potential anti-inflammatory, hypolipidemic, and anticancer agent. The topical niosomal formulation of BA was developed by Mehta *et al.* (2016) to improve its poor oral bioavailability. The developed formulation revealed spherical vesicles with a particle size of 707.9 nm and high encapsulation efficiency. The anti-inflammatory activity revealed higher inhibition than the marketed gel, consequently indicating that BA-loaded niosomal gel was better in terms of absorption, bioavailability, and release kinetics.

14.3.5 Hyalurosomes

Hyalurosomes are the next generation phospholipid vesicles, prepared by mixing drug dispersion and sodium hyaluronate in water, followed by the addition of lipids (Castangia *et al.*, 2016). Sodium hyaluronate acts as a penetration enhancer and immobilizes the lipid-based vesicles, thus imparting strength and preventing leakage at the site of application. These have been explored as potential carriers to enhance the residence time of the drug, leading eventually to enhanced local bioavailability of the drug (Kong *et al.*, 2011; Witting *et al.*, 2015).

Phycocyanin is a phycobiliprotein obtained from Klamath algae, which possesses significant biological properties, but is characterized by low bioavailability. Castangia *et al.* (2016) incorporated phycocyanin into hyalurosomes and was studied for permeation and retention characteristics. The results of permeation studies concluded that hyalurosomes assisted phycocyanin deposition in the deep skin layers due to formation of hyaluronan-phospholipid structure. Moreover, hyalurosomes are biocompatible with improved antioxidant activity *vis-à-vis* the drug solution. Manca *et al.* (2015) developed curcumin-loaded vesicles employing polyanion sodium hyaluronate and phospholipids. The *in vitro* permeation studies revealed improved capacity of hyalurosomes to favor faster drug deposition, thereby promoting tissue growth through cellular proliferation and migration. The liquorice extract was incorporated into hyalurosomes and the protective effect was investigated against oxidative damage. The anti-inflammatory activity on mice proved the ability of nanosystems to improve the local efficacy of the extract, thus favoring the re-epithelialization (Castangia *et al.*, 2015).

14.3.6 Microemulsions

Microemulsions (MEs) are clear, isotropic monophasic systems, composed of two immiscible liquids and characterized by low interfacial tension. MEs have the ability to improve topical delivery of both hydrophilic and lipophilic moieties compared with the conventional vehicles (Date and Patravale, 2007; Boonme *et al.*, 2016). The presence of a permeation enhancer as a formulation component and small globule size make MEs as exceptional carriers for promoting *in vitro* skin permeation (Benigni *et al.*, 2018; Wang *et al.*, 2018). The ethanolic extract of *Glycyrrhiza glabra* root and rhizome was incorporated into the ME carrier system, and the results suggested that the MEs possesses 13-fold higher *ex vivo* antioxidant capacity than with the extract solution (Mostafa *et al.*, 2014). Hesperetin, a flavonoid, possesses anti-inflammatory, UV-protective, and antioxidant effects. Hesperetin-loaded MEs were formulated by Tsai *et al.* (2010) to overcome its permeation issues. The hesperetin-loaded

MEs exhibited enhanced skin permeation *vis-à-vis* hesperetin solution. During pharmacodynamic studies, significant topical whitening effect was observed, indicating that the hesperetin-loaded ME could be an efficient whitening agent (Tsai *et al.*, 2010).

Multiple emulsions (MLEs) have been studied as carriers to enhance the bioavailability of active lipophilic antioxidant, such as kojic dipalmitate (KDP). The results revealed that incorporation of KDP into MLEs improved its safety profile and antioxidant activity, suggesting that MLEs presented higher capacity for maintaining the antioxidant activity of KDP (Goncalez *et al.*, 2015). Lycopene was incorporated in two different MEs containing BRIJ-propylene glycol, but with different oil phases. Both the ME systems significantly enhanced lycopene penetration in the SC layers compared with control solution. The results demonstrated that monoglycerides (MGs) of capric and caprylic acids containing MEs is a potential and safe carrier to improve lycopene delivery to the skin and antioxidant activity (Lopes *et al.*, 2010).

14.3.7 Nanoemulsions

Nanoemulsions (NEs) are non-equilibrium systems, having a droplet size of 50–500 nm with the added advantage of reduced amount of surfactant employed during the preparation of NEs in comparison to MEs (Gupta *et al.*, 2012; Thakur *et al.*, 2018).

NEs, composed of palm kernel oil, were loaded with *P. urinaria* extract and were characterized in terms of particle size, zeta potential and rheological properties. The *in vitro* release of the extract from the formulation showed 2,2-diphenyl-1-picrylhydrazyl (DPPH) radical scavenging activity and counteracting oxidative injury. The anti-inflammatory property of curcumin, in combination with emu oil, was investigated by developing a nanoemulgel formulation. NEs were prepared using emu oil, Cremophor RH 40 and Labrafil M2125CS as oil phase, surfactant, and co-surfactant. The arthritic scoring, paw volume, biochemical, molecular, radiological, and histological examination revealed significant improvement in anti-inflammatory activity using NEs compared with plain solutions of both curcumin and emu oil (Jeengar *et al.*, 2016).

14.3.8 Miscellaneous: Archaeosomes, Lecithin Organogel, and Self-Emulsifying Drug-Delivery Systems

Archaeosomes are the latest generation vesicular carriers, composed of polar lipids extracted from microorganisms belonging to the domain *Archaea*. The archaeal membranes are composed of polar ether lipids with branched isoprenoid chains attached to the glycerol, whereas the conventional phospholipids are composed of unsaturated fatty acids. Due to the polar ether lipids, the archaeosomes are reported to be stable to oxidative degradation, high temperature and pH conditions, thus making them suitable alternative carriers to liposomes (Eskandar *et al.*, 2014). Gonzalez-Paredes *et al.* (2011) explored archaeosomes as a topical vehicle for natural antioxidant compounds recovered from olive mill waste. The study concluded that archaeosomes possess significant potential for topical delivery of phenolic compounds due to their inherent characteristic like improvement in stability, high entrapment efficiency and antioxidant activity.

Lecithin organogel (LO) is a suitable carrier for topical delivery of many bioactive agents used in aging treatment. Lecithin is a cellular component isolated from soy beans/eggs and purified to obtain gelation in nonpolar solvents. It acts as an organic medium to improve dermal penetration of poorly permeable drugs by effectively partitioning into the skin, thus making it a dynamic vehicle, which can be explored as a carrier for antiaging agents (Kumar and Katare, 2005; Raut *et al.*, 2012).

Quercetin, a natural antioxidant, was effectively loaded into non-aqueous self-double-emulsifying (SDEDDS) formulation for topical delivery. The antioxidant study revealed that SDEDDS exhibited much higher permeation capacity and significantly increased accumulation of quercetin in the skin tissue. These studies demonstrated that the solid non-aqueous SDEDDS might be a promising carrier for topical application of poorly soluble bioactives (Wang *et al.*, 2016).

14.4 PATENTS ON VESICULAR CARRIERS FOR TOPICAL DELIVERY OF BIOACTIVES

The literature is really scarce on the patents, which specifically use the term "topical delivery of natural bioactives" in their title. Table 14.3 furnishes a detailed account of such patents related to topical delivery of bioactives in recent years. Kurzrock *et al.* (2008) have been granted a patent in 2008 for efficient loading of curcumin, comprising a liposomal complex effective to load curcumin into the vesicular system. Roederer (2006) developed a topical pain-relief composition containing CAP or capsaicinoids to achieve controlled release over time and minimize the side effect of burning, associated with CAP. McCook *et al.* (2008) developed a topical liposomal composition containing phospholipids and one lipophilic bioactive agent. Streeper and Singh (2004) reported the method of preparation and application of a variety of pharmaceutical compositions containing digitalis glycoside such as oleandrin, odoroside-A, neriifolin, proscillaridin-A, methyl-proscillaridin-A, digitoxin, for treating skin diseases in a human or non-human animal. Kunz *et al.* (1997) developed a composition for transporting a bioactive agent across a biological membrane comprising the bioactive agent, oil, an oil-immiscible compound, and a cationic surface active agent.

14.5 CONCLUSIONS

In the last decade, several insights into the structural properties and challenges of SC encouraged the development of newer vesicular systems. The objective was to understand the mechanism(s) of interaction between the lipidic systems and SC barrier, which would ultimately lead to the development of an efficient drug-delivery system for enhanced permeation and improved therapeutic efficacy. The delivery of drug molecules through the skin using conventional formulations does not result in the permeation through the SC barrier in sufficient amounts. A number of approaches have been employed to deliver the desired bioactives at the desired site. The low solubility and high molecular weight are the major challenging factors for natural bioactives to pass across the biological membrane to be absorbed following oral or topical administration. The potential advantages of vesicular systems for topical delivery include the ability to enhance drug penetration, biodegradability, and biocompatibility, sustained release profile, miniscule vesicle size (generally in nanometric size),

TABLE 14.3
Some Patents Based on Vesicular Technology for Topical Delivery of Plant Bioactives

Patent No.	Title of the Patent	Inventors	Publication Date	Reference
US20080138400A1	Liposomal curcumin for treatment of diseases	R Kurzrock, L Li, K Mehta, B B Aggarawal, L Helson	2008–06–12	Kurzrock *et al.* (2008)
US20080107747A1	Pain relief composition	J Roederer	2008–05–08	Roederer (2006)
US20080233183A1	Topical formulations having enhanced bioavailability	J P McCook, I S Persaud, N R Narain	2008–09–25	McCook *et al.* (2008)
US20060205679A1	Topical and oral formulations of cardiac glycosides for treating skin diseases	R Streeper, C Singh	2006–09–14	Streeper and Singh (2004)
US20020169138A1	Delivery vehicles for bioactive agents and uses thereof	L Kunz, T Tice, R Libby, C McDaniel	2002–11–14	Kunz *et al.* (1997)
CN106074365A	Capsaicin ethosomes gel and preparation method thereof	–	2016–11–09	–

and improved safety. These vesicles possess a great potential in the efficient noninvasive delivery of a variety of bioactives, with poor permeation due to unfavorable physicochemical characteristics. The nanometric size of the carriers allows close contact to the skin barrier, leading to occlusion and decrease of moisture loss, thus leading to enhanced permeation or penetration. Further, simple and robust methods of preparation have enhanced the laboratory production and scale-up opportunities. The biodegradability and biocompatibility further add to the nonirritating and nontoxic characteristics of these carriers.

Besides the safety of these vesicular carriers, these certainly improved the stability of the incorporated bioactives to a great extent. The wide spectrum of dermal and transdermal applications of vesicular carriers has certainly revolutionized the therapeutic strategies. The major challenges ahead for the formulation scientists is the search of newer and better raw materials to meet the growing demands. A wider choice of lipidic raw materials would certainly intensify the technological research in this field, thus facilitating the translation of further vesicular-based pharmaceutical products, from laboratories to the clinics. Apart from this, an improved understanding of skin and vesicle interaction is ultimately needed to investigate the influence of vesicular composition on the skin barrier. The future is anticipated to witness the creation of mixed carriers to explore several positive aspects of each carrier, along with improvements in their properties, especially for the delivery of biological drugs through the skin.

REFERENCES

Almeida FM, Silva CN, de Araujo Lopes SC, Santos DM, Torres FS, Cardoso FL. Physicochemical characterization and skin permeation of cationic transfersomes containing the synthetic peptide PnPP-19. *Curr Drug Deliv.* 2018; doi: 10.2174/1567201815666180108170206.

Ascenso A, Pinho S, Eleuterio C, Praca FG, Bentley M, Oliveira H. Lycopene from tomatoes: Vesicular nanocarrier formulations for dermal delivery. *J Agric Food Chem.* 2013;61:7284–93.

Basnet P, Hussain H, Tho I, Skalko-Basnet N. Liposomal delivery system enhances anti-inflammatory properties of curcumin. *J Pharm Sci.* 2012;101:598–609.

Benigni M, Pescina S, Grimaudo MA, Padula C, Santi P, Nicoli S. Development of microemulsions of suitable viscosity for cyclosporine skin delivery. *Int J Pharm.* 2018;545:197–205.

Benson HA. Elastic liposomes for topical and transdermal drug delivery. *Methods Mol Biol.* 2017;1522:107–17.

Boonme P, Kaewbanjong J, Amnuaikit T, Andreani T, Silva AM, Souto EB. Microemulsion and microemulsion-based gels for topical antifungal therapy with phytochemicals. *Curr Pharm Des.* 2016;22:4257–63.

Burgeson RE, Christiano AM. The dermal-epidermal junction. *Curr Opin Cell Biol.* 1987;9:651–8.

Caddeo C, Nacher A, Vassallo A, Armentano MF, Pons R, Fernandez-Busquets X. Effect of quercetin and resveratrol co-incorporated in liposomes against inflammatory/oxidative response associated with skin cancer. *Int J Pharm.* 2016;513:153–63.

Candido TM, De Oliveira CA, Ariede MB, Velasco MVR, Rosado C, Baby AR. Safety and antioxidant efficacy profiles of rutin-loaded ethosomes for topical application. *AAPS PharmSciTech.* 2018;19:1773–80.

Carrer DC, Higa LH, Tesoriero MV, Morilla MJ, Roncaglia DI, Romero EL. Structural features of ultradeformable archaeosomes for topical delivery of ovalbumin. *Colloids Surf B Biointerfaces.* 2014;121:281–9.

Castangia I, Caddeo C, Manca ML, Casu L, Latorre AC, Diez-Sales O. Delivery of liquorice extract by liposomes and hyalurosomes to protect the skin against oxidative stress injuries. *Carbohydr Polym.* 2015;10:657–63.

Castangia I, Manca ML, Catalan-Latorre A, Maccioni AM, Fadda AM, Manconi M. Phycocyanin-encapsulating hyalurosomes as carrier for skin delivery and protection from oxidative stress damage. *J Mater Sci Mater Med.* 2016;27:75.

Celia C, Cilurzo F, Trapasso E, Cosco D, Fresta M, Paolino D. Ethosomes and transfersomes containing linoleic acid: Physicochemical and technological features of topical drug delivery carriers for the potential treatment of melasma disorders. *Biomed Microdevices.* 2012;14:119–30.

Cevc G, Vierl U. Nanotechnology and the transdermal route: A state of the art review and critical appraisal. *J Control Release.* 2010;141:277–99.

Chen G, Li D, Jin Y, Zhang W, Teng L, Bunt C. Deformable liposomes by reverse-phase evaporation method for an enhanced skin delivery of (+)-catechin. *Drug Dev Ind Pharm.* 2014;40:260–5.

Chen JG, Liu YF, Gao TW. Preparation and anti-inflammatory activity of triptolide ethosomes in an erythema model. *J Liposome Res.* 2010;20:297–303.

Chen YY, Lu YH, Ma CH, Tao WW, Zhu JJ, Zhang X. A novel elastic liposome for skin delivery of papain and its application on hypertrophic scar. *Biomed Pharmacother.* 2017;87:82–91.
Choi JH, Cho SH, Yun JJ, Yu YB, Cho CW. Ethosomes and transfersomes for topical delivery of ginsenoside rhl from red ginseng: Characterization and *in vitro* evaluation. *J Nanosci Nanotechnol.* 2015;15:5660–2.
Choi MJ, Maibach HI. Elastic vesicles as topical/transdermal drug delivery systems. *Int J Cosmet Sci.* 2005;27:211–21.
Cosco D, Celia C, Cilurzo F, Trapasso E, Paolino D. Colloidal carriers for the enhanced delivery through the skin. *Expert Opin Drug Deliv.* 2008;5:737–55.
Date AA, Patravale VB. Microemulsions: Applications in transdermal and dermal delivery. *Crit Rev Ther Drug Carrier Syst.* 2007;24:547–96.
Desai P, Patlolla RR, Singh M. Interaction of nanoparticles and cell-penetrating peptides with skin for transdermal drug delivery. *Mol Membr Biol.* 2010;27:247–59.
Doppalapudi S, Mahira S, Khan W. Development and *in vitro* assessment of psoralen and resveratrol co-loaded ultradeformable liposomes for the treatment of vitiligo. *J Photochem Photobiol B.* 2017;174:44–57.
El Maghraby GM, Barry BW, Williams AC. Liposomes and skin: From drug delivery to model membranes. *Eur J Pharm Sci.* 2008;34:203–22.
Escarleth da CRR, Adriana MF, Jessica CEV, Fernanda BA, Rodrigo ASC, Alexandro CF. Development of a larvicidal nanoemulsion with Copaiba (*Copaifera duckei*) oleoresin. *Revista Brasileira de Farmacognosia.* 2014;24:699–705.
Eskandar M, Mohammad K, Somayeh H. Archaeosomes as means of nano-drug delivery. *Reviews in Medical Microbiology.* 2014;25:40–5.
Fan C, Li X, Zhou Y, Zhao Y, Ma S, Li W. Enhanced topical delivery of tetrandrine by ethosomes for treatment of arthritis. *Biomed Res Int.* 2013;2013:161943.
Franze S, Marengo A, Stella B, Minghetti P, Arpicco S, Cilurzo F. Hyaluronan-decorated liposomes as drug delivery systems for cutaneous administration. *Int J Pharm.* 2018;535:333–39.
Goncalez ML, Marcussi DG, Calixto GM, Correa MA, Chorilli M. Structural characterization and *in vitro* antioxidant activity of kojic dipalmitate loaded w/o/w multiple emulsions intended for skin disorders. *Biomed Res Int.* 2015;2015:1–8.
Gonzalez-Paredes A, Clares-Naveros B, Ruiz-Martinez MA, Durban-Fornieles JJ, Ramos-Cormenzana A, Monteoliva-Sanchez M. Delivery systems for natural antioxidant compounds: Archaeosomes and archaeosomal hydrogels characterization and release study. *Int J Pharm.* 2011;421:321–31.
Gray GM, White RJ, Williams RH, Yardley HJ. Lipid composition of the superficial stratum corneum cells of pig epidermis. *Br J Dermatol.* 1982;106:59–63.
Gupta M, Agrawal U, Vyas SP. Nanocarrier-based topical drug delivery for the treatment of skin diseases. *Expert Opin Drug Deliv.* 2012;9:783–804.
Gupta R, Gupta M, Mangal S, Agrawal U, Vyas SP. Capsaicin-loaded vesicular systems designed for enhancing localized delivery for psoriasis therapy. *Artif Cells Nanomed Biotechnol.* 2016;44:825–34.
Hatahet T, Morille M, Hommoss A, Devoisselle JM, Muller RH, Begu S. Liposomes, lipid nanocapsules and smart Crystals(R): A comparative study for an effective quercetin delivery to the skin. *Int J Pharm.* 2018;542:176–85.
Iqbal B, Ali J, Baboota S. Recent advances and development in epidermal and dermal drug deposition enhancement technology. *Int J Dermatol.* 2018;57:646–60.
Jain S, Sapre R, Tiwary AK, Jain NK. Proultraflexible lipid vesicles for effective transdermal delivery of levonorgestrel: Development, characterization and performance evaluation. *AAPS Pharm SciTech.* 2005;6:513–22.
Jeengar MK, Rompicharla SV, Shrivastava S, Chella N, Shastri NR, Naidu VG. Emu oil based nano-emulgel for topical delivery of curcumin. *Int J Pharm.* 2016;506:222–36.
Jøraholmen MW, Skalko-Basnet N, Acharya G, Basnet P. Resveratrol-loaded liposomes for topical treatment of the vaginal inflammation and infections. *Eur J Pharm Sci.* 2015;79:112–21.
Jose A, Labala S, Ninave KM, Gade SK, Venuganti VVK. Effective skin cancer treatment by topical co-delivery of curcumin and STAT3 siRNA using cationic liposomes. *AAPS PharmSciTech.* 2018;19(1):166–175.
Kanikkannan N, Singh J, Ramarao P. Transdermal iontophoretic delivery of timolol maleate in albino rabbits. *Int J Pharm.* 2000;197:69–76.
Katare OP, Raza K, Singh B, Dogra S. Novel drug delivery systems in topical treatment of psoriasis: Rigors and vigors. *Indian J Dermatol Venereol Leprol.* 2010;76:612–21.
Kianvash N, Bahador A, Pourhajibagher M, Ghafari H, Nikoui V, Rezayat SM. Evaluation of propylene glycol nanoliposomes containing curcumin on burn wound model in rat: Biocompatibility, wound healing, and anti-bacterial effects. *Drug Deliv Transl Res.* 2017;7:654–663.

Kong M, Chen XG, Kweon DK, Park HJ. Investigations on skin permeation of hyaluronic acid based nanoemulsion as transdermal carrier. *Carbohydr Polym.* 2011;86:837–43.

Kumar R, Katare OP. Lecithin organogels as a potential phospholipid-structured system for topical drug delivery: A review. *AAPS PharmSciTech.* 2005;6:E298–10.

Kunz L, Tice T, Libby R, McDaniel C, inventors. Delivery vehicles for bioactive agents and uses thereof. 1997.

Kurzrock R, Li L, Mehta K, Aggarawal BB, Helson L, inventors. Liposomal curcumin for treatment of diseases. 2008.

Lademann J, Otberg N, Richter H, Weigmann HJ, Lindemann U, Schaefer H. Investigation of follicular penetration of topically applied substances. *Skin Pharmacol Appl Skin Physiol.* 2001;14:17–22.

Lopes LB, VanDeWall H, Li HT, Venugopal V, Li HK, Naydin S. Topical delivery of lycopene using microemulsions: Enhanced skin penetration and tissue antioxidant activity. *J Pharm Sci.* 2010;99:1346–57.

Madni A, Rahim MA, Mahmood MA, Jabar A, Rehman M, Shah H. Enhancement of dissolution and skin permeability of pentazocine by proniosomes and niosomal gel. *AAPS PharmSciTech.* 2018;19:1544–53.

Mahdi ES, Noor AM, Sakeena MH, Abdullah GZ, Abdulkarim MF, Sattar MA. Formulation and *in vitro* release evaluation of newly synthesized palm kernel oil esters-based nanoemulsion delivery system for 30% ethanolic dried extract derived from local *Phyllanthus urinaria* for skin antiaging. *Int J Nanomedicine.* 2011;6:2499–512.

Manca ML, Castangia I, Zaru M, Nacher A, Valenti D, Fernandez-Busquets X. Development of curcumin loaded sodium hyaluronate immobilized vesicles (hyalurosomes) and their potential on skin inflammation and wound restoring. *Biomaterials.* 2015;71:100–09.

Manosroi A, Chankhampan C, Manosroi W, Manosroi J. Transdermal absorption enhancement of papain loaded in elastic niosomes incorporated in gel for scar treatment. *Eur J Pharm Sci.* 2013;48:474–83.

McClements DJ, Decker EA, Weiss J. Emulsion-based delivery systems for lipophilic bioactive components. *J Food Sci.* 2007;72:R109–24.

McCook JP, Persaud I, Narain NR, inventors. Pathfinder Management, Inc., assignee. Topical formulations having enhanced bioavailability. 2008.

Mehta M, Dureja H, Garg M. Development and optimization of boswellic acid-loaded proniosomal gel. *Drug Deliv.* 2016;23:3072–81.

Mostafa DM, Ammar NM, Abd El-Alim SH, El-anssary AA. Transdermal microemulsions of *Glycyrrhiza glabra* L.: Characterization, stability and evaluation of antioxidant potential. *Drug Deliv.* 2014;21:130–9.

Mostafa DM, Ammar NM, Basha M, Hussein RA, El Awdan S, Awad G. Transdermal microemulsions of *Boswellia carterii* Bird: Formulation, characterization and *in vivo* evaluation of anti-inflammatory activity. *Drug Deliv.* 2015;22:748–56.

Negi P, Aggarwal M, Sharma G, Rathore C, Singh B, Katare OP. Niosome-based hydrogel of resveratrol for topical applications: An effective therapy for pain related disorder(s). *Biomed Pharmacother.* 2017;88:480–87.

Pando D, Caddeo C, Manconi M, Fadda AM, Pazos C. Nanodesign of olein vesicles for the topical delivery of the antioxidant resveratrol. *J Pharm Pharmacol.* 2013;65:1158–67.

Paolino D, Lucania G, Mardente D, Alhaique F, Fresta M. Ethosomes for skin delivery of ammonium glycyrrhizinate: *In vitro* percutaneous permeation through human skin and *in vivo* anti-inflammatory activity on human volunteers. *J Control Release.* 2005;106:99–110.

Patel D, Shukla S, Gupta S. Apigenin and cancer chemoprevention: Progress, potential and promise (review). *Int J Oncol.* 2007;30:233–45.

Pathak K, Vaidya A, Sharma V. Confronting penetration threshold via fluidic terpenoid nanovesicles. *Curr Drug Deliv.* 2018; doi: 10.2174/1567201815666180108162110.

Pierre MB, Dos Santos Miranda Costa I. Liposomal systems as drug delivery vehicles for dermal and transdermal applications. *Arch Dermatol Res.* 2011;303:607–21.

Pirvu CD, Hlevca C, Ortan A, Prisada R. Elastic vesicles as drugs carriers through the skin. *Farmacia.* 2010;58:128–35.

Plangsombat N, Rungsardthong K, Kongkaneramit L, Waranuch N, Sarisuta N. Anti-inflammatory activity of liposomes of *Asparagus racemosus* root extracts prepared by various methods. *Exp Ther Med.* 2016;12:2790–96.

Prasanthi D, Lakshmi PK. Terpenes: Effect of lipophilicity in enhancing transdermal delivery of alfuzosin hydrochloride. *J Adv Pharm Technol Res.* 2012;3:216–23.

Priprem A, Damrongrungruang T, Limsitthichaikoon S, Khampaenjiraroch B, Nukulkit C, Thapphasaraphong S. Topical niosome gel containing an anthocyanin complex: A potential oral wound healing in rats. *AAPS PharmSciTech.* 2018a;19:1681–92.

Priprem A, Nukulkit C, Johns NP, Laohasiriwong S, Yimtae K, Soontornpas C. Transmucosal delivery of melatonin-encapsulated niosomes in a mucoadhesive gel. *Ther Deliv.* 2018b;9:343–57.

Puglia C, Bonina F. Lipid nanoparticles as novel delivery systems for cosmetics and dermal pharmaceuticals. *Expert Opin Drug Deliv.* 2012;9:429–41.

Raut S, Bhadoriya SS, Uplanchiwar V, Mishra V, Gahane A, Jain SK. Lecithin organogel: A unique micellar system for the delivery of bioactive agents in the treatment of skin aging. *Acta Pharma Sin B.* 2012;2:8–15.

Raza K, Kumar M, Kumar P, Malik R, Sharma G, Kaur M. Topical delivery of aceclofenac: Challenges and promises of novel drug delivery systems. *Biomed Res Int.* 2014;2014:1–11.

Raza K, Singh B, Mahajan A, Negi P, Bhatia A, Katare OP. Design and evaluation of flexible membrane vesicles (FMVs) for enhanced topical delivery of capsaicin. *J Drug Target.* 2011;19:293–302.

Roederer JE, inventor. Pain relief composition. 2006.

Sarwa KK, Mazumder BF, Rudrapal M, Verma VK. Potential of capsaicin-loaded transfersomes in arthritic rats. *Drug Deliv.* 2015;22:638–46.

Schreier H, Bouwstra J. Liposomes and niosomes as topical drug carriers: Dermal and transdermal drug delivery. *J Control Rel.* 1994;30:1–15.

Sharma G, Goyal H, Thakur K, Raza K, Katare OP. Novel elastic membrane vesicles (EMVs) and ethosomes-mediated effective topical delivery of aceclofenac: A new therapeutic approach for pain and inflammation. *Drug Deliv.* 2016b;23:3135–45.

Sharma G, Kaur M, Raza K, Thakur K, Katare OP. Aceclofenac–β-cyclodextrin-vesicles: A dual carrier approach for skin with enhanced stability, efficacy and dermatokinetic profile. *RSC Adv.* 2016a;6:20713–27.

Sharma G, Saini MK, Thakur K, Kapil N, Garg NK, Raza K. Aceclofenac cocrystal nanoliposomes for rheumatoid arthritis with better dermatokinetic attributes: A preclinical study. *Nanomedicine (Lond).* 2017b;12:615–38.

Sharma G, Singh J, Anand D, Kumar M, Raza K, Pareek A. Aceclofenac: Species-dependent metabolism and newer paradigm shift from oral to non-oral delivery. *Curr Top Med Chem.* 2017a;17:107–19.

Shen LN, Zhang YT, Wang Q, Xu L, Feng NP. Enhanced *in vitro* and *in vivo* skin deposition of apigenin delivered using ethosomes. *Int J Pharm.* 2014;460:280–8.

Singh D, Dubey P, Pradhan M, Singh MR. Ceramic nanocarriers: Versatile nanosystem for protein and peptide delivery. *Expert Opin Drug Deliv.* 2013;10:241–59.

Singh D, Pradhan M, Nag M, Singh MR. Vesicular system: Versatile carrier for transdermal delivery of bioactives. *Artif Cells Nanomed Biotechnol.* 2015;43:282–90.

Streeper R, Singh C, inventors. Topical and oral formulations of cardiac glycosides for treating skin diseases. 2004.

Tavano L, Alfano P, Muzzalupo R, de Cindio B. Niosomes vs microemulsions: New carriers for topical delivery of Capsaicin. *Colloids Surf B Biointerfaces.* 2011;87:333–9.

Teixeira MC, Carbone C, Souto EB. Beyond liposomes: Recent advances on lipid based nanostructures for poorly soluble/poorly permeable drug delivery. *Prog Lipid Res.* 2017;68:1–11.

Thakur K, Sharma G, Singh B, Jain A, Tyagi R, Chhibber S. Cationic-bilayered nanoemulsion of fusidic acid: An investigation on eradication of methicillin-resistant Staphylococcus aureus 33591 infection in burn wound. *Nanomedicine (Lond).* 2018;13:825–47.

Tong X, Van Dross RT, Abu-Yousif A, Morrison AR, Pelling JC. Apigenin prevents UVB-induced cyclooxygenase 2 expression: Coupled mRNA stabilization and translational inhibition. *Mol Cell Biol.* 2007;27:283–96.

Touitou E, Dayan N, Bergelson L, Godin B, Eliaz M. Ethosomes - novel vesicular carriers for enhanced delivery: Characterization and skin penetration properties. *J Control Release.* 2000;65:403–18.

Tsai YH, Lee KF, Huang YB, Huang CT, Wu PC. *In vitro* permeation and *in vivo* whitening effect of topical hesperetin microemulsion delivery system. *Int J Pharm.* 2010;388:257–62.

Ubbink J, Kruger J. Physical approaches for the delivery of active ingredients in foods. *Trends Food Sci Technol.* 2006;17:244–54.

Wang C, Zhu J, Zhang D, Yang Y, Zheng L, Qu Y. Ionic liquid - microemulsions assisting in the transdermal delivery of Dencichine: Preparation, in-vitro and in-vivo evaluations, and investigation of the permeation mechanism. *Int J Pharm.* 2018;535:120–31.

Wang Q, Hu C, Zhang H, Zhang Y, Liu T, Qian A. Evaluation of a new solid non-aqueous self-double-emulsifying drug-delivery system for topical application of quercetin. *J Microencapsul.* 2016;33:785–94.

Wertz PW, Swartzendruber DC, Kitko DJ, Madison KC, Downing DT. The role of the corneocyte lipid envelopes in cohesion of the stratum corneum. *J Invest Dermatol.* 1989;93:169–72.

Witting M, Boreham A, Brodwolf R, Vavrova K, Alexiev U, Friess W. Interactions of hyaluronic Acid with the skin and implications for the dermal delivery of biomacromolecules. *Mol Pharm.* 2015;12:1391–401.

Yadav NK, Nanda S, Sharma G, Katare OP. Systematically optimized coenzyme q10–loaded novel proniosomal formulation for treatment of photo-induced aging in mice: Characterization, biocompatibility studies, biochemical estimations and anti-aging evaluation. *J Drug Target*. 2016;24:257–71.
Zhang YT, Shen LN, Zhao JH, Feng NP. Evaluation of psoralen ethosomes for topical delivery in rats by using *in vivo* microdialysis. *Int J Nanomedicine*. 2014;9:669–78.
Zorec B, Zupancic S, Kristl J, Pavselj N. Combinations of nanovesicles and physical methods for enhanced transdermal delivery of a model hydrophilic drug. *Eur J Pharm Biopharm*. 2018;12:387–97.

15 Nanotechnology-Based Biopharmaceutical Systems

From Laboratory Research to Industrial Development

Rajashree Gude, Sarwar Beg, Harmanjot Kaur, Teenu Sharma, Bhupinder Singh, and Umesh Banakar**

CONTENTS

15.1 INTRODUCTION

Nanobiomedicine explores applied scientific applications and benefits of nanotechnology in healthcare, taking into ambit the molecular and clinical pharmacology, genetics, diagnostics, among others, including targeted therapeutics. Applications of nanotechnology in pharmaceutical dosage forms have been steadily evolving in the past two decades with varying degrees of success. However, the advantages offered by this technology are being recognized quite recently. Nanotechnology has also been explored in treating numerous diseased states and physiological conditions. In many of such cases, the development of a product for the end user seems to be in the distant future, primarily because the scientific basis of such pursuits is yet to be clearly established and understood. Albeit it is often a quite lengthy and cumbersome exercise, the need for such commercial pursuits has been quite immediate. This continuous on-going competition of time-consuming scientific enquiry and the immediate need (i.e., commercial interest) is fueling the growth of nanotechnology applications in such pharmaceutical systems.

One such application is by exploiting the advantage of enhanced bioavailability of nanopharmaceutical systems. For long term treatments, the oral solid pharmaceutical dosage forms (i.e., tablets and capsules) or oral suspensions can reduce the overall daily consumption of the total mass of drug (i.e., dosage levels) by as much as 50%. Such applications can be widely foreseen

* Corresponding author.

in long-term treatments such as antiretroviral therapy (Smans and Jans, 2010; Delaet *et al.*, 2013), among others. Currently, the scientific and commercial interests of the pharmaceutical community are centered on understanding the physicochemical principles of surface modifications, solubility improvement, bioefficacy improvement, and ultimately, clinical applications of nanotechnology to pharmaceutical systems, resulting in safe, stable, and clinically efficient pharmaceutical dosage forms.

The scientific literature, since the advent of the principles of nanotechnology and their potential bioefficacious use in delivering drug substance, is replete with such reports over the past two decades. During the period of 2000 through 2018, it is estimated that there have been over 90 conferences, over 3000 scientific findings reported, over 1200 reports and publications, over 250 patents, and billions of dollars spent in the pursuit of developing nanotechnology-based drug-delivery systems. These reports range from feasibility studies and patent applications, as well as published patents, abstracts presented at numerous conferences and other scientific publishing formats, several theses and dissertations, and the likes. Considering such proliferation of activity in the scientific community, it has resulted in gaining improved understanding of various physical, chemical, mechanical, and mechanistic functioning of nanotechnology-enabled drug-delivery systems, that is, pharmaceutical formulations. In this process, while there is substantial creation of new knowledge, the number as well as the frequency of approved nanotechnology-based pharmaceutical systems is quite limited. It is seen that the drug substance used to formulate nanoparticulate-based pharmaceutical systems is often well-known and the processes employed for the preparation/manufacture of such formulations are also fairly common. Additionally, on a strict time-scale comparison basis, the overall time required in developing a new chemical entity (NCE) subsequent to a confirmed lead compound status to its approval to market through the various stages of safety and efficacy evaluation is over 10 years. While such a comparison may not be completely fair, time has come to assess the potential reasons for such disparity—that is, unseemingly lengthy development time spent on bringing forth successful and so-called "approvable" nanotechnology-based drug-delivery systems.

Translation of research findings into practice is quite a challenging and arduous process. The question that often surfaces is, "Is some observations' evidence missed and/or overlooked during the product development stage(s)" Absence of evidence is not verily the evidence of absence! In this context, a thorough assessment of the steps and stages of development of a typical nanotechnology-based drug-delivery system, that is, a product is warranted that will potentially set the tone within the drug product development scientific community, leading to the realization of translating exploratory research findings into commercially viable nanoparticulate pharmaceutical systems.

15.1.1 Nanotechnology-Driven Drug Delivery Systems

Micromeritics has been exploited with numerous objectives in the context of diverse drug-delivery systems. These exploits extend from micronization to nanonization of the drug substance in designing the pharmaceutical formulation system to accomplish the desired biological (i.e., bioavailability) and clinical (i.e., therapeutic) efficacy. Thus, the utility and application of micromeritics to nanostructured drug-delivery systems harvest the potential advantages in the two crucial phases of the drug-development process: the discovery phase and development phase. Nanotechnology, in fact, plays significant role in both phases of drug development.

In the *Discovery Phase*, nanotechnology offers the following advantages:

- Simplicity of formulation methods
- Amenability to its direct use in screening
- Reduction in dose of drug formulation
- Production of sample formulations with GRAS materials
- Administration of drug formulations by various routes
- Easy scale-up through development and manufacture
- Cost-effectiveness

Similarly, in the *Development Phase*, nanotechnology offers several advantages, as follows:

- Enhancement of drug solubility and bioavailability
- Avoidance of increased length of manufacture process
- Applicability to novel dosage forms/delivery systems
- Plausibility of utilization in preclinical, clinical, and marketed products
- Possibility of being manufactured from milligram to kilogram levels
- Suitability for GLP and GMP, and commercial supplies
- Cost-effectiveness

Nanotechnology has practically revolutionized drug delivery in healthcare, which has witnessed its culmination in most kinds of pharmaceutical systems, including immediate release (IR), delayed release (DR), controlled release (CR), and extended release (ER), as follows:

- Tablets: IR, DR, CR, ER, fast-melt, and others
- Capsules: IR, DR
- Powders
- Topicals, dermals, and transdermals
- Concentrated liquid dispersions
- Aerosols/hnhalational products
- Others

The advantage of greater surface area-to-volume ratio of particles can be related to the physical, chemical, and pharmacological properties of the drug. Healthcare professionals worldwide have recognized significant compensation and opportunities offered by nanotechnology when applied to pharmaceutical systems. These can be summarized, as follows (Liversidge and Cundy, 1995; Setler, 1999):

- Improved bioavailability by enhancing drug solubility
- Rapid onset of therapeutic action
- Improved dose proportionality
- Reduced fed/fasted state variability
- Reduced intrasubject variability
- Enhanced absorption rate
- Others

These salient meritorious visages have been successfully exploited to develop, manufacture, and market clinically efficient pharmaceutical products. In doing so, several opportunities along with corresponding challenges have been recognized, while exploring the utility of nanostructured drug-delivery systems:

- Improving the solubility and bioavailability of hydrophobic drugs
- Eliminating or minimizing the toxicity
- Increasing target specificity to particular cells or tissues
- Developing delivery systems for slow/controlled release
- Accumulating at the target site and thus reducing adverse effects
- Decreasing toxicity using biocompatible nanomaterials

A bird's eye view of literature reveals that the early, perhaps the first, nanoparticulate compositions of poorly soluble therapeutic or diagnostic agents were described in US patents (Liversidge *et al.*, 1992), while the methods for making nanoparticulate compositions have been reported in several patents (Bosch *et al.*, 1996; Bruno *et al.*, 1996; Czekai and Seaman, 1998, 1999; Ruddy and Roberts,

1998). Nanoparticulate pharmaceutical compositions have been developed and evaluated in various therapeutic categories for various drug substances, namely, oncology drugs (Kellar *et al.,* 1995), NSAIDs (Eickhoff *et al.,* 1996a; Osifo, 1996; Ruddy *et al.,* 1997; Wiedmann *et al.,* 1998; Liversidge *et al.,* 2001), digestible oils and fatty acids (Illig, 1996), human immunodeficiency virus (HIV) protease inhibitors (Wong, 1996), rapidly disintegrating oral solid-dosage forms (Ryde and Ruddy, 2002), bioadhesive and targeted drug-delivery systems (Kato *et al.,* 2002; Liversidge *et al.,* 2002), antimicrobial agents (Violante and Steigbigel, 1988), controlled-release pharmaceutical compositions (Jain *et al.,* 2002), and others. Various nanotechnology-based pharmaceutical products that are in various stages of development, or that have been approved, are presented in Table 15.1.

15.1.2 Million-Dollar Mistake or Multimillion-Dollar Opportunity

The authors' experience while developing a nanotechnology-based oral pharmaceutical dosage form provides a compelling story necessitating in-depth attention that needs to be provided to determine the potential cause(s) for the unusually long duration observed while translating research findings (early prototype formulation) into practice. A BCS Class II (poorly soluble; highly permeable) drug substance was nanosized through a seemingly well-controlled nanofluidization process. The resulting slurry of the nanosized suspension was granulated. The granules were then blended using a rate-controlling polymer and then compressed into the final dosage form. This prototype formulation was subjected to a pilot bioefficacy evaluation, with price tag of several hundreds of thousands of US dollars, and the bioefficacy met the expected targets. Subsequently, the prototype formulation was scaled-up to manufacture a "pivotal" batch, so-called a biobatch, which was subjected to definitive bioefficacy evaluation (price tag of approximately 1 million USD), which failed to meet the expected bioefficacy targets. The root-cause-analysis revealed that the physicochemical surface characteristics of the nanoparticulate system of the prototype formulation needed better characterization and control during their manufacture, which failed to reproduce themselves during the scale-up and manufacture of the pivotal biobatch, thus failing bioefficacy expectations in the final stage. Following appropriate characterization and applying adequate controls during the scale-up, the biobatch was once again manufactured and subjected to definitive bioefficacy assessment, which ultimately met the expected targets of bioefficacy.

15.1.3 Development of Robust Nanotechnology-Driven Pharmaceutical Dosage Forms

15.1.3.1 Mechanistic Approach

Development of robust nanotechnology-based pharmaceutical formulations that can withstand the rigors of various stages of the drug development process requires an in-depth understanding of the mechanisms involved in their manufacturing process(es) and the corresponding impact or influence of the outcome-related efficacy, both bioefficacy and therapeutic efficacy. It stands to reason that such an understanding will provide the necessary clues and the related information, which will decrease the potential pitfalls resulting in development delays and economically expensive exigencies. As a result, a fresh look at the pharmaceutical and biopharmaceutical considerations relating to the development of nanotechnology-based pharmaceuticals is essential.

In general, a list (not all inclusive) of potential difficulties encountered, while transforming exploratory research findings into viable product include

- Micromeritic characterization
- Industrial scale-up considerations
- Processing variables
- Release profiling
- Understanding of cross-disciplinary functions
- Biopharmaceutical considerations
- Others

TABLE 15.1
An Update on Nanotechnology-Based Pharmaceutical Products at Various Stages of Development, under Clinical Trials, and/or Approved, as of 2018

Type of Nanostructured System(s)	Product	Hydrophobic Drug	Indication	Route of Delivery	Technology by/Licensed to	Year of Approval/ Progress in Clinical Trials
Polymeric micelles	NK105	Paclitaxel	Metastatic breast cancer	IV	Nippon Kayaku	Phase III (2016)
Nanocochelates	MAT2203	Amphotericin	Antifungal	Oral	Matinas BioPharma	Phase IIa (2016)
Nanocapsules	–	Lapatinib and Paclitaxel combination	Metastatic Breast Cancer	IV	–	Phase III (2014)
Thermally Sensitive Liposomes	ThermoDox	Doxorubicin	Hepatocellular Carcinoma	IV infusion	Celsion corporation	Phase III (2014-2019, ongoing)
Nanocrystals	Emend	Aprepitant	Chemotherapy-induced nausea and vomiting	Oral	Merck/Elan nanocrystals	2013
Liposomes	Marquibo	Vincristine	Adult acute myelogenous leukemia	IV	Spectrum Pharmaceuticals	2012
Nanoparticles	Sienna+	Dextran-coated Iron oxide	Sentinal Llmphnode mapping	Subcutaneously	Sysmex Europe GmbH	2011
Nanocrystals	Invega Sustena	Paliperidone Palmitate	Tteatment of schizophrenia	IV	Elan/Johnson & Johnson	2009
Polymeric nanoparticles	Abraxane	Paclitaxel	Metastatic breast cancer	IV infusion	AbraxisLife Science, USA	2005
Nano-crystals	Triglide	Fenofibrate	Treatment of high cholesterol and high triglycerides	Oral	Skye Pharma/First Horizon Pharmaceuticals	2005
Nanocrystals	Rapamune	Sirolimus	Immune regression	Oral	Wyeth/Elan nanocrystals	2004
Nanocrystals	Tricor	Fenofibrate	Treatment of high cholesterol and high triglycerides	Oral	Elan/Abbott	2004
Nanocrystals	Emend	Aprepitant	Antiemetic	Oral	Elan/Merck	2003
Liposomes	Depocyt	Cytarabine	Neoplastic meningitis and lymphomatous Meningitis	Epidural	Napp	2002

(*Continued*)

TABLE 15.1 (*Continued*)
An Update on Nanotechnology-Based Pharmaceutical Products at Various Stages of Development, under Clinical Trials, and/or Approved, as of 2018

Type of Nanostructured System(s)	Product	Hydrophobic Drug	Indication	Route of Delivery	Technology by/Licensed to	Year of Approval/ Progress in Clinical Trials
Liposomes	Lipo-Dox	Doxorubicin	Breast and ovarian cancer	IV	Taiwan Liposome	2001
Liposomes	Daunoxome	Daunorubicin	Kaposi'vs sarcoma	IV	Galen	1996
Liposomes	Abelcet	Amphotericin B	Aspergillosis	IV infusion	Enzon	1995
Liposomes	Ambisome	Amphotericin B	Severe fungal infections	IV	Gilead	1990
Polymeric micelles	NK-911	Doxorubicin	Cancer targeting	IV	Nippon Kayaku Co. Japan	Phase II
Polymeric micelles	NK-012	SN-38	Cancer targeting	IV	Nippon Kayaku Co. Japan	Phase II
Polymeric micelles	Genexol-PM	Paclitaxel	Solubilization	IV	Samyang Corp. Korea	Phase II

Source: Adapted from Osifo IK. Nanoparticulate Iodipamide derivatives for use as x-ray Contrast agents. US 5,521,218; 1996; Beg S et al. *Pharma Times*. 2018; 50(2): 63–67; Jain S et al. *Pharma Times*. 2018; 50(2): 13–19; Singh B et al. *The Pharma Review*. 2018; 16(93): 37–45.

A complete knowledge of the biopharmaceutics and the resulting clinical implications associated with nanopharmaceutical systems is vital to maximize their potential utility in developing stable and effective pharmaceutical systems. The biopharmaceutical model that can represent the pharmaceutics (e.g., solubility) and the bioavailability of such systems, in general, is presented in Figure 15.1.

The model represents that *in vitro* and *in vivo* dissolution are the critical stages that are influenced due to the decrease in the particle size of the drug and subsequently the formulation, thereby increasing the surface area for faster dissolution of the drug. As a result, the dissolution rate-limited absorption of the drug will be enhanced, if not optimized, in the formulation of poorly soluble drugs from their respective formulations. The ultimate result is the bioavailability of the drug from such nanopharmaceutical systems, wherein the absorption rate, in particular, gets increased, as reflected in the increased values of observed C_{max} for a dose administered, compared to that observed of its counterpart, microparticulate systems. Increase in *in vitro* as well as *in vivo* dissolution rate can be explained by the augmented dissolution of the nanoparticles. The rate of dissolution is dependent on the concentration gradient, i.e., the difference between the saturation concentration of the solute in the solvent (i.e., medium) and the concentration of the dissolved solute in the medium. Additionally, the diffusional distance between the site of dissolution and the concentration gradient contributes to the apparent drug dissolution rate from the dosage form. The combined effect of decrease in the diffusional distance and the increase in the saturation concentration, results in an apparent increase in the dissolution rate (Fromming, 1998; Muller *et al.*, 1998). Simply stated, reduction in particle size of the solute (i.e., drug substance) increases surface area in contact with the dissolution medium, resulting eventually in increased dissolution pressure. Further, reduction in the diffusional distance ensures substantial increase in the concentration gradient, thus enhancing the apparent dissolution rate of the solute in the dissolution medium. Potential positive fallout of nanoparticles in drug formulations is that these particles with increased surface area are charged and tend to adhere to the site, where they ultimately dissolve in the fluid available at the site (Durrer *et al.*, 1994). This increased adhesiveness of these nanoparticulates present further advantages of reduction in erratic absorption and increase in their bioavailability.

It becomes abundantly clear from the above that the biopharmaceutics model must be appropriately modified to accommodate the micromeritic processing, i.e., nanosizing considerations, as presented below Figure 15.2. Additionally, on closer look, one realizes that the processes of nanosizing, combined with their appropriate characterization, and adequate stabilization, will be crucially essential for predictive and reproducible biopharmaceutical properties, and thereby for ultimate bioefficacy and clinical therapeutic efficacy of the nanotechnology-oriented pharmaceutical dosage forms.

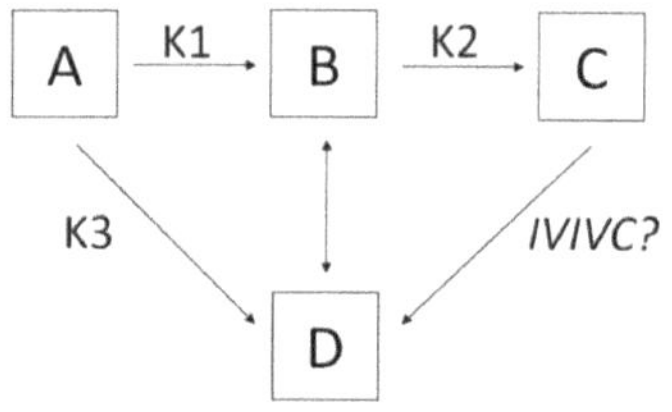

A: Drug in dosage form

B: Drug dissolved in the gut

C: Drug in blood/ systemic circulation

D: Drug dissolved in *in vitro* dissolution system

K1: *in vivo* dissolution rate constant

K2: *in vivo* drug absorption rate constant

K3: *in vitro* dissolution rate constant

FIGURE 15.1 IVIVC model representing the *in vivo* dissolution of the drug from the formulation (A→B), *in vivo* absorption of the dissolved drug from the formulation, (B→C) and *in vitro* dissolution of the drug from the formulation (A→D).

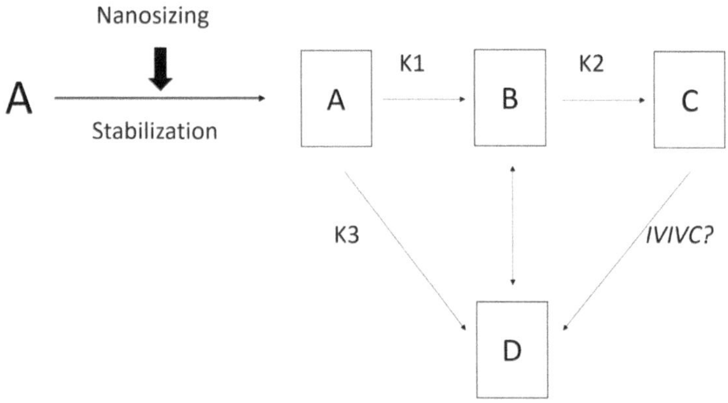

FIGURE 15.2 IVIVC model representing the *in vivo* dissolution of the drug from the formulation (A→B), *in vivo* absorption of the dissolved drug from the formulation, (B→C) and *in vitro* dissolution of the drug from the formulation (A→D).

The pharmaceutics of the process of nanosizing and the resulting changes in the micromeritics of the drug particle will result in changes in the inherent property of the solid particulate systems. Such changes include in size, shape, morphology, surface area, and surface properties, such as surface charges, among others. These changes can potentially lead to physical instability of the drug particle in the composition (i.e., formulation), which in turn can lead to apparent changes in the properties of the composition and the final pharmaceutical dosage unit. Reduction in particle size, resulting in changes in surface characteristics, is schematically presented in Figure 15.3. Such changes in surface properties of the solid particle can become a potential source of instability in the pharmaceutical system, which can further result in unexpected variability in the biopharmaceutical outcomes, such as bioefficacy (or bioavailability) and variability in clinical therapeutics arising from there. There are ample number of literature reports available on enhancing the biopharmaceutical performance of the formulations of many poorly water-soluble drugs.

Studies have also been carried out in our laboratories evaluating the plausible enhancement of absorption potential of rosuvastatin in rats using various kinds of solid lipidic nanoparticles (SLNs) (Beg *et al.*, 2017). Manifold improvement in drug absorption parameters (i.e., 12–25-folds in C_{max}, 16–25-folds in AUC, and 1.5–2.8-folds in T_{max}) was observed during *in vivo* pharmacokinetic studies, with the PEGylated phospholid-based SLNs (PL-PEG-SLNs), phospholipid based-SLNs (PL-SLNs), and naïve SLNs, over the pure drug, in that decreasing order (Figure 15.4). This can

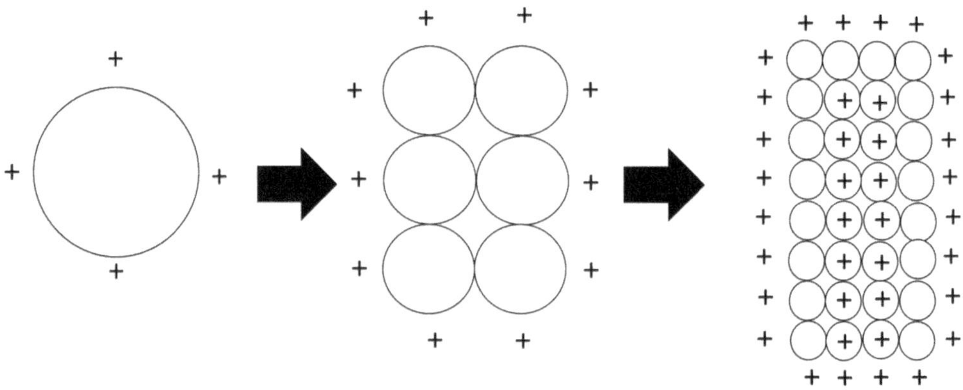

FIGURE 15.3 Schematic representation of the result of changes in size of a solid particle.

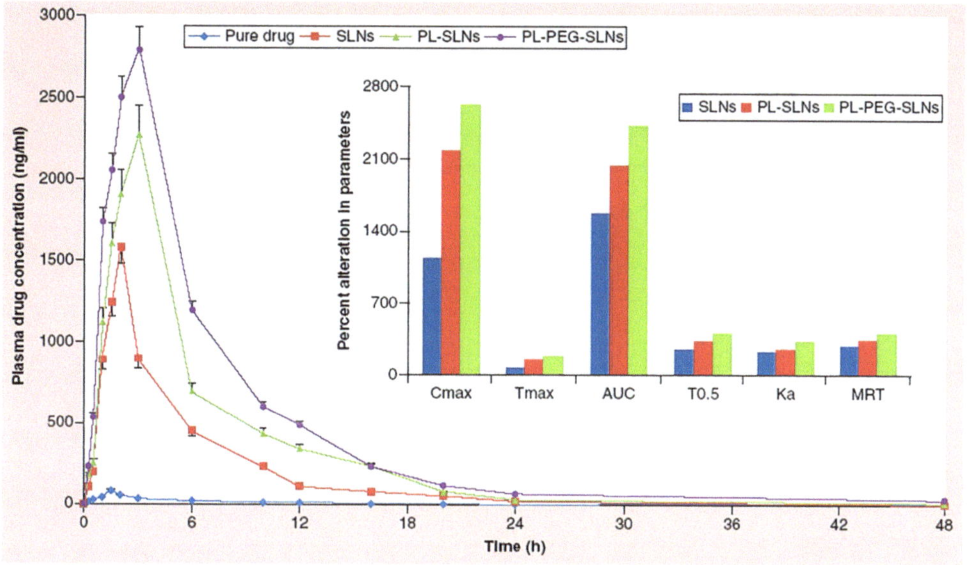

FIGURE 15.4 Plasma-concentration versus time profile of pure drug (rosuvastatin), its solid lipid nanoparticles (SLNs), phospholipid-conjugated SLNs and PEGylated phospholipid-conjugated SLNs after oral administration in fasted Wistar rats. (Adapted with permission from Beg S et al. *Nanomedicine.* 2017; 12(4): 333–356.)

ostensibly be explained by the internalization of SLNs through low-density lipoprotein receptor-mediated pathways owing to the high affinity of phospholipids with apolipoprotein B100 binding site and the role of PEGylated phospholipid in the potentiation of drug absorption through intestinal lymphatic pathways. This is in close agreement with findings construing that SLNs tend to get absorbed through paracellular and transcellular routes due to their miniscule size and improved permeability characteristics. Besides, phospholipids and polyethylene glycol (PEG) also significantly helped in improving the drug absorption from the gastrointestinal tract, while the latter also aided in prolonging the circulation, as is clearly apparent from mild extension in the profile of the SLNs. For investigating possible *in vitro*/*in vivo* correlation (IVIVC) between the *in vitro* drug release and *in vivo* drug absorbed, high magnitudes of linear model coefficient (i.e., $R > 0.97$) for plain SLNs, PL-SLNs, PL-PEG-SLNs, and for pure drug suspension were observed, confirming the prevalence of significant level A IVIVC.

Likewise, Figure 15.5 portrays much pronounced plasma level profiles of lopinavir from the cationic nanoemulsions and plain nanoemulsions *vis-à-vis* that of the conventional marketed oral formulations (Garg *et al.,* 2016). Distinct augmentation of about 4-fold in the values of absorption parameters like C_{max}, $AUC_{24}h$, K_a, and 2-fold reduction in T_{max} were discerned with OPT-SNEOF tablets and OPT-L-SNEOFs, respectively, *vis-à-vis* the pure drug ($p < .001$), thus vouching enhanced rate and extent of drug absorption from these two formulations. Significant reduction in the AUC values in SNEOFs administered to cycloheximide-treated group ($p < .001$) could be ascribed to the circumnavigation of hepatic first-pass effect and involvement of lymphatic pathways, besides of course, the enhanced solubility. This augmentation in bioavailability of lopinavir via lymphatic pathways results in the site-specific inhibition of the HIV for the suppression of the acquired immune deficiency syndrome (AIDS) symptoms and further biodistribution of the virus to other organs. Moreover, enhancement in the values of AUC for the OPT-SNEOF tablets and OPT-L-SNEOFs in case of cycloheximide-treated group *vis-à-vis* the pure drug could be explained by inhibition of the P-gp efflux and opening of the tight junctions owing to the presence of the emulgent, Tween 80.

Nonetheless, for such impressive biopharmaceutical performance of the nanoformulations, as above, to be extrapolated to the commercial scale, several other issues like their production using

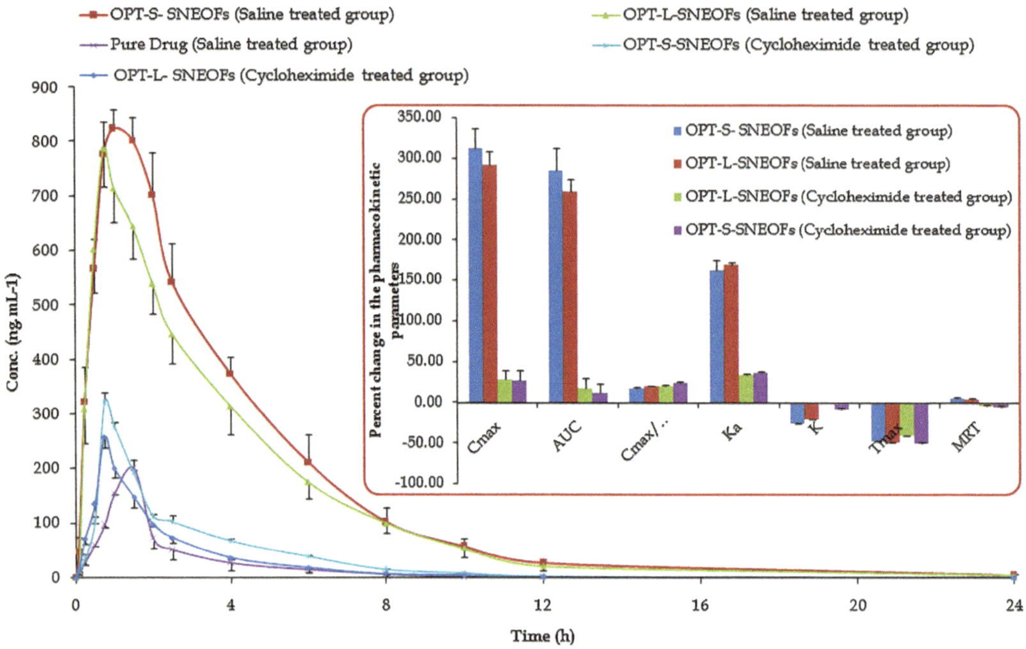

FIGURE 15.5 Mean plasma-concentration-time profile for conventional and nanostructured formulations; CSNEOFs-Cationic nanoemulsion formulation; SNEOFs-Plain nanoemulsion formulation; MKT-Marketed formulation. (Adapted with permission from Garg B et al. *Colloids and Surfaces B*. 2016; 141: 611–622.)

geometrically and kinematically analogous equipment, key issues of stability and reproducibility, etc., must be surmounted before their sojourn from bench-to-bedside gets completed and the marketed formulation sees the light of the day.

Numerous nanopharmaceuticals are in various stages to development across the globe. However, the headway in translating these innovative nanomedicines into commercial successes has been limited by the inability to manufacture their batches at the required scale, in line with stringent quality control measures as per the federal agencies. The quantities required for clinical studies are modest, but pilot processes do not fit easily into the existing manufacturing plants. This demands special attention and training to avoid any deviations between the outcomes of the products developed at lab-scale and shop-floor.

15.1.4 Vital Scale-Up Considerations

A number of nanocarrier delivery technologies have not been found to be scalable due to lack of robustness of the method and process of production, and the exorbitant cost of materials employed. It is, in fact, a lot easier to modify nanomaterials at laboratory scale for improved performance than at larger scale. Major unmet challenges related to the scale-up manufacturing of nanocarriers include reproducibility in product quality attributes, such as drug encapsulation efficiency, drug-lipid ratio, drug payload, size and shape, charge, rheological properties and physico-chemical stability. Identification of these potential scale-up issues invariably coerces the formulator to consider "dimensional analysis" as the potential tool to resolve these plausible tech-transfer issues regarding nanoformulations of bioactives. Some of the widely used manufacturing techniques include hot-homogenization, high pressure homogenization, microemulsification, ultrasonication, microfluidization, a blend of microprecipitation and homogenization, and super-critical fluid (SCF) technology. A brief account of such vital processes (Shah and Khan, 2009; Katare *et al.,* 2018) is presented here.

15.1.4.1 High-Pressure Homogenization (HPH)

This is a fluid mechanical process that involves subdivision of particles or droplets into micron- or nano-size range to create a stable biphasic system for further processing. The fluid passes through minute gap in the special homogenizing valve, with its design being the most important part of the homogenizing equipment. This creates conditions of high turbulence and shear, combined with compression, acceleration, pressure drop and impact, which results in the size reduction of particles. The major parameters that need to be considered during HPH are selection of nozzle, pressure, flow rate, solid concentration, and number of cycles. Proper selection of these parameters and availability of geometrically and kinematically similar equipment from lab scale to commercial scale contribute for successful process scale-up of a nanostructured carrier-based drug-product.

15.1.4.2 Hot Melt Extrusion (HME)

An archetypal process of HME includes (a) feeding of the extruder, (b) conveying of mass (mixing, melting, kneading, and reduction of particle size), (c) flow through the die, and (d) exit from the die and downstream processing (i.e., cooling, size reduction, and further processing). Various critical input variables and system parameters of HME process, which must be considered during pilot-plant scale-up, are enumerated in Table 15.2.

15.1.4.3 Microfluidics

A recently introduced innovative technique for formation of nanoparticles is micro-fluidic "lab-on-a-chip" technology, based on the application of well-controlled forces, which control fluid movements and other processing conditions in the microfluidizer. The technique offers several potential benefits for the production of nanoparticles, such as miniaturization, integration, and automation.

15.1.4.4 Bubbling Technique

Developing an effective scale-up method for large-scale production of diverse vesicular carrier systems is associated with numerous specific problems and technical restrictions. The developed

TABLE 15.2
Plausible HME Process Variables during Scale-Up of Nanocarrier Formulations of Bioactives

Extrusion Input Variables		Extrusion System Parameters	
Configuration	Screw design Barrel design Die geometry	**Material**	Product temperature Viscosity Product temperature at stage on screw Mixing quality
Adjustable	Feed rate Screw speed Barrel temperature	**Process**	Specific energy input (mechanical & thermal) Residence time distribution Shear stress time distribution Torque at screws Die pressure Pressure Melt temperature variation
Material	Melting point Glass transition Molecular weight Density Specific heat capsity Miscibility/interactions		

manufacturing technique should allow economic scale-up and evasion of hazardous chemicals (e.g., chloroform, methylene chloride, N-methylglucoside, etc.). Several preparatory techniques for large scale production of vesicular carrier systems include sonication, thin-film hydration, high-shear homogenization, solvent injection, bubbling, and extrusion. "Bubbling technique" is a relatively recent technique with a very high potential of process scale-up. Liposomal dispersions with an average diameter of about 300 nm have been reported to be prepared by bubbling a stream of nitrogen through a dispersion of lipid materials. An evaluation of the scale-up feasibility of this process has also yielded promising results. Removal of unentrapped free drug from the liposomal dispersions is another vital issue, addressed by techniques like ultracentrifugation and dialysis for large-scale preparation.

Thus, with the advent of newer technologies and advances in the engineering field, novel nanopharmaceutical products are being introduced. Successful scale-up depends upon systematically and judiciously evaluated pilot plant scale-up studies, which should be based on methodical understanding of various formulation and process parameters throughout various developmental stages of a nanopharmaceutical drug product.

15.1.5 Potential Solution

It has now been recognized that the decrease in particle size leads to substantial and often detrimental changes in the inherent surface properties of the solid particle, i.e., the drug substance. These changes in surface properties include, but are not limited to, the nanoparticles acquiring surface charges, which impart unexpected and unwanted physical instability of the particle in the composition during the processing of the formulation. This instability can lead to segregation of particles, complexing of particles with excipients in the formulation, adhesion of the particle to the wall of the container, among others. One of the potential ways to address and overcome this physical instability is to neutralize the surface charges (i.e., the surface energy) associated with the surface of the particle. Such neutralization is accomplished by adsorbing the nanoparticle on an inert excipient. Such excipients are often referred to as surface modifiers or surface stabilizers. They often work through the processes of adsorption and complexation, independently or in combination, to overcome the potential possibility of any physical instability due to surface energy-associated with nanosizing of the solid particle.

It has further been recognized that the primary mechanisms through which physical stability of the nanoparticles is accomplished, is when the surface stabilizer physically adheres to the surface of the drug. Additionally, sequestration of the drug by wetting agent occurs by association complexation. Furthermore, surface stabilizer adsorbed onto the drug substance maintains an effective particle size of less than 1 micron. The sum total of these mechanisms helps to render the process of nanosizing of the drug substance, along with the surface stabilizer and thereby the composition through which various processes of transition into a final finished dosage form that is stable, both physically and chemically.

One of the potential physical properties exhibited by nanoparticles is their tendency to aggregate, thereby inducing physical instability, and more importantly, forfeiting the objective of nanosizing the drug substance to the avail of the numerous advantages of increased efficacy, both physical and biological. With the inclusion of a single, or more than one, surface-modifier, such physical instability is overcome through the measurement of redispersibility of the composition containing nanoparticles. Several studies in the literature report the use of surfactants like sodium lauryl sulfate (SLS), dioctyl sodium sulfosuccinate (DOSS) among others, as a cloud point modifier for nanoparticulate surface stabilizers (George and Rajagopalan, 1994a,b; Hollister *et al.,* 1994; Bacon *et al.,* 1995; Wood *et al.,* 2001) and numerous polymers and clays, individually or as synergistic combinations (Ruddy *et al.,* 1995, 1996a,b; Eickhoff *et al.,* 1996b; Wong *et al.,* 1996; Ryde and Ruddy, 2003). Ryde and Ruddy (2003) reported stabilizing solid dose nanoparticulate compositions for several drug substances employing a synergistic combination of one surfactant-based surface stabilizer and a polymeric surface stabilizer through measurement of their redispersibility potential

(Ryde and Ruddy, 2003). The redispersion properties of solid-dose nanocrystalline dispersion (NCD) of a mitogen-activating protein (MAP) kinase inhibitor composition, comprising DOSS and a polymeric stabilizer in electrolyte solutions, and both of the acidic solutions (0.001–0.1 M HCl and 0.001–0.1 M NaCl) are presented in Table 15.3. The term "small" particles were defined as those below 1 micron and the term "large" particles were defined as those above 1 micron.

This very approach of utilizing the potential of a synergistic combination of surface stabilizers was efficiently explored, while translating these findings into a commercial embodiment, namely, Megestrol (nanoparticulate) Oral Suspension (Megace^R), which is physically stable and effective, biologically as well as clinically.

In order to illustrate the utility of nanotech-oriented drug delivery formulations, two studies reported from our laboratories are cited here. In a research work published by Kapil and Singh (2012), the authors have described the development of solid lipid nanoparticles (SLNs) of quercetin, envisaged to facilitate the entry of the phytobioactive across the blood–brain barrier (BBB) into the CNS, thus paving a way for novel potential therapy for Alzheimer's disease (AD). Quercetin, a natural bioactive flavonoid, is known to act as a novel neuroprotectant by mitigating the levels of reactive oxygen species. Albeit more potent antioxidant and radical scavenger than even vitamins C, E, and β-carotene, limited permeability of quercetin across the BBB is the major impediment.

A micro-emulsification technique was adopted for formulating SLNs using different ratios of lipids (Compritol 888 ATO) and surfactants (Polysorbate 80). To design SLNs for delivering quercetin across the BBB into the CNS, various quality attributes such as particle size (<200 nm), drug entrapment efficiency, drug release characteristics and zeta potential were selected. For assessing the *in vivo* efficacy of SLNs of quercetin, cognitive performance of rats was evaluated using spatial navigation, maze acquisition and maze retention through Morris water maze test, and elevated plus maze paradigm. The rats treated with SLN-encapsulated quercetin showed better memory-retention *vis-à-vis* other animal groups. These animals also exhibited better gross behavioral activity than the test animals, as measured using actophotometer. Nearly four-fold augmentation in the brain levels of quercetin was observed in animals treated with the SLNs *vis-à-vis* those treated with pure quercetin. Biochemical estimations of rat brain homogenates indicated highly significant better regulation of glutathione, MDA, and nitrite levels. The combined results of biochemical and behavioral studies of SLNs are depicted graphically as Figure 15.6, corroborating successful development and evaluation

TABLE 15.3
Redispersion Properties of a Solid Dose Nanocrystalline Dispersion (NCD) of a MAP Kinase Inhibitor Composition Comprising DOSS and a Polymeric Stabilizer in Electrolyte Solutions

Electrolyte Concentration (M)	Type	No Sonication Mean	No Sonication Small %	No Sonication Large %	1 min. Sonication Mean	1 min. Sonication Small %	No Sonication Large %
0	–	99	100	0	99	100	0
0.001	HCl	100	100	0	100	100	0
0.01	HCl	105	100	0	106	100	0
0.1	HCl	4708	23	77	1901	52	48
0.001	NaCl	103	100	0	103	100	0
0.001	NaCl	101	100	0	101	100	0
0.1	NaCl	105	100	0	105	100	0

Source: Adapted with permission from Ryde NP, Ruddy SB. Nanoparticulate dispersions comprising a synergistic combination of a polymeric surface stabilizer and dioctyl sodium sulfosuccinate. US 6,592,903; 2003.

Note: All sizes are in nanometers (nm). Small particles: <1 micron; Large particles: ≥1 micron.

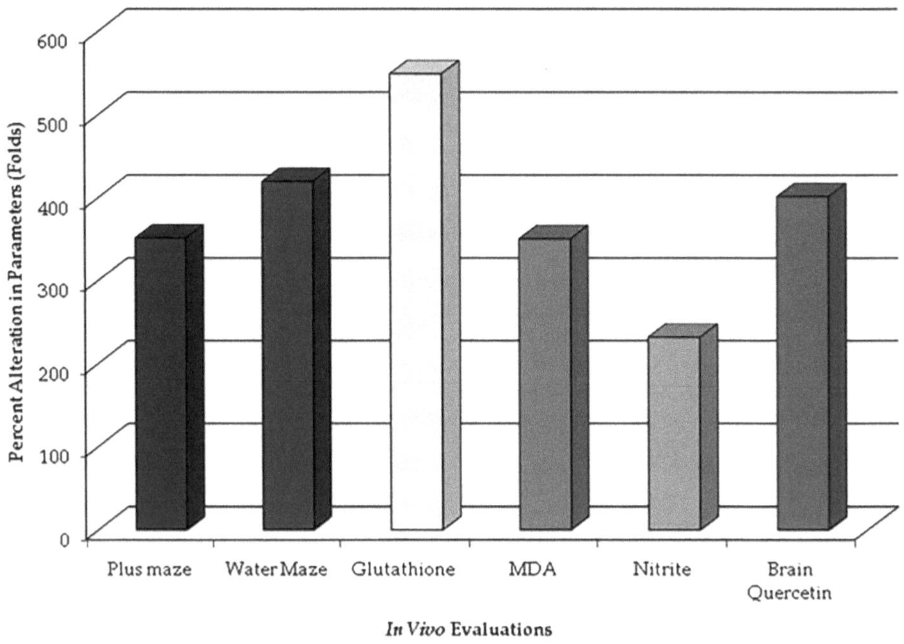

FIGURE 15.6 Biochemical and behavioral studies of quercetin SLNs with improved brain targeting potential, bioefficacy and potential therapeutic performance. (From Kapil R, Singh B. *QbD and Product Performance Focus Group Newsletter.* 2012; 4: 4–7.)

of nanoformulations of quercetin with improved brain targeting potential, bioefficacy and potential therapeutic performance.

In another research work from our laboratories, by Sandhu et al. (2017), the authors have demonstrated improved oral bioavailability of tamoxifen (TMX) nanoformulations, following its co-administration with the plant bioflavonoid naringenin (NG). Since tamoxifen exhibits low oral bioavailability and its anticancer bioefficacy tends to be limited for the purpose, it was envisioned to enhance the same using the P-gp inhibitory effect of NG, and incorporating the drug and/or bioactive in various kinds of lipid nanoformulations.

In this regard, self-nanoemulsifying drugdelivery systems (SNEDDS) were developed using polyunsaturated fatty acid (PUFA) along with NG in a synergistic manner, that is, TMX-NG-SNEDDS using optimized blends of the lipid (i.e., corn oil), surfactant (i.e., Labrasol), and cosolvent (i.e., Transcutol P), along with drug and/or NG. In vivo oral pharmacokinetic studies were performed in Wistar rats, which indicated enhanced oral bioavailability of the developed TMX-NG–SNEDDS over TMX-NG–Suspension, TMX–SNEDDS, and TMX-suspension (Figure 15.7). Overall, nanoformulations exhibited highly significant (i.e., nearly 7 to 11-fold improvement, $p < .001$) in the biopharmaceutical attributes for the purpose.

15.2 CONCLUSIONS AND FUTURE PROSPECTS

It is an undisputed fact that the knowledge and control of the size of particles, and surface area of a particle, can be related to the physical, chemical, and pharmacologic properties of drugs. Harvesting distinctive advantages offered by nanotechnology, a plethora of opportunities to develop nanotechnology-based drug-delivery systems that are more effective and stable, biologically as well as clinically, have been realized. The chapter encompasses a few case studies demonstrating phenomenal potential of such systematically developed nanoformulations in augmenting the bioefficacy and potential therapeutic outcomes of a couple of phytobioactives and drugs. While

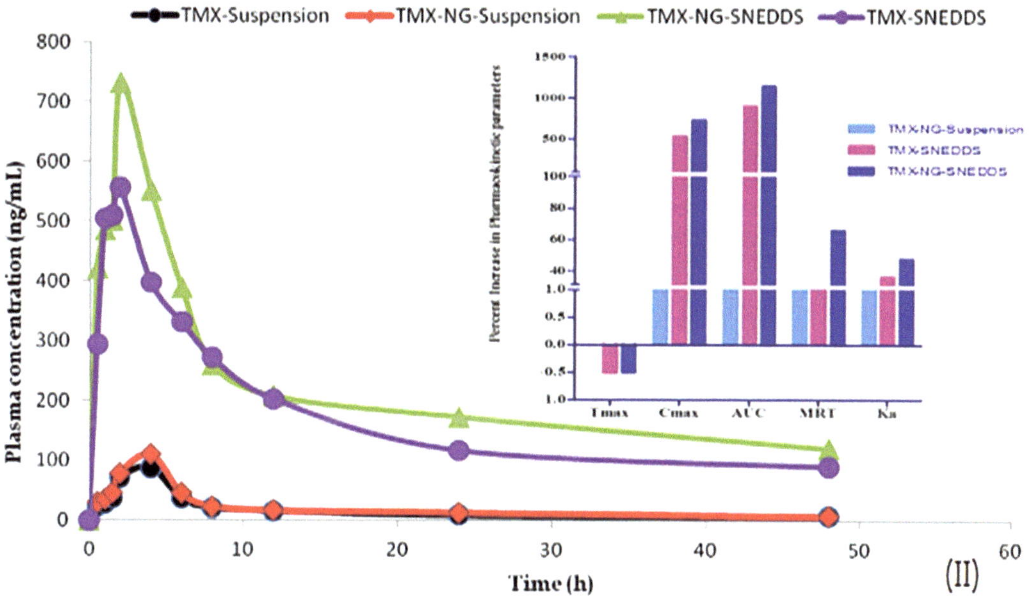

FIGURE 15.7 In vivo oral pharmacokinetic studies performed in Wistar rats indicated enhanced oral bioavailability of the developed TMX-NG–SNEDDS over TMX-NG–Suspension, TMX–SNEDDS and TMX-suspension. (Adapted with permission from Sandhu PS et al. *Nanomedicine: Nanotechnology, Biology and Medicine*. 2017; 13: 1703–1713.)

nanotechnology in the product development of various drugs and bioactives has been maturing at rapid pace, as is evidenced by the proliferation of research reports and findings in literature, yet it is time to translate research findings into a tangible embodiment of physic-chemically stable, and biologically and clinically effective commercial products. This certainly needs attention with the perspective of reducing the time involved, while translating research findings to practice. Beyond a doubt, "absence of evidence" can never be taken as the "evidence of absence"! In this context, a renewed focus is essential to improve our understanding of the essential requirements, while harvesting the "vigors" of nanotechnology-enabled robust pharmaceutical systems that can withstand the "rigors" of various stages of drug development. Already, a humble endeavor has been made, while scribbling the current chapter, in delineating the vital manufacturing process considerations while successfully translating the laboratory level studies for developing nanostructured formulations into marketable commodities. Besides, sufficient attention should be provided to various elements of processing of drug delivery systems and their implications on their respective biopharmaceutical and clinical outcomes. According to the adage, "proof is in the pudding, and the devil is in the details," substantial attention to details at the "nano level," such as elements of pharmaceutical processing of such formulations, with the "macro" objective of reducing the time between research to practice. should be the guiding principles.

REFERENCES

Bacon ER, Daum SJ, Spara PP. Polyiodinated Aroyloxy Esters. US 5,466,433; 1995.

Beg S, Jain S, Kushwah V, Bhatti GK, Sandhu PS, Katare OP, Singh B. Novel surface-engineered solid lipid nanoparticles of rosuvastatin calcium for low-density lipoprotein-receptor targeting: A Quality by Design-driven perspective. *Nanomedicine*. 2017; 12(4): 333–356.

Beg S, Khurana RK, Teenu, Singh B. Lipid and polymer-based nanostructured systems for enhanced oral bioavailability. *Pharma Times*. 2018; 50(2): 63–67.

Bosch WH, Mawr B, Marcera DM, Mueller RL, Swanson JR, Mishra DS. Process for preparing therapeutic compositions containing Nanoparticles. US 5,510,118; 1996.

Bruno JA, Doty BD, Gustow E, Illig KJ, Rajagopalan N, Sarpotdar P. Method of grinding pharmaceutical substances. US 5,518,187; 1996.

Czekai DA, Seaman LP. Continuous method of grinding pharmaceutical substances. *US Pat.*, US 5,718,388; 1998.

Czekai DA, Seaman LP. Method of grinding pharmaceutical substances. US 5,862,999; 1999.

Delaet UAF, Heyns PEH, Jans EMJ. WIPO/PCT, WO 2013/004816A1; 2013.

Durrer C, Irache JM, Puisieux F, Duchene D, Ponchel G. Mucoadhesion of latexes. I. analytical methods & kinetic studies. *Pharm Res.* 1994; 11(5): 674–679.

Eickhoff WM, Engers DA, Mueller KR. Nanoparticulate NSAID compositions. US 5,518,738; 1996a.

Eickhoff WM, Mueller KR, Engers DA. Formulations of compounds as nanoparticulate dispersions in digestible oils or fatty acids. US 5,560,93; 1996b.

Fromming KH. *Pharmazeutische Technologie, Moderne Arzeiformen*, 2nd Ed. WVG Stuttgart; 1998.

Garg B, Katare OP, Beg S, Lohan S, Singh B. Systematic development of solid self-nanoemulsifying oily formulations (S-SNEOFs) for enhancing the oral bioavailability and intestinal lymphatic uptake of lopinavir. *Colloids and Surfaces B.* 2016; 141: 611–622.

George CN, Rajagopalan N. Use of Ionic cloud point modifiers to prevent particle aggregation during sterilization. US 5,298,262; 1994a.

George CN, Rajagopalan N. Formulations for nanoparticulate X-ray blood pool contrast agents using high molecular weight nonionic surfactants. US 5,326,552; 1994b.

Hollister KR, Ladd D, McIntire GL, Na GC, Rajagopalan N, Yuan BO. Use of purified surface modifiers to prevent particle aggregation during sterilization. US 5,352,459; 1994.

Illig CR. Nanoparticulate diagnostic mixed carboxylic anhydrides as x-ray contrast agents for blood pool and lymphatic system imaging. US 5,573,749; 1996.

Jain RA, Swanson J, Hontz R, Devane J, Cumming KI, Clancy MJA et al. Controlled release nanoparticulate compositions. *US Patent Appln.* No. 2002/0012675A1; 2002.

Jain S, Saini S, Teenu, Singh B. Nanopharmaceuticals: Global perspectives and national prospective. *Pharma Times.* 2018; 50(2): 13–19.

Kapil R, Singh B. "Optimized" novel drug delivery systems for the treatment of Alzheimer's disease: A QbD-based Developmental Approach. *QbD and Product Performance Focus Group Newsletter.* 2012; 4: 4–7.

Katare OP, Singh B, Nirbhavane P, Kumar R. Pharmaceutical pilot-plant and scale-up in product development. In: *Systematic Optimization of Pharmaceutical Products and Processes Using Modern Approaches in Pharmaceutical Product Development.* NK Jain, ed., CBS Publishers, New Delhi; 2018: pp. 549–595.

Kato J, Fujimoto K, Takahashi T. Polyester resin combination. US 6,423,814B1; 2002.

Kellar KE, Karpinski KE, Liversidge E, Gunther WHH, McIntire GL, Orman BV. Water insoluble non-magnetic manganese particles as magnetic resonance contract enhancement agents. *US Pat.*, US 5,401,492; 1995.

Liversidge GG, Cundy KC. Particle size reduction for improvement of oral bioavailability of hydrophobic drugs: Absolute oral bioavailability of nanocrystalline danazol in beagle dogs. *Int J Pharm.* 1995; 125: 91–97.

Liversidge GG, Cundy KC, Bishop JF, Czekai DA. Surface modified drug nanoparticles. US 5,145,684; 1992.

Liversidge GG, Eickhoff MW, Illig KJ, Sarpotdar P, Ruddy SB. Methods for targeting drug delivery to the upper and/or lower gastrointestinal tract. US6432381B2; 2002.

Liversidge GG, Engers DA, Roberts ME, Rudy SB, Wong SM, Xu S. Methods of treating mammals using nanocrystalline formulations of human immunodeficiency virus(HIV) Protease inhibitors. US 6,221,400B1; 2001.

Muller RH, Benita S, Bohm B, eds. *Emulsions and Nanosuspensions for the Formulation of Poorly Soluble Drugs.* Medpharm Scientific Publishers, Stuttgart, Germany; 1998: pp. 68–74.

Osifo IK. Nanoparticulate Iodipamide derivatives for use as X-ray Contrast agents. US 5,521,218; 1996.

Ruddy SB, Eickhoff WM, Liversidge G. Site specific adhesion within the GI tract using nanoparticles stabilized by high molecular weight, linear poly (ethylene oxide) polymers. US 5,580,579; 1996a.

Ruddy SB, Eickhoff WM, Liversidge G, Cooper ER. Formulations of oral gastrointestinal diagnostic contrast x-ray agents in combination with pharmaceutically acceptable clays. US 5,466,440; 1995.

Ruddy SB, Eickhoff WM, Liversidge G, Cooper ER. Formulations of oral therapeutic agents in combination with pharmaceutically acceptable clays. US 5,585,108; 1996b.

Ruddy SB, Mark WM, Liversidge G, Roberts ME. Barium salt formulations stabilized by non-ionic and anionic stabilizers. US 5,593,657; 1997.

Ruddy SB, Roberts ME. Nanoparticles containing the R(-) enantiomer of Ibuprofen. US 5,718,919; 1998.

Ryde NP, Ruddy SB. Solid dose nanoparticulate compositions comprising a synergistic combination of a polymeric surface stabilizer and dioctyl sodium sulfosuccinate. US 6,375,986B1; 2002.

Ryde NP, Ruddy SB. Nanoparticulate dispersions comprising a synergistic combination of a polymeric surface stabilizer and dioctyl sodium sulfosuccinate. US 6,592,903; 2003.

Sandhu PS, Kumar R, Beg R, Jain S, Kushwah V, Katare OP, Singh B. Natural lipids enriched self-nano-emulsifying systems for effective co-delivery of tamoxifen and naringenin: Systematic approach for improved breast cancer therapeutics. *Nanomedicine.* 2017; 13: 1703–1713.

Setler P. Identifying new oral technologies to meet drug delivery needs for delivery of peptides and proteins and poorly soluble drugs, Ch.12, HR Limited Drug Delivery Systems, London; 1999: pp. 71–91.

Shah RB, Khan MA. Nanopharmaceuticals: Challenges and regulatory perspective. In: *Nanotechnology in Drug Delivery.* Springer, New York; 2009: pp. 621–646.

Singh B, Beg S, Saini S, Kaur R, Teenu. Vital drug delivery nanoconstructs for enhanced oral bioavailability of bioactives. *The Pharma Review.* 2018; 16(93): 37–45.

Smans GF, Jans EMJ. Relating to anti-HIV tablet formulations. *US Pat. Appln. Publn.*, US2010/0189783; 2010.

Violante MR, Steigbigel RT. Particulate composition and use thereof as antimicrobial agent. US 4,783,484; 1988.

Wiedmann TS, Wood RW, De Castro L. Aerosols containing Beclomethazone nanoparticle dispersions. US 5,747,001; 1998.

Wong SM. Butylene oxide-ethylene oxide block copolymer surfactant as stabilizer coatings for nanoparticle compositions. US 5,587,143; 1996.

Wong SM, Cooper ER, Xu S. Polyalkylene block copolymers as surface modifiers for nanoparticles. US 5,565,188; 1996.

Wood RW, De Castro L, Bosch HW. Nebulized aerosols containing nanoparticle dispersions. US 6,264,922; 2001.

Index

C

D

E

F

M

N

O

P

Q

R

S